# Aromatic and Heteroaromatic Chemistry

## Volume 7

A Specialist Periodical Report

# Aromatic and Heteroaromatic Chemistry
Volume 7

A Review of the Literature Abstracted
between July 1977 and June 1978

Senior Reporters
**H. Suschitzky and O. Meth-Cohn,** *Department of Chemistry and Applied Chemistry, University of Salford*

*Reporters*
**R. S. Atkinson,** *University of Leicester*
**G. V. Boyd,** *Chelsea College, London*
**G. M. Brooke,** *University of Durham*
**D. J. Chadwick,** *University of Liverpool*
**A. H. Jackson,** *University College, Cardiff*
**G. R. Proctor,** *University of Strathclyde, Glasgow*
**S. M. Roberts,** *University of Salford*
**R. K. Smalley,** *University of Salford*
**A. W. Somerville,** *University of Salford*
**R. C. Storr,** *University of Liverpool*

The Chemical Society
Burlington House, London, W1V 0BN

**British Library Cataloguing in Publication Data**

Aromatic and heteroaromatic chemistry.–
  (Chemical Society. Specialist periodical reports).
  Vol. 7
  1. Aromatic compounds
  I. Suschitzky, Hans  II. Meth-Cohn, Otto  III. Series
  547′.6    QD331    72-95095

  ISBN 0-85186-600-X
  ISSN 0305-9715

Set in Times on Linotron and printed offset by
J. W. Arrowsmith Ltd., Bristol, England

Made in Great Britain

# Introduction

In this seventh Report the literature coverage is based essentially on volumes 87 and 88 of *Chemical Abstracts*. We have arranged the chapters as in Volume 6 except that the topics treated in the last two sections of the previous volume, namely 'Naturally Occurring Aromatic Oxygen-ring Compounds' and 'Other Naturally Occurring Aromatic Compounds' are now omitted. This, it was felt, was justified in view of the liberal treatment and accessibility of current reviews on natural products elsewhere.

We have again asked our contributors to comment on highlights in their subject. The synthesis of a stable tetrahedrane, one of the most attractive recent challenges, has now been achieved by a group of German workers (Chapter 1, ref. 40). Evidence for the existence of other elusive molecules, namely thiiren (Chapter 1, ref. 32), seleniren (Chapter 2, ref. 391), and azete (Chapter 1, ref. 80), the latter during the photolysis of an oxazinone, has been obtained. The interpretation of new i.r. data for cyclobutadiene as a non-square structure removes a troubling inconsistency between experiment and theory (Chapter 1, ref. 39). Five-membered heterocyclic systems continue to provide starting materials for fascinating chemistry: irradiation of tetrakis(trifluoromethyl)thiophen gives a Dewar-thiophen (Chapter 2, ref. 115) while that of a tetrazoline at low temperature provides the first example of a nitrogen analogue of trimethylenemethane (Chapter 2, ref. 419). 2-Chloro-3-ethylbenzoxazolium salts proved remarkably versatile reagents for the synthesis of ketones, isocyanates, isocyanides, and alkyl chlorides (Chapter 2, refs. 319—322). Among six-membered heterocycles, pyridine is still a favourite target for studying basic reactions. Thus, the first examples of direct alkylation, acylation, and regioselective metallation at the 3-position have been reported (Chapter 4, refs. 29, 30, and 33, respectively). A new non-cyclic cryptate derived from 8-hydroxyquinoline has been prepared which, in some respects, is superior to 18-crown-6 and is claimed to have the greatest complexation ability observed so far for neutral ligands (Chapter 4, ref. 93). Phase-transfer catalysts continue to provide useful synthetic aids (Chapter 4, refs. 91, 105, and 111).

There is an interesting claim pertaining to seven-membered rings (Chapter 5, ref. 43), namely that azatropolones have been prepared. If substantiated, this would represent a significant contribution to the heterocyclic field.

The first example of a medium-sized cyclic cumulene (Chapter 6, ref. 35) and of a bridged [10] annulene analogue of azulene (Chapter 6, ref. 57) are noted amongst a wealth of new annulenes.

Noteworthy points in the chapter on Electrophilic Substitution are the surprising findings, based on n.m.r. parameters of a model system, that the polar inductive effect of alkyl groups attached to $sp^2$-carbon is zero (ref. 12). The correlation of the *ortho*-directing effect in the metallation of aromatics (refs. 76—79) and a detailed study of the acylation of the pyrrole anion rationalizing the results on the HSAB principle (refs. 265, 266) are worthy of mention.

Noteable features described in the chapter on porphyrins include (i) the synthesis of porphyrin-*a* (ref. 22), (ii) increasing interest in synthesizing models for haemoglobin (ref. 31), and (iii) the biosynthetic work in the haem and vitamin $B_{12}$ fields (Section 6).

We thank all our present Reporters for delivering and promptly dealing with their chapter within the required time, and our previous reviewers for their help. As always, the professional advice and invaluable assistance by the Editorial Staff of The Chemical Society is much appreciated by us.

H. SUSCHITZKY & O. METH-COHN

# Contents

Contents                                                                    xi

# 1
# Three- and Four-membered Ring Systems

BY R. C. STORR

## 1 Three-membered Carbocyclic Systems

Calculations of the potential surfaces of the planar cyclopropenyl radical[1,2] and anion[1] indicate that both are subject to Jahn–Teller distortions such that the lowest energy configuration of the radical has one short and two long bonds whereas that of the anion has two short and one long bond. For the carbene (1), INDO calculations reveal a large singlet–triplet separation, the singlet state being stabilised by $2\pi$ aromatic character of the ring.[3]

Symmetry-allowed extrusion of the cyclopropenyl cation is not observed on loss of nitrogen from (2). Instead the ether (3) is formed in methanol, possibly *via* the pyramidally bridged cation (4).[4]

(1) 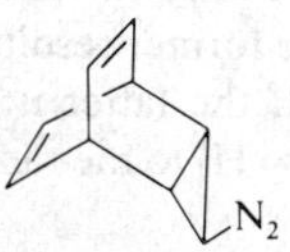          (2) 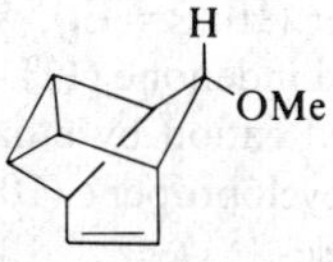          (3) 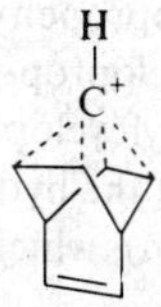          (4)

Yoshida and his co-workers have continued to contribute greatly to the field of cyclopropenium ion chemistry. In addition to describing a large number of cyclopropenium ions, cyclopropen-ones, and -thiones in the patent literature[5] they have produced the first triafulvalene dication (5).[6] This was obtained by treatment of (6; X = H) with butyl-lithium and then allowing the resulting carbenoid (6; X = Li) to react with (6; X = Cl). The transformation illustrates

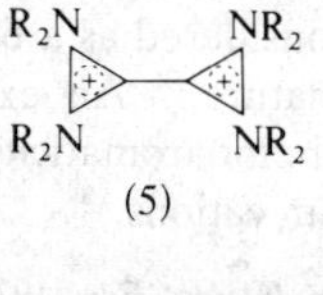

(5) 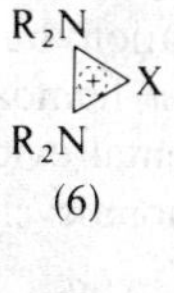                              (6)

[1] E. R. Davidson and W. T. Borden, *J. Chem. Phys.*, 1977, **67**, 2191.
[2] D. Poppinger, L. Radom, and M. A. Vincent, *Chem. Phys.*, 1977, **23**, 437.
[3] C.-K. Lee and W.-K. Li, *J. Mol. Structure*, 1977, **38**, 253.
[4] R. M. Magid and G. W. Whitehead, *Tetrahedron Letters*, 1977, 1951; see also R. M. Coates and E. R. Fretz, *ibid.*, p. 1955.
[5] Z. Yoshida, *Chem. Abs.*, 1976, **87**, 102 020, 134 027; 1978, **88**, 74 104, 74 107, 74 114, 104 762.
[6] Z. Yoshida, H. Konishi, S. Sawada, and H. Ogoshi, *J.C.S. Chem. Comm.*, 1977, 850.

the use of intermediates such as (6; X = Li) to effect formal electrophilic substitution in the cyclopropenium ion system.[7] The related (6; X = MgBr) produced from the iodocyclopropenium salt (6; X = I) and PhMgBr behaves as a typical Grignard reagent with electrophiles. In contrast to the iodo-cation (6; X = I), the chloro-cation (6; X = Cl) gives the aryldiamino-cyclopropenium cation (6; X = Ph) with PhMgBr.[8]

Hydride abstraction from (7) with trityl perchlorate does not give the dication (8), but rather the monocation (9) resulting from deprotonation of (8). [13]C n.m.r. spectroscopy reveals that the central carbon possesses a considerable amount of electron density, indicating that the dipolar form (9a) makes a significant contribution.[9] There is evidence that formation of dicyclopropene from cyclopropenyl cations under conditions of two-electron reduction proceeds through coupling of cyclopropenyl radicals produced by electron transfer from anion to cation.[10]

The cyclopropenyl ether (10; R = Et), with FeCl₃, in refluxing ethanol, gives triphenylcyclopropene and indenone (11). The former results from reduction of the triphenylcyclopropenyl cation by ethanol; the latter is thought to involve oxidation of the hydroxy-cyclopropene (10; R = H) to the radical (12) (or further to the cation), which cyclises.[11]

The cation (13), prepared from 11-methyl-11-bromotricyclo[4.4.1.0¹·⁶]-undecane in SbF₅–SO₂ClF at −120 °C, is neither a symmetrical cyclopropyl cation nor an allylic cation, but is best considered as a bent cyclopropyl cation showing significant $2\pi$-homoaromatic nature.[12] An excellent review of the concept and experimental evidence for homoaromaticity includes sections on mono-, bis-, and tris-homocyclopropenium cations.[13]

[7] R. Weiss, C. Priesner, and H. Wolf, *Angew. Chem. Internat. Edn.*, 1978, **17**, 446.
[8] Z. Yoshida, H. Konishi, Y. Miura, and H. Ogoshi, *Tetrahedron Letters*, 1977, 4319.
[9] K. Komatsu, K. Matsumoto, and K. Okamoto, *J.C.S. Chem. Comm.*, 1977, 232.
[10] R. W. Johnson, T. Widlanski, and R. Breslow, *Tetrahedron Letters*, 1976, 4685.
[11] A. S. Monahan, J. D. Frielich, J. J. Fong, and D. Kronenthal, *J. Org. Chem.*, 1978, **43**, 232.
[12] G. A. Olah, G. Liang, D. B. Ledlie, and M. G. Costopoulos, *J. Amer. Chem. Soc.*, 1977, **99**, 4196.
[13] L. A. Paquette, *Angew. Chem. Internat. Edn.*, 1978, **17**, 106.

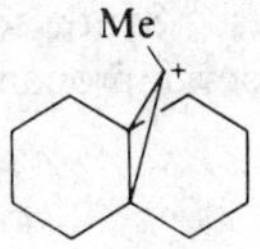

(13) 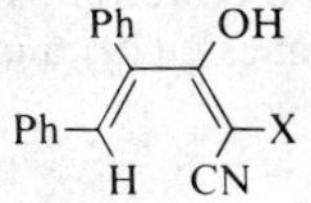  (14)  X = CN or CO₂Et

Full details of the preparation of cyclopropenone have appeared in *Organic Syntheses*.[14] Under basic conditions, diphenylcyclopropenone reacts with malononitrile and ethyl cyanoformate at the carbonyl group to give the butadienes (14).[15] The key step in the oxidation of cyclopropenones (15) by peracid

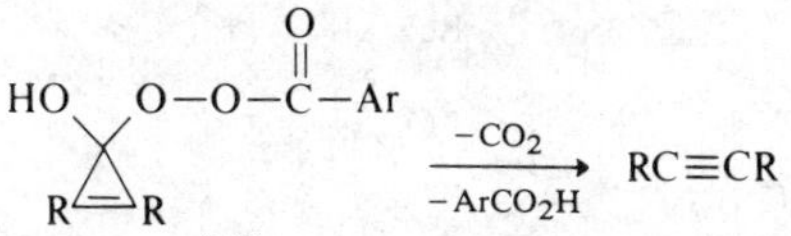

(15)

appears to be formation of the acetylene from attack by peracid at the carbonyl group, as shown, rather than formation of an oxabicyclobutanone. The acetylene then undergoes further oxidation.[16] CNDO/2 calculations have been reported for a number of cyclopropenones and cyclopropenethiones.[17] The first e.s.r. spectrum of a cyclopropenone radical anion has been observed for diphenylcyclopropenone. The anion is relatively unstable and undergoes loss of CO to give the radical anion of diphenylacetylene.[18] No spectra were obtainable for dialkylcyclopropenones.

A novel activation of diphenylcyclopropenone involving ring cleavage of the C–C double bond has been observed with $Pt_3(Me_3CNC)_6$.[19]

The reaction of cyclopropenones (16) with compounds containing active methylene groups leads to methylenecyclopropenes (17), which on oxidation give a new family of alkylidenequinocyclopropanes (18). These are highly coloured,

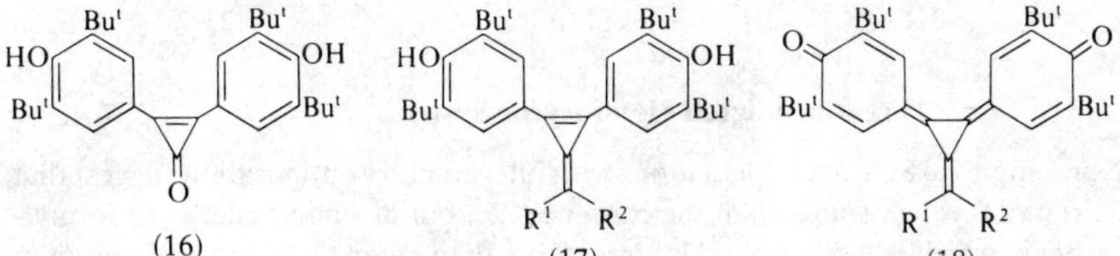

(16)  (17)  (18)

$R^1 = R^2 = CN$ or $COMe$; $R^1 = CN$, $R^2 = CO_2Et$

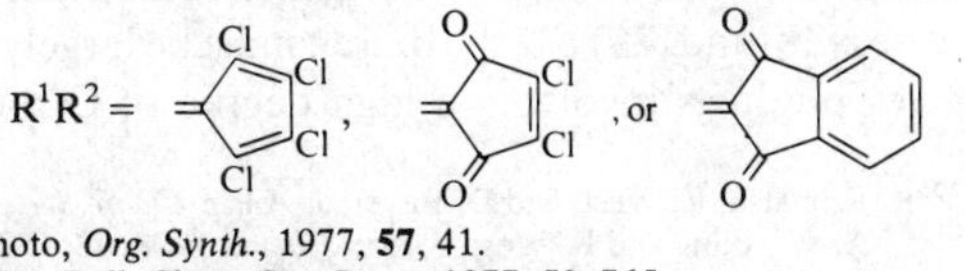

[14] R. Breslow, J. Pecararo, and T. Sugimoto, *Org. Synth.*, 1977, **57**, 41.
[15] I. Agranat, G. V. Boyd, and M. A. Wirt, *Bull. Chem. Soc. Japan*, 1977, **50**, 765.
[16] J. K. Crandall and W. W. Conover, *J. Org. Chem.* 1978, **43**, 1323.
[17] N. J. Fitzpatrick and M. O. Fanning, *J. Mol. Structure*, 1977, **42**, 261.
[18] P. Fürderer, F. Gerson, and A. Krebs, *Helv. Chim. Acta*, 1977, **60**, 1226.
[19] W. E. Carroll, M. Green, J. A. K. Howard, M. Pfeffer, and F. G. A. Stone, *Angew. Chem. Internat. Edn.*, 1977, **16**, 793.

strongly dichroic, and powerful oxidising agents.[20] Syntheses and properties of the quinocyclopropanes[21] (19) and quinoiminocyclopropanes[22] (20) have also been reported.

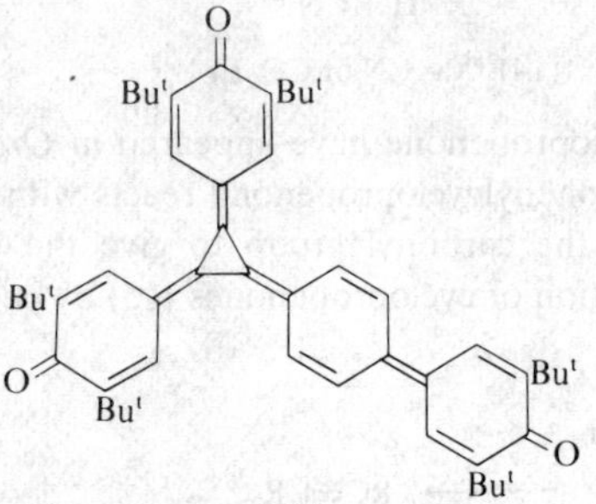

(19) 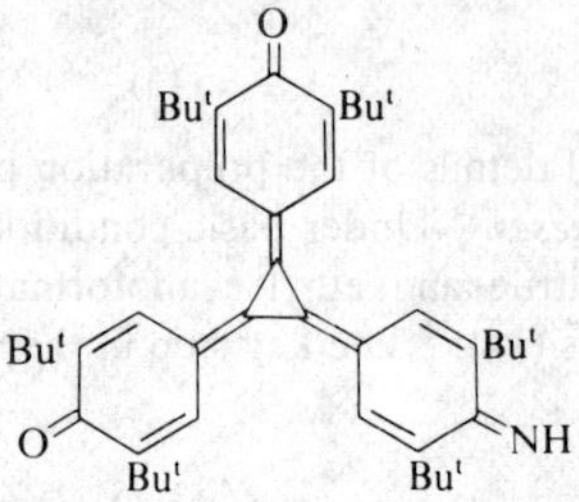                                                          (20)

The *o*-tropoquinonecyclopropenide (21) is somewhat less stable than the previously prepared *para*-isomer. Spectral data indicate that it is nearly planar; however, the mono- and di-cations appear to be non-planar, owing to steric hindrance and electronic repulsion.[23] In agreement with previous *X*-ray studies, [35]Cl quadrupole resonance data for the cyclopentadiene cyclopropenide (22) indicate that the cyclopentadiene ring is largely dienoid.[24] The [13]C n.m.r. spectra of the highly stabilised thiocarbonyl ylides (23) suggest that they possess appreciable C=S character, as shown.[25]

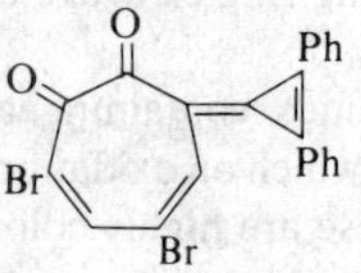

(21) 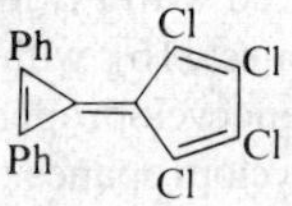                          (22) 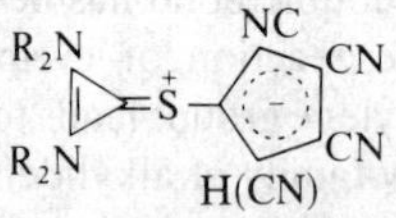                          (23)

## 2 Three-membered Heterocyclic Systems

Non-empirical SCF MO calculations with full geometry optimisation suggest that oxiren (24) is less stable than the carbene (25), but of similar energy to formyl-carbene, which is ~80 kcal mol$^{-1}$ less stable than keten.[26] The role of oxiren in oxocarbene rearrangements is well established. Use of the labelled diazo-compounds (26) and (27) indicates that the equilibrium between the two oxocar-benes (28) and (29) *via* the oxiren must be largely displaced to the right, since (26) gives products involving a high degree of O migration on photolysis, whereas

[20] K. Komatsu, R. West, and D. Beyer, *J. Amer. Chem. Soc.*, 1977, **99**, 6290.
[21] L. A. Wendling and R. West, *J. Org. Chem.*, 1978, **43**, 1573.
[22] L. A. Wendling and R. West, *J. Org. Chem.*, 1978, **43**, 1577.
[23] K. Takahashi, K. Marita, and K. Takase, *Tetrahedron Letters*, 1977, 1511.
[24] I. Agranat, M. Hayek, and D. Gill, *Tetrahedron*, 1977, **33**, 239.
[25] K. Nakasuji, K. Nishino, I. Murata, H. Ogoshi, and Z. Yoshida, *Angew, Chem. Internat. Edn.*, 1977, **16**, 866.
[26] A. C. Hopkinson, M. Lien, K. Yates, and I. G. Csizmadia, *Progr. Theor. Org. Chem.*, 1977, **2**, 230.

(24)    (25)    (26)    (27)

(28)    (29)

those from (27) involve practically no migration.[27] Labelling studies with formyl-carbene that had been produced by photolysis of formyldiazomethane revealed that formation of oxiren is a minor process compared with direct Wolff rearrangement.[28] Attempts to generate oxirens by retro-Diels–Alder reactions in such ideal precursors as (30)[29,30] were largely unsuccessful, although some evidence for the formation of small amounts of keten was found from flash pyrolysis of (30) at 600–800 °C.[29]

(30)

*Ab initio* calculations for the $C_2H_2S$ system indicate that thiiren (35) is the least stable species in the series (31)–(35). However, clear evidence for the presence of

(31)    (32)    (33)    (34)    (35)

thiiren in the photolysis of matrix-isolated 1,2,3-thiadiazole (and, more surprisingly, isothiazole) has appeared.[31] The distribution of the label in products derived from labelled thiadiazole established that a species with thiiren symmetry was involved, and further support came from direct spectroscopic observation. Seleniren was generated similarly.[32] The formation of both 2,7- and 2,8-dicarbomethoxythianthrenes in the pyrolysis of 6-carbomethoxybenzo-1,2,3-thiadiazole is consistent with the intervention of a benzothiiren intermediate.[35] A thiiren intermediate has also been suggested in the thermolysis of the thiadiazole (36).[34] Examples of stable thiirenium salts have appeared.[35]

[27] K. P. Zeller, *Angew. Chem. Internat. Edn.*, 1977, **16**, 781.
[28] K. P. Zeller, *Tetrahedron Letters*, 1977, 707.
[29] E. G. Lewars and G. Morrison, *Tetrahedron Letters*, 1977, 501.
[30] E. G. Lewars and G. Morrison, *Canad. J. Chem.* 1977, **55**, 966; H. Hart, J. B.-C. Jiang, and M. Sasaoka, *J. Org. Chem.*, 1977, **42**, 3840.
[31] O. P. Strausz, R. K. Gosavi, F. Bernardi, P. Mezey, J. D. Goddard, and I. G. Csizmadia, *Chem. Phys. Letters*, 1978, **53**, 211.
[32] A. Krantz and J. Laureni, *J. Amer. Chem. Soc.*, 1977, **99**, 4842; *Ber. Bunsengesellschaft. Phys. Chem.*, 1978, **82**, 13.
[33] T. Wooldridge and T. D. Roberts, *Tetrahedron Letters*, 1977, 2643.
[34] K. Senga, M. Ichiba, and S. Nishigaki, *J. Org. Chem.* 1978, **43**, 1677.
[35] R. Destro, T. Pilati, and M. Simonetta, *J.C.S. Chem. Comm.*, 1977, 576; G. Capozzi, V. Lucchini, G. Modena, and P. Scrimin, *Tetrahedron Letters*, 1977, 911.

MeN

(36)

## 3 Four-membered Carbocyclic Systems

New calculations[36-38] for the potential surfaces of the singlet and triplet states of cyclobutadiene are essentially in agreement with the generally accepted picture, namely that the square triplet is of higher energy than the rectangular singlet molecule. The rectangular singlet is more stable than the square singlet, which corresponds to a transition state between the two possible rectangular configurations. More surprisingly, the square singlet is estimated to be lower in energy than the square triplet, this apparent violation of Hund's rule being explained by dynamic spin polarisation.[37]

One of the great problems of recent years has been to reconcile the theoretical picture of parent cyclobutadiene with the observed i.r. spectrum of the matrix-isolated species, which suggested that the molecule is square. One explanation advanced was that the species actually observed was a metastable triplet. MINDO/3 has been used to calculate the vibrational frequencies for singlet and triplet cyclobutadiene and their mono-deuteriated derivatives (the method was shown to be reliable for several simple systems) and the spectrum calculated for the triplet is consistent with the experimental spectrum.[38] However, it now appears that this whole problem has been due to the failure to observe certain bands in the original i.r. spectrum. Generation of cyclobutadiene in a matrix from several precursors, under conditions where the spectrum can be more clearly observed, reveals new absorptions which mean that the symmetry of the species is *less* than $D_{4h}$ and that it is therefore very probably rectangular.[39]

Tetra-t-butyltetrahedrane has been prepared by irradiation of tetra-t-butyl-cyclopentadienone as the first example of this long-sought system.[40] It is remarkably stable (m.pt. 135 °C) because lengthening of any of the four C–C bonds causes increased compression of the t-butyl groups, and it is only transformed into tetra-t-butylcyclobutadiene on heating to 130 °C. Some evidence for the formation of tetralithiotetrahedrane has also been claimed.[41] Several other attempts to generate tetrahedranes led mostly to cyclobutadienes.[42,43] Interestingly, irradiation of the ozonide (37) gave the symmetrical tetra(trifluoromethyl)cyclo-

[36] W. T. Borden, E. R. Davidson, and P. Hart, *J. Amer. Chem. Soc.* 1978, **100**, 388.

[37] H. Kollmar and V. Staemmler, *J. Amer. Chem. Soc.*, 1977, **99**, 3583.

[38] M. J. S. Dewar and A. Komornicki, *J. Amer. Chem. Soc.*, 1977, **99**, 6174.

[39] S. Masamune, F. A. Souto-Bachiller, T. Machiguchi, and J. E. Bertie, *J. Amer. Chem. Soc.*, 1978, **100**, 4889.

[40] G. Maier, S. Pfrem, U. Schäfer, and R. Matusch, *Angew, Chem. Internat. Edn.*, 1978, **17**, 520.

[41] G. Rauscher, T. Clark, D. Poppinger, and P. von R. Schleyer, *Angew. Chem. Internat. Edn.*, 1978, **17**, 276.

[42] S. Masamune, T. Machiguchi, and M. Aratani, *J. Amer. Chem. Soc.*, 1977, **99**, 3524.

[43] G. Maier, H. P. Reisenauer, and H. A. Freitag, *Tetrahedron Letters*, 1978, 121; Y. Kobayashi, I. Kumadaki, A. Ohsawa, Y. Hanzawa, M. Honda, W. Migashita, and Y. Iitaka, *Tetrahedron Letters*, 1977, 1795.

butadiene, which had a half-life of several hours at 145 K. Significantly, its i.r. spectrum is consistent with a rectangular rather than a square geometry.[42]

Cyclobutenylidene (38) rearranges to give vinylacetylene with no evidence for the formation of cyclobutadiene or methylenecyclopropane. This preferred pathway also emerges in MINDO/3 calculations for the system.[44]

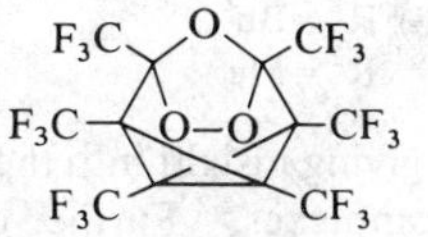

(37)      (38)

All five oxidation states in the reversible redox system involving the [4]radialene (39) and the cyclobutadiene (40) have been observed. As expected,

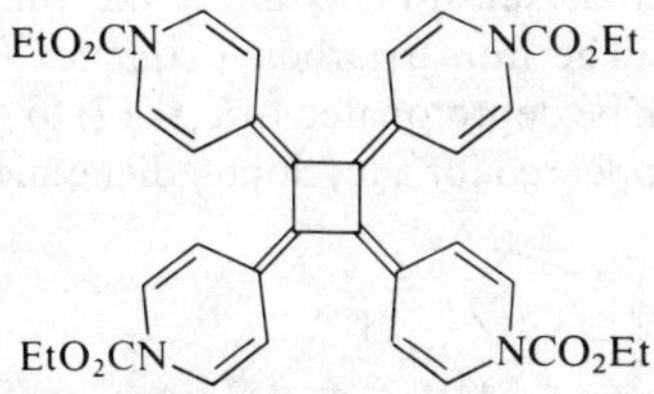

(39) 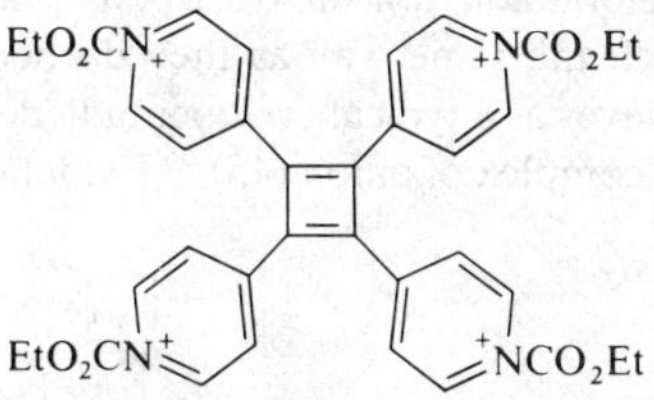     (40)

introduction of anti-aromatic character is reflected in the potential for the final electron transfer.[45] The high reactivity of cyclobutadiene towards sterically hindered dienophiles has been noted[46] and cyclobutadiene has been utilised in a synthesis of *cis,trans*-octa-1,5-dienes.[47] Fragmentation to benzene and cyclobutadiene is a minor process in the photolysis of *syn*-tricyclo[4.4.0.0²,⁵]deca-3,7,9-triene.[48]

Further evidence that butalene (41) is produced from the chloro-Dewarbenzene (42) with base comes from the use of a methyl-labelled analogue. The final distribution of the label reveals that about half of the reaction proceeds through a butalene.[49]

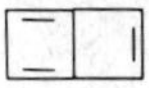

(41) 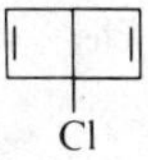     (42)

A comprehensive review of cyclobutadiene–metal complexes has appeared.[50] The first bi(cyclobutadiene)nickel complex of the sandwich type has been reported.[51] σ-Bonded intermediates (43) have been isolated at low temperature from

[44] S. F. Dyer, S. Kammula, and P. B. Shevlin, *J. Amer. Chem. Soc.*, 1977, **99**, 8104.
[45] M. Horner and S. Hünig, *Angew. Chem. Internat. Edn.*, 1977, **16**, 410.
[46] E. G. Georgescu and M. D. Gheorghia, *Rev. Roumaine Chim.*, 1977, **22**, 907.
[47] H. D. Martin, M. Hekman, G. Rist, A. Sauter, and D. Bellus, *Angew. Chem. Internat. Edn.*, 1977, **16**, 406.
[48] E. L. Allred, B. R. Beck, and N. A. Mumford, *J. Amer. Chem. Soc.*, 1977, **99**, 2694.
[49] R. Breslow and P. L. Khanna, *Tetrahedron Letters*, 1977, 3429.
[50] A. Efraty, *Chem. Rev.*, 1977, **77**, 691.
[51] H. Hoberg, R. Krause-Goeing, and R. Mynott, *Angew. Chem. Internat. Edn.*, 1978, **17**, 123.

$$(43) \quad R^1 = Bu^t$$
$$R^2 = Me$$

$[PdCl_2(PhCN)_2]$ and acetylene, thus giving insight into the mechanism of formation of palladium–cyclobutadiene complexes.[52] Further examples of formation
of ($\pi$-cyclobutadiene)cobalt complexes from $[(\pi\text{-}C_5H_5)CoL_2]$ and acetylenes[53]
and further examples of 'conventional aromatic chemistry' for cyclobutadiene
that is co-ordinated to iron tricarbonyl[54] have appeared. $^1$H n.m.r. spectra of
(cyclobutadiene)iron tricarbonyls indicate that substituents affect the ring in
much the same way as they do benzene.[55] The iron tricarbonyl complex (44)
behaves as a typical cyclopentadiene and can be deprotonated ($pK_a \approx 17$) to give
the complex of anion (45).[56] Evidence has appeared for a (cyclobutadiene)nickel

(44)                    (45)

complex in the $NiBr_2$-catalysed polymerisation of hex-3-yne,[57] and the structure
of cyclobutadienedicobalt hexacarbonyl has been established by $X$-ray
diffraction.[58] Complex (46), incorporating the common type of $\eta^4$-co-ordinated
cyclobutadiene, can be readily converted into the less common type of $\eta^2$-
complex (47).[59] The newly reported di-t-butoxyethyne can be transformed into

(46)                         (47)

squaric and deltic acids.[60] Full details of a new synthesis of semisquaric acid,
involving hydrolysis of the photodimers of chlorovinylene carbonate, have

[52] E. A. Kelly, P. M. Bailey, and P. M. Maitlis, *J.C.S. Chem. Comm.*, 1977, 289.
[53] A. Clearfield, R. Gopal, M. D. Rausch, E. F. Tokas, F. A. Higbie, and I. Bernal, *J. Organometallic Chem.*, 1977, **135**, 229; K. Yusufuku and H. Yamazaki, *ibid.*, 1977, **127**, 197.
[54] P. Marcincal and E. Cuingnet, *Bull. Soc. Chim. France*, 1977, 489; I. G. Dinulescu, E. G. Georgescu, and M. Avram, *J. Organometallic Chem.*, 1977, **127**, 193.
[55] P. L. Pruitt, E. R. Biehl, and P. C. Reeves, *J.C.S. Perkin II*, 1977, 907.
[56] J. T. Bamberg and R. G. Bergman, *J. Amer. Chem. Soc.*, 1977, **99**, 3173.
[57] P. Mouret, G. Guerch, and S. Martin, *Compt. rend.*, 1977, **248**, *C*, 747.
[58] P. E. Riley and R. E. Davis, *J. Organometallic Chem.*, 1977, **137**, 91.
[59] W. Winter and J. Strähle, *Angew. Chem. Internat. Edn.*, 1978, **17**, 128.
[60] M. A. Pericás and F. Serratosa, *Tetrahedron Letters*, 1977, 4437.
[61] H. D. Scharf, H. Frauenrath, and W. Pinske, *Chem. Ber.*, 1978, **111**, 168.

appeared.[61] A number of papers have been concerned with systems related to squaric acid[62] and cyclobutenedione.[63] The synthesis of 1,3,4-triamino-2-oxacyclobutenylium cations (48) has been described.[64]

(48)

## 4 Fused Four-membered Carbocyclic Systems

An $X$-ray structure determination for the stable benzocyclobutadiene (49) indicates, as expected, that the canonical form shown best describes the structure.[65] Sophisticated calculations have been carried out for a wide range of benzo-fused cyclobutadiene derivatives. The agreement between calculated and experimental transition energies and moments for the few known compounds is impressive enough to suggest that predictions for the (as yet) unknown compounds are reliable.[66] Because of the discrepancy between the calculated spectra and the data reported for the tentatively claimed 1,2-diphenylphenanthro[1]cyclobutene (50),[67] the alternative structure (51) has been proposed.[66]

distances in Å
(49)          (50)          (51)

Some $^{13}$C n.m.r. spectra for benzo- and dibenzo-cyclobutadiene dications indicate that these species are fully delocalised six- and ten-electron aromatic systems respectively.[68] Flash-pyrolytic extrusion of nitrogen from benzocinnoline analogues has been employed in the generation of the fused cyclobutadiene derivatives (52),[69] (53), and (54).[70] 1,2-Dimethylbenzocyclobutadiene gives the

[62] H. Ehrhardt, S. Hünig, and H. Pütter, *Chem. Ber.*, 1977, **110**, 2506; S. Hünig and H. Pütter, *ibid.*, p. 2524; G. Seitz, R. Schmiedel, and K. Mann, *Arch. Pharm.*, 1977, **310**, 549, 991; G. Seitz, R. Matusch, and K. Mann, *Chem. Ztg.*, 1977, **101**, 557; W. Reid and M. Vogl, *Annalen*, 1977, 101; H. Knorr, W. Reid, G. Oremek, and P. Pustoslemsek, *ibid.*, p. 948; G. Farina, G. Capobianco, and B. Lunelli, *Chimica e Industria*, 1977, **59**, 215.
[63] N. Obata and T. Takizawa, *Bull. Chem. Soc. Japan*, 1977, **50**, 2017; M. E. Jung and J. A. Lowe, *J. Org. Chem.*, 1977, **42**, 2371; A. Roedig, G. Bonse, E. M. Ganns, and H. Heinze, *Tetrahedron*, 1977, **33**, 2437.
[64] G. Seitz, R. Schmiedel, and R. Sutrisno, *Synthesis*, 1977, 845.
[65] W. Winter and H. Straub, *Angew. Chem. Internat. Edn.*, 1978, **17**, 127.
[66] H. Volger and G. Ege, *J. Amer. Chem. Soc.*, 1977, **99**, 4599.
[67] E. D. Bergmann and I. Agranat, *Israel J. Chem.*, 1965, **3**, 197.
[68] G. A. Olah and G. Liang, *J. Amer. Chem. Soc.*, 1977, **99**, 6045.
[69] J. W. Barton and R. B. Walker, *Tetrahedron Letters*, 1978, 1005.
[70] S. Kanoktanaporn and J. A. H. MacBride, *Tetrahedron Letters*, 1977, 1817.

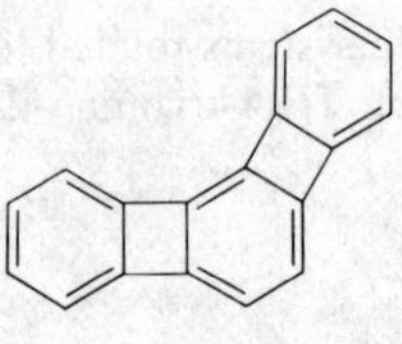

(52) 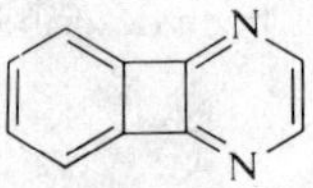                          (53) 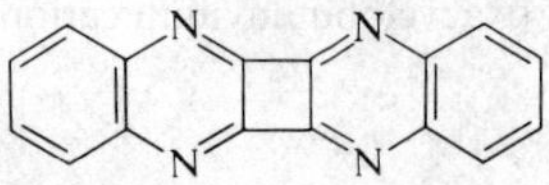                          (54)

expected angular dimer (55), which rearranges to the semibullvalene (56) on heating at 250 °C.[71] A bathochromic shift in the electronic spectrum of (57) reflects the increased strain in this molecule compared with 2,3-dimethyl-biphenylene.[72] 'Mixed' biphenylenes have been prepared successfully from simultaneous flash pyrolysis of different benzyne precursors.[73]

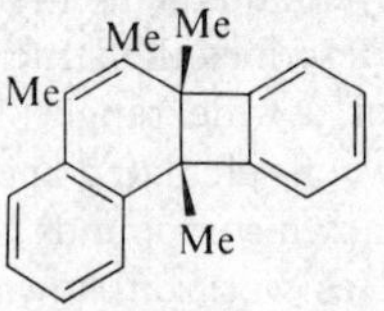

(55) 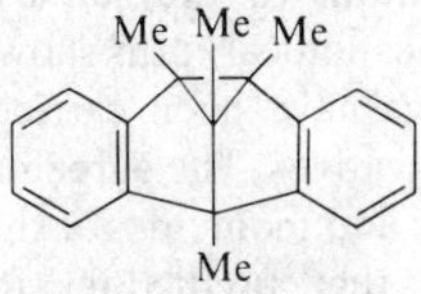                          (56) 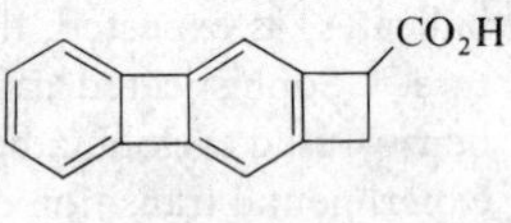                          (57)

Full details of the formation of 3,4-cyclobuta[1,2]cyclohepten-6-one (58) derivatives from the dichlorocarbene ring-expansion reaction of 1- and 2-methoxybiphenylenes have appeared. Spectral data suggest that there is considerable bond fixation in the seven-membered ring, to minimise cyclo-butadienoid character in the four-membered ring.[74] Similar considerations apply to the fused tropolones (59)[75] and (60),[76] which exist in the tautomeric forms shown and undergo methylation only at the points indicated.

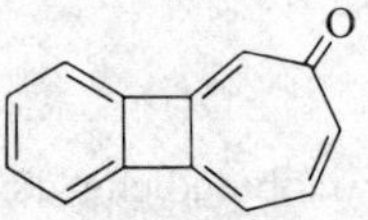

(58) 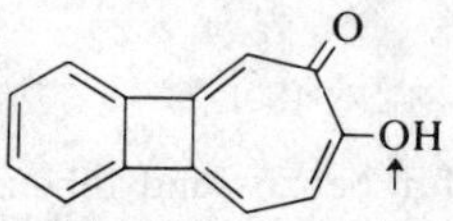                          (59) 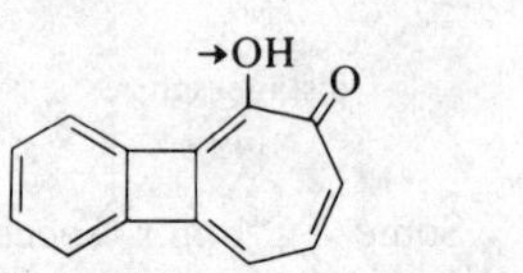                          (60)

A new synthesis of 9,10-diphenylbicyclo[6.2.0]decapentaenes (61) has been reported. Preliminary spectral data suggest a nearly planar structure.[77] The

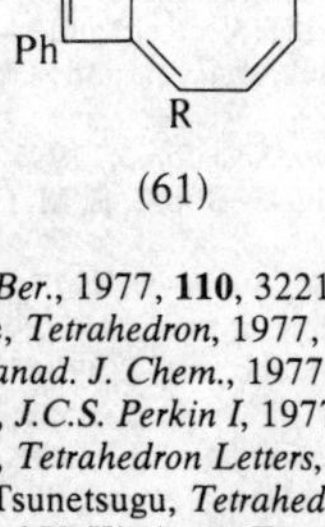

(61)

[71] H. Straub and J. Hambrecht, *Chem. Ber.*, 1977, **110**, 3221.
[72] P. R. Buckland and J. F. W. McOmie, *Tetrahedron*, 1977, **33**, 1797.
[73] A. Martineau and D. C. DeJongh, *Canad. J. Chem.*, 1977, **55**, 34.
[74] M. Sato, S. Ebine, and J. Tsunetsugu, *J.C.S. Perkin I*, 1977, 1282.
[75] M. Sato, S. Ebine, and J. Tsunetsugu, *Tetrahedron Letters*, 1977, 855.
[76] M. Sato, H. Fujino, S. Ebine, and J. Tsunetsugu, *Tetrahedron Letters*, 1978, 143.
[77] M. Oda, H. Oikawa, N. Fukazawa, and Y. Kitahara, *Tetrahedron Letters*, 1977, 4409.

benzocyclobutene oxide (63) can be isolated from the reaction of (62) with $^3O_2$ in solution. With further $O_2$ this gives ozonide (64), which on heating gives (65), (66), and (67), so providing some evidence that the oxidation of benzo-cyclobutene proceeds through analogous intermediates. In the solid state, (62)

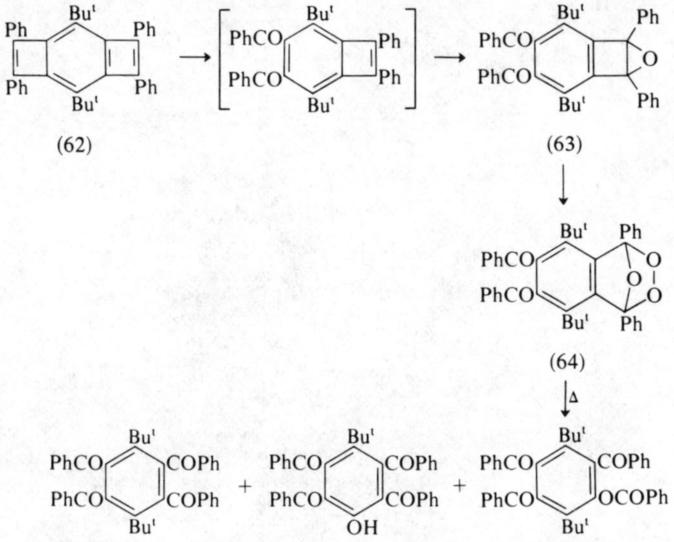

(62)            (63)

(64)

(67)      (66)      (65)

gives (68) with oxygen, cyclisation of this *o*-quinonemethide being inhibited by the development of cyclobutadienoid character.[78]

Cycloaddition of TCNE occurs at the benzene nucleus of (62; *p*-halogeno-phenyl for Ph) to give the propellane (69).[79]

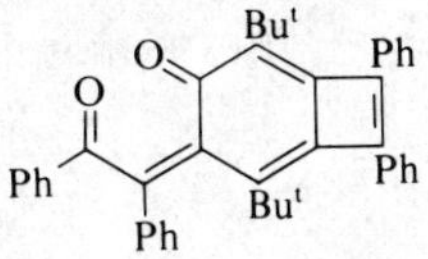

(68) 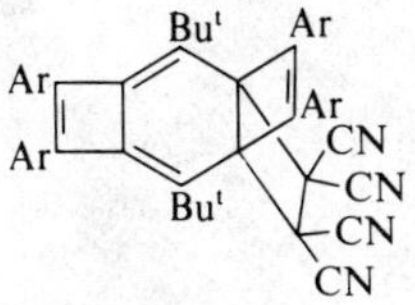          (69)

## 5 Four-membered Heterocyclic Systems

Photolysis (>270 nm) of oxazinone (70) at 7 K gives the β-lactone (71), which undergoes further fragmentation to give MeCN, Bu$^t$CN, propyne, and t-butyl-acetylene. That such products can be isolated strongly suggests the intermediate formation of the azete (72), which undergoes spontaneous cycloreversion *via* the two modes indicated overleaf.[80]

[78] F. Toda, N. Dan, K. Tanaka, and Y. Takehira, *J. Amer. Chem. Soc.*, 1977, **99**, 4529.
[79] F. Toda, and T. Yoshioka, *Chem. Letters*, 1977, 561.
[80] G. Maier and U. Schaefer, *Tetrahedron Letters*, 1977, 1053.

(70) → (71) → (72)

# 2

# Five-membered Ring Systems

BY G. V. BOYD

## 1 Introduction

This chapter deals with aromatic and heteroaromatic compounds containing five-membered rings. Monocyclic systems, their benzo-analogues, and compounds containing two or more five-membered rings, whether fused or linked, are reviewed, but those with an annelated six-membered heterocycle, such as purines and indolizines, are omitted. These will be found in Chapter 4. Syntheses, physical properties, and reactions other than substitutions, such as thermolysis, photolysis, and additions and cycloadditions, are discussed. As in previous volumes, the term 'aromatic' has been interpreted rather liberally, the criterion for inclusion being full conjugation rather than compliance with Hückel's rule. Accordingly, compounds such as fulvenes, pyrazolinones, oxazolinones, and meso-ionic compounds, and even the outright anti-aromatic cyclopentadienones and pentalenes, are dealt with.

## 2 Carbocyclic Compounds

**Cyclopentadiene Radicals, Anions, and Ylides.**—The $^{13}$C n.m.r. spectra of lithium cyclopentadienide and indenide have been determined in several solvents[1] and the e.s.r. spectra of the cyclopentadienyl radical and several of its silicon derivatives, *e.g.* (1), have been recorded.[2] *X*-Ray crystallography shows that lithium 7b*H*-indeno[1,2,3-*jk*]fluorenide has the sandwich structure (2).[3]

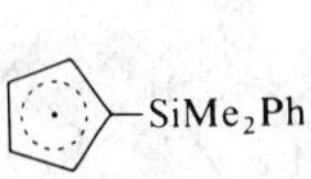

(1) 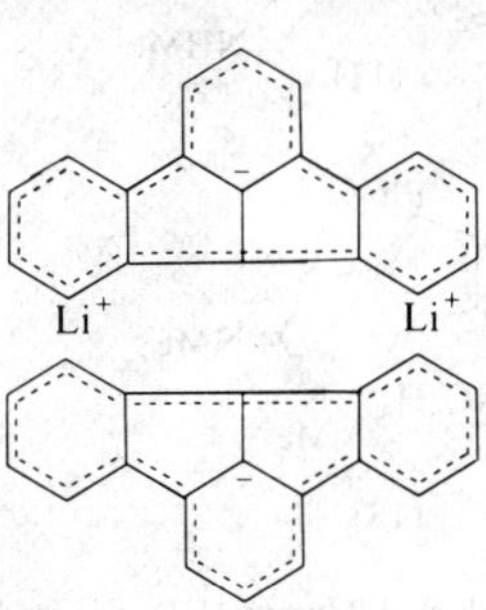          (2)

[1] U. Edlund, *Org. Magn. Resonance*, 1977, **9**, 593.
[2] M. Kira, M. Watanabe, and H. Sakurai, *J. Amer. Chem. Soc.*, 1977, **99**, 7780.
[3] D. Bladauski, H. Dietrich, H. J. Hecht, and D. Rewicki, *Angew. Chem. Internat. Edn.*, 1977, **16**, 474.

The iodonium ylide (3) reacts with the cyclopropenethione (4) to form the stable thiocarbonyl ylide (5);[4] with thioureas, the analogues (6; $R^1$—$R^4$ = H or alkyl) are obtained.[5] Bis(cyclopentadienyliodonium) ylides, such as (8), have been prepared by the action of the appropriate potassium cyclopentadienides on compound (7).[6] The fluorene compounds (9; $R^1$—$R^3$ = H or alkyl) exist as equilibrium mixtures, containing only minor amounts of the dipolar tautomers (9b).[7,8] They react, however, as ylides: thus compound (9; $R^1R^2 = CH_2CH_2$, $R^3$ = H) gives the anil oxide (10) on treatment with nitrosobenzene, and the cyclopentadiene analogue (11) forms a mixture of the fulvene (12) and the cyclopentathiazine derivative (13) by the action of *p*-nitrobenzaldehyde.[9] The guanidino- and selenourea-fluorene compounds (14; X = NH or Se) exist entirely in the non-ylidic forms shown.[8,9]

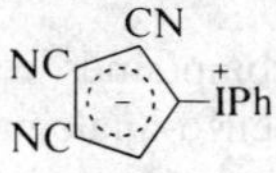

(3) 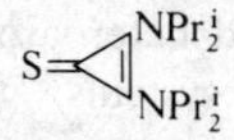          (4) 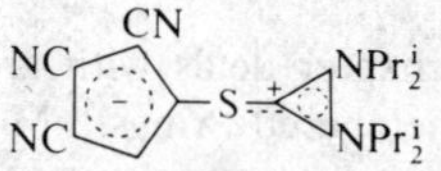          (5)

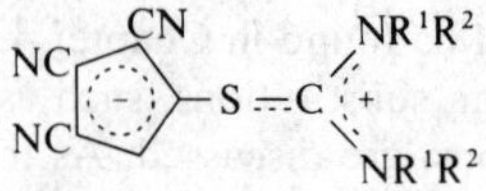

(6) 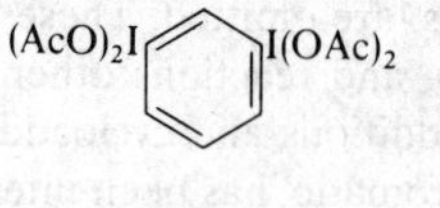          (7) 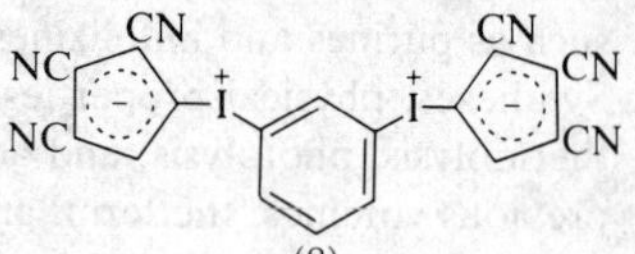          (8)

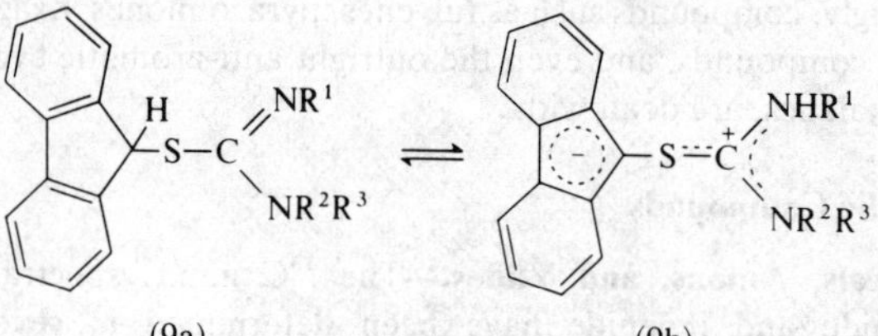

(9a)          (9b) 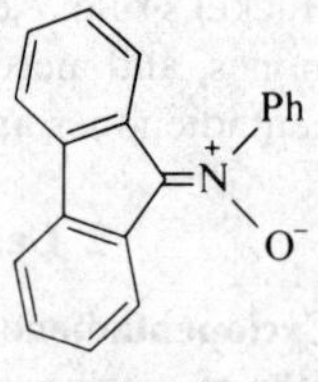          (10)

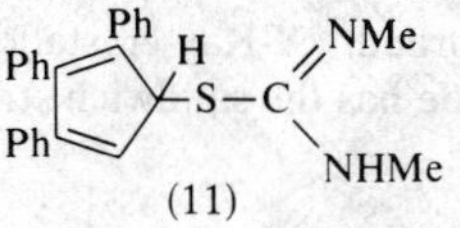

(11) 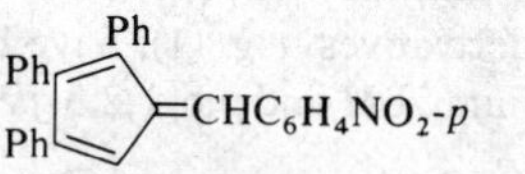          (12)

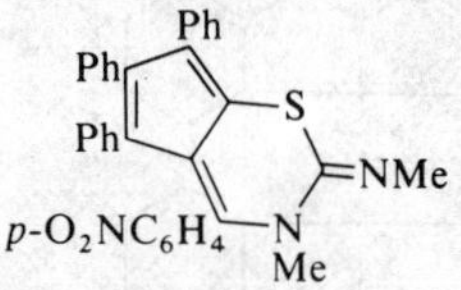

(13) 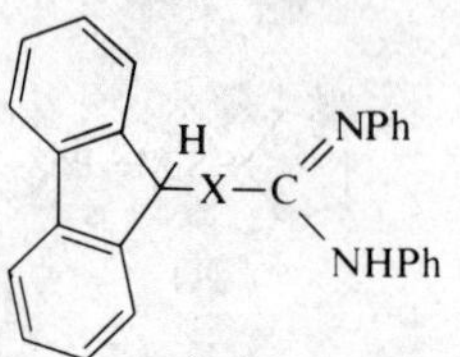          (14)

[4] K. Nakasuji, K. Nishino, I. Murata, H. Ogoshi, and Z. Yoshida, *Angew. Chem. Internat. Edn.*, 1977, **16**, 866.
[5] P. Gronski and K. Hartke, *Chem. Ber.*, 1978, **111**, 272.
[6] K. Friedrich and W. Amann, *Tetrahedron Letters*, 1977, 2885.
[7] K. Hartke, H. Burzlaff, R. Böhme, and A. Shaukat, *Chem. Ber.*, 1977, **110**, 3689.
[8] H. Lumbroso, C. Liegeois, and D. Lloyd, *Tetrahedron*, 1977, **33**, 2583.
[9] D. Lloyd, R. W. Millar, H. Lumbroso, and C. Liegeois, *Tetrahedron*, 1977, **33**, 1379.

Treatment of the sulphonium chloride (15) with thallium cyclopentadienide gives the bis(dimethylsulphonio)cyclopentadienide (16), isolated as the per-chlorate; the tri-substituted salt (17) has also been obtained.[10]

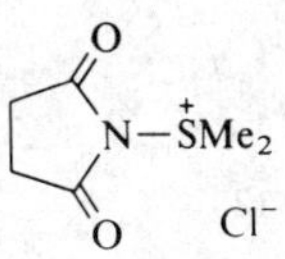

(15) 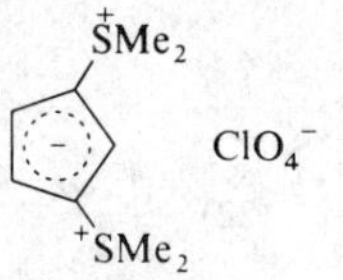  (16) 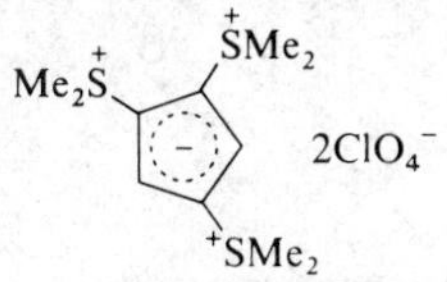  (17)

**Fulvenes.**—The $^{35}$Cl nuclear quadrupole resonance frequencies of a series of tetrachlorofulvenes (18; $R^1,R^2 = H$ or Ar) indicate bond localization in these compounds.[11] The e.s.r. spectra of the radicals (19; X = CH or N) have been reported.[12]

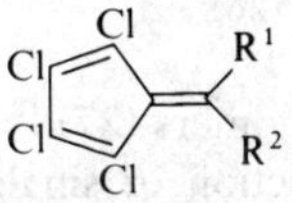

(18) 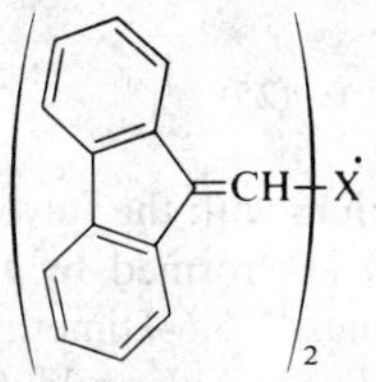  (19)

Fulvenes are formed by the action of sodium cyclopentadienide, indenide, and fluorenide on chloroacetates RCHClOAc.[13] Diaryl thioketones react with cyclopentadienyliron dicarbonyl and its molybdenum and tungsten analogues to yield 6,6-diaryl-fulvenes.[14] The ester (20) dimerises stereoselectively to compound (21).[15] Pyrolysis of phthalide at 740 °C gives unstable fulveneallene

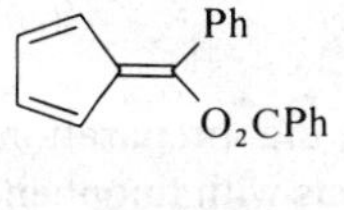

(20) 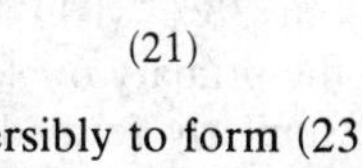  (21)

(22), which adds tetracyanoethylene reversibly to form (23).[16] Compound (24) decomposes on flash vacuum pyrolysis to 1,1-diphenylethylene and benzo[a]fulvene (25) by successive [1,3] migration of the ethano-bridge, [1,5] hydrogen shift, and $10\pi + 2\pi$ cycloreversion.[17] Vinyl-indenes (26; $R^1$, $R^2 = H$ or

[10] K. H. Schlingensief and K. Hartke, *Tetrahedron Letters*, 1977, 1269.
[11] I. Agranat, M. Hayek, and D. Gill, *Tetrahedron*, 1977, **33**, 239.
[12] F. A. Neugebauer and H. Fischer, *Z. Naturforsch.*, 1977, **32b**, 904.
[13] (a) M. Neuenschwander and R. Iseli, *Helv. Chim. Acta*, 1977, **60**, 1061; (b) M. Neuenschwander, R. Vögeli, H. P. Fahrni, H. Lehmann, and J. P. Ruder, *ibid.*, p. 1073.
[14] H. Alper and H.-N. Paik, *J. Amer. Chem. Soc.*, 1978, **100**, 506.
[15] O. Wennerstrom, *Acta. Chem. Scand. (B)*, 1977, **31**, 915.
[16] R. Botter, J. Jullien, J. M. Pechine, J. J. Piada, and D. Solgadi, *J. Electron Spectrosc. Relat. Phenom.*, 1978, **13**, 141 (*Chem. Abs.*, 1978, **88**, 169 224).
[17] R. N. Warrener, K. I. Gell, and M. N. Paddon-Row, *Tetrahedron Letters*, 1977, 53.

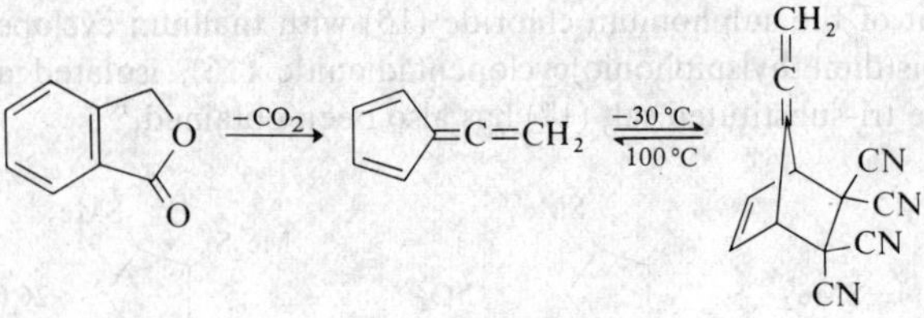

(22)                                                (23)

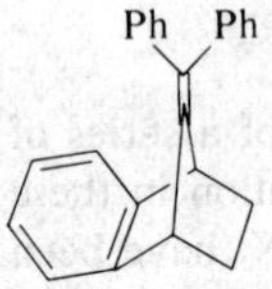

(24) 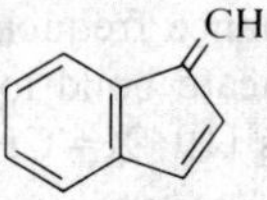                    (25) 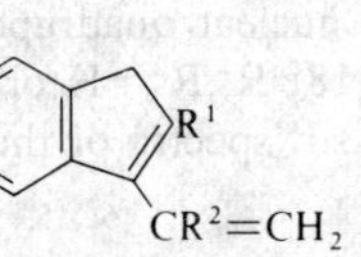                    (26) 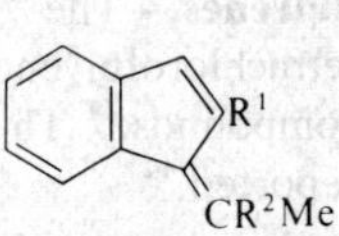                    (27)

Me) exist in equilibrium with the fulvene tautomers (27).[18] Unstable peroxides (28; R = Me or Ph) are formed by the reaction of singlet oxygen with the corresponding fulvenes.[19] 6,6-Dimethylfulvene and cyclopentadienone give a 1:1 mixture of the Diels–Alder adduct (29) and the product (30) of its Cope

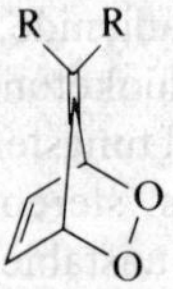

(28) 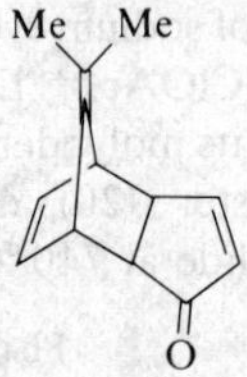                    (29) 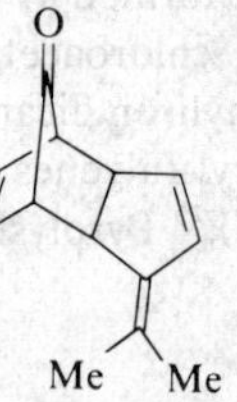                    (30)

rearrangement.[20] Three articles describe the preparation of azulenes from fulvenes: 6-dimethylaminofulvene (31) reacts with thiophen *SS*-dioxide to yield azulene (33; R = H) in 33% yield by elimination of sulphur dioxide and dimethylamine from the primary $6\pi + 4\pi$ cycloadduct (32);[21] 5,6-dichloroazulene (33; R = Cl) is similarly formed (60%) from 3,4-dichlorothiophen *SS*-

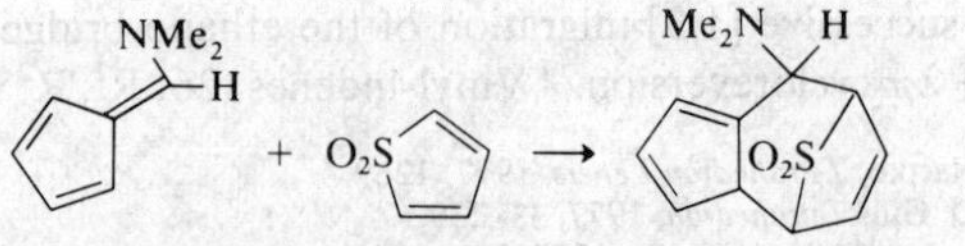

(31)                          +  O₂S          →          (32) 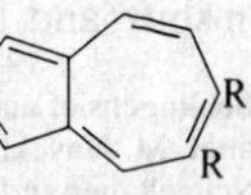                          (33)

[18] R. Begamasco and Q. N. Porter, *Austral. J. Chem.*, 1977, **30**, 1051.
[19] W. Adam and I. Erden, *Angew. Chem. Internat. Edn.*, 1978, **17**, 210.
[20] M. N. Paddon-Row, H. K. Patney, and R. N. Warrener, *Austral. J. Chem.*, 1977, **30**, 2307.
[21] D. Copland, D. Leaver, and W. B. Menzies, *Tetrahedron Letters*, 1977, 639.

dioxide,[22] and the perchlorate (34) condenses with 2-cyanomethylazulene to give the azuleno[1,2-*f*]azulene derivative (35).[23] The crystalline fulvalene (36) is produced by treatment of tetraphenylcyclopentadienone with cyclopentadienylmagnesium bromide, followed by dehydration; it behaves as a diene towards dimethyl acetylenedicarboxylate, giving (37), and as a dienophile with cyclopentadiene to form the adduct (38).[24]

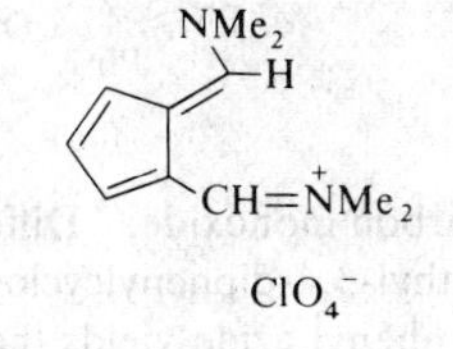

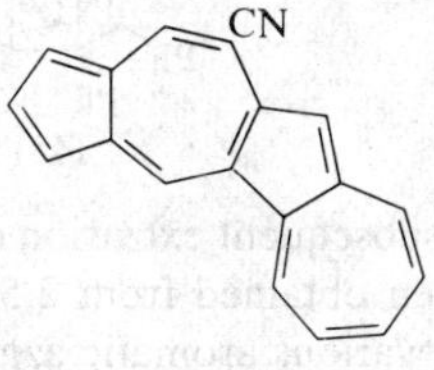

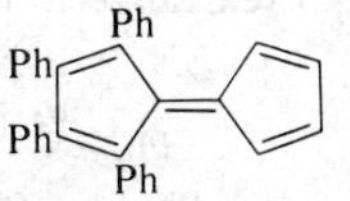

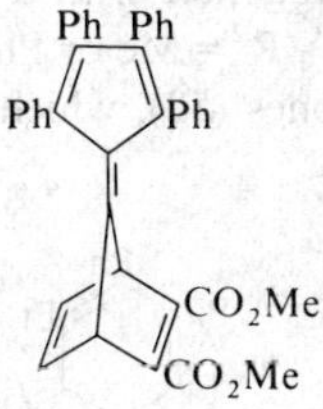

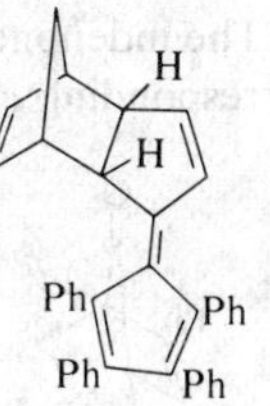

**Cyclopentadienones.**—The stable perchlorocyclopentadienone derivative (40) results from the thermal electrocyclic reaction of the dione (39).[25] Indenone (42) is conveniently prepared by heating the sulphoxide (41);[26] 2,3-diphenyl-indenones (43; R = H, Me, Cl, OMe, or NHAc) are produced in low yields by

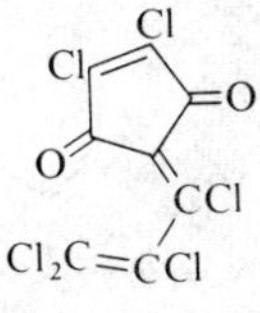

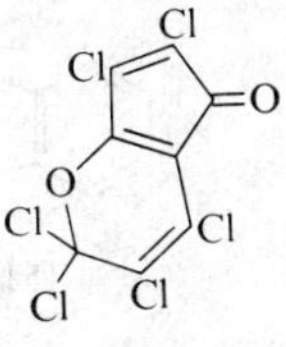

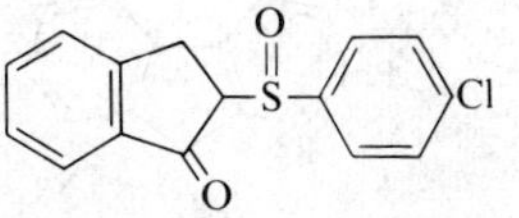

[22] S. E. Reiter, L. C. Dunn, and K. N. Houk, *J. Amer. Chem. Soc.*, 1977, **99**, 4199.
[23] C. Jutz, H. G. Peuker, and W. Kosbahn, *Synthesis*, 1976, 673.
[24] H. Prinzbach, H. Sauter, H. G. Hörster, H. H. Limbach, and L. Knothe, *Annalen*, 1977, 869.
[25] K. Kusada, A. Roedig, and G. Bonse, *Chem. Letters*, 1977, 819.
[26] H. H. Szmant and R. Nanjundiah, *J. Org. Chem.*, 1978, **43**, 1835.

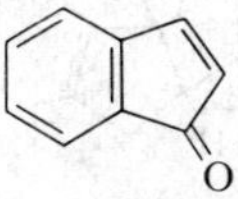

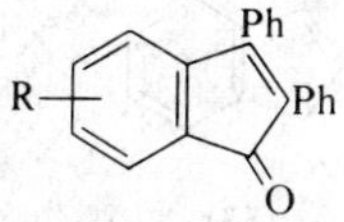

treatment of diphenylacetylene with aromatic aldehydes in the presence of di-t-butyl peroxide.[27] Irradiation of the unstable cyclopentadienone (44; $R^1 =$ Me) gives the cage compound (45).[28] Tetraphenylcyclopentadienone reacts with bis(methoxycarbonyl)carbene to yield the cyclopentadiene derivative (46) by

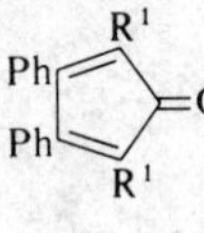

(44) 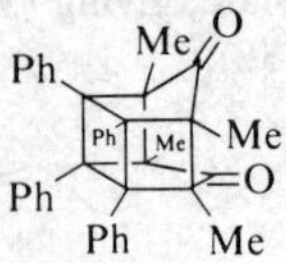                              (45) 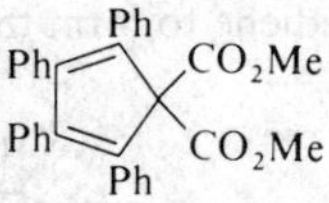                              (46)

1,4-addition and subsequent extrusion of carbon monoxide.[29] Different types of products have been obtained from 2,5-diethyl-3,4-diphenylcyclopentadienone (44; $R^1 =$ Et) and various aromatic azides: phenyl azide yields the cycloadduct (47), 3,5-dinitrophenyl azide the α-pyridone (48), and α-azidostyrene, $N_3CPh=CH_2$, gives compound (49), together with a trace of the Diels–Alder product (50).[30] The indenones (51; $R^1$, $R^2 =$ Me or Ph) are reduced electrochemically to the corresponding *cis*-indanones (52), which at pH 9.3 rearrange to the *trans*-

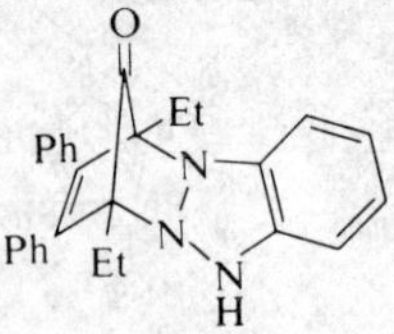

(47) 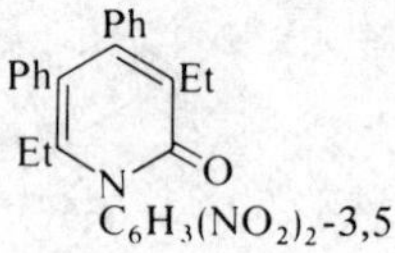                              (48) 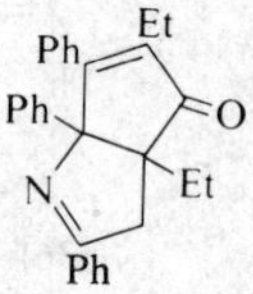                              (49)

isomers.[31] The cyclopentadienones (44; $R^1 =$ Me, Et, or Ph) react with 3-chloroindene (53) to form fluorenes (55) *via* the Diels–Alder adducts (54); the

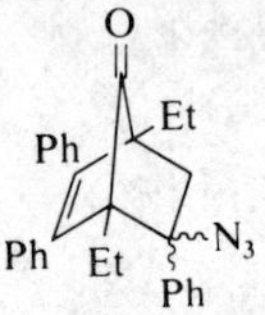

(50) 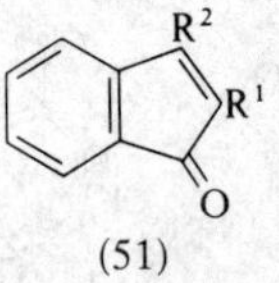                              (51) 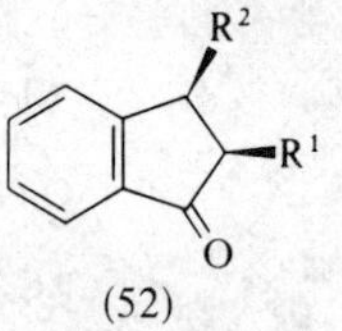                              (52)

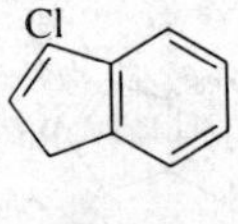

(53) 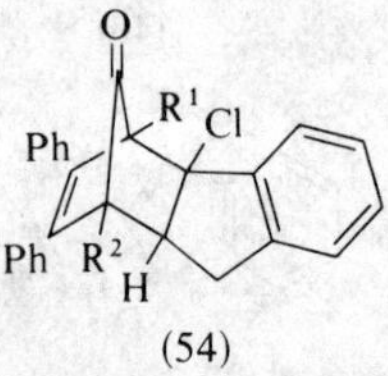                              (54) 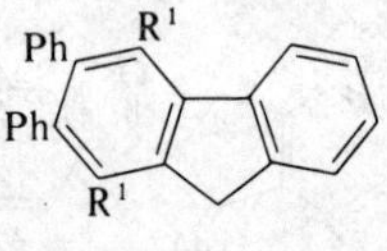                              (55)

[27] A. Monahan, P. Campbell, S. Cheh, J. Fong, S. Grossman, J. Miller, P. Rankin, and J. Vallee, *Synth. Comm.*, 1977, **7**, 553.
[28] B. Fuchs and M. Pasternak, *J.C.S. Chem. Comm.*, 1977, 537.
[29] W. Lilienblum and R. W. Hoffmann, *Chem. Ber.*, 1977, **110**, 3405.
[30] A. Hassner, D. J. Anderson, and R. H. Reuss, *Tetrahedron Letters*, 1977, 2463.
[31] J. Sarrazin and A. Tallec, *Tetrahedron Letters*, 1977, 1579.

condensed cyclopentadienones (56; $R^1$, $R^2 =$ alkyl) similarly yield the indenofluoranthenes (57),[32] and a series of dialkyl-acenaphtho[1,2-*k*]fluoranthenes (59) has been prepared from compounds (56) and chloroacenaphthylene (58).[33]

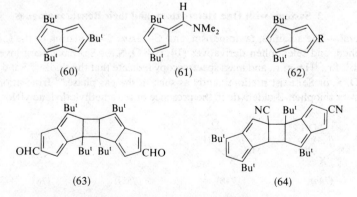

**Pentalenes.**—The *X*-ray structure of the tri-t-butylpentalene (60) has been determined; the compound is planar and the double bonds are localised.[34] Pentalenes (62), together with dimers, are produced by the action of acetylenes $RC{\equiv}CH$ (R = $CO_2Me$, CHO, or CN) on 1,3-di-t-butyl-6-dimethylamino-fulvene (61). The aldehyde (62; R = CHO) dimerized to yield mainly compound (63), whereas the 'head-to-tail' dimer (64) was isolated from the cyano-derivative (62; R = CN).[35] The latter adds water under acidic conditions to give the alcohol (65), while dimethylamine affords the rearranged adduct (66). Cycloaddition reactions of the pentalene di-ester (67) proceed differently with dicyanoacetylene and cyclopentadiene: the former yields the cyclobutene derivative (68), the latter the Diels–Alder product (69).[36]

[32] K. Ghosh and A. J. Bhattacharya, *Indian J. Chem.*, 1977, **15B**, 32.
[33] P. K. Banerjee and A. J. Bhattacharya, *Indian J. Chem.*, 1977, **15B**, 953.
[34] B. Kitschke and H. J. Lindner, *Tetrahedron Letters*, 1977, 2511.
[35] M. Suda and K. Hafner, *Tetrahedron Letters*, 1977, 2449.
[36] M. Suda and K. Hafner, *Tetrahedron Letters*, 1977, 2453.

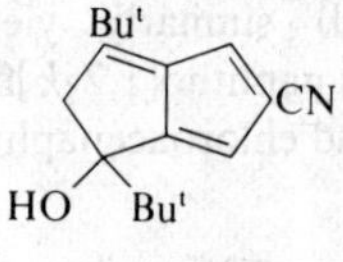

(65) 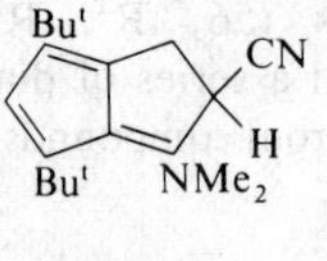                    (66) 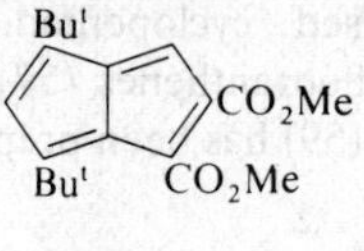                    (67)

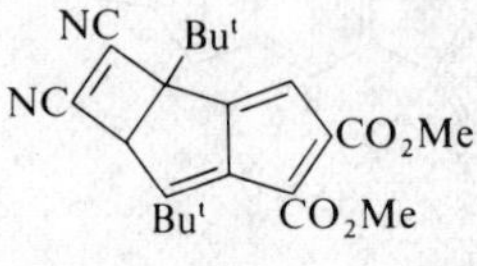

(68) 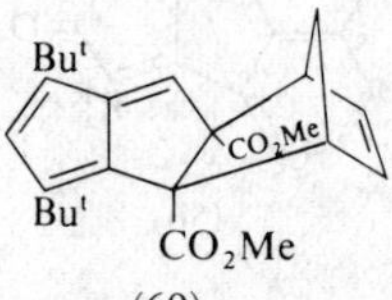                    (69)

The $^1$H n.m.r. spectra of the dibenzopentalene dianions (70)[37] and (71; R = H or Me)[37,38] indicate that there is slight peripheral delocalisation in these formal $18\pi$-systems *cf.* (71a). The dibenzopentalene (72) has been oxidised by SbF$_5$ plus SO$_2$FCl to the $14\pi$-pentalene dication (73).[38]

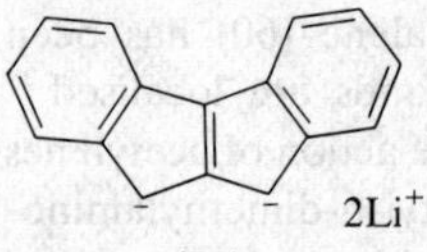

(70) 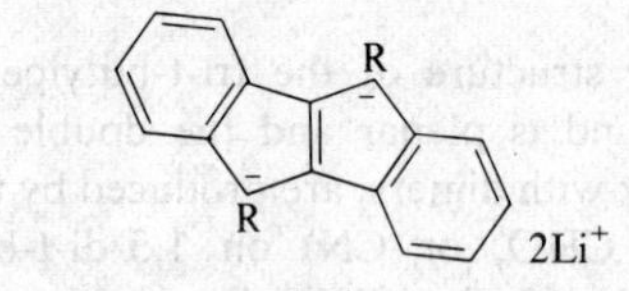                    (71) 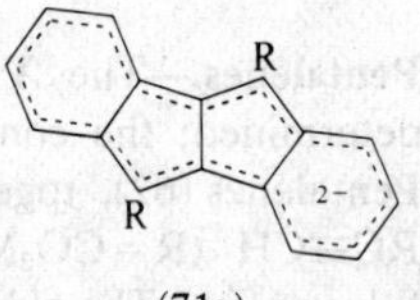                    (71a)

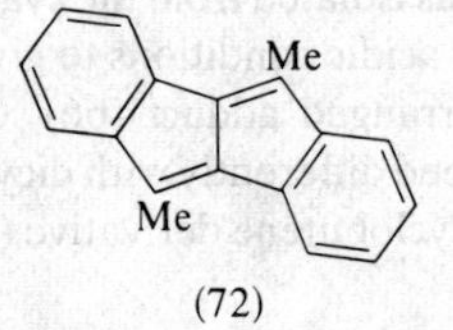

(72) 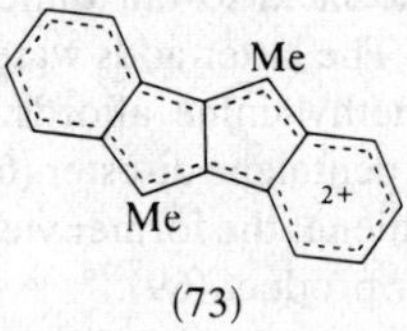                    (73)

### 3 Systems with One Heteroatom, and their Benzo-analogues

**General.**—The acidities, tautomerism, and *C-* *versus* *O*-alkylation of the furan, thiophen, and selenophen derivatives (74; X = O, S, or Se) have been investigated.[39] I.r., $^1$H n.m.r., and mass spectroscopy indicate that the thiols (75 and 76; X = O, S, or Se) exist predominantly as such in the gas phase.[40] Irradiation of furan- or thiophen-2-aldehyde in the presence of tetramethylethylene yields the

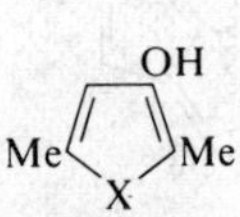

(74a) 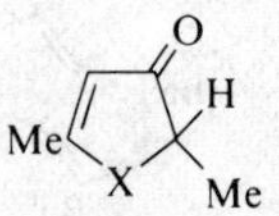                   (74b) 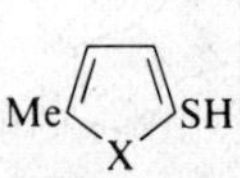                   (75) 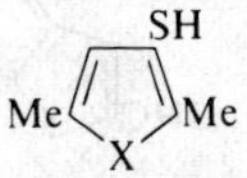                    (76)

[37] T. Uyehara, T. Honda, and Y. Kitahara, *Chem. Letters*, 1977, 1233.
[38] I. Willner and M. Rabinovitz, *J. Amer. Chem. Soc.*, 1978, **100**, 337.
[39] R. Lantz and A. B. Hornfeldt, *Chemica Scripta*, 1976, **10**, 126.
[40] B. Cederlund, R. Lantz, A. B. Hornfeldt, O. Thorstad, and K. Undheim, *Acta Chem. Scand.* (*B*), 1977, **31**, 198.

oxetans (77; X = O or S, respectively) cleanly; furan-3-aldehyde similarly gives (78; X = O), but thiophen-3-aldehyde affords a mixture of the adduct (79) and compound (80), the break-down product of the oxetan (78; X = S).[41]

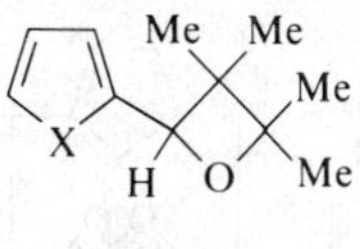

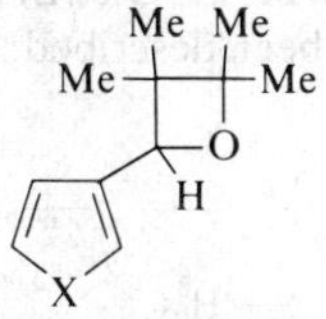

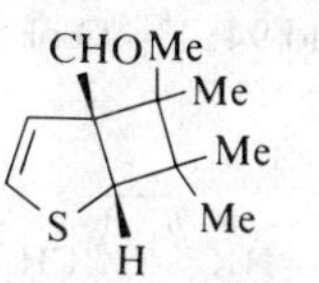

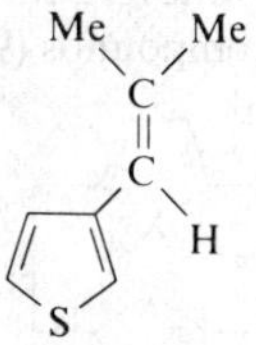

(77)          (78)          (79)          (80)

Gas-phase vacuum pyrolysis of the substituted butadienes (81; X = O, S, or NMe) leads to the bicyclic heterocycles (82) *via* electrocyclization, followed by [1,5] hydrogen shifts.[42] Benzofuran, benzothiophen, and *N*-methylindole react with ethyl diazoacetate to give the corresponding carbene adducts (83; X = O, S, or NMe) as mixtures of geometrical isomers; with benzothiophen, the insertion product (84) is also obtained.[43]

The electronic structures of benzo[*c*]furan (85) and its sulphur, selenium, and nitrogen analogues have been discussed in the light of their $^1$H n.m.r. spectra; for each system, bond fixation in the six-membered ring is indicated, independent of the nature of the heteroatom.[44] Treatment of the rhodium complex (86) with *m*-chloroperbenzoic acid, sulphur, selenium, or nitrosobenzene leads to the quinones (87; X = O, S, Se, or NPh, respectively).[45]

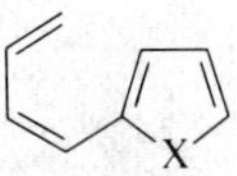

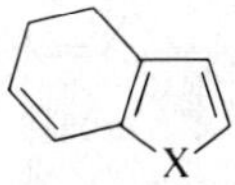

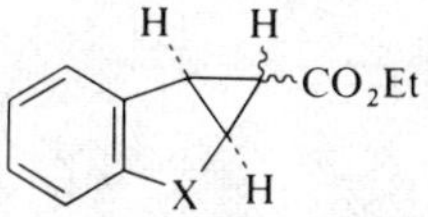

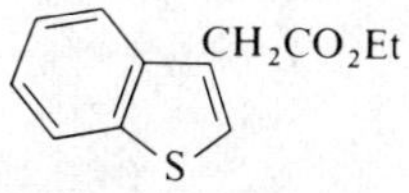

(81)          (82)          (83)          (84)

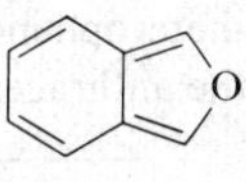

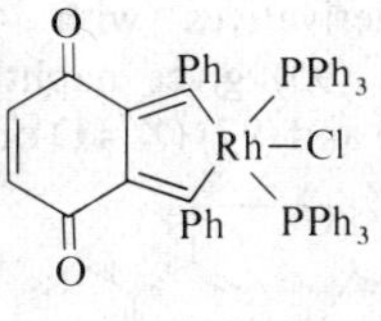

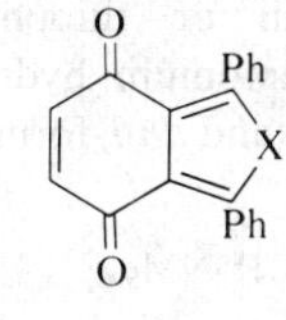

(85)          (86)          (87)

The crystal structures of [2,2](2,5)furanophane (88; X = Y = O) and of the corresponding thienophane (88; X = Y = S) have been determined.[46] The 'cross-breeding' dimerisation of the dimethylene compounds (90; Z = O) and (90; Z = S), produced by heating a mixture of the quaternary furyl- and thienyl-ammonium hydroxides (89; Z = O) and (89; Z = S), yields the mixed compound

[41] T. S. Cantrell, *J. Org. Chem.*, 1977, **42**, 3774.
[42] B. I. Rosen and W. P. Weber, *Tetrahedron Letters*, 1977, 151.
[43] E. Wenkert, M. E. Alonso, H. E. Gottlieb, E. L. Sanchez, and R. Pellicciari, *J. Org. Chem.*, 1977, **42**, 3945.
[44] E. Chacko, J. Bornstein, and D. J. Sardella, *J. Amer. Chem. Soc.*, 1977, **99**, 8248.
[45] J. Hambrecht and E. Müller, *Annalen*, 1977, 387.
[46] N. B. Pahor, M. Calligaris, and L. Randaccio, *J.C.S. Perkin II*, 1978, 42.

(88; X = O, Y = S). Hydrolysis of the latter gives 3,6-dioxo[8](2,5)thienophane (91), whose dynamic properties, *i.e.* ring *versus* chair flipping, have been investigated by variable-temperature $^1$H n.m.r. spectroscopy.[47] The preparation of paracyclo-furano- and -thieno-phanes (92; X = O or S) and of the multi-layered compounds (93 and 94; X = O or S) has been described; the electronic spectra of

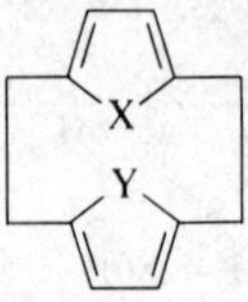

(88) 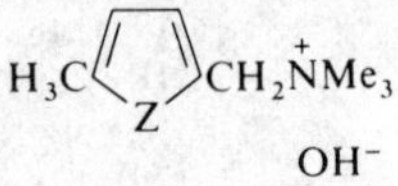                    (89) 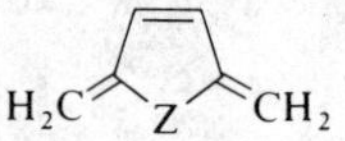                    (90) 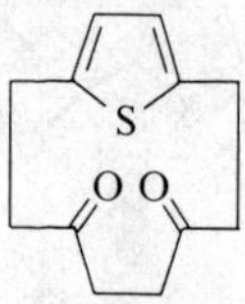                    (91)

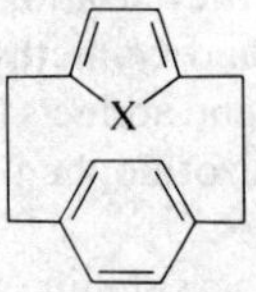

(92)

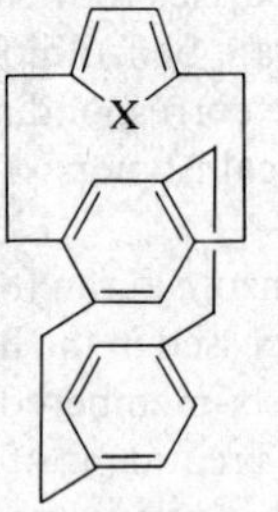

(93)

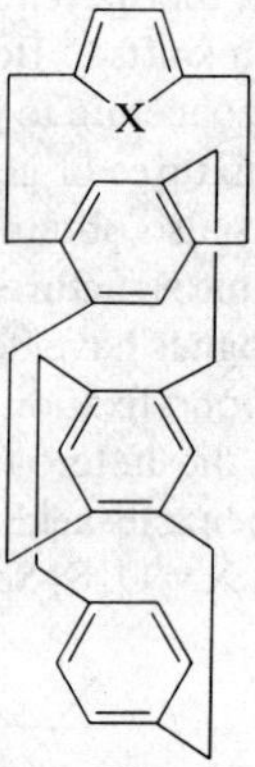

(94)

these compounds reveal stronger trans-annular interactions in the thienophanes than in the furan analogues.[48] Cross-breeding 'co-dimerisation' of the appropriate furan or thiophen derivatives with 4-(1-methylnaphthyl)methyl-trimethylammonium hydroxide (95) gives naphthaleno-heterophanes, which exist in *syn*- and *anti*-forms (96) and (97) (X = O or S). In the anthraceno-series,

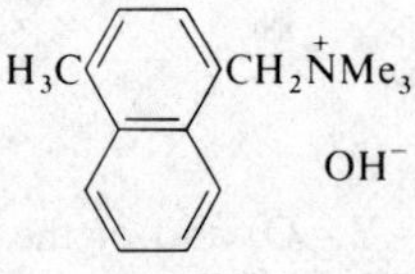

(95) 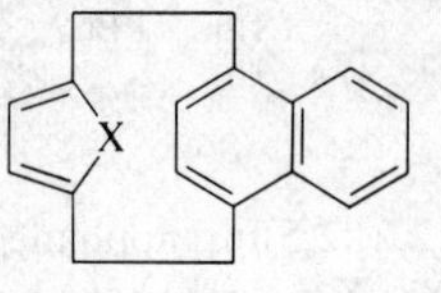                    (96) 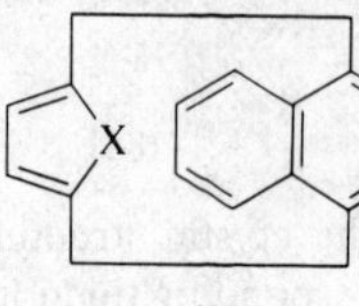                    (97)

the 9,10-bridged compounds (98; X = O or S) show stronger trans-annular electronic interactions than the [1,4]-analogues (99; X = O or S).[49] The geometries of the furanophanes (97 and 98; X = O) have been studied.[50] The

[47] A. W. Lee, P. M. Keehn, S. M. Ramos, and S. M. Rosenfeld, *Heterocycles*, 1977, **7**, 81.
[48] T. Otsubo, S. Mizogami, N. Osaka, Y. Sakata, and S. Misumi, *Bull. Chem. Soc. Japan*, 1977, **50**, 1841.
[49] T. Otsubo, S. Mizogami, N. Osaka, Y. Sakata, and S. Misumi, *Bull. Chem. Soc. Japan*, 1977, **50**, 1858.
[50] C. B. Shana, S. M. Rosenfeld, and P. M. Keehn, *Tetrahedron*, 1977, **33**, 1081.

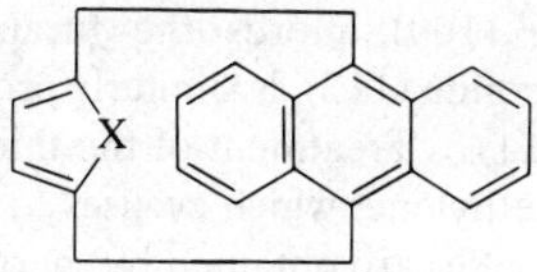

(98) 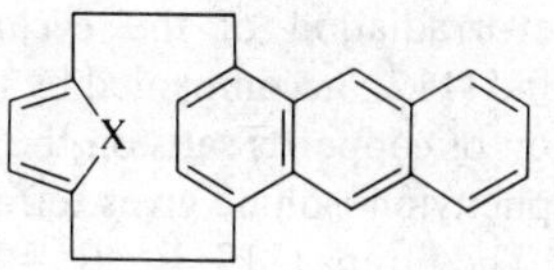             (99)

synthesis of macrocyclic compounds containing furan, thiophen, and pyridine rings has been reviewed.[51]

**Furans and Benzofurans.**—*Formation.* The acid-catalysed condensation of 2-chlorocyclopentanone with dimethyl β-oxoglutarate leads to the furan ester (100).[52] Compound (101) is obtained by the action of nitrotrichloroethylene on acetylacetone in the presence of bases.[53] 1-Diethylaminopropyne, $MeC \equiv CNEt_2$, and benzoin undergo three competing reactions to give a mixture of the furans (102) and (103) and the butenolide (104).[54] The sulphonium ylide (105) and

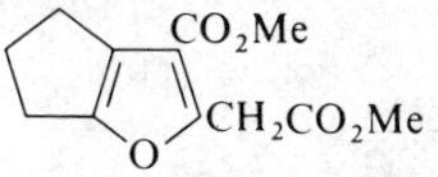

(100) 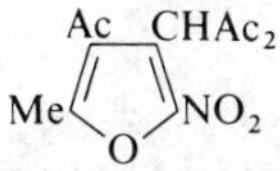          (101)

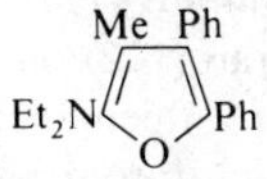

(102) 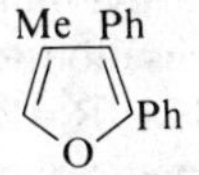      (103) 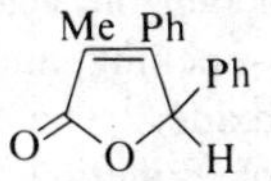      (104)

dimethyl acetylenedicarboxylate form the tri-ester (108); the authors suggest initial formation of the adduct (106), which undergoes a [1,3] shift of the acetyl group to yield the rearranged ylide (107), which collapses to the product as shown in Scheme 1.[55]

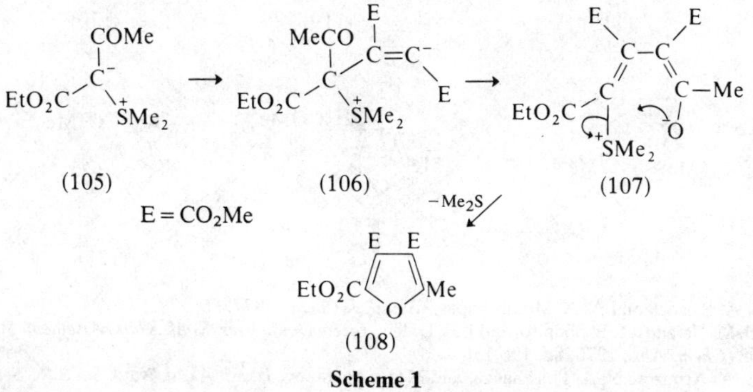

(105)          (106)          (107)

$E = CO_2Me$        $-Me_2S$

(108)

**Scheme 1**

[51] G. R. Newkome, J. D. Sauer, J. M. Roper, and D. C. Hager, *Chem. Rev.*, 1977, **77**, 513.
[52] O. Campos and J. M. Cook, *J. Heterocyclic Chem.*, 1977, **14**, 711.
[53] V. A. Buevich, L. I. Deiko, and V. V. Perekalin, *Zhur. org. Khim.*, 1977, **13**, 972.
[54] S. I. Pennanen, *Heterocycles*, 1977, **6**, 701.
[55] H. Saikachi and T. Kitagawa, *Chem. and Pharm. Bull (Japan)*, 1977, **25**, 809.

Singlet-irradiation of the cyclopropene (109) affords the furan (110);[56] compound (112), accompanied by the butenolide (113), is similarly produced by the action of copper(I) salts on the ester (111).[57] Treatment of the thiiran (114) with triphenylphosphine gives tetrabenzoylethylene, which cyclises to the furan (115).[58] The furans (117; $R^1$, $R^2$ = H, Me, or Ph) are obtained by thermolysis of the cyclobutenediols (116); intermediate 1,4-diketones were isolated in two

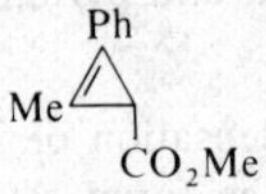

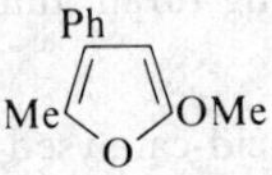

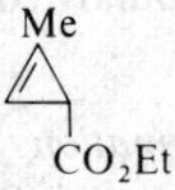

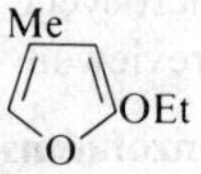

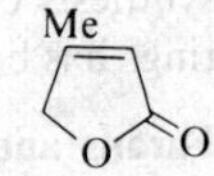

(109)          (110)          (111)          (112)          (113)

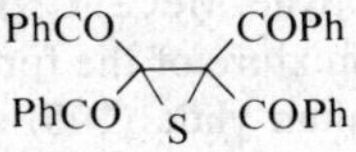

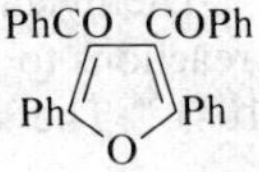

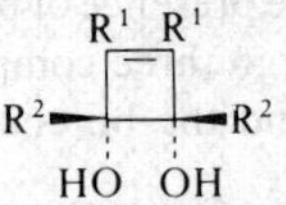

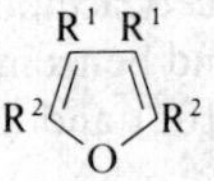

(114)               (115)               (116)               (117)

cases.[59] 1,3-Dioxolan-4-yl benzoates (118; $R^1-R^3$ = H or alkyl) decompose on heating to the dioxolanylium salts (119), which yield furans (120) by elimination of water and benzoic acid.[60] The pyrylium perchlorate (121) undergoes hydrolysis, ring-opening, and re-cyclisation to the furan (122) on treatment with sodium hydroxide.[61] The esters (123; R = 2-furyl or 2-thienyl) are produced by the action of dimethyl acetylenedicarboxylate on furoin or its thiophen analogue.[62]

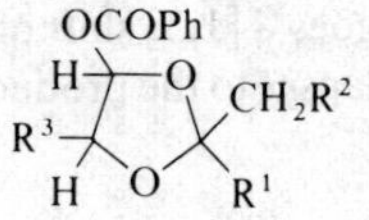

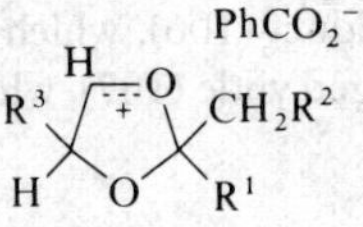

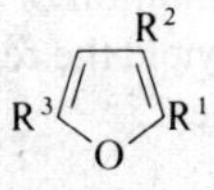

(118)               (119)               (120)

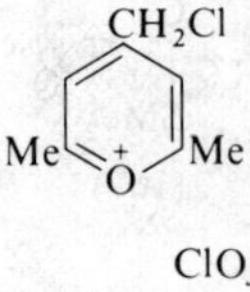

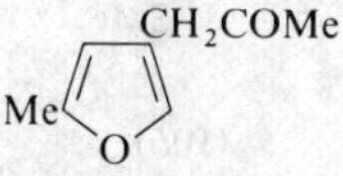

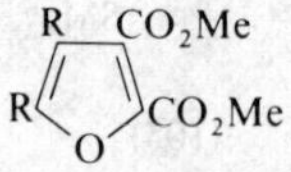

(121)               (122)               (123)

[56] J. A. Pincock and A. A. Moutsokapas, *Canad. J. Chem.*, 1977, **55**, 979.
[57] O. M. Nefedov, I. E. Shapiro, and I. E. Dalgii, *Tezisy Dokl.-Vses. Konf. Khim. Atsetilena 5th*, 1975, 68 (*Chem. Abs.*, 1978, **88**, 190 136).
[58] B. A. Arbuzov, N. A. Polezhaeva, and M. N. Agafonov, *Izvest. Akad. Nauk S.S.S.R., Ser. khim.*, 1977, 1399.
[59] J. Hambrecht, *Synthesis*, 1977, 280.
[60] H. D. Scharf and E. Wolters, *Chem. Ber.*, 1978, **111**, 639.
[61] V. I. Dulenko, N. N. Alekseev, V. M. Golyak, and L. V. Dulenko, *Khim. geterotsikl. Soedinenii*, 1977, 1135.
[62] S. I. Pennanen, *J. Heterocyclic Chem.*, 1977, **14**, 745.

Photolysis of the benzocyclopropenes (124; R = H or Ph) affords fulvene-allenes (125) and minor amounts of the benzofurans (126).[63] Dibenzofuran is produced by the action of pyridinium bromide on 2,2'-dihydroxybiphenyl.[64]

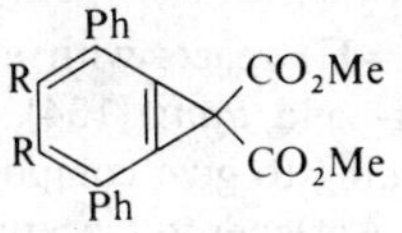

(124) 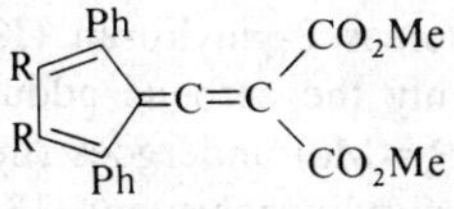          (125) 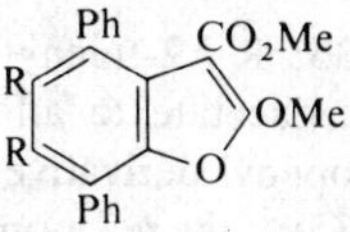          (126)

Irradiation of chlorinated diphenyl ethers, possessing at least one *ortho*-chlorine atom, yields mono-, di-, and tri-chlorobenzofurans by reductive cyclization.[65] The naphtho[1,8-*bc*]furylium salt (128) is prepared by treatment of the hydroxy-ketone (127) with acetic anhydride and perchloric acid.[66] The synthesis of

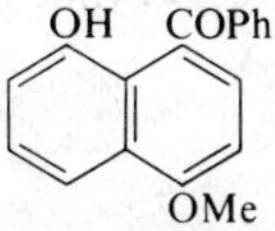

(127) 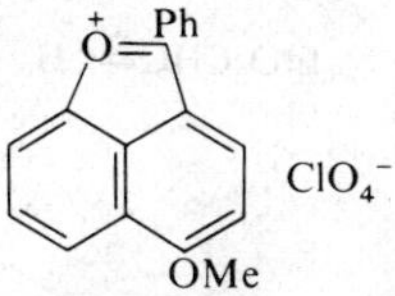          (128)

phenanthro[4,5-*bcd*]furan (129) has been described.[67] 2,3-Diphenylbenzofuran, on irradiation, forms the dihydrobenzo[*b*]phenanthro[9,10-*d*]furan (130), which has been converted into the fully aromatic compound by bromination–dehydrobromination.[68]

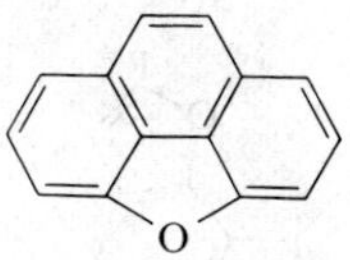

(129) 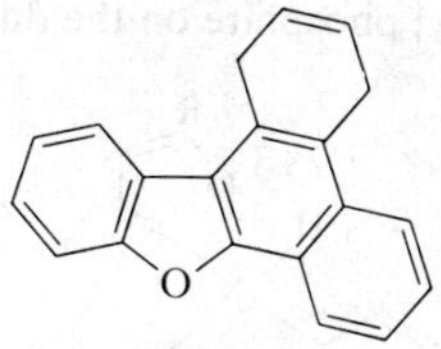          (130)

*Reactions.* Birch reduction of 2-furoic acid with lithium in liquid ammonia gives the 2,5-dihydro-compound (131) in 90% yield.[69] Hydrogenation of the vinyl-furans (132; R = CO₂H, CO₂Me, Ac, or COPh) in the presence of nickel bromide occurs selectively at the exocyclic double bond.[70] Anodic methoxylation of 2-(2-thienylmethyl)-furan yields a mixture of geometrically isomeric adducts

[63] H. Duerr and H. J. Ahr, *Tetrahedron Letters*, 1977, 1991.

[64] J. P. Bachelet, P. Demerseman, and R. Royer, *J. Heterocyclic Chem.*, 1977, **14**, 1409.

[65] (a) G. G. Choudhry, G. Sundstrom, F. W. M. Van der Wielen, and O. Hutzinger, *Chemosphere*, 1977, **6**, 327; (b) A. Norstrom, K. Andersson, and C. Rappe, *ibid.*, p. 241.

[66] V. V. Mezheritskii, O. N. Zhukovskaya, V. V. Tkachenko, and G. N. Dorofeenko, *Khim. geterotsikl. Soedinenii*, 1977, 1693.

[67] T. Horaguchi, *Bull. Chem. Soc. Japan*, 1977, **50**, 3329.

[68] A. Couture, A. Lablache-Combier, and H. Ofenberg, *Org. Photochem. Synth.*, 1976, **2**, 7.

[69] H. R. Divanford and M. M. Joullie, *Org. Prep. Proced. Internat.*, 1978, **10**, 94.

[70] M. Bartok, K. Lakos-Lang, L. G. Bogatskaya, G. L. Kamalov, and A. V. Bogatskii, *Doklady Akad. Nauk S.S.S.R.*, 1977, **234**, 590.

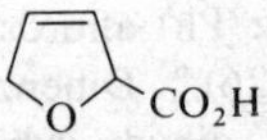

(131) 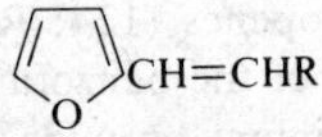                    (132) 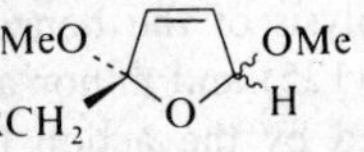                    (133)

(133; R = 2-thienyl).[71] Whereas 2-vinylfuran (132; R = H) reacts with ethyl diazoacetate to afford mainly the carbene adducts *cis*- and *trans*-(134), the propenyl derivative (132; R = Me) undergoes ring-opening to give compound (135);[72] the reaction with furfurylideneacetone (132; R = Ac) leads to 1-acetyl-2-(2-furyl)-cyclopropane (136).[73] Ring expansion of the alcohols (137; R = alkyl) to the dihydropyranones (138) on treatment with pyridinium chlorochromate has been observed.[74] The photochemical reaction of furan with *p*-dicyanobenzene in methanol to yield compound (139) is sensitized by naphthalene or phenanth-

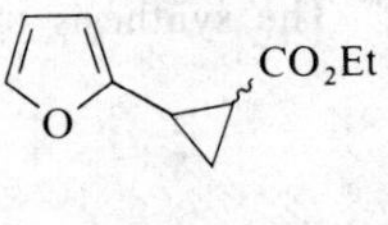

(134) 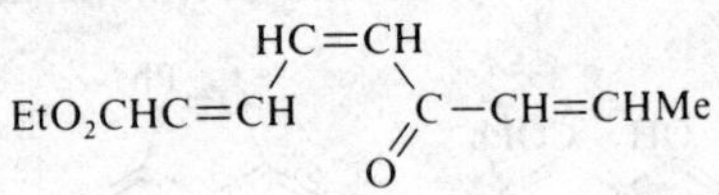                    (135) 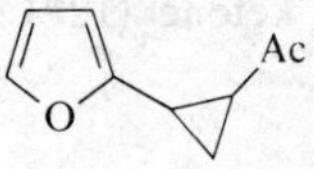                    (136)

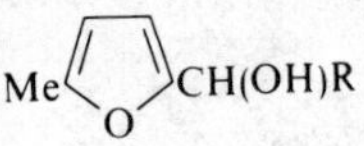

(137) 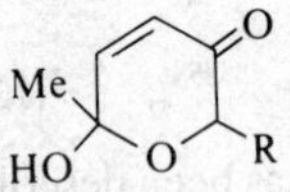                    (138) 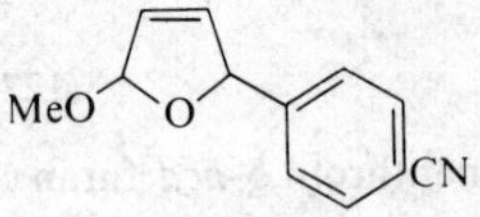                    (139)

rene.[75] Furo[3,2-*c*]carbazolephosphonates (141; R = Me, Et, or Bu^t), represen-tatives of a new ring system, are obtained in an unexpected way, namely by the action of triethyl phosphite on the difuryl-*o*-nitrophenylmethanes (140).[76]

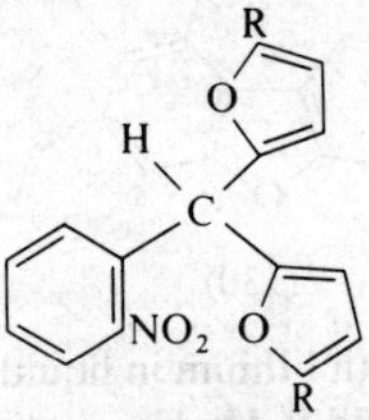

(140) 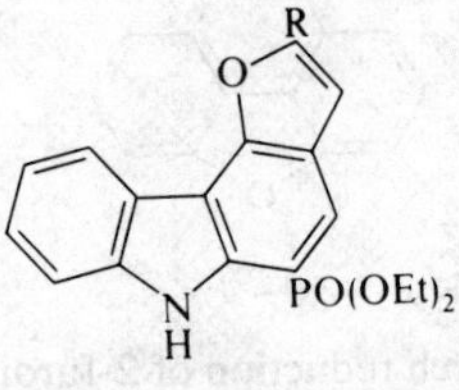                    (141)

The addition of the nitrile oxide (142) to furan is regiospecific, giving solely compound (143); with benzofuran, on the other hand, both possible adducts, (144) and (145), are obtained.[77] The Diels–Alder reaction of furan with maleic anhydride has been re-investigated. Both *endo*- and *exo*-adducts are formed; the

[71] J. Srogl, M. Janda, I. Stibor, and Z. Salajka, *Coll. Czech. Chem. Comm.*, 1977, **42**, 1361.
[72] O. M. Nefedov, M. Shostakovskii, and A. E. Vasilvitskii, *Angew. Chem. Internat. Edn.*, 1977, **16**, 646.
[73] V. A. Zefirova and R. A. Karakhanov, *Khim. geterotsikl. Soedinenii*, 1977, 1032.
[74] G. Piancatelli, A. Scettri, and M. D'Aurie, *Tetrahedron Letters*, 1977, 2199.
[75] C. Pac, A. Nakasone, and H. Sakurai, *J. Amer. Chem. Soc.*, 1977, **99**, 5806.
[76] G. Jones and W. H. McKinley, *Tetrahedron Letters*, 1977, 2457.
[77] P. L. Beltrame, M. G. Cattania, V. Redaelli, and G. Zecchi, *J.C.S. Perkin II*, 1977, 706.

ArCNO

(142)    (143) 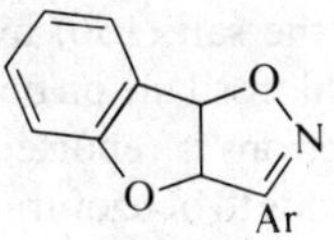    (144) 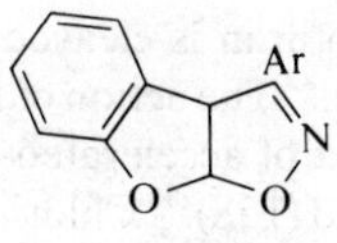    (145)

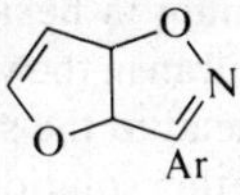

$$Ar = \text{Me Cl} \diagup \text{Me} \diagdown \text{Me Cl}$$

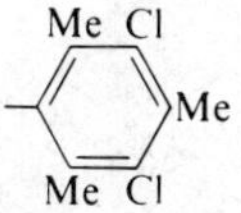

former is favoured kinetically to the extent of 3.8 kcal mol$^{-1}$, but the latter is thermodynamically more stable by 1.9 kcal mol$^{-1}$. Both isomers are formed reversibly; the *exo*-compound (146) is eventually isolated.[78] Cyclo-butano[*c*]benzyne (148), generated by the action of butyl-lithium on compound (147), has been trapped by furan as the cycloadduct (149).[79] Treatment of the

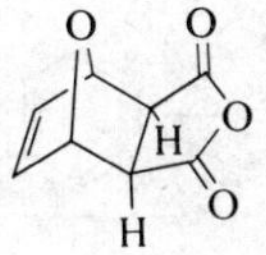

(146) 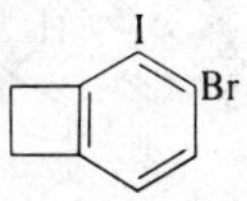    (147) 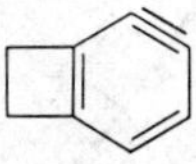    (148) 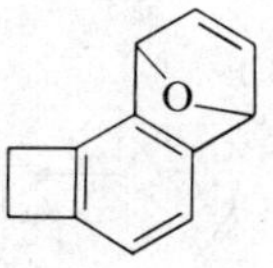    (149)

chloro-benzocycloheptadienes (150) or (151) with potassium t-butoxide in the presence of furan yields compound (153), the cycloadduct of furan to the transient allene (152).[80] Benzhydroxamic acid, PhCONHOH, reacts with 2,5-dimethyl-furan in the presence of silver oxide to yield the dioxazoline (155); it has been suggested that the reaction involves Diels–Alder addition of the unstable nitroso-compound PhCONO to give (154), which rearranges to the product as shown.[81]

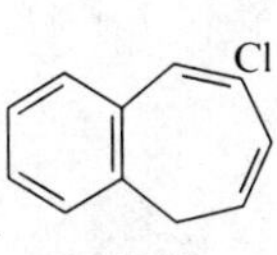

(150) 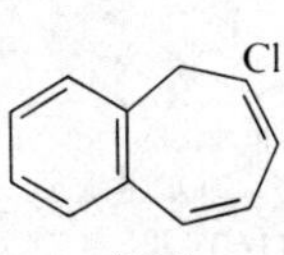    (151) 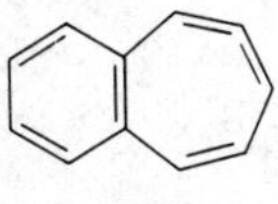    (152) 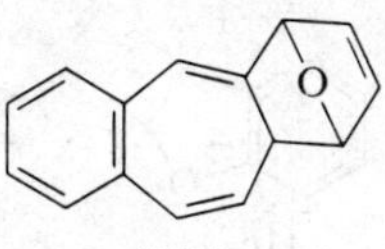    (153)

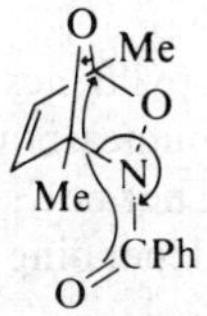

(154) 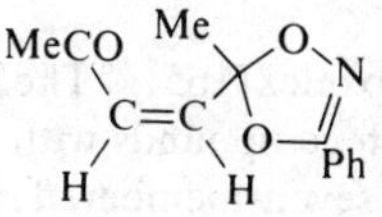    (155)

[78] M. W. Lee and W. C. Herndon, *J. Org. Chem.*, 1978, **43**, 518.
[79] R. L. Hillard III and K. P. C. Vollhardt, *Angew. Chem. Internat. Edn.*, 1977, **16**, 399.
[80] E. E. Waali, J. M. Lewis, D. E. Lee, E. W. Allen III, and A. K. Chappell, *J. Org. Chem.*, 1977, **42**, 3460.
[81] C. J. B. Dobbin, D. Mackay, M. R. Penney, and L. H. Dao, *J.C.S. Chem. Comm.*, 1977, 703.

Benzofuran is cleaved to the salt (156) by lithium in hexamethylphosphoric triamide.[82] The action of trichlorosilane on benzofuran in the vapour phase under the effect of accelerated electrons is reported to lead to the silicon heterocycles (157) and (158).[83] Chlorine adds to benzofuran in ether solution to form a mixture of *cis-* and *trans-*dichlorides (159), which, on treatment with sodium ethoxide,

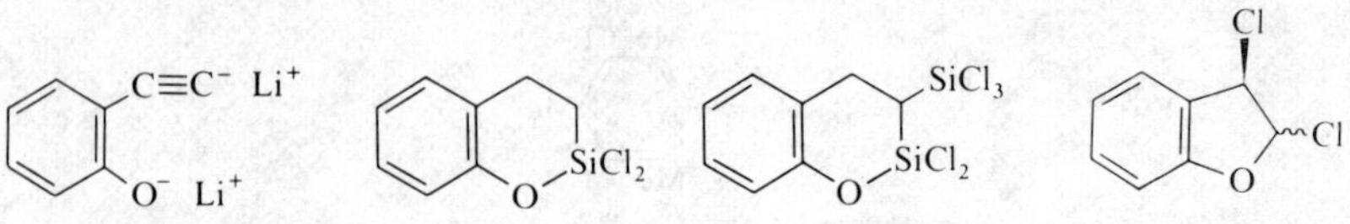

(156)                    (157)                    (158)                    (159)

eliminate hydrogen chloride to yield 3-chlorobenzofuran.[84] A mixture of four products (160)–(163) is obtained by the sensitised photo-addition of dimethyl acetylenedicarboxylate to benzofuran. These are thought to be formed as shown in Scheme 2.[85] Singlet oxygen adds to 2-vinylbenzofurans (164; $R^1$—$R^3$ = H, Me,

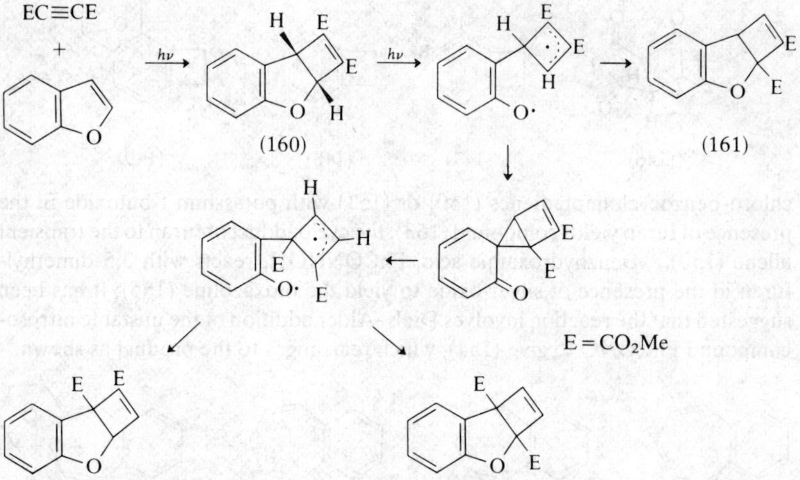

(160)                                                                (161)

E = CO₂Me

(162)                                        (163)

**Scheme 2**

or Ph) to yield the peroxides (165).[86] The furotropylidenes (166) and (167) (R = H or Me) form sandwich compounds with chromiumcarbonyl, in which the metal atom is bonded to the seven-membered ring.[87] Thermolysis of the 2-(azidovinyl)-benzofurans (168) gives different products, depending on the nature of the

[82] R. A. Karakhanov, S. K. Sharipova, M. I. Rozhkova, and E. A. Viktorova, *Khim. geterotsikl. Soedinenii*, 1978, 22.
[83] B. I. Vainshtein, I. P. Tokareva, T. L. Krasnova, V. V. Stepanov, and E. A. Chernyshev, *Khim. geterotsikl. Soedinenii*, 1977, 554.
[84] E. Baciocchi, S. Clementi, and G. V. Sebastiani, *J. Heterocyclic Chem.*, 1977, **14**, 359.
[85] A. H. Tinnemans and D. C. Neckers, *J. Org. Chem.*, 1977, **42**, 2374.
[86] M. Matsumoto, S. Dobashi, and K. Kondo, *Bull. Chem. Soc. Japan*, 1977, **50**, 3026.
[87] M. El Borai, R. Guilard, P. Fournari, J. Dusausoy, and J. Protas, *Bull. Soc. chim. France*, 1977, 75.

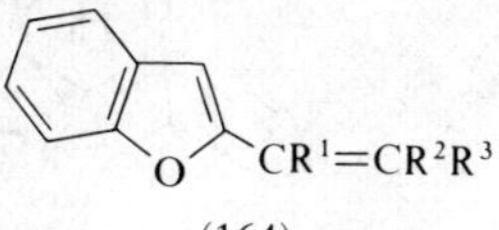

(164)

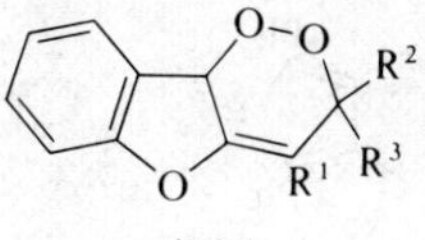

(165)

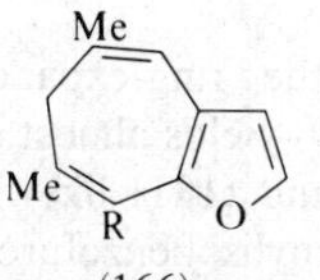

(166)

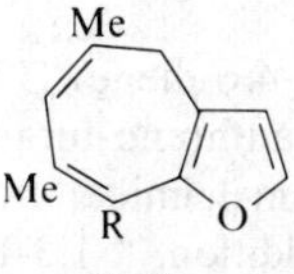

(167)

substituent at C-3. Compound (168; R = H) yields the benzofuro-pyrrole (169), the methyl analogue (168; R = Me) the benzofuro-pyridine (170), and the phenyl compound (168; R = Ph) the condensed azepine (171).[88] 3-Acetyl-2-(1-naph-thyl)-benzofuran (172) has been converted by standard methods into the benzo-phenaleno-furan (173), a member of a new ring system.[89]

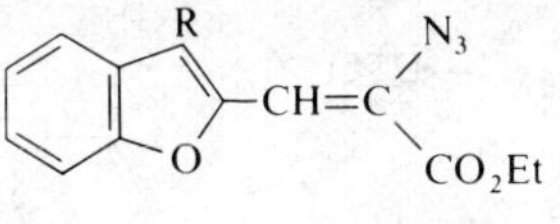

(168)

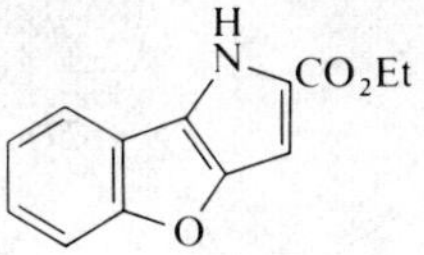

(169)

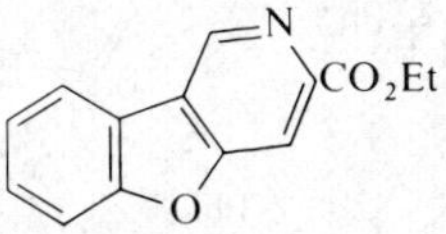

(170)

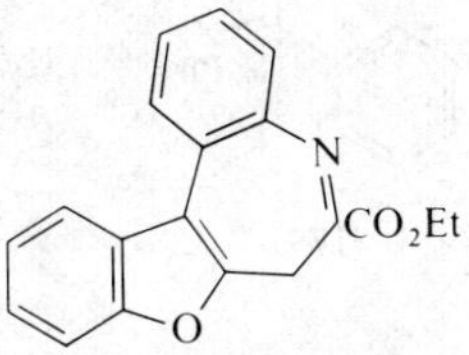

(171)

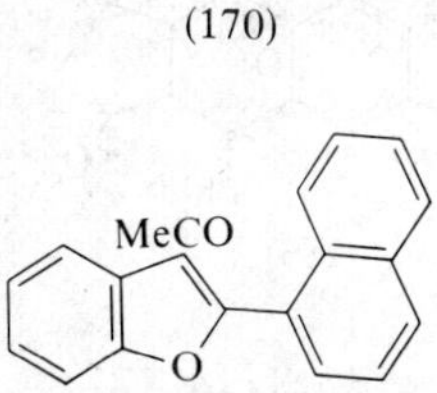

(172)

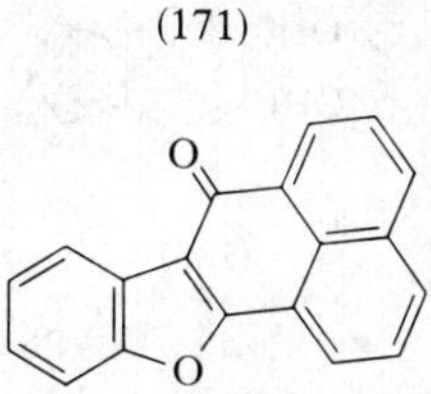

(173)

1,3-Diphenylisobenzofuran (174) undergoes stereospecific Diels–Alder reactions with the cyclopropenes (175; $R^1$, $R^2$ = H or Me) to form compounds (176);[90]

[88] K. Isomura, H. Taguchi, T. Tanaka, and H. Taniguchi, *Chem. Letters*, 1977, 401.
[89] M. Brenner and C. Brush, *Tetrahedron Letters*, 1977, 419.
[90] I. G. Bolesov, L. G. Zaitseva, V. V. Plemenkov, I. B. Avezov, and L. S. Surmina, *Zhur. org. Khim.*, 1978, **14**, 71.

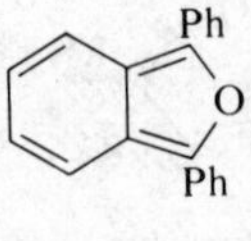

(174) 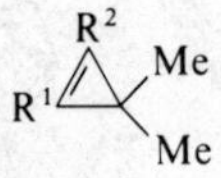                        (175) 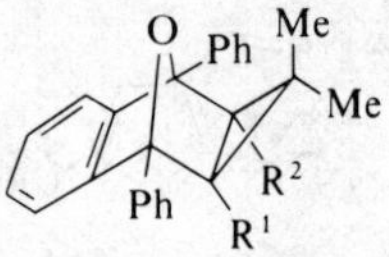                              (176)

with spiro[2,4]hepta-4,6-diene (177), the ring-expanded adduct (178) is obtained.[91] The phenanthreno-furan (179) yields almost exclusively the adduct (180) with *N*-phenylmaleimide; compound (181) likewise undergoes stereospecific *endo*-cycloaddition.[92] 1,3-Diphenylisobenzofuro[5,6-*d*]tropone (185) has been prepared from the dialdehyde (182) by the following sequence: reaction with *N*-phenylmaleimide gave a cycloadduct, which existed in the cyclic hydrated form (183); condensation of the latter with acetone yielded (184), which on heating underwent Diels–Alder fragmentation to the product.[93]

(177) 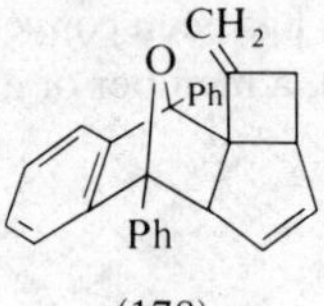                    (178)

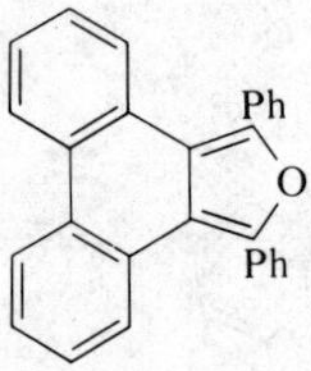

(179) 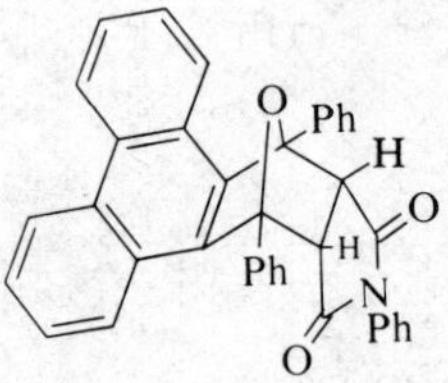                  (180) 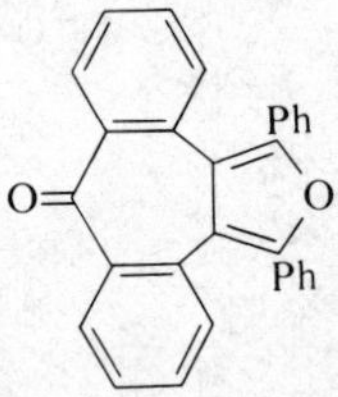                  (181)

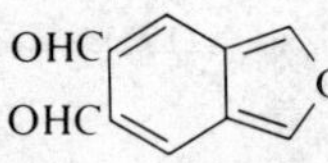

(182) 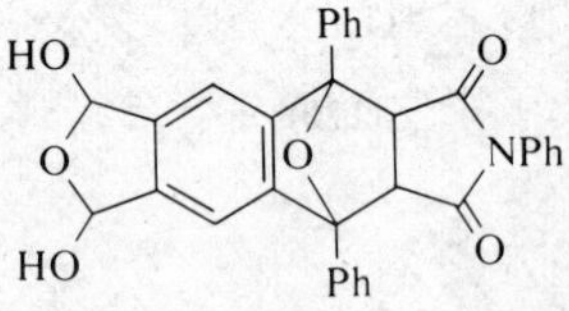                        (183)

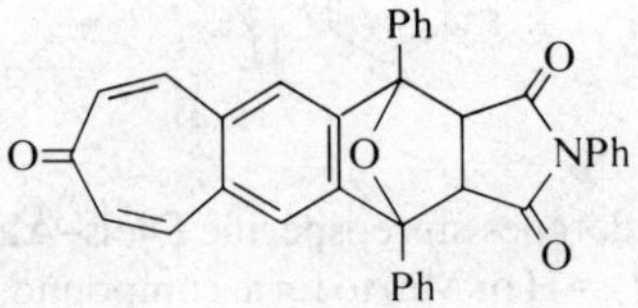

(184) 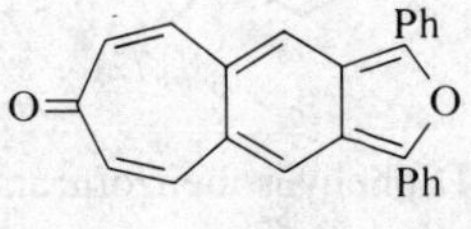                              (185)

[91]  R. D. Miller, D. Kaufmann, and J. J. Mayerle, *J. Amer. Chem Soc.*, 1977, **99**, 8511.
[92]  T. Sasaki, K. Kanematsu, K. Iizuka, and N. Izumichi, *Tetrahedron*, 1976, **32**, 2879.
[93]  D. Villessot and Y. Lepage, *Tetrahedron Letters*, 1977, 1495.

**Thiophens and Benzothiophens.**—*General.* Tris-(5-acetyl-3-thienyl)methane
(186) forms 2:1 clathrates with a wide variety of organic compounds.[94] The
magnetic circular dichroism of thiophen has been measured.[95] The [1]H n.m.r.
spectra of thiophen-2-aldehyde and of thiophen-2,5-dialdehyde, partially orient-
ed in anisotropic soap solutions, have been reported.[96] The c.d. spectra of
(*S*)-(−)-α-hydroxy-2-thienylacetic acid (187) and of (*R*)-(−)-α-hydroxy-3-

(186)          (187)

thienylacetic acid have been obtained.[97] *X*-Ray crystallography and [1]H n.m.r.
studies on the geometry of bridged thiophens, *e.g.* compound (188), have been
published.[98] The barriers to inversion in the bridged bithienyls (189; X = O, S,
NMe, or NPh; R = H or Br) have been investigated by variable-temperature [1]H
n.m.r. spectroscopy;[99] the photoelectron spectra of these compounds and of the
isomers (190) indicate a twisted conformation in the vapour phase.[100]

(188)          (189)          (190)

*Formation.* The aluminium-chloride-catalysed reaction of cyclopentanecarbonyl
chloride with allyl chloride, followed by treatment with phosphorus sulphide,
results in 2-cyclopentylthiophen (191).[101] Thermolysis of the triazolines (192;
$NR_2$ = morpholino) yields a mixture of the elimination products (193) and the

(191)          (192)          (193)          (194)

thiophens (194).[102] The nitroketenaminals (195; R = alkyl, Ar, or acyl) react with
phenacyl bromide to afford the complex thiophen derivatives (196).[103] The

[94] L. B. Din and O. Meth-Cohn, *J.C.S. Chem. Comm.*, 1977, 741.
[95] R. Hakansson, B. Norden, and E. W. Thulstrup, *Chem. Phys. Letters*, 1977, **50**, 305.
[96] A. Amanzi, D. Silvestri, C. A. Varacini, and P. L. Barili, *Chem. Phys. Letters*, 1977, **51**, 116.
[97] R. Hakansson and S. Gronowitz, *Tetrahedron*, 1976, **32**, 2973.
[98] Ya. L. Goldfarb and S. Z. Taits, *Tezisy Dokl.-Vses. Soveshch. Org. Kristallokhim. 1st*, 1974, 52 (*Chem. Abs.*, 1977, **87**, 101 823).
[99] (*a*) A. Almqvist and R. Hakansson, *Chemica Scripta*, 1976, **10**, 117; (*b*) *ibid.*, p. 120.
[100] P. Meunier and G. Pfister-Guillouzo, *Canad. J. Chem.*, 1977, **55**, 2867.
[101] A. G. Ismailov, E. I. Mamedov, and V. G. Ibragimov, *Zhur. org. Khim.*, 1977, **13**, 2612.
[102] D. Pocar, L. M. Rossi, R. Stradi, and P. Trimarco, *J.C.S. Perkin I*, 1977, 2337.
[103] S. Rajappa and R. Sreenivasan, *Indian J. Chem.*, 1977, **15B**, 301.

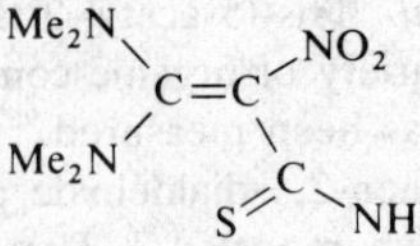

(195)

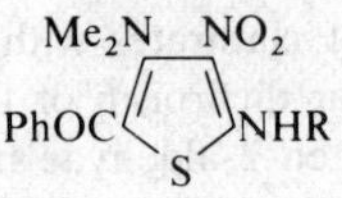

(196)

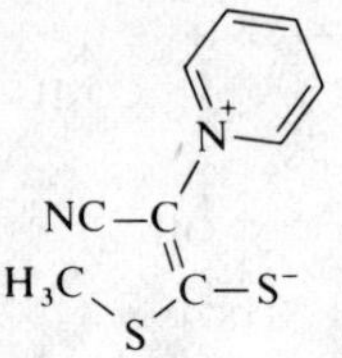

(197)

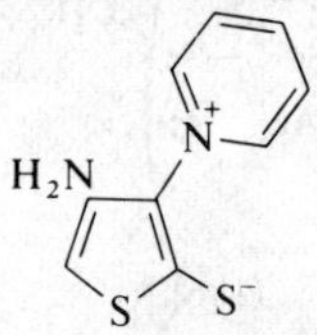

(198)

pyridinium betaine (197) cyclises to compound (198) in the presence of tri-ethylamine.[104] A 3 : 1 mixture of diphenylthiophens (200) and (201) results from the action of phenylacetylene on dimorpholinodisulphide (199); the proportions of isomeric thiophens obtained from other acetylenes have been determined.[105]

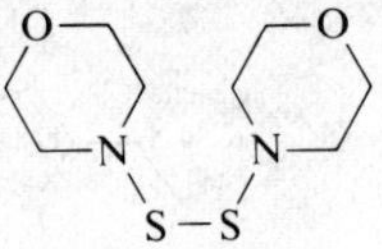

(199)

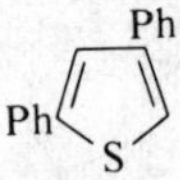

(200)

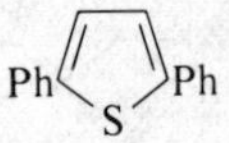

(201)

Oxidative coupling of the enol ester (202), prepared from 2-acetylthiophen and trimethylsilyl chloride, yields the diketone (203), which is transformed into 2,2′,5′,2″-terthienyl (204) by the action of hydrogen sulphide.[106] The macrocycle (206) has been synthesized from thiophen-2,5-dialdehyde and the bis(phos-phonium) ylide (205).[107]

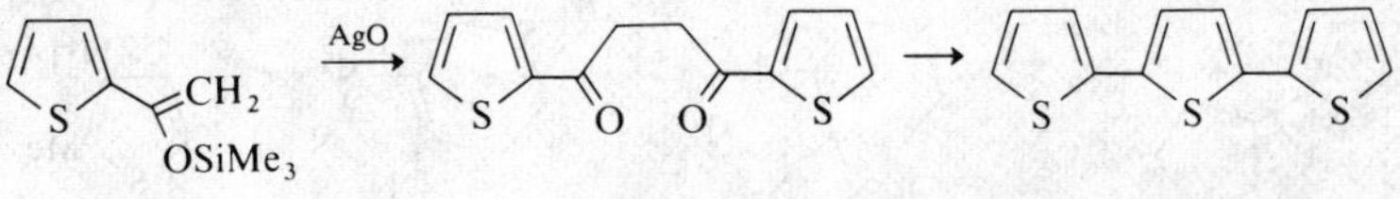

(202)

(203)

(204)

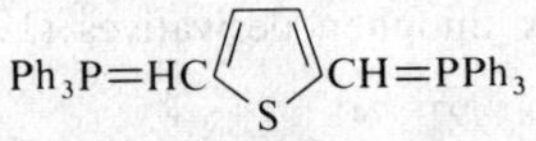

(205)

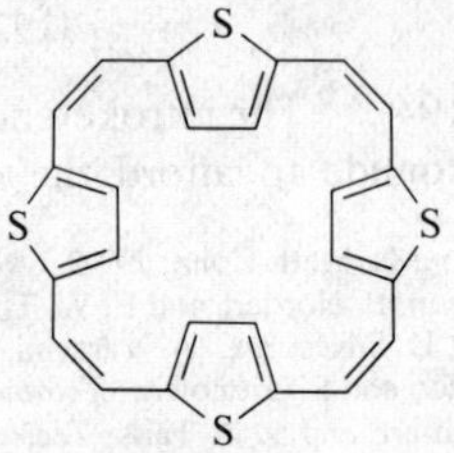

(206)

[104] Y. Tominaga, H. Fujito, Y. Matsuda, and G. Kobayashi, *Heterocycles*, 1977, **6**, 1871.
[105] F. M. Benitez and J. R. Grunwell, *Tetrahedron Letters*, 1977, 3413.
[106] T. Asano, S. Ito, N. Saito, and K. Hatakeda, *Heterocycles*, 1977, **6**, 317.
[107] A. Strand, B. Tulin, and O. Wennerstrom, *Acta Chem. Scand.* (*B*), 1977, **31**, 521.

The action of thionyl chloride on cinnamic acid affords 3-chloro-benzothiophen-2-carbonyl chloride (207);[108] phenylpropiolic acid and its methyl ester likewise yield 3-chlorobenzothiophen derivatives on treatment with sulphur dichloride, disulphur dichloride, and thionyl chloride, but not with sulphuryl chloride.[109] The bromo-acid (208) is obtained by the combined action of hydrogen bromide and sulphur dioxide on *m*-methoxyphenylpropiolic acid.[110]

(207)            (208)

Irradiation of a mixture of the naphthoquinone (209) and the thiophen (210) yields compound (211), which has been reduced to the anthra[2,1-*b*]thiophen (212).[111] The synthesis of naphtho[1,2-*b*:8,7-*b'*]dithiophen (213) has been described.[112]

(209)            (210)            (211)

(212)            (213)

*Reactions.* In the low-temperature plasma of glow discharges, phenyl-thiophens yield mainly naphthalene.[113] The action of ozone on thiophen results in substantial quantities of oxygen, less of carbon dioxide and sulphur dioxide, and what is probably a polymeric hydrocarbon.[114] Irradiation of tetrakis(trifluoromethyl)thiophen (214) gives the Dewar-thiophen (215).[115] In the presence of bases, 3-acyl-2-amino-thiophens (216; $R^1$ = H, alkyl, or $CO_2Et$; $R^2$ = alkyl or CHO; $R^3$ = H, Me, or Ph) are converted into the nitriles (217) by a process involving ring-opening, recyclisation, and dehydration; esters (216; $R^3$ = OEt), on the other hand, yield thiolactones (218) in this reaction.[116] The cleavage of 3-lithio-thiophens (219; $R^1$—$R^3$ = Me, $Bu^t$, or Ph) is hindered by the presence of

[108] R. Moll and B. Hesse, *Z. Chem.*, 1977, **17**, 133.
[109] W. N. Lok and A. D. Ward, *Austral. J. Chem.*, 1978, **31**, 605.
[110] I. V. Smirnov-Zamkov and Yu. L. Zborovskii, *Zhur. org. Khim.*, 1977, **13**, 667.
[111] K. Maruyama, K. Mitsui, and T. Otsuki, *Chem. Letters*, 1977, 853.
[112] D. Muller, J. F. Muller, and D. Cagniant, *J. Chem. Res. (S)*, 1977, 328.
[113] J. Skramstad and B. Smedsrud, *Acta Chem. Scand. (B)*, 1977, **31**, 625.
[114] A. B. Kaduk and S. Toby, *Internat. J. Chem. Kinetics*, 1977, **9**, 829.
[115] (*a*) Y. Kobayashi, I. Kumadaki, A. Ohsawa, Y. Sekine, and A. Ando, *Heterocycles*, 1977, **6**, 1587; (*b*) *J.C.S. Perkin I*, 1977, 2355.
[116] O. Meth-Cohn and B. Narine, *J. Chem. Res. (S)*, 1977, 294.

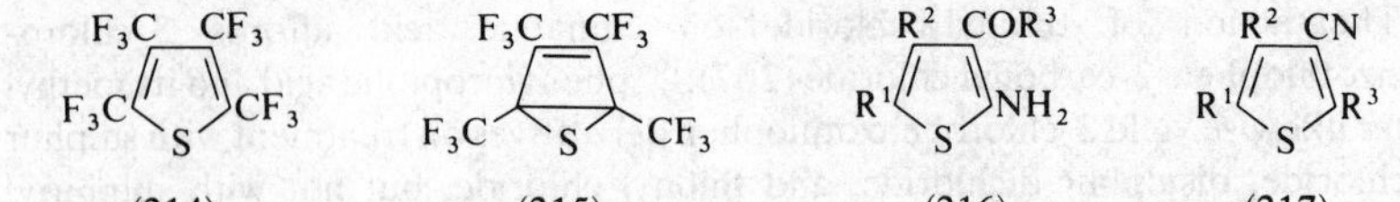

(214)                    (215)                    (216)                    (217)

(218)                    (219)

bulky substituents.[117] Treatment of the disulphonate (220) with lithium di-isopropylamide, followed by methylation with dimethyl sulphate, yields (221), presumably by way of an intermediate di-lithium compound.[118] 2,2'-Bithienyl (222) and the 2,3'-isomer (223) are produced by the action of a mixture of palladium(II) chloride, iron(III) chloride, and manganese dioxide on thiophen; the proportions of the products depend on the composition of the catalyst.[119]

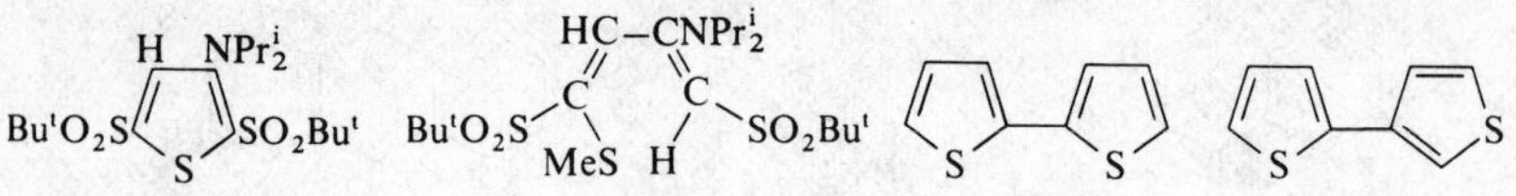

(220)                    (221)                    (222)                    (223)

Thermolysis of allyl 2-thienyl sulphide (224) yields a mixture of the thiol (225), the diallyl compound (226), and the bicyclic heterocycles (227) and (228).[120] In the base-induced Truce–Smiles rearrangement of 5-mesitylenesulphonyl-2-methylthiophen (229) to yield (230), the thienyl group migrates with a change in

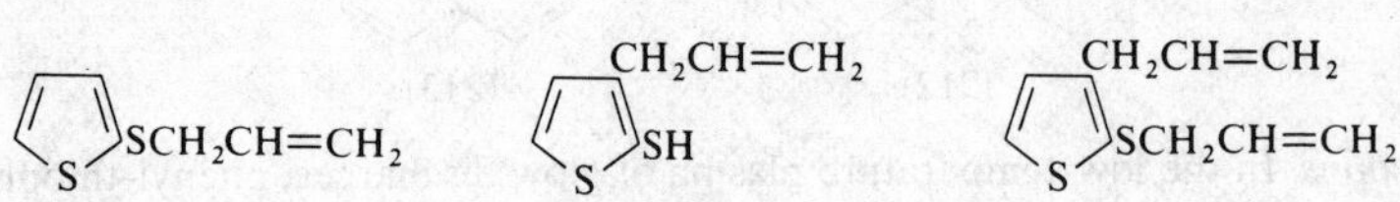

(224)                    (225)                    (226)

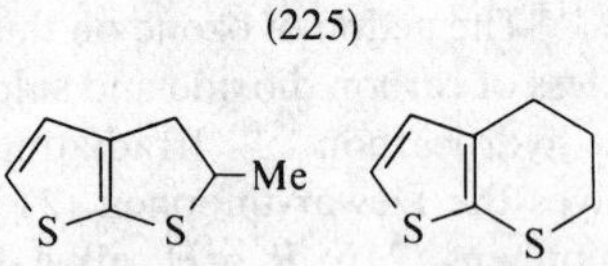

(227)                    (228)

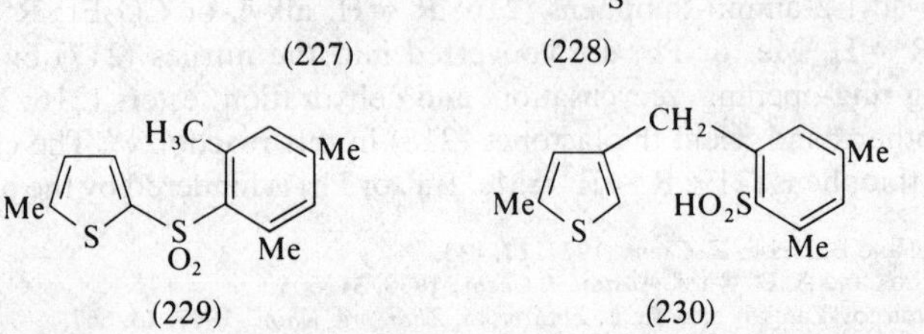

(229)                    (230)

[117] T. Frejd, *J. Heterocyclic Chem.*, 1977, **14**, 1085.
[118] F. M. Stoyanovich, G. B. Chermanova, and Ya. L. Goldfarb, *Izvest. Akad. Nauk S.S.S.R., Ser. khim.*, 1977, 1367.
[119] E. S. Rudakov and R. I. Rudakova, *Dopovidi Akad. Nauk Ukrain. R.S.R., Ser. B.*, 1977, 815 (*Chem. Abs.*, 1978, **88**, 22 514).
[120] A. V. Anisimov, V. F. Ionova, and E. A. Viktorova, *Zhur. org. Khim.*, 1977, **13**, 2624.

orientation.[121] The Friedel–Crafts reaction of thiophen-2,3-dicarbonyl chloride (231) with benzene affords a mixture of 2,3-dibenzoylthiophen (232), the keto-acid (233), the corresponding acid chloride, the quinone (234), and the lactone (235).[122] The dihydro- and tetrahydro-thiophens (237) and (238) have been isolated as intermediates in the reaction of 3,4-dinitrothiophen (236) with sodium 2,4,6-trimethylthiophenate to give the *cine*-substitution product (239).[123] In the

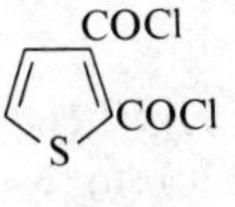

(231) 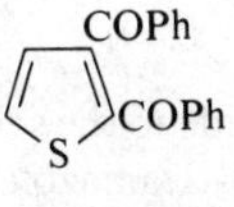  (232) 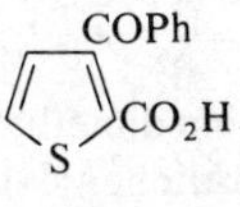  (233) 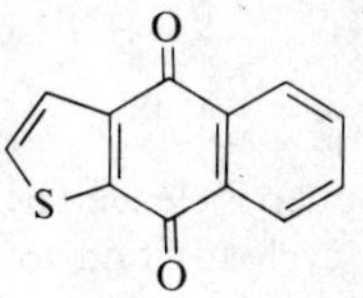  (234) 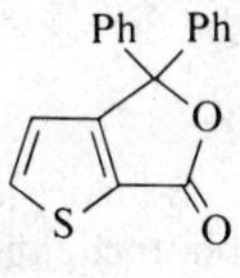  (235)

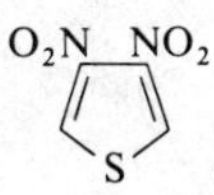

(236) 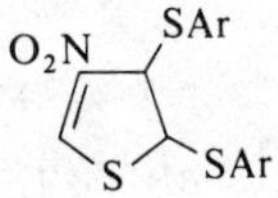  (237) 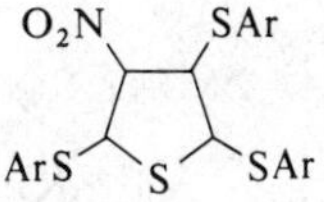  (238) 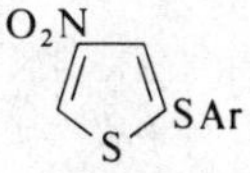  (239)

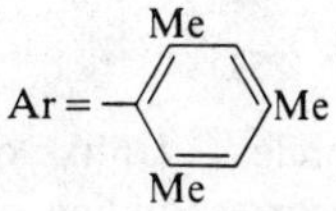

$$Ar = -\underset{Me}{\overset{Me}{\bigcirc}}Me$$

addition of dimethyl acetylenedicarboxylate to ethyl 2-aminothiophen-3-carbo-xylate (240), the expected product (241) is accompanied by small amounts of the 1:1 cycloadduct (242), the 2:1 cycloadduct (243), and the benzothiophen (244).[124] 2,4-Diphenylthiophen (245) reacts with dimethyl acetylene-dicarboxylate in boiling *o*-dichlorobenzene to afford a mixture of the naphtho-thiophen (246) and the desulphurized Diels–Alder adduct (247).[125] The

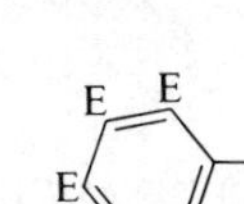

(240) 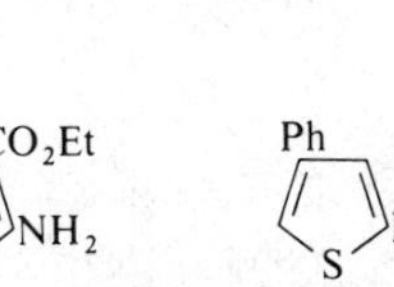  (241) 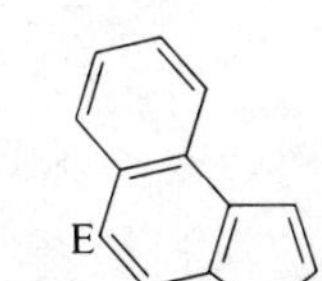  (242) 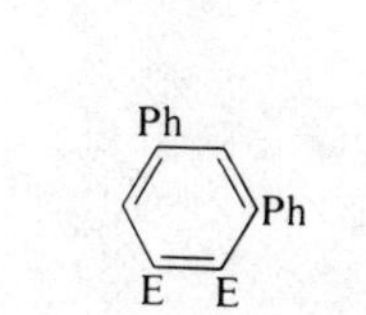  (243)

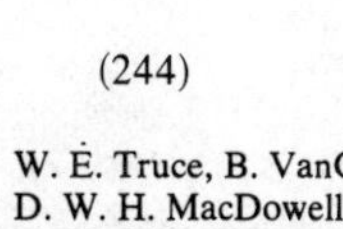

(244) 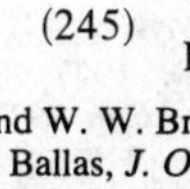  (245) 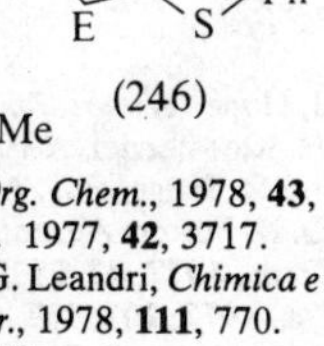  (246) 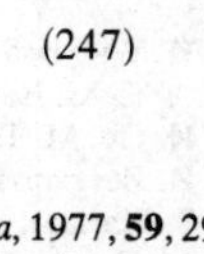  (247)

E = CO₂Me

[121] W. E. Truce, B. VanGemert, and W. W. Brand, *J. Org. Chem.*, 1978, **43**, 101.
[122] D. W. H. MacDowell and F. L. Ballas, *J. Org. Chem.* 1977, **42**, 3717.
[123] M. Novi, F. Sancassan, G. Guanti, C. Dell'Erba, and G. Leandri, *Chimica e Industria*, 1977, **59**, 299.
[124] H. Biere, C. Herrmann, and G. A. Hoyer, *Chem. Ber.*, 1978, **111**, 770.
[125] K. Kobayashi and K. Mutai, *Chem. Letters*, 1977, 1149.

photochemical reaction of thiophen with *N*-methyldibromomaleimide leads to the complex product (248).[126]

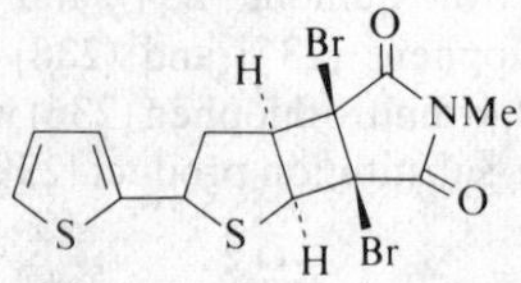

(248) 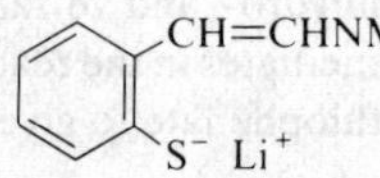                                                    (249)

Benzo[*b*]thiophen is cleaved by lithium dimethylamide to the salt (249).[127] The photochemical cycloaddition of 2,5-diphenyl-1,3,4-oxadiazole (250) to 3-methylbenzothiophen yields compound (251), but in the presence of iodine the addition is reversed, yielding compound (252).[128] Benzothiophen *SS*-dioxide

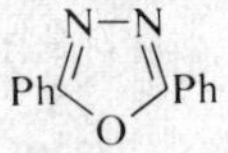

(250) 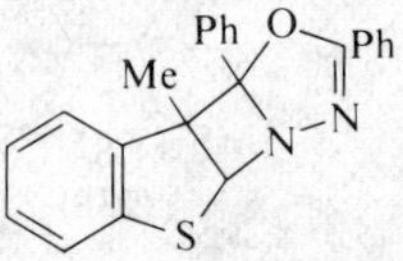                        (251) 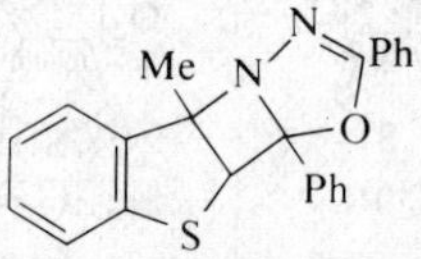                        (252)

(253; R = H) is a potent dienophile.[129] Unlike open-chain $\alpha\beta$-unsaturated sulphones, it does not form the corresponding epoxide when treated with hydroperoxide anion, but instead yields the ketone (254); 3-substituted benzothiophen dioxides (253; R = Me, Et, or Ph) give alcohols (255) in this reaction.[130]

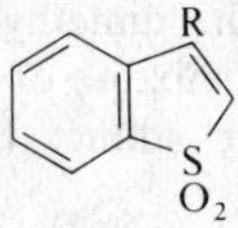

(253) 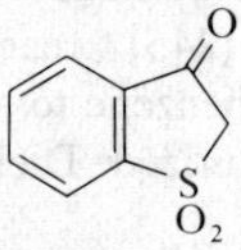                        (254) 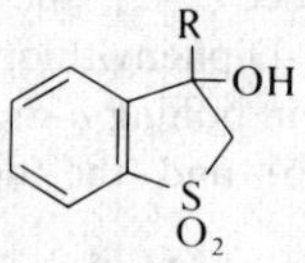                        (255)

The combined action of maleic anhydride and chloranil on 2,2'-bibenzothiophen (256) leads to the benzothienodibenzothiophen (257).[131] Dibenzothiophen *SS*-dioxide (258) is cleaved by aqueous alkali at 300 °C to the biphenyl derivative (259).[132]

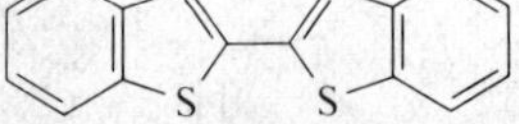

(256) 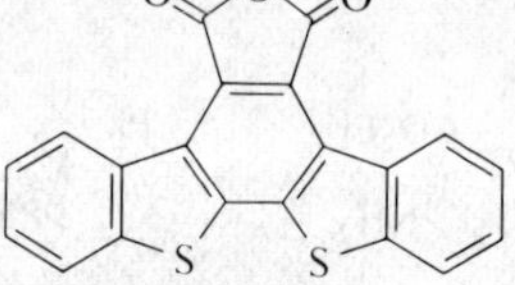                                              (257)

[126] H. Wamhoff and H. J. Hupe, *Tetrahedron Letters*, 1978, 125.
[127] A. E. M. Beyer and H. Kloosterziel, *Rec. Trav. chim.*, 1977, **96**, 178.
[128] K. Oe, M. Tashiro, and O. Tsuge, *Bull. Chem. Soc. Japan*, 1977, **50**, 3281.
[129] R. Bergamasco and Q. N. Porter, *Austral. J. Chem.*, 1977, **30**, 1523.
[130] S. Marmar, *J. Org. Chem.*, 1977, **42**, 2927.
[131] M. Zander, *Chem.-Ztg.*, 1977, **101**, 507.
[132] R. B. LaCount and S. Friedman, *J. Org. Chem.*, 1977, **42**, 2751.

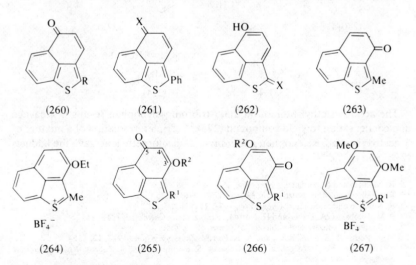

Further studies (see these Reports, Vol. 6, p. 27) on the chemistry of the 5-oxo-5$H$-naphtho[1,8-$bc$]thiophen system (260; R = Ph or Br), which is iso-$\pi$-electronic with phenalenone, include the reaction of the phenyl derivative with benzenesulphonyl isocyanate, isothiocyanate, or sulphinimide to yield the benzenesulphonylimine (261; X = PhSO$_2$N) and with diphenylketen to yield the diphenylmethylene compound (261; X = CPh$_2$),[133] and an investigation of the action of nucleophiles on the bromo-derivative, which results in the formation of imino- or methylene-compounds [262; X = NR, NNHR, or C(CN)$_2$] in the case of amines, hydrazines, or malononitrile anion;[134,135] however, with sodium methoxide, the ether (260; R = OMe) is produced.[135] Aqueous alkali gives a mixture containing (262; X = O) as the predominant tautomer.[135] Triethyloxonium fluoroborate transforms the 3-oxo-naphthothiophen (263) into the salt (264), which possesses a reactive methyl group.[136] Condensation of 2-methyl- or 2-ethyl-benzothiophen with malonyl chloride gives an equilibrium mixture of the tautomeric hydroxy-oxo-naphthothiophens (265) and (266) (R$^1$ = Me or Et; R$^2$ = H), which on treatment with dimethyl sulphate yields a mixture of the ethers (265) and (266) (R$^2$ = Me). Both ethers are converted into the fluoroborate (267) by trimethyloxonium fluoroborate.[137] The ethyl compounds (265) $\rightleftharpoons$ (266) (R$^1$ = Et; R$^2$ = H) react with toluene-$p$-sulphonyl azide to yield 4-diazo-derivatives;

[133] R. Neidlein and A. D. Krämer, *J. Heterocyclic Chem.*, 1977, **14**, 1369.
[134] R. Neidlein and H. Seel, *J. Heterocyclic Chem.*, 1977, **14**, 1379.
[135] R. Neidlein and K. F. Cepera, *Chem. Ber.*, 1977, **110**, 2388.
[136] R. Neidlein, K. F. Cepera, and M. H. Salzl, *Chem.-Ztg.*, 1977, **101**, 558.
[137] R. Neidlein and M. H. Salzl, *J. Chem. Res. (S)*, 1977, 118.

heating with aromatic aldehydes produces the cyclic ethers (268).[138] Benzenesulphonyl isocyanate and diphenylketen attack the methyl ethers at the carbonyl group to form the corresponding benzenesulphonyl-imines and diphenylmethylene derivatives, respectively;[139,140] in contrast, t-butylcyanoketen reacts at the alkyl side-chain to yield ketones, *e.g.* (269).[140]

(268)                                           (269)

**Selenophens.**—The preparation of selenophen by the action of sodium hydrogen selenide on bis(trimethylsilyl)butadiyne, $Me_3SiC{\equiv}C{-}C{\equiv}CSiMe_3$, has been described.[141] The reaction of dimethyl acetylenedicarboxylate with the potassium salt $PhC{\equiv}CSe^- K^+$ yields the ester (270), together with a little of the diselenole (271).[142,143] Treatment of dibenzylideneacetone with selenium tetrabromide gives compound (272), which is converted into the benzoselenophen (273; R = COCH=CHPh) in the presence of pyridine.[144] The ester (273; R = $CO_2Et$) has been obtained by cyclization of ethyl *o*-methylselenocinnamate (274) by means of bromine and pyridine.[145]

(270)                     (271)                          (272)

(273)                          (274)

The action of ethyl-lithium on 2,3,5-tribromoselenophen results in cleavage and condensation to yield compound (275).[146] Triplet irradiation of a mixture of 3-acetoxybenzo[*b*]selenophen and *trans*-1,2-dichloroethylene gave the adducts (276) and (277).[147]

[138] R. Neidlein and M. H. Salzl, *Chem.-Ztg.*, 1977, **101**, 357.
[139] R. Neidlein and M. H. Salzl, *Arch. Pharm.*, 1977, **310**, 685.
[140] R. Neidlein and G. Schäfer, *Chem.-Ztg.*, 1977, **101**, 509.
[141] P. M. Jacobs, M. A. Davis, and H. Norton, *J. Heterocyclic Chem.*, 1977, **14**, 1115.
[142] A. Shafiee, I. Lalezari, and F. Savabi, *Synthesis*, 1977, 765.
[143] M. L. Petrov, L. S. Rodionova, and A. A. Petrov, *Zhur. org. Khim.*, 1977, **13**, 1564.
[144] V. G. Lendel, Yu. V. Migalina, S. V. Galla, A. S. Kozmin, and N. S. Zefirov, *Khim. geterotsikl. Soedinenii*, 1977, 1340.
[145] G. Marechal, L. Christiaens, M. Renson, P. Jacquignon, and A. Croisy, *Bull. Soc. chim. France*, 1977, 157.
[146] T. Frejd, *Chemica Scripta*, 1976, **10**, 133.
[147] Tran Quang Minh, L. Christiaens, P. Grandclaudon, and A. Lablache-Combier, *Tetrahedron*, 1977, **33**, 2225.

(275)          (276)          (277)

**Tellurophen.**—The chemistry of tellurophen has been reviewed.[148] The [125]Te n.m.r. spectra of the tellurophens (278; R = Cl, $CH_2OH$, CHO, $CO_2H$, *etc.*) show large shifts caused by the substituents, 2.44 times larger than in the [77]Se n.m.r. spectra of the corresponding selenophens.[149] The electric dipole moments of tellurophen and its derivatives have been determined; the mesomeric moments of the unsubstituted heterocycles follow the order of their aromatic character: thiophen > selenophen > tellurophen > furan.[150]

(278)

**Pyrroles.**—*Formation.* Toluene-*p*-sulphonyl-alkyl isocyanides react with electron-poor olefins to furnish pyrroles; thus compound (279) and methyl cinnamate give the ester (280).[151] Treatment of the isocyano-ester $MeCH{=}C(NC)CO_2Et$ with $Li^+$ $CH(NC)CO_2Et$ affords the pyrrole derivative (281).[152] A series of amino-nitriles (282; R = alkyl, Ph, *etc.*) has been prepared by the action of

TosylCHMeNC

(279)          (280)          (281)          (282)

malononitrile on the acetamido-ketones AcNHCHRCOMe.[153] The pyrrole derivative (283) was isolated when an aqueous solution of D-glucose and glycine was kept at 56 °C for two weeks; valine gave an analogous product.[154] The formation of pyrroles from oximes and acetylenes has been reported: acetoxime and phenylacetylene yield 2-methyl-5-phenylpyrrole;[155] from methyl ethyl ketoxime and acetylene, a mixture of compounds (284) and (285) is obtained,[156] and the oxime of 2-acetylthiophen reacts with acetylene to give 2-(2-thienyl)-pyrrole or its *N*-vinyl-derivative, depending on the proportions of the

[148] F. Fringuelli, G. Marino, and T. Gianlorenzo, *Adv. Heterocyclic Chem.*, 1977, **21**, 119.
[149] T. Drakenberg, F. Fringuelli, S. Gronowitz, A. B. Hornfeldt, I. Johnson, and A. Taticchi, *Chemica Scripta*, 1976, **10**, 139.
[150] H. Lumbroso, D. M. Bertin, F. Fringuelli, and A. Taticchi, *J.C.S. Perkin II*, 1977, 775.
[151] O. Possel and A. Van Leusen, *Heterocycles*, 1977, **7**, 77.
[152] U. Schöllkopf and R. Meyer, *Annalen*, 1977, 1174.
[153] R. W. Johnson, R. J. Mattson, and J. W. Sowell, *J. Heterocyclic Chem.*, 1977, **14**, 383.
[154] H. Kato, H. Sonobe, and M. Fujimaki, *Agric. and Biol. Chem. (Japan)*, 1977, **41**, 711 (*Chem. Abs.*, 1977, **87**, 84 755).
[155] B. A. Trofimov, A. I. Mikhaleva, S. E. Korostova, and G. A. Kalabin, *Khim. geterotsikl. Soedinenii*, 1977, 994.
[156] B. A. Trofimov, A. I. Mikhaleva, A. N. Vasilev, and M. V. Sigalov, *Khim. geterotsikl. Soedinenii*, 1978, 54.

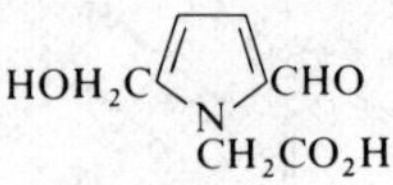

HOH₂C / CHO
N
CH₂CO₂H

(283)

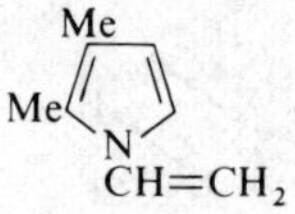

Me
Me / N
CH=CH₂

(284)

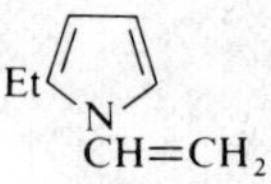

Et / N
CH=CH₂

(285)

Ar / N / CO₂Et
H

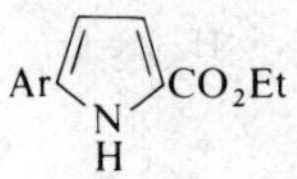

(286)

reagents.[157] The azido-butadienes $ArCH=CHCH=C(N_3)CO_2Et$ are transformed into the pyrroles (286) on heating.[158] Acetylacetone reacts with 2-phenylazirine (287) in the presence of bis-(2,4-pentanedionato)-nickel to give the pyrrole ketone (288) quantitatively.[159] Contrary to a previous report, the thermal rearrangement of 3-butadienyl-azirines produces pyrroles rather than azepines; thus compound (289) yields the pyrrole (290).[160] The azirine (291) rearranges photochemically to the aldehyde (292).[161] Treatment of *N*-methyl-hexahydroazepine (293) with sulphur in hexamethylphosphoric triamide results in ring contraction and dehydrogenation to yield a polysulphide (294).[162]

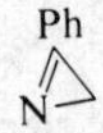

Ph
N

(287)

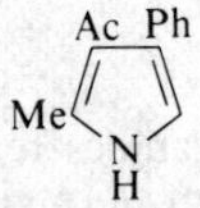

Ac Ph
Me / N
H

(288)

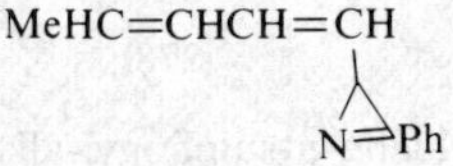

MeHC=CHCH=CH
N=Ph

(289)

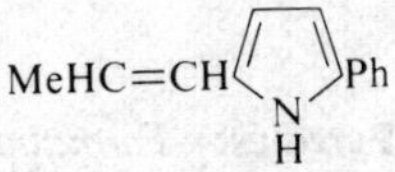

MeHC=CH / N / Ph
H

(290)

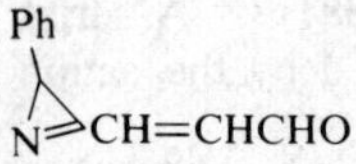

Ph
N=CH=CHCHO

(291)

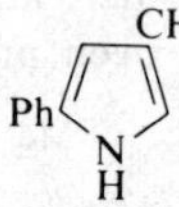

CHO
Ph / N
H

(292)

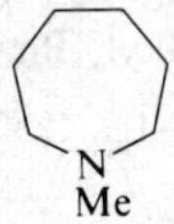

N
Me

(293)

Sₓ
Et / N         N / Et
Me           Me

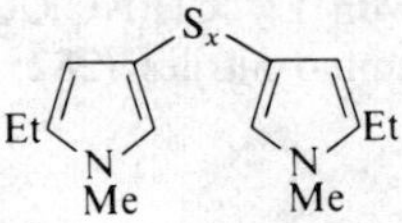

(294)

*Reactions.* The first direct evidence for hydration of amides has been obtained in the case of *N*-trifluoroacetylpyrrole, which forms compound (295) in aqueous acetonitrile.[163] From the oxidation of 2,4,5-triphenylpyrrole with dichromate, the novel products (296)–(298) have been isolated.[164] The action of allyl bromide on

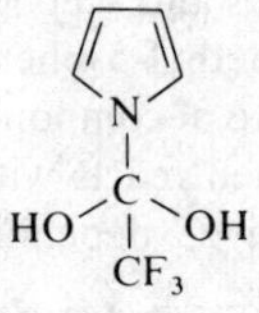

N
C
HO   OH
CF₃

(295)

[157] B. A. Trofimov, A. I. Mikhaleva, R. N. Nesterenko, A. N. Vasilev, A. S. Nakhmanovich, and M. G. Voronkov, *Khim. geterotsikl. Soedinenii*, 1977, 1136.
[158] H. Hemetsberger, I. Spira, and W. Schönfelder, *J. Chem. Res. (S)*, 1977, 247.
[159] P. Faria dos Santos Filho and U. Schuchardt, *Angew. Chem. Internat. Edn.*, 1977, **16**, 647.
[160] K. Isomura, T. Tanaka, and H. Taniguchi, *Chem. Letters*, 1977, 397.
[161] T. Mukai, T. Kumagai, and O. Seshimoto, *Pure Appl. Chem.*, 1977, **49**, 287.
[162] J. Perregaard, S. Scheibye, H. J. Meyer, I. Thomsen, and S. O. Lawesson, *Bull. Soc. chim. belges*, 1977, **86**, 679.
[163] A. Cipiciani, P. Linda, and G. Savelli, *J.C.S. Chem. Comm.*, 1977, 857.
[164] V. Sprio, S. Petruso, L. Ceranlo, and L. Lamartino, *J. Heterocyclic Chem.*, 1977, **14**, 797.

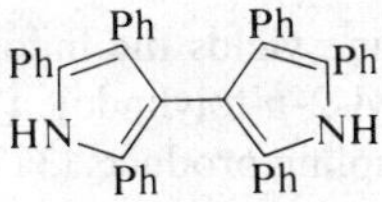

(296) 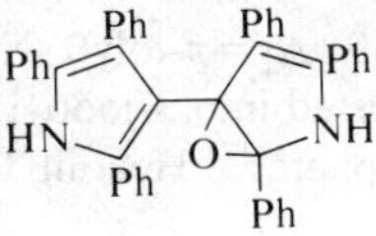　　　　　　　(297) 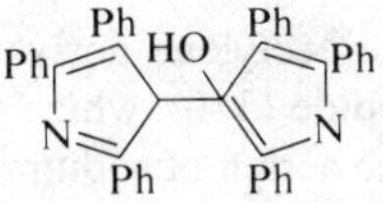　　　　　　　(298)

the hydroxy-pyrrole (299) results in a mixture of the *O*-allyl ether and the pyrrolinone (300).[165]

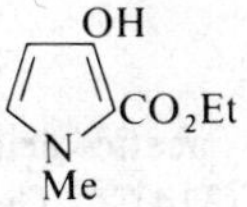

(299) 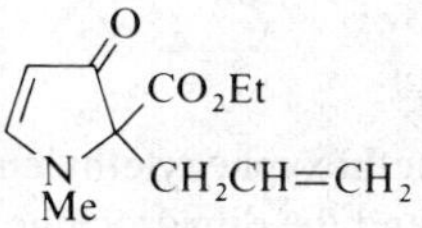　　　　　(300) 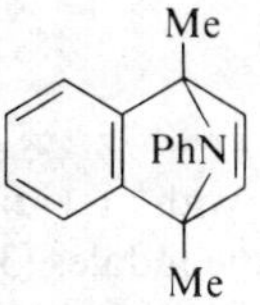　　　　　(301) 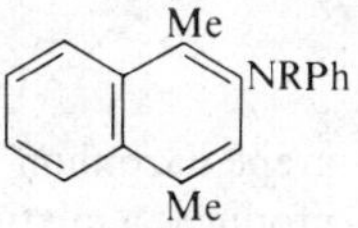　　　　　(302)

The reaction of benzyne with 2,5-dimethyl-1-phenylpyrrole yields not only the Diels–Alder product (301), but also the naphthalene derivatives (302; R = H) and (302; R = NO).[166] The adduct (303; R = SiMe₃) of tetrafluorobenzyne to *N*-trimethylsilylpyrrole was hydrolysed to compound (303; R = H), which added benzonitrile phenylimine, $PhC\equiv\overset{+}{N}-\bar{N}Ph$, to yield the condensed pyrazoline (304). Thermal fragmentation of the latter gave 1,3-diphenylpyrazole and tetrafluoroisoindole (305).[167]

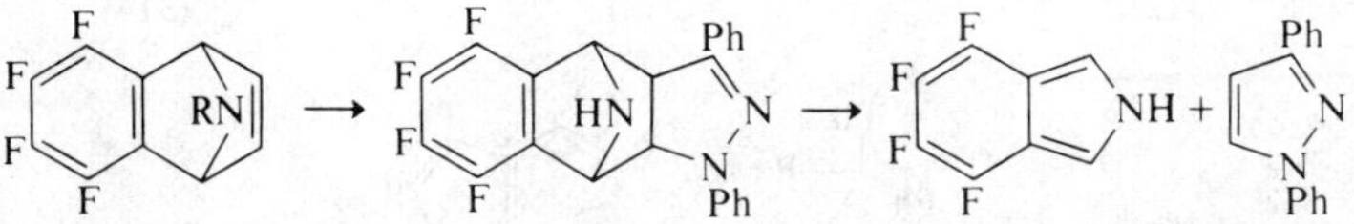

(303)　　　　　　　(304)　　　　　　　(305)

**Indoles.**—*Formation.* Phenacyltriphenylarsonium bromide, $Ph_3\overset{+}{As}CH_2COPh$ Br⁻, reacts with aniline to give 2-phenylindole.[168] *o*-Tolyl isocyanide (306) is selectively lithiated at the methyl group by lithium di-isopropylamide at −78°C; warming the product to room temperature, followed by aqueous work-up, yields indole almost quantitatively.[169] The ester (307) cyclises to the indole (308) in the presence of palladium acetate and triphenylphosphine.[170] Treatment

(306) 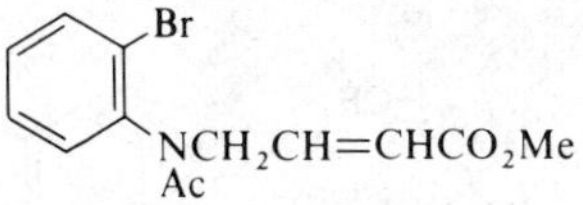　　　　　　　(307) 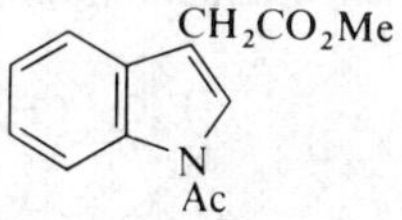　　　　　　　(308)

[165] T. Momose, T. Tanaka, and T. Yokota, *Heterocycles*, 1977, **6**, 1821.
[166] J. M. Vernon, M. Ahmed, and J. M. Moran, *J.C.S. Perkin I*, 1977, 1084.
[167] P. S. Anderson, M. E. Christy, E. L. Engelhardt, G. F. Lundell, and G. S. Ponticello, *J. Heterocyclic Chem.*, 1977, **14**, 213.
[168] R. K. Bansal and K. S. Sharma, *Tetrahedron Letters*, 1977, 1923.
[169] Y. Ito, K. Kobayashi, and T. Saegusa, *J. Amer. Chem. Soc.*, 1977, **99**, 3532.
[170] M. Mori, K. Chiba, and Y. Ban, *Tetrahedron Letters*, 1977, 1037.

of the acetylene derivative (309; Ar = *p*-tolyl) with iodine yields the indolium tri-iodide (310), which is converted into 3-iodo-1-methyl-2-*p*-tolylindole (311) by the action of sodium thiosulphate.[171] Heating the coupling products (312) of

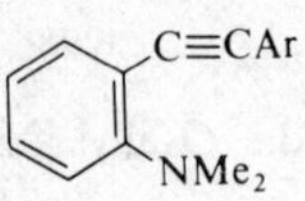

(309) 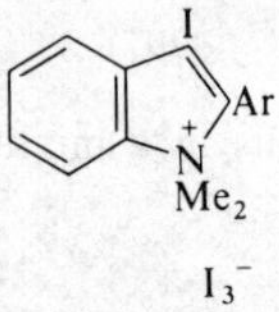                           (310) 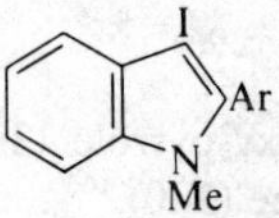                           (311)

arenediazonium salts with 1,1-di-*p*-methoxyphenylethylene in acetic acid produces a mixture of the indoles (313) and the dihydrocinnolines (314).[172] The azirines (315; R = H, Me, or Ph) rearrange thermally or under the influence of palladium chloride to the indoles (316).[173] The primary adducts (317) of nitrones

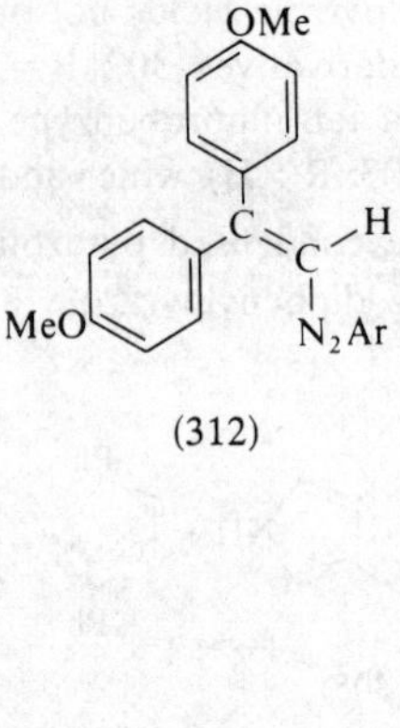

(312) 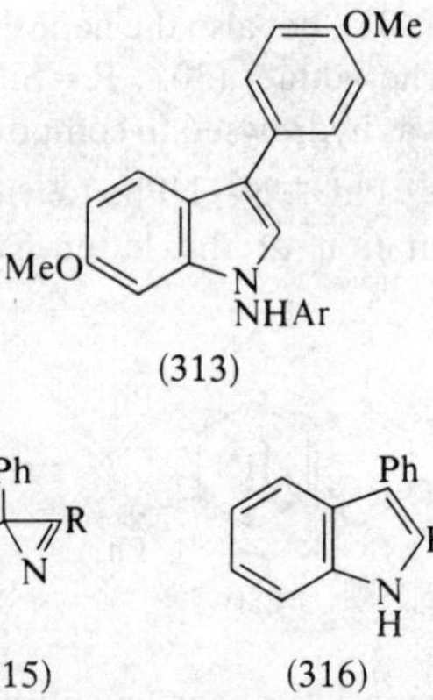                           (313) 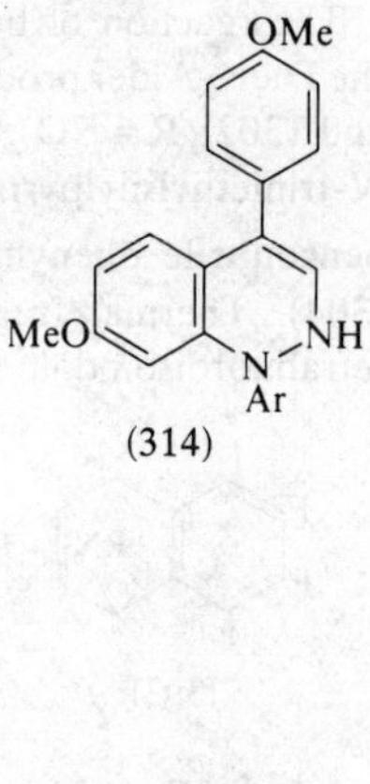                           (314)

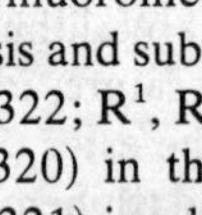

(315) 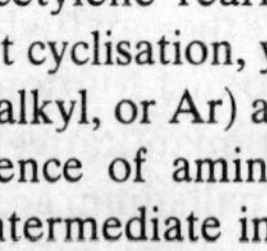                           (316)

ArCH=NPhO to bis-(trifluoromethyl)-acetylene rearrange to the oxazolines (318), which, on hydrolysis and subsequent cyclisation, yield the indole (319).[174] Indole-3-carboxamides (322; $R^1$, $R^2$ = H, alkyl, or Ar) are produced by irradiating the diazo-ketones (320) in the presence of amines $HNR^1R^2$; the Wolff rearrangement product (321) is a likely intermediate in this reaction.[175] Crossover experiments show that the thermal rearrangement of the 3*H*-indole (323) to

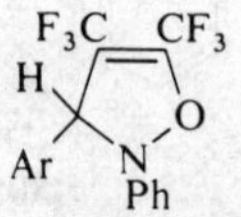

(317) 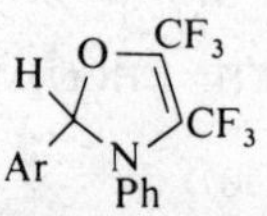                           (318) 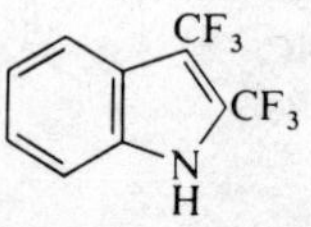                           (319)

[171] R. W. M. Ten Hoedt, G. Van Koten, and J. G. Noltes, *Synth. Comm.*, 1977, **7**, 61.
[172] A. B. Sakla, G. Aziz, S. A. El-Sayed, and N. S. Ibrahim, *Indian J. Chem.*, 1976, **14B**, 742.
[173] K. Isomura, K. Uto, and H. Taniguchi, *J.C.S. Chem. Comm.*, 1977, 664.
[174] Y. Kobayashi, I. Kumadaki, and T. Yoshida, *Heterocycles*, 1977, **8**, 387.
[175] J. T. Carlock, J. S. Bradshaw, B. Stanovnik, and M. Tišler, *J. Org. Chem.*, 1977, **42**, 1883.

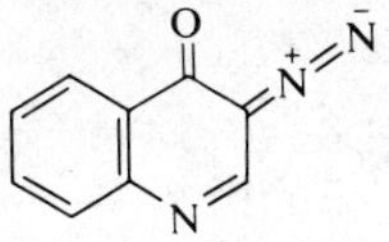

(320) 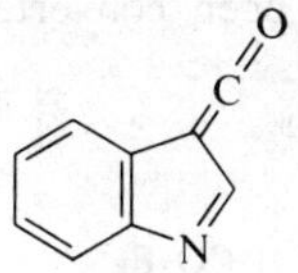 (321) 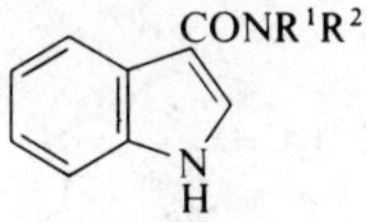 (322)

the indole (324) is intermolecular.[176] The toluene-*p*-sulphonylhydrazone (325) decomposes to the indolo-quinone (326) on heating.[177] The formation of the

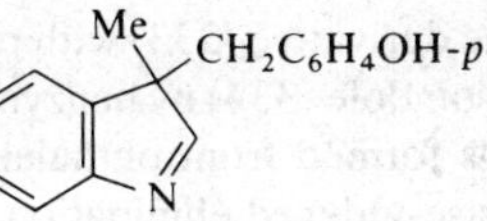

(323) 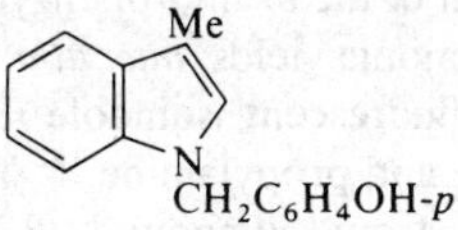 (324)

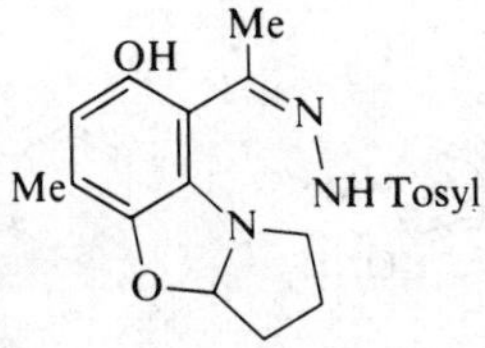

(325) 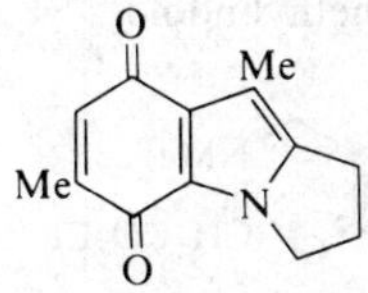 (326)

tetrahydrocarbazole (328) by Fischer indolization of the hydrazone (327) is thought to proceed *via* successive [1,2] shifts of the ethyl group.[178] The benz-

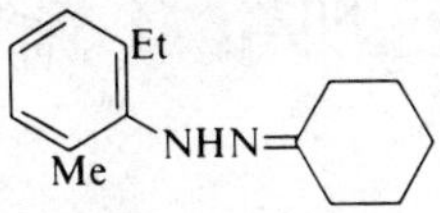

(327) 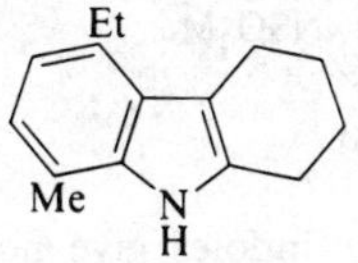 (328)

indoles (329) and (330) have been prepared from 1- and 2-naphthylamine, respectively, by a multi-step sequence, starting with specific *ortho*-alkylation.[179]

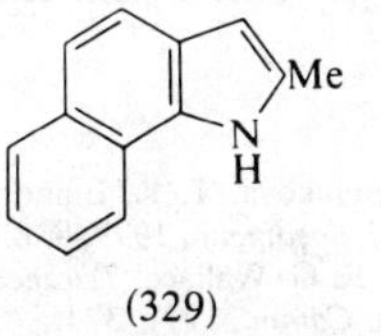

(329) 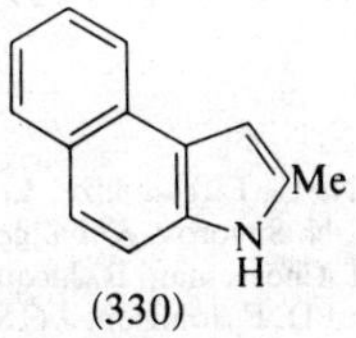 (330)

[176] G. Decodts, *Bull. Soc. chim. France*, 1976, 1839.
[177] (a) M. Akiba, Y. Kosugi, and T. Takada, *Heterocycles*, 1977, **6**, 1125; (b) T. Takada, Y. Kosugi, and M. Akiba, *Chem. and Pharm. Bull. (Japan)*, 1977, **25**, 543.
[178] B. Miller and E. R. Matjeka, *Tetrahedron Letters*, 1977, 131.
[179] P. G. Gassman and W. N. Schenk, *J. Org. Chem.* 1977, **42**, 3240.

The bis-hydrazone (331) has been converted into the pyrrolo[2,3-*e*]indole (332).[180]

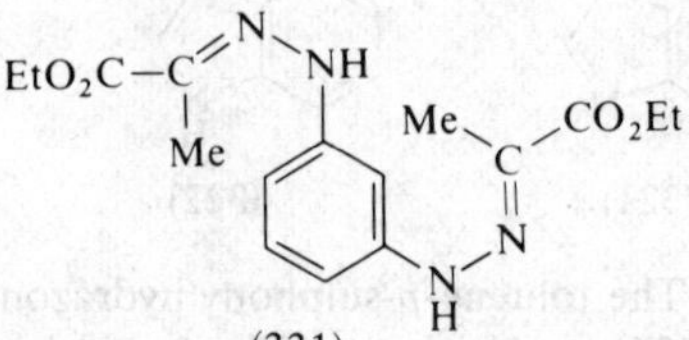

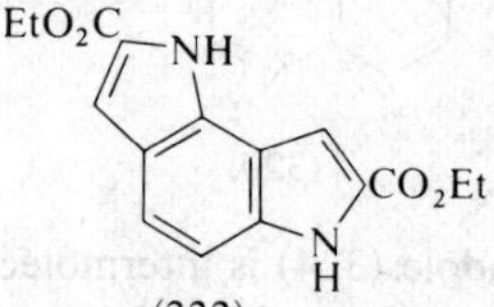

(331)                                                          (332)

Treatment of the *o*-chlorobenzylamine derivative (333) with potassium amide in liquid ammonia yields, *inter alia*, the isoindole (334) by benzyne cyclization.[181] The stable fluorescent isoindole (335) is formed from phthalaldehyde, t-butyl mercaptans, and propylamine.[182] The base-induced elimination of methanesulphinic acid from compound (336) affords the bis-isoindole (337);[183] the benzo[1,2-*b*:4,5-*c'*]dipyrrole (338) has been prepared in two steps from 5,6-dibenzoyl-1-methylindole.[184]

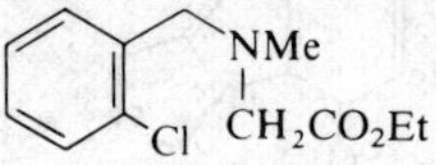

(333) 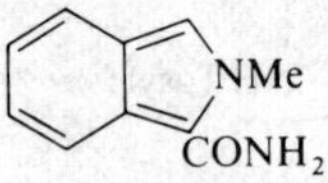                        (334) 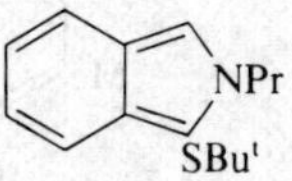                        (335)

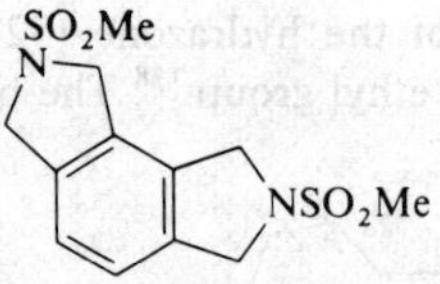

(336) 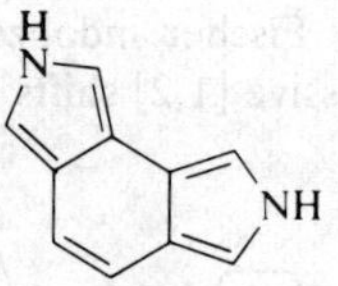                        (337) 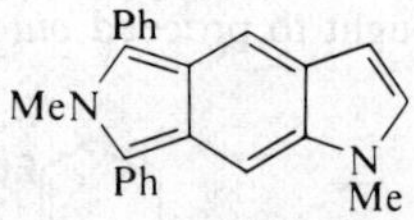                        (338)

*Reactions.* Vinyl-indoles give normal Diels–Alder products with tetracyano-ethylene, such as compound (339) from 1-methyl-3-vinylindole, but in the case of the 2-methyl derivative (340), steric hindrance causes formation of the cyclo-butane adduct (341).[185] Irradiation of 1,3-diacetylindole produces the dimer (342), as shown by *X*-ray crystallography.[186] The arylazo-indoles (343) give products (344) of the semidine type on treatment with zinc and acetic anhydride.[187]

[180] S. A. Samsoniya, N. L. Targamadze, L. G. Tretyakova, T. K. Efimova, K. F. Turchin, I. M. Gverdtsiteli, and N. N. Suvorov, *Khim. geterotsikl. Soedinenii*, 1977, 938.
[181] I. Ahmed, G. W. H. Cheeseman, B. Jacques, and R. G. Wallace, *Tetrahedron*, 1977, **33**, 2255.
[182] S. S. Simons, Jr., and D. F. Johnson, *J.C.S. Chem. Comm.*, 1977, 374.
[183] R. Kreher and K. J. Herd, *Angew. Chem. Internat. Edn.*, 1978, **17**, 68.
[184] J. Duflos, G. Queguiner, and P. Pastour, *J. Chem. Res. (S)*, 1978, 39.
[185] R. Bergamasco, Q. N. Porter, and C. Yap, *Austral. J. Chem.*, 1977, **30**, 1531.
[186] T. Hino, M. Taniguchi, T. Date, and Y. Iidaka, *Heterocycles*, 1977, **7**, 105.
[187] G. N. Kurilo, N. I. Tostova, A. A. Chekasova, and A. N. Grinev, *Khim. geterotsikl. Soedinenii*, 1977, 353.

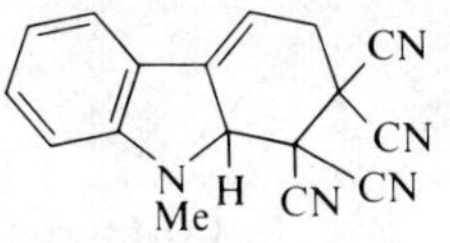

(339) 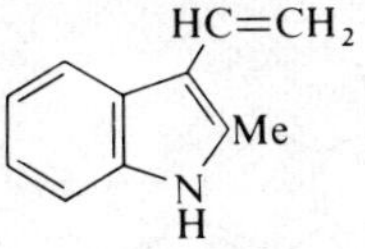          (340) 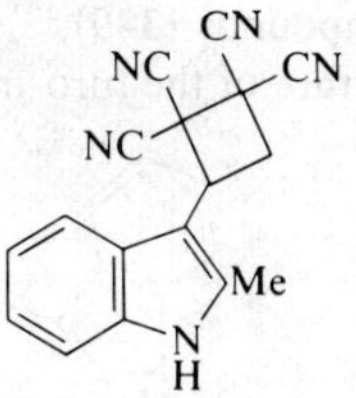          (341)

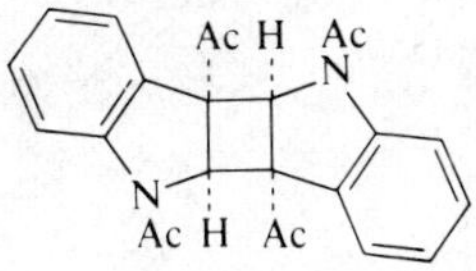

(342) 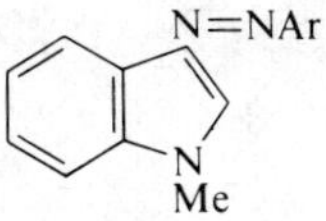          (343) 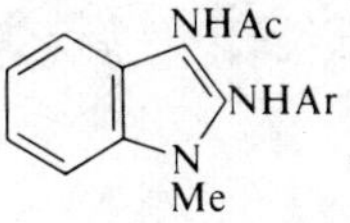          (344)

A new trimer (345) has been obtained by the action of toluene-*p*-sulphonic acid on indole.[188] The sodium-catalysed addition of benzylidenebenzylamine to indole yields the tricyclic compound (346).[189] 1-Methylindole reacts with 1 mole

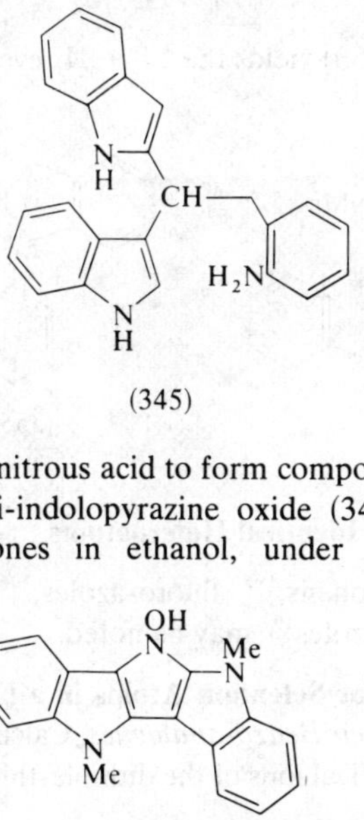

(345) 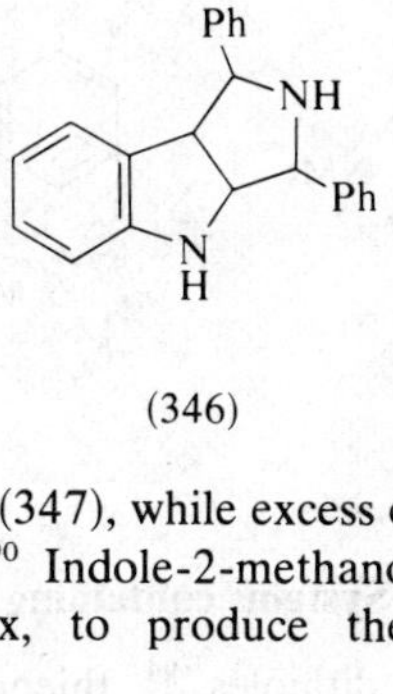                          (346)

equivalent of nitrous acid to form compound (347), while excess of nitrous acid affords the di-indolopyrazine oxide (348).[190] Indole-2-methanols react with α-bromo-ketones in ethanol, under reflux, to produce the unexpected

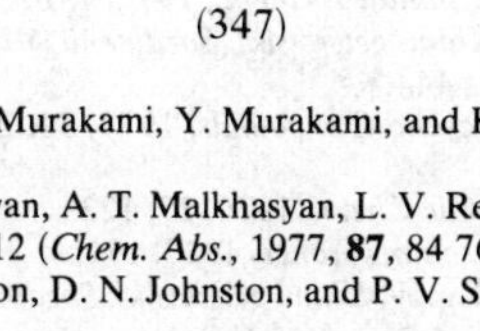

(347) 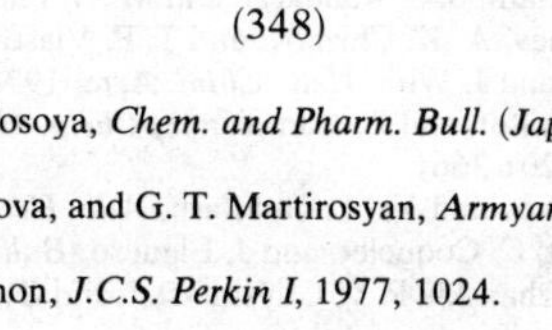                          (348)

[188] H. Ishii, K. Murakami, Y. Murakami, and K. Hosoya, *Chem. and Pharm. Bull. (Japan)*, 1977, **25**, 3122.
[189] L. V. Asratyan, A. T. Malkhasyan, L. V. Revazova, and G. T. Martirosyan, *Armyan. Khim. Zhur.*, 1977, **30**, 312 (*Chem. Abs.*, 1977, **87**, 84 762).
[190] A. H. Jackson, D. N. Johnston, and P. V. Shannon, *J.C.S. Perkin I*, 1977, 1024.

compounds (349).[191] Heating the indolyloxyacetic acid (350) in DMF gives a mixture of the furo-indole derivatives (351) and (352).[192]

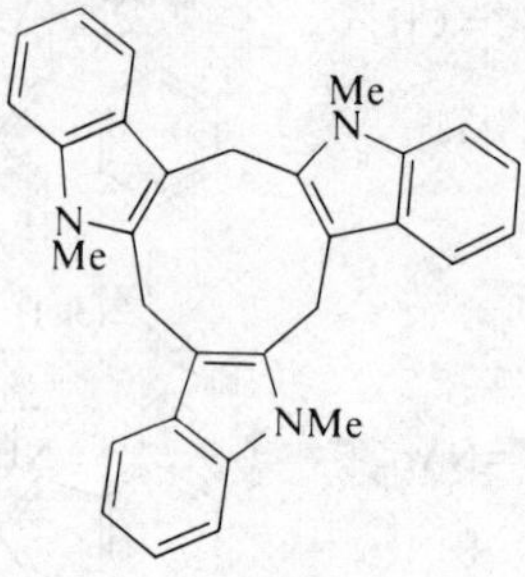

(349)

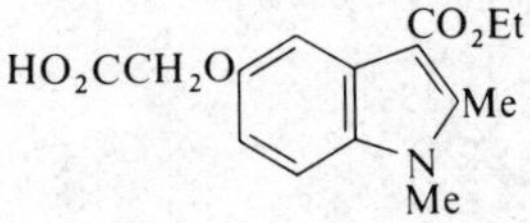

(350)

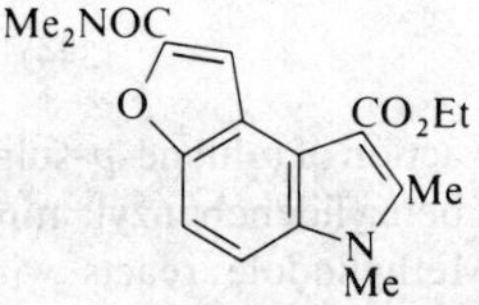

(351) 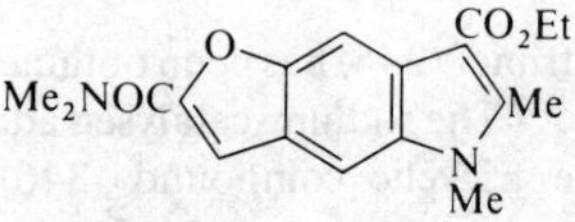                                        (352)

Irradiation of *N*-methylisoindole (353) yields the $_\pi 4_s + _\pi 4_s$ cyclodimers (354) and (355).[193]

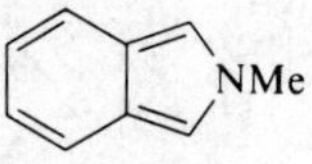

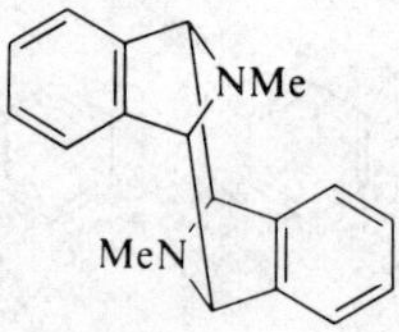

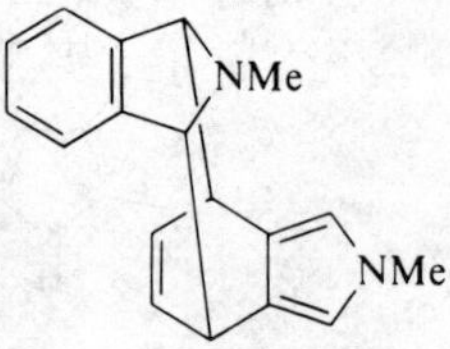

(353)                    (354)                    (355)

## 4 Systems containing Two Identical Heteroatoms

Reviews on dithioles,[194] thienothiophens,[195] fluoro-azoles,[196] and on the luminescence and photochemistry of azoles[197] may be noted.

**Compounds containing Two Sulphur or Selenium Atoms in a Five-membered Ring.**—*Monocyclic Compounds and their Benzo-analogues.* Calculated coupling constants in the e.s.r. spectra of the radical ions of the dithiole-thiones (356) and

[191] S. P. Hiremath, S. S. Kaddargi, and M. G. Purohit, *Indian J. Chem.*, 1977, **15B**, 1103.
[192] A. N. Grinev, A. K. Chizhov, and T. F. Vlasova, *Khim. geterotsikl. Soedinenii*, 1977, 766.
[193] W. Rettig and J. Wirz, *Helv. Chim. Acta*, 1978, **61**, 444.
[194] R. D. Hamilton and E. Campaigne, *Chem. Heterocyclic Compounds*, 1977, **30**, 271 (*Chem. Abs.*, 1977, **87**, 201 366).
[195] V. P. Litvinov and Ya. L. Goldfarb, *Adv. Heterocyclic Chem.*, 1976, **19**, 123.
[196] P. Bouchet, C. Coquelet, and J. Elguero, *Bull. Soc. chim. France*, 1977, 171.
[197] M. I. Knyazhanskii, P. V. Gilyanovskii, and O. A. Osipov, *Khim. geterotsikl. Soedinenii*, 1977, 1455.

(357) agree well with the experimental values.[198] Protonation of compound (357) and its benzo-analogue in $HSO_3F$–$SbF_5$–$SO_2$ has been studied by means of $^1H$ and $^{13}C$ n.m.r. spectroscopy.[199] The $X$-ray structure of the quinone methide (358) has been determined; $^1H$ n.m.r. spectroscopy reveals the existence of a weak barrier to rotation around the inter-annular bond due to the contribution of the dipolar canonical form (358b).[200]

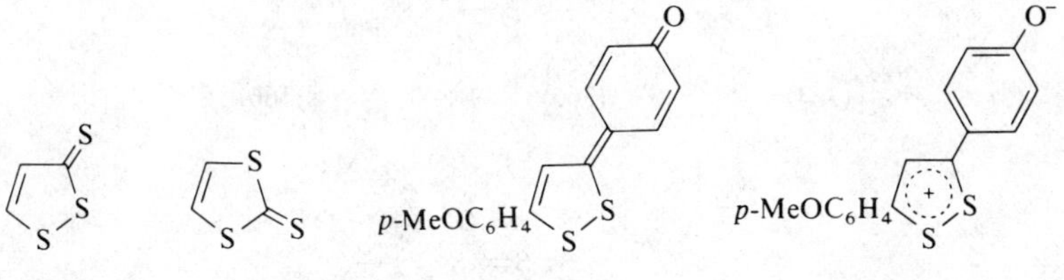

(356)    (357)    (358a)    (358b)

The 1,3-dithiolan derivative (359) is transformed into the 1,2-dithiole-3-thione (360) by the action of phosphorus sulphide.[201] Treatment of *o*-benzenedithiol with the esters $RCO_2Me$ (R = Pr, Bu$^t$, n-$C_{15}H_{31}$, Ph, or Ar) in the presence of fluoroboric acid affords the benzodithiolium salts (361).[202] The dithiocarbamate $Me_2NCS_2^-$ $H_2Me_2N^+$ reacts with chloranil to give the salt (362), which is converted into the dithione (363) by hydrogen sulphide.[203]

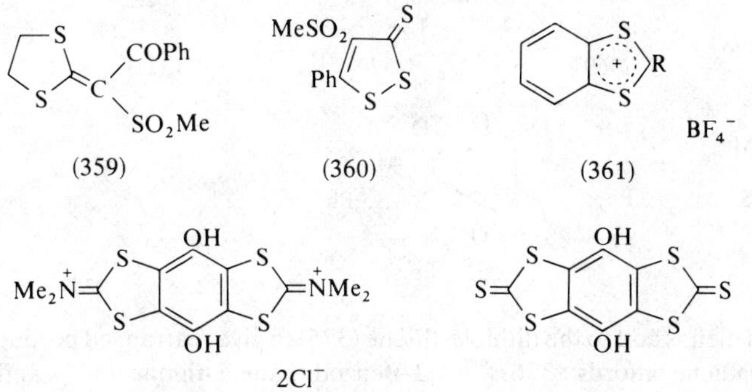

(359)    (360)    (361)

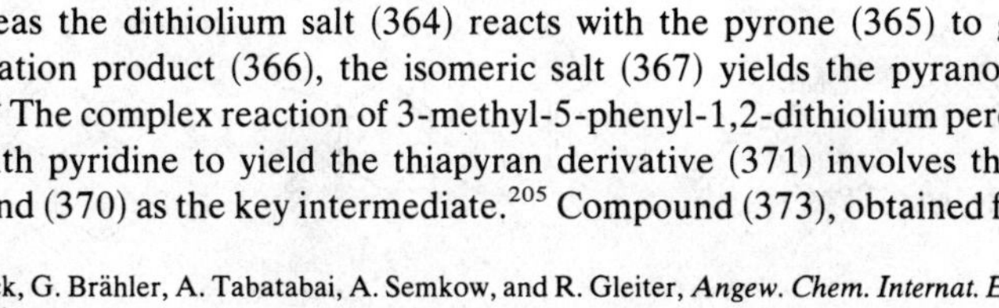

(362)    (363)

Whereas the dithiolium salt (364) reacts with the pyrone (365) to give the condensation product (366), the isomeric salt (367) yields the pyrano-pyrone (368).[204] The complex reaction of 3-methyl-5-phenyl-1,2-dithiolium perchlorate (369) with pyridine to yield the thiapyran derivative (371) involves the spiro-compound (370) as the key intermediate.[205] Compound (373), obtained from the

[198] H. Bock, G. Brähler, A. Tabatabai, A. Semkow, and R. Gleiter, *Angew. Chem. Internat. Edn.*, 1977, **16**, 724.
[199] G. A. Olah and J. L. Grant, *J. Org. Chem.*, 1977, **42**, 2237.
[200] K.-T. H. Wei, I. C. Paul, G. Le Coustumer, R. Pinel, and Y. Mollier, *Tetrahedron Letters*, 1977, 2717.
[201] D. Laduree, D. Paquer, and P. Rioult, *Rec. Trav. chim.*, 1977, **96**, 254.
[202] I. Degani and R. Fochi, *Synthesis*, 1977, 263.
[203] N. G. Demetriadis, S. J. Huang, and E. T. Samulski, *Tetrahedron Letters*, 1977, 2223.
[204] E. G. Frandsen, *Tetrahedron*, 1977, **33**, 869.
[205] E. I. G. Brown, D. Leaver, and D. M. McKinnon, *J.C.S. Perkin I*, 1977, 1511.

salt (372) and acetonedicarboxylic anhydride, has been converted into the thione (374) by acid hydrolysis and subsequent heating with phosphorus sulphide.[206]

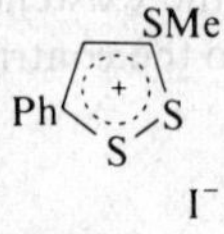

(364) 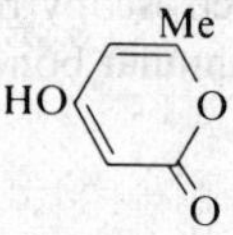                    (365) 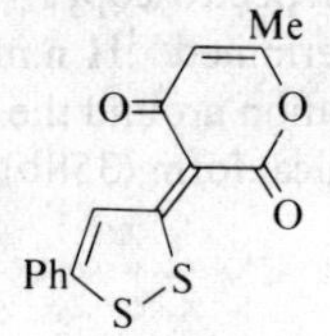                    (366)

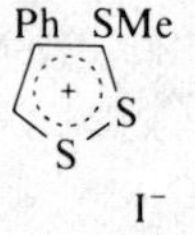

(367) 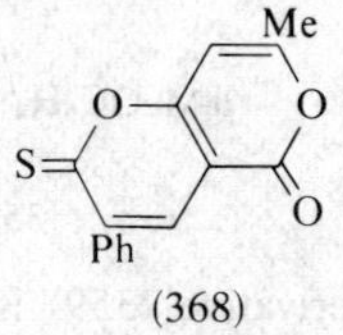                    (368)

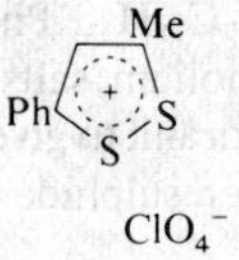

(369) 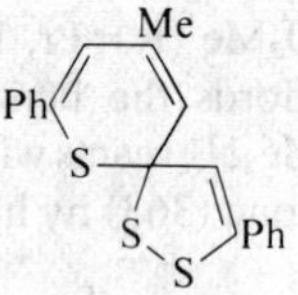                    (370) 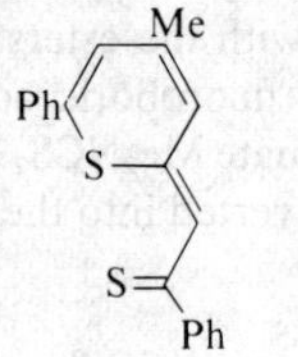                    (371)

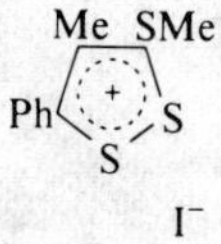

(372) 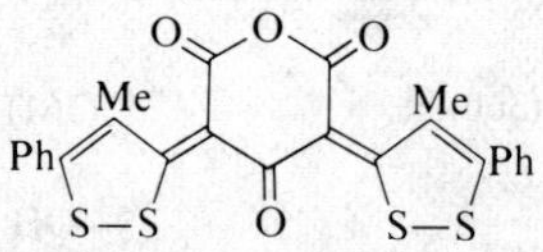                    (373) 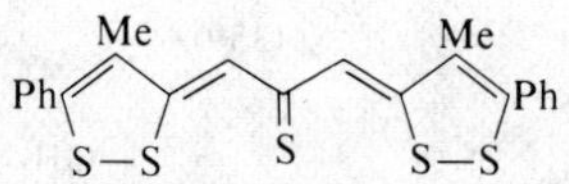                    (374)

Strained olefins add to the dithiole-thione (375) to give rearranged products; thus norbornadiene affords (376).[207] 1,2-Benzodithiole-3-thione (377) and cyclo-pentene form the photoadduct (378).[208]

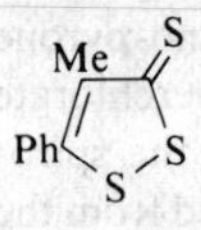

(375) 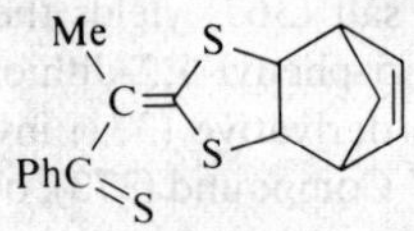                    (376) 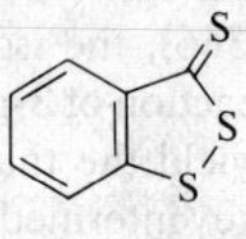                    (377) 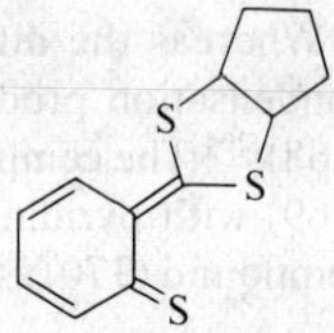                    (378)

1,3-Dithiole-2-thione (379; R = H) reacts with the keten (380) to form the heptafulvalene analogue (381), whose $^1$H n.m.r. spectrum suggests that it is

[206] E. G. Frandsen, *J.C.S. Chem. Comm.*, 1977, 851.
[207] V. N. Drozd and G. S. Bogomolova, *Zhur. org. Khim.*, 1977, **13**, 2012.
[208] P. De Mayo and H. Y. Ng, *Canad. J. Chem.*, 1977, **55**, 3763.

polyolefinic.[209] The related diketone (383) is produced by the condensation of 3-hydroxytropone with the dithiolium salt (382).[210] The action of pyrrolidino-cyclopentene on the thione (379; R = Ph) leads to the 1,3-dithionin-2-thione (384); with pyrrolidinocyclohexene, on the other hand, the thiophen derivative (385) is obtained.[211] The quinone methides (387; R = alkyl) are produced by the action of benzo-1,3-dithiolium fluoroborate (386) on 2,6-dialkyl-phenols.[212]

(379)     (380)     (381)     (382)

(383)     (384)     (385)

(386)     (387)

*Tetrathiafulvalenes and Selenium Analogues.* A review on 'organic metals' has been published.[213] The photoelectron spectra of tetrathiafulvalene (388) and the corresponding di- and tetra-selenium compounds have been determined[214] and the $^1$H n.m.r. spectra of tetrathiafulvalene in nematic liquid-crystalline solvents have been measured.[215] Synthetic procedures for symmetrically substituted tetrathiafulvalenes include the reduction of 1,3-dithiolium hexafluorophosphate with zinc dust,[216] the reductive coupling of 1,3-dithiole-2-thiones [*cf.* (379)] with derivatives of tervalent phosphorus,[217] and a five-stage synthesis of tetraethyl-tetrathiafulvalene, starting from the dithiocarbamate (389).[218] Treatment of a mixture of the dithiolium salts (390; R$^1$, R$^2$ = H, Me, or Et) and (391) with triethylamine results in three products, two symmetrical tetrathiafulvalenes and the 'crossed' compounds (392).[219] The conversion of the bisdithiolone (393) into

[209] K. Kato, Y. Kitahara, N. Morita, and T. Asao. *Chem. Letters*, 1977, 873.
[210] J. Nakayama, M. Ishihara, and M. Hoshino, *Chem. Letters*, 1977, 287.
[211] F. Ishii, R. Okazaki, and N. Inamoto, *Heterocycles*, 1977, **6**, 313.
[212] J. Nakayama, K. Yamashita, M. Hoshino, and T. Takemasa, *Chem. Letters*, 1977, 789.
[213] J. H. Perlstein, *Angew. Chem. Internat. Edn.*, 1977, **16**, 519.
[214] A. Schweig, N. Thon, and E. M. Engler, *J. Electron Spectroscopy Related Phenomena*, 1977, **12**, 335.
[215] T. C. Wong, E. E. Burnell, and L. Weiler, *Chem. Phys. Letters*, 1977, **50**, 243.
[216] A. Kruger and F. Wudl, *J. Org. Chem.*, 1977, **42**, 2778.
[217] S. Yoneda, T. Kawase, M. Inaba, and Z. Yoshida, *J. Org. Chem.*, 1978, **43**, 595.
[218] A. Mas, J. M. Fabre, E. Torreilles, L. Giral, and G. Brun, *Tetrahedron Letters*, 1977, 2579.
[219] J. M. Fabre, E. Torreilles, J. P. Gilbert, M. Chanaa, and L. Giral, *Tetrahedron Letters*, 1977, 4033.

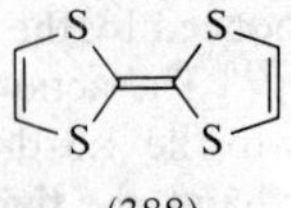

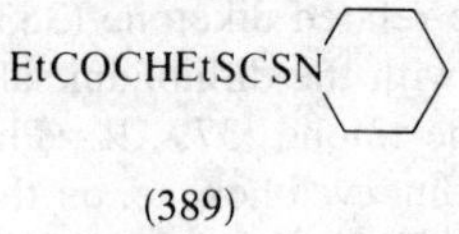

EtCOCHEtSCSN

(388)                (389)

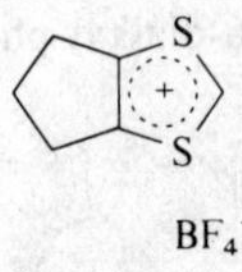

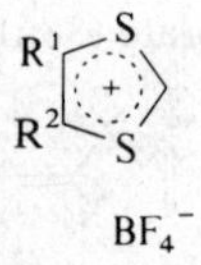

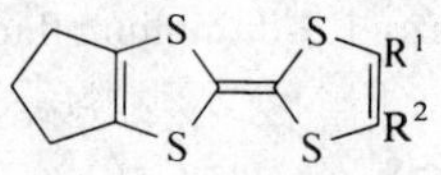

(390)        (391)           (392)

BF$_4^-$       BF$_4^-$

the tetrathiafulvalene derivative (394) has been described.[220] Dithiolium salts (391) react with the phosphonium ylide (395) in the presence of triethylamine to yield benzotetrathiafulvalenes (396).[221] 'Transthiolation' of tetrakis(methoxy-carbonyl)tetrathiafulvalene with *o*-benzenedithiol results in a mixture of compounds (396; R$^1$ = R$^2$ = CO$_2$Me) and (397).[222] The bistetrathiafulvalene

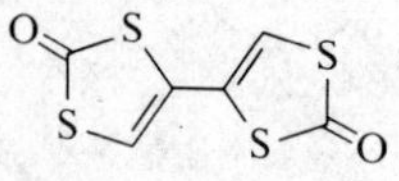

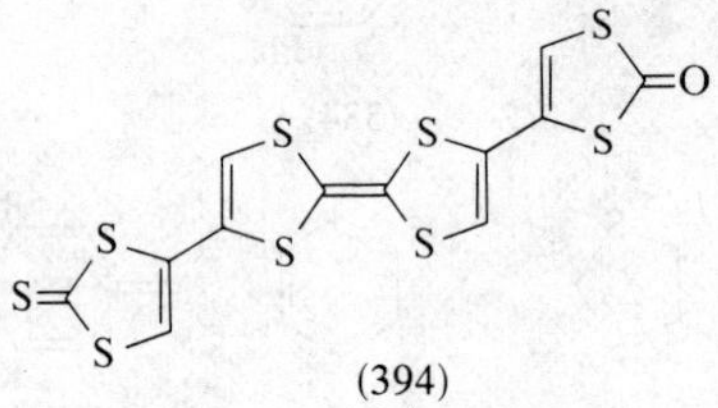

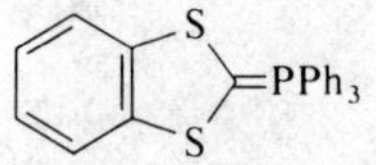

(393)           (394)          (395)

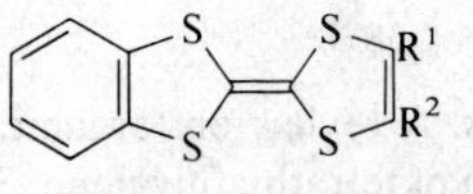

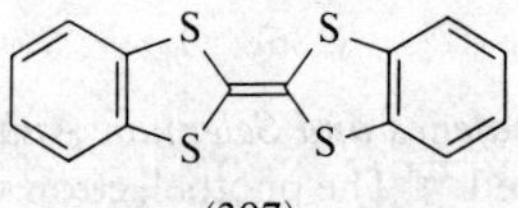

(396)                (397)

(398)[223] and the 'push–pull'-stabilised purple-black *p*-quinodimethane (399)[224] have been synthesized. 'Dithiapendione' (400) yields salts of the anion (401) on treatment with strong bases.[225]

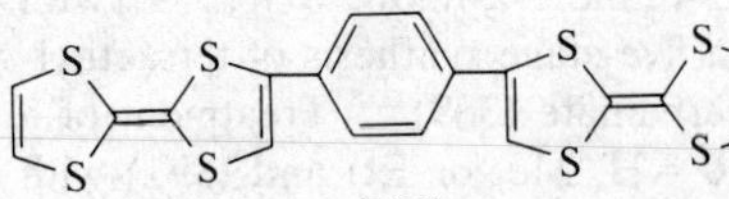

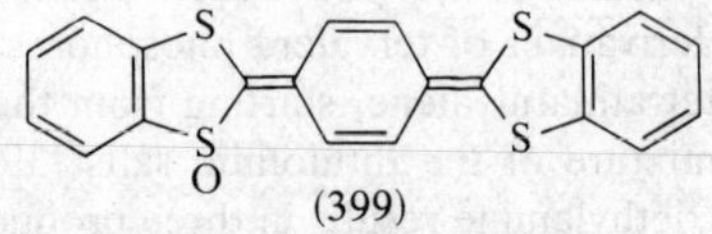

(398)                (399)

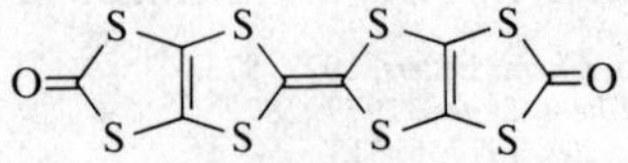

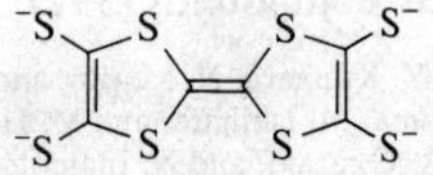

(400)                (401)

[220] J. R. Andersen, V. V. Patel, and E. M. Engler, *Tetrahedron Letters*, 1978, 239.
[221] N. C. Gonnella and M. P. Cava, *J. Org. Chem.*, 1978, **43**, 369.
[222] M. Mizuno and M. P. Cava, *J. Org. Chem.*, 1978, **43**, 416.
[223] M. L. Kaplan, R. C. Haddon, and F. Wudl, *J.C.S. Chem. Comm.*, 1977, 388.
[224] M. V. Lakshmikantham and M. P. Cava, *J. Org. Chem.*, 1978, **43**, 82.
[225] R. R. Shumaker and E. M. Engler, *J. Amer. Chem. Soc.*, 1977, **99**, 552.

The charge-transfer salt (402) is a better conductor than the corresponding sulphur compound.[226] The action of trimethyl phosphite on the diselenolethione

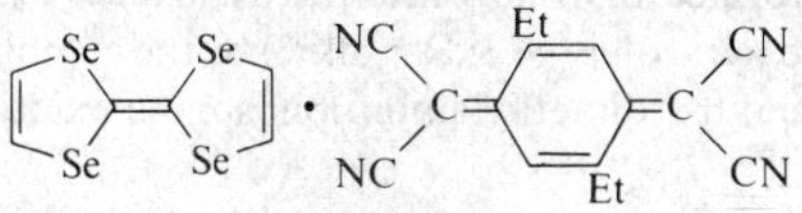

(402)

(403) gives a 6:1 mixture of the tetraselenafulvalene (404) and the mixed analogue (405).[227] The diselenadithiafulvalene (406) has been obtained by successive treatment of 2-chlorocyclopentanone with tetramethylthiourea, hydrogen selenide, sulphuric acid, hydrogen selenide, and, finally, triethyl phosphite.[228]

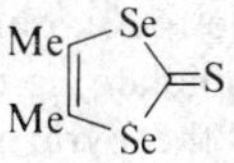

(403) 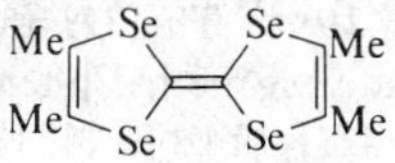        (404) 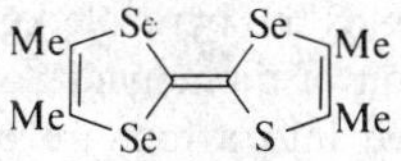        (405)

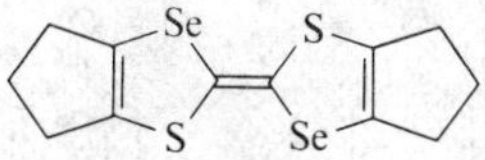

(406)

**Pyrazoles.**—*Formation.* Butadiyne reacts with methylhydrazine to yield either 1,3-dimethylpyrazole (407) or the 1,5-isomer, depending on conditions.[229] 2,2-Diacylketen *SS*-dimethyl acetals and related compounds yield methylthio-pyrazoles on treatment with hydrazine; thus compound (408) gives the ketone (409),[230] and the nitrile (410) gives the amine (411).[231] The azine (412) is converted into the pyrazole (413) in refluxing xylene; the reaction involves

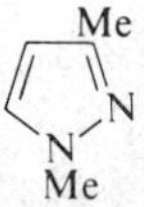

(407) 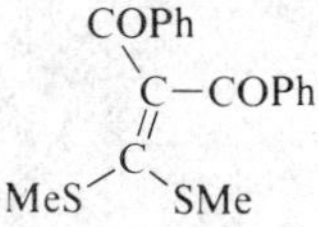        (408) 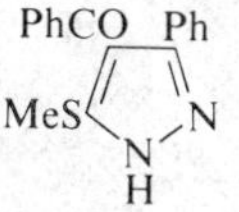        (409) 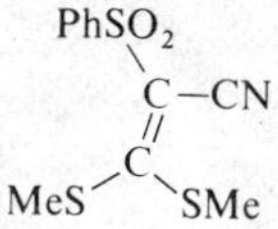        (410) 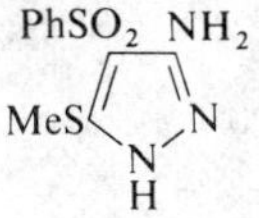        (411)

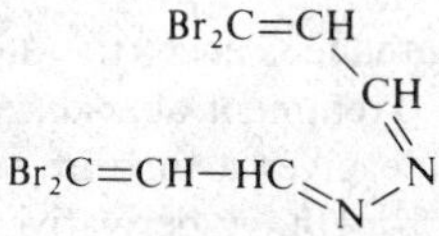

(412) 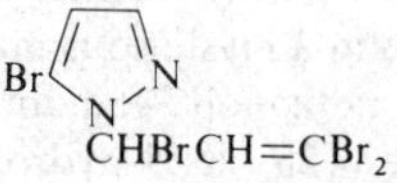        (413)

[226] J. R. Andersen, R. A. Craven, J. E. Weidenborner, and E. M. Engler, *J.C.S. Chem. Comm.*, 1977, 526.
[227] E. M. Engler, V. V. Patel, and R. R. Shumaker, *J.C.S. Chem. Comm.*, 1977, 835.
[228] P. Shu, A. N. Bloch, T. F. Carruthers, and D. O. Cowan, *J.C.S. Chem. Comm.*, 1977, 505.
[229] E. G. Darbinyan, Yu. B. Mitardzhyan, A. A. Saakyan, and S. G. Matsoyan, *Armyan. Khim. Zhur.*, 1977, **30**, 332 (*Chem. Abs.*, 1977, **87**, 117 811).
[230] E. C. Taylor and W. R. Purdum, *Heterocycles*, 1977, **6**, 1865.
[231] M. Augustin, R. Schmidt, and W. D. Rudorf, *Z. Chem.*, 1977, **17**, 289.

intramolecular addition, followed by migration of a bromine atom.[232] Mixed azines (414; $R^1$—$R^5$ = H, alkyl, or Ar) of $\alpha$-diketones and $\alpha\beta$-unsaturated aldehydes are transformed thermally into the pyrazoles (415).[233] In the remarkable formation of the pyrazoles (417; R = alkyl) by the combined action of lithium di-isopropylamide and tris(dimethylamino)phosphine oxide on the azines (416),

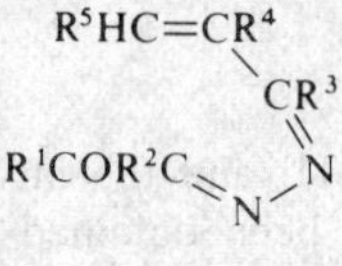

(414) 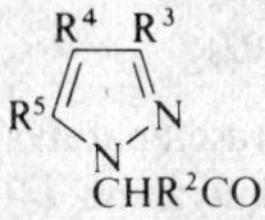                    (415) 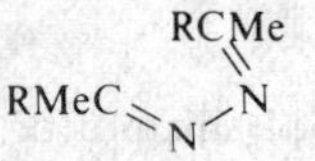                    (416) 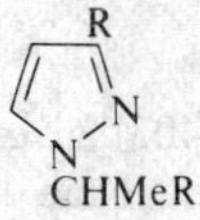                    (417)

the additional carbon atom is probably supplied by the latter reagent.[234] The oxidative condensation of benzaldehyde methylhydrazone with the diketone (418) yields the pyrazole ketone (419).[235] The 3$H$-pyrazole (420), obtained by the action of diphenyldiazomethane on methyl phenylpropiolate, is thermally converted into a mixture of the 4$H$-pyrazole (421) and the pyrazole (422), whereas photolysis leads to the cyclopropene (423) and the indene (424).[236]

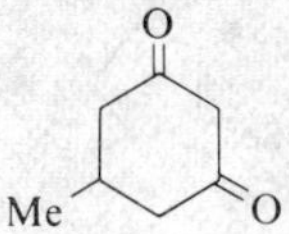

(418) 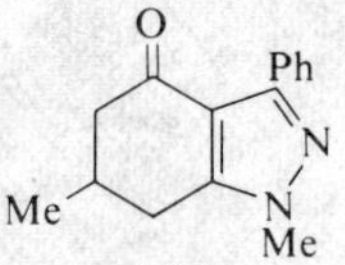                    (419)

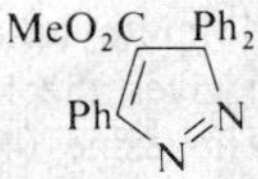

(420) 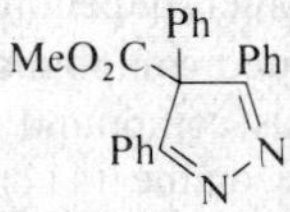                    (421) 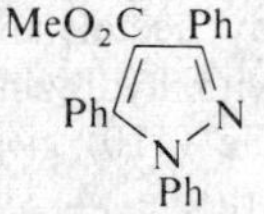                    (422) 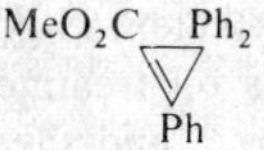                    (423)

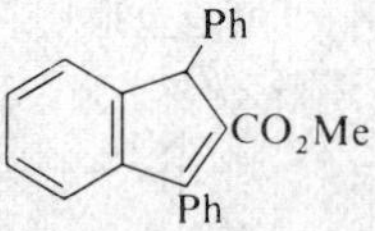

(424)

$\alpha\beta$-Dinitrostyrene and diphenyldiazomethane react to afford 3,4,5-triphenylpyrazole in a mysterious manner.[237] Treatment of $\omega$-diazoacetophenone with aqueous methanolic sodium hydroxide gives a mixture of 3-benzoyl-4-phenylpyrazole (425), its 5-hydroxy- and 5-methoxy-derivatives, 3-benzoyl-4-hydroxy-5-phenylpyrazole, the tetrazole (426), and the dihydrotetrazine

[232] P. Freche, A. Gorgues, and E. Levas, *Tetrahedron*, 1977, **33**, 2069.
[233] T. A. Albright, S. Evans, C. S. Kim, C. S. Labaw, A. B. Russiello, and E. E. Schweizer, *J. Org. Chem.*, 1977, **42**, 3691.
[234] Y. Tamaru, T. Harada, and Z. Yoshida, *Tetrahedron Letters*, 1977, 4323.
[235] W. Sucrow and V. Sandmann, *Chimia (Switz.)* 1977, **31**, 49.
[236] M. I. Komendantov and R. R. Bekmukhametov, *Tezisy Dokl.-Vses. Konf. Khim. Atsetilena 5th*, 1975, 374 (*Chem. Abs.*, 1978, **88**, 190 677).
[237] F. A. Gabitov, O. B. Kremleva, and A. L. Fridman, *Zhur. org. Khim.*, 1977, **13**, 1117.

(427).[238] Pyrazole *N*-oxides (429) are formed by the action of nitrile oxides RCNO (R = Ph, Ac, or CO$_2$Et) on the imidoyl-substituted oxosulphonium ylide (428).[239] Extrusion of sulphur from 6*H*-1,3,4-thiadiazines (430; R$^1$ = H or Ph; R$^2$ = Ph, SMe, NH$_2$, NMe$_2$, *etc*.) by the action of sodium ethoxide or triethyl phosphite yields the pyrazoles (431).[240] The diazatricycloundecatriene (433) is formed by thermolysis of the sodium salt of the toluene-*p*-sulphonylhydrazone (432).[241]

(425)   (426)   (427)

(428)   (429)

(430)   (431)   (432)   (433)

*Reactions.* The tautomerism of 5-hydroxy-pyrazoles (434) has been investigated by $^1$H[242] and $^{13}$C[243] and $^{15}$N[243] n.m.r. spectroscopy; the thiols (435) exist predominantly as such in neutral solvents.[244] Intermolecular migration of chlorine from the side-chain to the nucleus of compound (436) has been observed.[245] The colour reaction of phenols with 4-dimethylaminoantipyrine (437) in the presence of oxidants is due to the formation of quinone imines, such as (438).[246] Pyrolysis of antipyrine-4-diazonium fluoroborate (439) yields the dimeric azo-compound (440).[247] Irradiation of 1,4,5-triphenylpyrazole (441) in the presence of iodine leads to the phenanthropyrazole (442); when benzophenone is used as sensitizer, ring fission occurs to yield the anilino-nitrile (443).[248]

[238] P. Yates and R. J. Mayfield, *Canad. J. Chem.*, 1977, **55**, 145.
[239] R. Faragher and T. L. Gilchrist, *J.C.S. Perkin I*, 1977, 1196.
[240] (a) W. D. Pfeiffer, E. Dilk, and E. Bulka, *Synthesis*, 1977, 196; (b) W. D. Pfeiffer and E. Bulka, *ibid.*, p. 485.
[241] T. Miyashi, Y. Nishizawa, T. Sugiyama, and T. Mukai, *J. Amer. Chem. Soc.*, 1977, **99**, 6109.
[242] L. N. Kurkovskaya, N. N. Shapetko, A. S. Vitvitskaya, and I. Ya. Kvitko, *Zhur. org. Khim.*, 1977, **13**, 1750.
[243] G. E. Hawkes, E. W. Randall, J. Elguero, and C. J. Marzin, *J.C.S. Perkin II*, 1977, 1024.
[244] A. Maquestiau, Y. van Haverbeke, and J. C. Vanovervelt, *Bull. Soc. chim. belges*, 1977, **86**, 949.
[245] A. N. Sinyakov, S. F. Vasilevskii, and M. S. Shvartsberg, *Izvest. Akad. Nauk S.S.S.R., Ser. khim.*, 1977, 2306.
[246] D. Svoboda, M. Fraenkl, and J. Gasparic, *Mikrochim. Acta*, 1977, 285.
[247] P. J. Robbins, *J. Heterocyclic Chem.*, 1977, **14**, 1107.
[248] J. Grimshaw and D. Mannus, *J.C.S. Perkin I*, 1977, 2096.

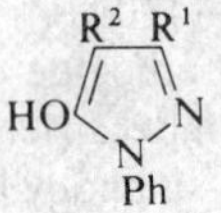

(434) 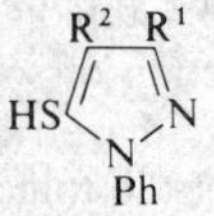        (435) 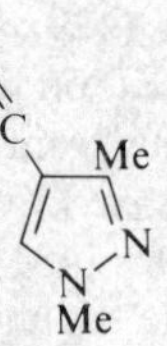        (436) 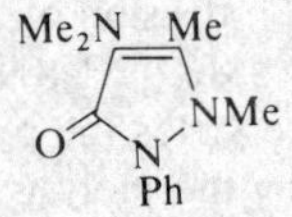        (437)

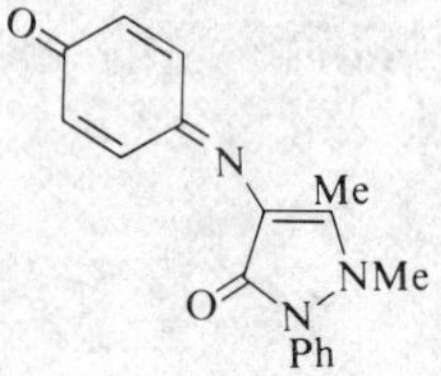

(438) 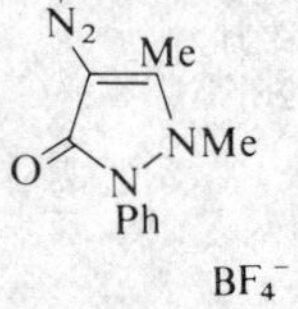                    (439) 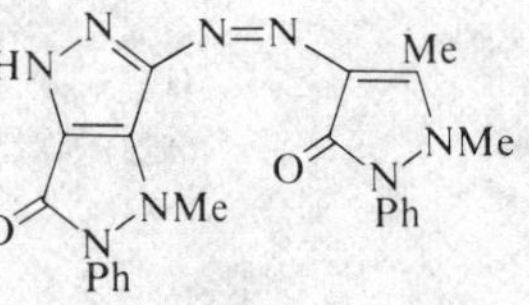                    (440)

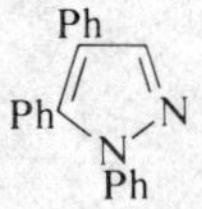

BF$_4^-$

(441) 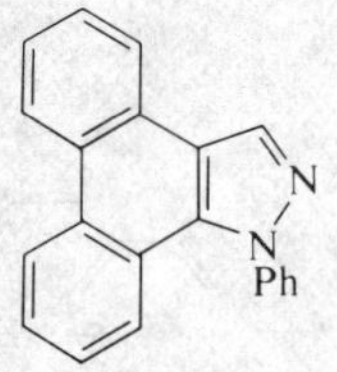                    (442) 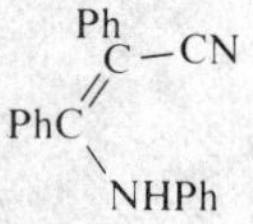                    (443)

The spiro-compound (445) is produced by treating β-bromo-β-nitrostyrene with the sodium salt of the pyrazolinone (444).[249] The kinetics of the Diels–Alder reaction of the pyrazolinone (446) with a series of alkyl vinyl ethers to yield the pyrazolo-pyrans (447) have been reported.[250] 1-Hydroxy-pyrazoles (448;

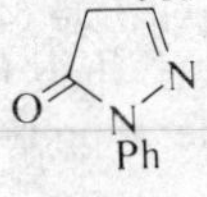

(444) 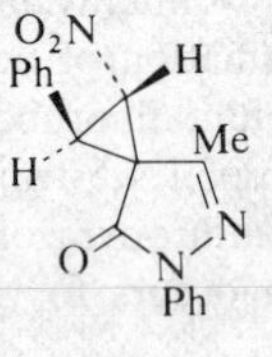                    (445) 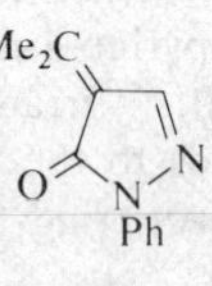                    (446) 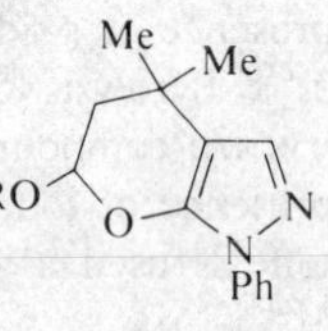                    (447)

R$^1$—R$^3$ = Me or Ph) are converted into the 4H-pyrazole derivatives (449) by the action of t-butyl hypochlorite; analogous products are obtained from 1-hydroxy-pyrazole-2-oxides.[251] 3,5-Dimethyl-1,4-dinitropyrazole (450) reacts with piperidine in primary alcohols ROH to yield mixtures of N nitrosopiperidine and the ethers (451).[252]

[249] T. G. Tkhor, A. S. Sopova, and B. I. Ionin, *Zhur. org. Khim.*, 1977, **13**, 851.
[250] G. Desimoni, P. Righetti, G. Tacconi, and A. Vigliani, *Gazzetta*, 1977, **107**, 91.
[251] J. P. Freeman, E. R. Janiga, and J. F. Lorenc, *J. Org. Chem.*, 1977, **42**, 3721.
[252] C. L. Habraken and S. M. Bonser, *Heterocycles*, 1977, **7**, 259.

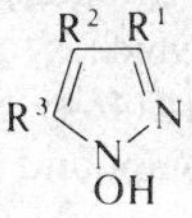

(448) 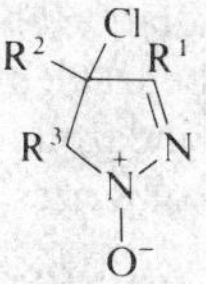    (449) 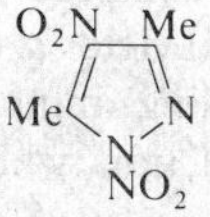    (450) 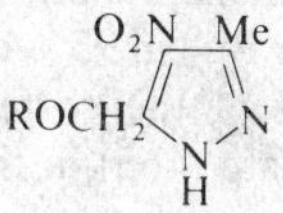    (451)

**Indazoles.**—The $^{13}C$ n.m.r. spectra of numerous indazoles have been reported[253a,b] and the tautomerism of nitro-indazoles has been investigated by dipole-moment measurements.[253c] Flash thermolysis of 2,5-diphenyltetrazole (452) or of 3,5-diphenyl-1,3,4-oxadiazolin-2-one (453) generates transient diphenyl-nitrile imine (454), which cyclises to 3-phenylindazole (455).[254] The production of

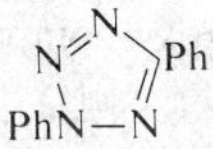

(452) 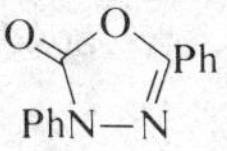    (453) 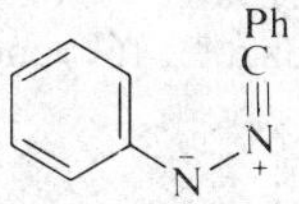    (454) 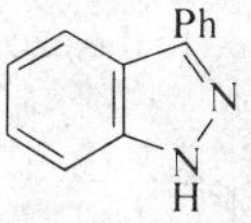    (455)

the racemic $3H$-indazole derivative (457) from the optically active diazonium salt (456) implies that the diazo-compounds (458) are general intermediates in

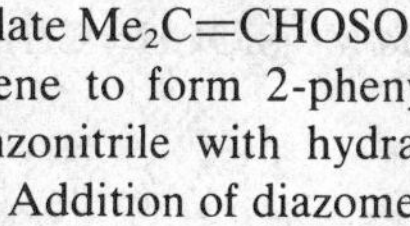

(456) 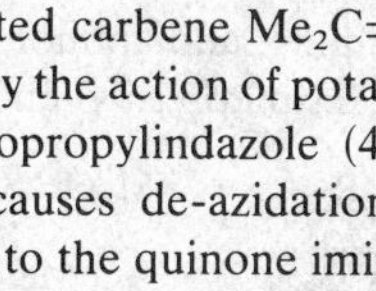    (457) 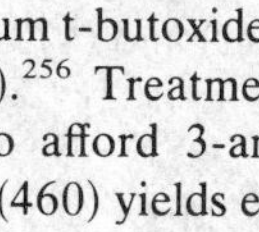    (458)

indazole synthesis.[255] The unsaturated carbene $Me_2C=C\colon$ (generated from the vinyl triflate $Me_2C=CHOSO_2CF_3$ by the action of potassium t-butoxide) adds to azobenzene to form 2-phenyl-3-isopropylindazole (459).[256] Treatment of *o*-azidobenzonitrile with hydrazine causes de-azidation to afford 3-aminoind-azole.[257] Addition of diazomethane to the quinone imine (460) yields either the adduct (461) or the de-alkylated product (462), depending on the absence or

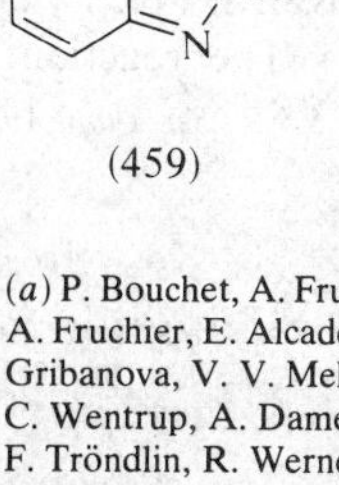

(459) 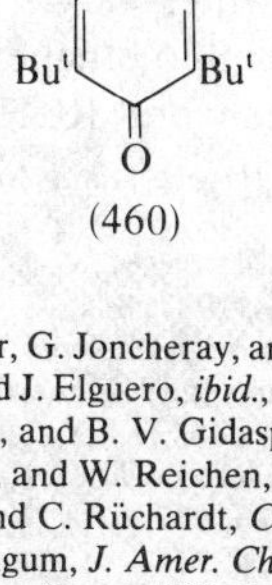    (460) 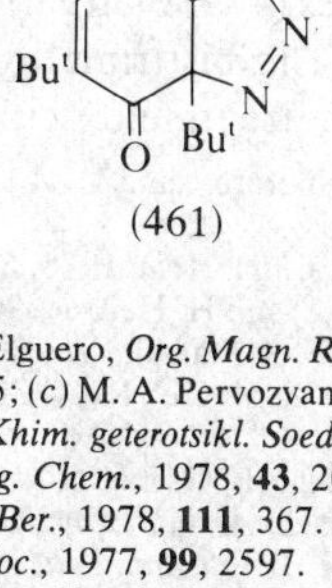    (461) 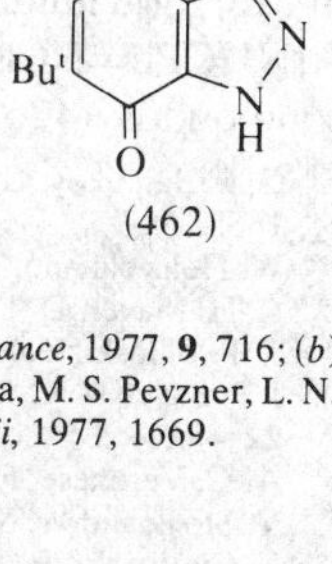    (462)

[253] (a) P. Bouchet, A. Fruchier, G. Joncheray, and J. Elguero, *Org. Magn. Resonance*, 1977, **9**, 716; (b) A. Fruchier, E. Alcade, and J. Elguero, *ibid.*, p. 235; (c) M. A. Pervozvanskaya, M. S. Pevzner, L. N. Gribanova, V. V. Melniko, and B. V. Gidaspov, *Khim. geterotsikl. Soedinenii*, 1977, 1669.

[254] C. Wentrup, A. Damerius, and W. Reichen, *J. Org. Chem.*, 1978, **43**, 2037.

[255] F. Tröndlin, R. Werner, and C. Rüchardt, *Chem. Ber.*, 1978, **111**, 367.

[256] P. J. Stang and M. G. Mangum, *J. Amer. Chem. Soc.*, 1977, **99**, 2597.

[257] T. McC. Paterson, R. K. Smalley, and H. Suschitzky, *Tetrahedron Letters*, 1977, 3973.

presence of methanol.[258] Low-temperature irradiation of 2-alkyl-indazoles (463; R = Me or Bu$^t$) in 95% ethanol yields *o*-aminobenzaldehyde, which is formed by hydrolysis of the corresponding imines; the indazolium salt (464) gives the intermediate (465) in this reaction, which tautomerizes to compound (466).[259]

(463)  (464)  (465)  (466)

Butyl isocyanate transforms the amino-indazole (467) into (468) with migration of the ethoxycarbonyl group.[260] The anion of 3-amino-1-methylindazole is autoxidised to the azo-indazole (469) and the corresponding azoxy-compound.[261]

(467)  (468)  (469)

**Imidazoles.**—Dipole-moment measurements suggest that *N*-acetyl-imidazole, -benzimidazole, and -indole exist as equilibrium mixtures of (*Z*)- and (*E*)-conformers.[262]

The imidazole ketone (470) is produced by the action of benzylamine on the oxime Ac$_2$C=NOH.[263] Toluene-*p*-sulphonyl-alkyl isocyanides TosylCHRNC (R = Me, Et, or CH$_2$Ph) react with benzylideneaniline to form the imidazoles (471).[151] Treatment of the bromo-ketone MeCOCHBrCH$_2$OMe with a mixture of formaldehyde, methylamine, and ammonia gives about equal amounts of the ethers (472) and (473).[264] Trimethylsilyl enol ethers Me$_3$SiOCH=CHR (R= H

(470)  (471)  (472)  (473)

or Me) form imidazoles (474) on treatment with *N*-aryl-*N'*-chloro-benzamidines ArNHCPh=NCl.[265] The bis(trimethylsilylimino)ethane derivative (475) yields compounds (476) on treatment with aldehydes RCHO.[266] The reaction of

[258] G. F. Bannikov, G. A. Nikiforov, and V. V. Ershov, *Izvest. Akad. Nauk S.S.S.R., Ser. khim.*, 1977, 1685.
[259] W. Heinzelmann, *Helv. Chim. Acta*, 1978, **61**, 618.
[260] K. H. Mayer, D. Lauerer, and H. Heitzer, *Synthesis*, 1977, 804.
[261] T. P. Filipskikh and A. F. Pozharskii, *Khim. geterotsikl. Soedinenii*, 1977, 952.
[262] J. P. Fayet, M. C. Vertut, P. Mauret, R. M. Claramunt, and J. Elguero, *Rev. Roumaine Chim.*, 1977, **22**, 471.
[263] A. C. Veronese, F. D.'Angeli, G. Zanotti, and A. Del Pra, *J.C.S. Chem. Comm.*, 1977, 443.
[264] V. Stöck and W. Schunack, *Arch. Pharm.*, 1977, **310**, 677.
[265] L. Citerio and R. Stradi, *Tetrahedron Letters*, 1977, 4227.
[266] I. Matsuda, T. Takahashi, and Y. Ishii, *Chem. Letters*, 1977, 1457.

cyanides RCN with 2,3-diphenylazirine (477) in the presence of boron trifluoride etherate affords the imidazoles (476).[267] 2,4,6-Triaryl-4$H$-1,3,5-thiadiazines

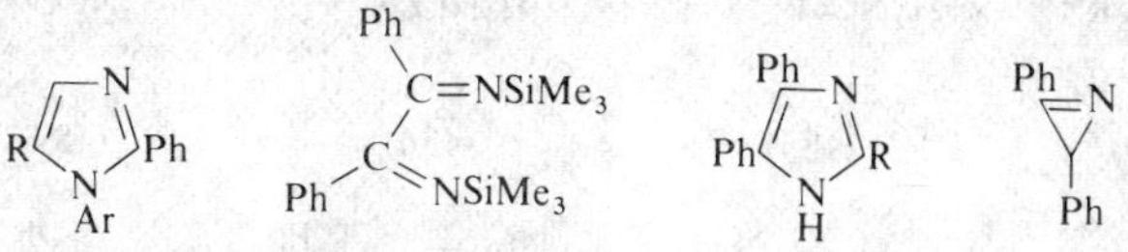

(474)      (475)      (476)      (477)

(478) lose sulphur in the presence of bases to form triaryl-imidazoles (481); the reaction probably proceeds by electrocyclization of the $8\pi$-anion (479) (see Scheme 3).[268] The heterocyclic prostaglandin analogue (482) has been synthesized.[269] Phthalaldehyde reacts with diaminomaleodinitrile to yield the condensed imidazole (483).[270]

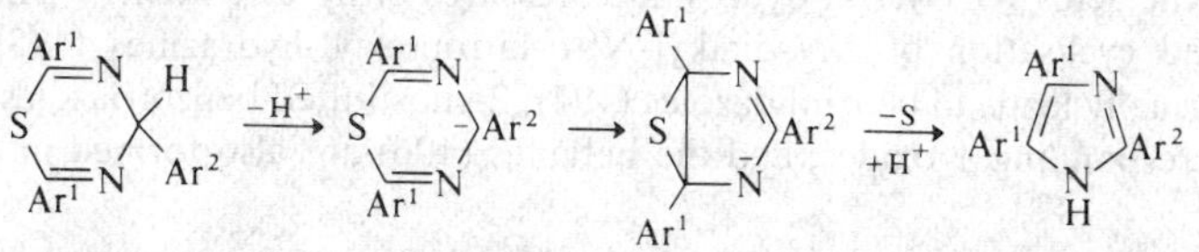

(478)      (479)      (480)      (481)

**Scheme 3**

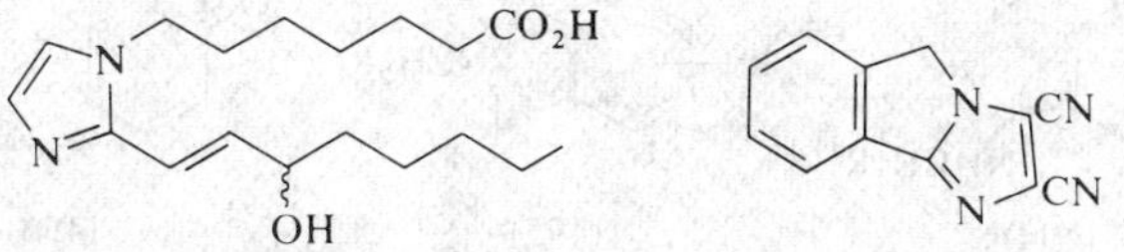

(482)      (483)

The rates of the thermal cleavage of $N$-alkyl groups from 1,3-dialkyl-imidazolium halides (484; $R^1 = R^3$ = alkyl, $R^2$ = H, $R^4$ = H, Ph, or $NO_2$) have been measured.[271] Deprotonation of 1,2,3-trimethylimidazolium salts (484; $R^1-R^3$ = Me; $R^4$ = H) at the 2-methyl group has been investigated in a model study of thiamine catalysis.[272] Photolysis of the 4-diazo-imidazole (485) in the presence of ethanol leads to a mixture of the carbene-insertion products (486) and (487).[273] Treatment of 2-diazo-4,5-dicyanoimidazole (488) with tetramethyl-thiourea gives the stable ylide (489).[5] The photochemical rearrangement of imidazole oxides (490) to 2-imidazolinones (491) has been described.[274]

[267] H. Bader and H. J. Hansen, *Helv. Chim. Acta*, 1978, **61**, 286.
[268] C. Giordano, L. Cassar, S. Panossian, and A. Belli, *J.C.S. Perkin II*, 1977, 939.
[269] M. Pailer and H. Gutwillinger, *Monatsh.*, 1977, **108**, 1059.
[270] W. Rasshofer and F. Vögtle, *J. Chem. Res. (S)*, 1977, 265.
[271] B. K. M. Chan, N.-H. Chang, and M. R. Grimmett, *Austral. J. Chem.*, 1977, **30**, 2005.
[272] J. A. Zoltewicz and J. K. O'Halloran, *J. Org. Chem.*, 1978, **43**, 1713.
[273] U. G. Kang and H. Shechter, *J. Amer. Chem. Soc.*, 1978, **100**, 651.
[274] R. Bartnik and G. Mloston, *Roczniki Chem.*, 1977, **51**, 1747.

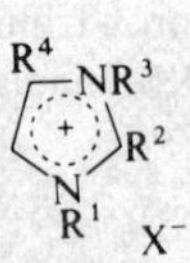

(484) 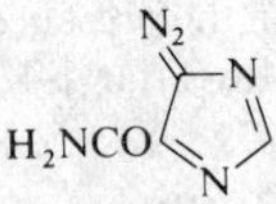          (485) 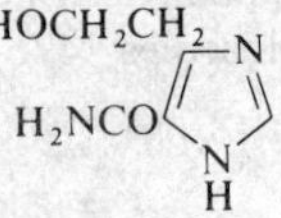          (486) 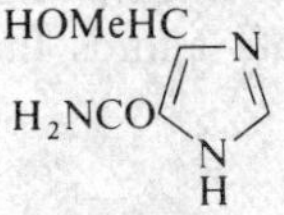          (487)

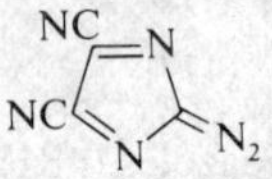

(488) 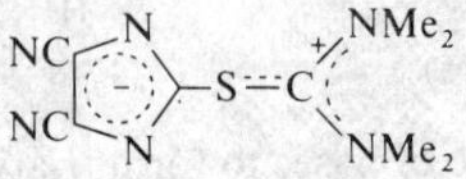              (489) 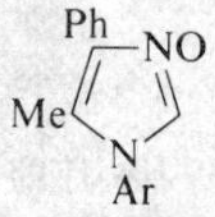              (490) 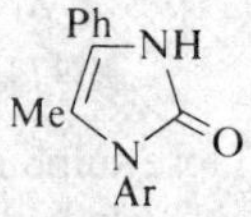      (491)

**Benzimidazoles.**—Benzimidazoles with bulky substituents in position 2, *e.g.* (492; R = Bu$^t$ or 1-adamantyl), are produced from *o*-phenylenediamines and carboxylic acids $RCO_2H$ at hydrostatic pressures of up to 8 kbar.[275] The acid-catalysed cyclisation of *NN*-dialkyl-*N'*-*o*-nitrophenyl-hydrazines (493; R$^1$ = alkyl) usually leads to benzimidazoles (494); 2-substituted benzotriazoles (495), the corresponding 1-oxides, and the betaines (496) are also formed in certain

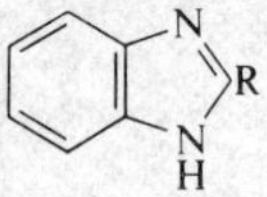

(492) 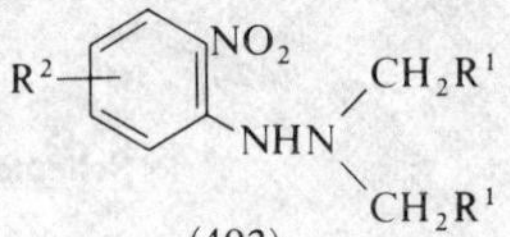                    (493)

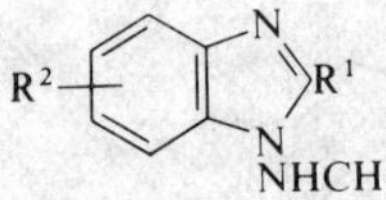

(494) 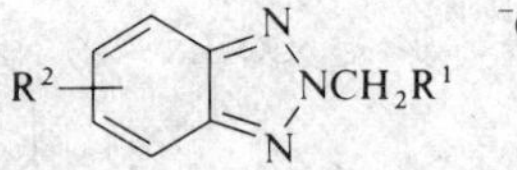              (495) 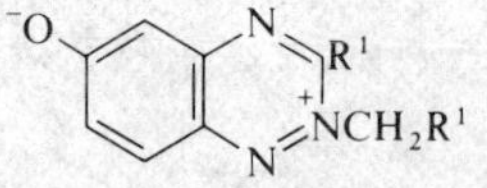              (496)

cases.[276] Benzimidazolium nitrites (498) are obtained by heating the *o*-nitrophenyl-benzamidines (497).[277] Two simple preparations of 1-hydroxy-benzimidazoles (499) have been reported: the 2-phenyl derivative is formed from

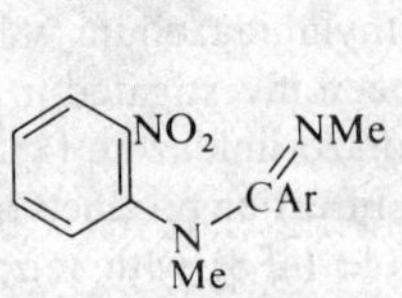

(497) 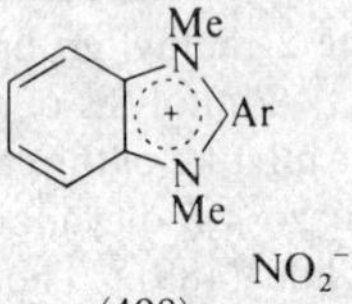            (498) 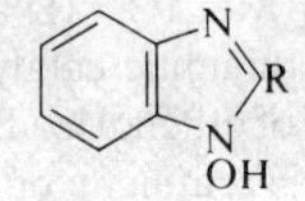            (499)

benzaldehyde and *o*-nitrosoaniline;[278] treatment of *N*-cyanomethyl-*o*-nitroaniline with sodium carbonate yields the 2-cyano-derivative.[279] Mesitonitrile oxide

[275] G. Holan, J. J. Evans, and M. Linton, *J.C.S. Perkin I*, 1977, 1200.

[276] D. W. S. Latham, O. Meth-Cohn, and H. Suschitzky, *J.C.S. Perkin I*, 1977, 478.

[277] H. M. Wolff and K. Hartke, *Tetrahedron Letters*, 1977, 3453.

[278] M. Z. Nazer, M. J. Haddadin, J. P. Petridou, and C. H. Issidorides, *Heterocycles*, 1977, **6**, 541.

[279] L. Konopski and B. Serafin, *Roczniki Chem.*, 1977, **51**, 1783.

and *N-p*-nitrophenyl-*SS*-dimethylsulphimide, $p$-$NO_2C_6H_4\bar{N}$—$\overset{+}{S}Me_2$, yield a mixture of the benzimidazole *N*-oxide (500) and the benzo-oxadiazine (501) in a general reaction of sterically stabilized benzonitrile oxides.[280]

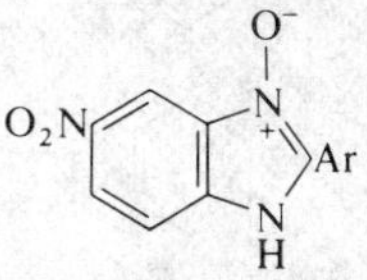

(500) 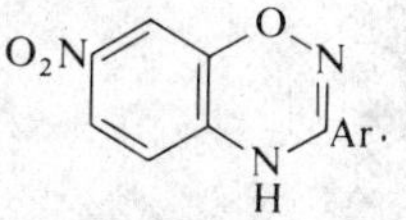          (501)          $Ar = 2,4,6\text{-}Me_3C_6H_2$

Irradiation of the bisazirine (502) in the presence of methyl acrylate affords the phenanthro-imidazole (503) in a complex intramolecular cycloaddition reaction.[281] The dibenzimidazole (505) is produced by the action of oxygen on the Schiff's base (504).[282]

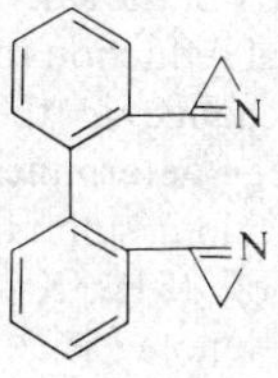

(502) 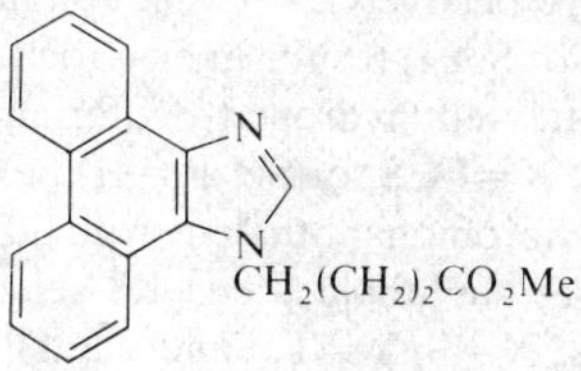          (503)

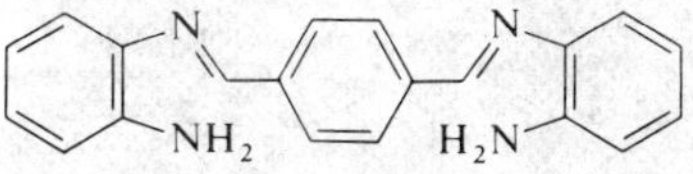

(504) 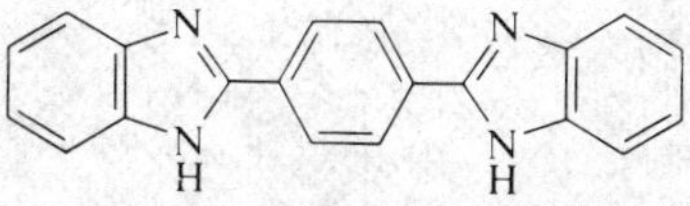          (505)

2-Methylbenzimidazole (492; R = Me) yields the benzimidazoline (506) in the Vilsmeier–Haack reaction with DMF–phosphorus oxychloride.[283] Treatment of 2-amino-1-benzhydryl-benzimidazole with sodium in liquid ammonia affords a mixture of 2-nitrobenzimidazole and the azo-compound (507) by an autoxidation process.[284] The benzimidazole derivative (508), prepared by the action of diphenylamine-2,2'-dicarbonyl chloride on benzimidazole, is 'dearomatized' on heating to give the condensed heterocycle (509).[285]

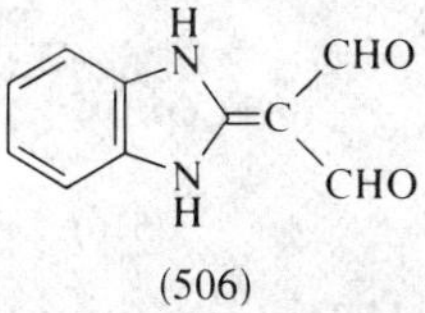

(506) 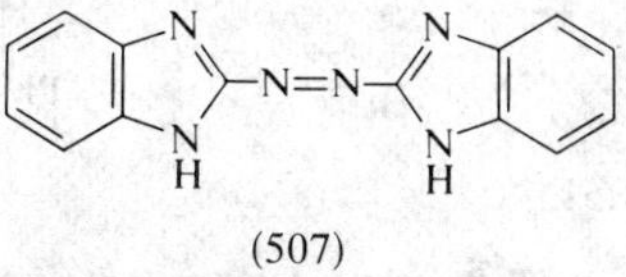          (507)

[280] S. Shiraishi, T. Shigemoto, and S. Ogawa, *Bull. Chem. Soc. Japan*, 1978, **51**, 563.
[281] A. Padwa and H. Ku, *J.C.S. Chem. Comm.*, 1977, 551.
[282] N. J. Coville and E. W. Neuse, *J. Org. Chem.*, 1977, **42**, 3485.
[283] H. A. Naik, V. Purnaprajna, and S. Seshadri, *Indian J. Chem.*, 1977, **15B** 338.
[284] A. F. Pozharskii, T. P. Filipskikh, and E. A. Zvezdina, *Tezisy Vses. Soveshch. Khim. Nitrosoedinenii, 5th*, 1974, 57 (*Chem. Abs.*, 1977, **87**, 135 191).
[285] A. Banerji, J. C. Cass, and A. R. Katritzky, *J.C.S. Perkin I*, 1977, 1162.

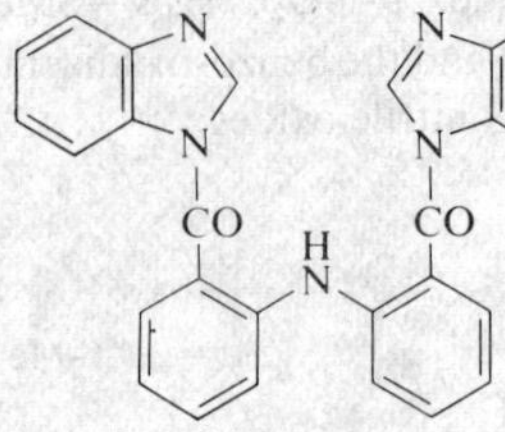

(508) 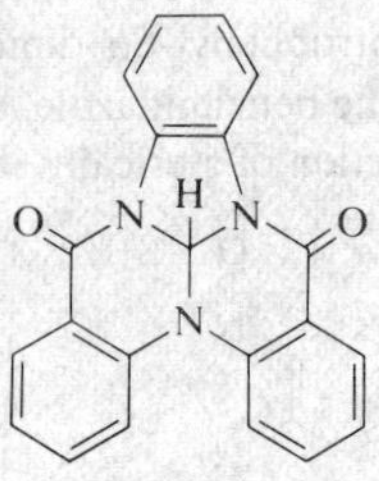                                    (509)

## 5 Systems containing Two Different Heteroatoms

**General.**—Aliphatic seleno-esters RCSeOEt condense with *o*-phenyl-enediamine, *o*-aminophenol, or *o*-aminothiophenol to yield the benzazoles (510; X = NH, O, or S, respectively).[286] Cyclic voltammetry of the salts (511; X, Y = O, S, or Se; R = SEt or SeEt) results in electrochemical reduction to dimers.[287] The rates of base-catalysed hydrogen–deuterium exchange in the 2-position of compounds (510; X = O, S, or Se; R = H) have been determined.[288] Successive treatment of 1,8-dibromonaphthalene with butyl-lithium, sulphur, butyl-lithium, and selenium gives naphtho[1,8-*cd*]-1,2-selenathiole (512; X = S, Y = Se); the tellurathiole (512; X = S, Y = Te) and the telluraselenole (512; X = Se, Y = Te) were prepared similarly.[289]

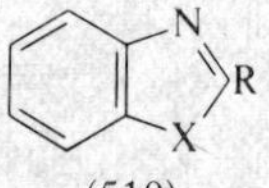

(510) 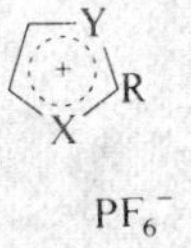                    (511) 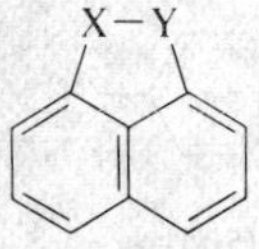                    (512)

**Isoxazoles.**—Isoxazolin-5-ones (513) exist in the gaseous phase in the form shown.[290] Dehydrogenation of isoxazolines (514; $R^1$, $R^2$ = alkyl or Ar) with active γ-manganese dioxide yields the corresponding isoxazoles quantitatively.[291] The oxazinone (515) undergoes ring contraction on treatment with hydroxylamine hydrochloride to give the isoxazole derivative (516).[292] The condensed isoxazoles

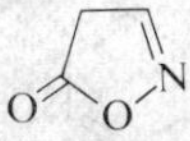

(513) 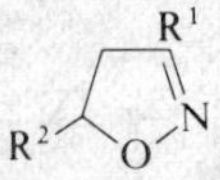            (514) 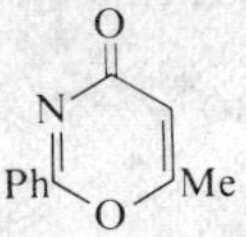            (515) 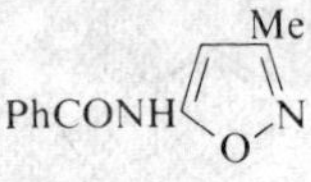            (516)

[286] V. I. Cohen and S. Pourabass, *J. Heterocyclic Chem.*, 1977, **14**, 1321.
[287] R. D. Braun and D. C. Green, *J. Electroanalyt. Chem. Interfacial Electrochem.*, 1977, **79**, 381 (*Chem. Abs.*, 1977, **87**, 133 493).
[288] O. Attanasi, *Gazzetta*, 1977, **107**, 359.
[289] J. Meinwald, D. Dauplaise, and J. Clardy, *J. Amer. Chem. Soc.*, 1977, **99**, 7743.
[290] J. L. Aubagnac and D. Bourgeon, *Org. Mass Spectrometry*, 1977, **12**, 65.
[291] A. Barco, S. Benetti, G. P. Pollini, and P. G. Baraldi, *Synthesis*, 1977, 837.
[292] Y. Yamamoto and Y. Azuma, *Heterocycles*, 1978, **9**, 185.

(518) are obtained by the action of the nitrile oxides ArCNO on the cyclohexenones (517; R = OH, MeO, Cl, morpholino, *etc.*).[293] The regioselectivity of 1,3-dipolar cycloaddition reactions of mesitonitrile oxide is opposite for electron-poor olefins to that of acetylenes; thus, with methyl acrylate the isoxazoline (519) is the main product, whereas methyl propiolate yields substantial amounts of the isoxazole (520).[294] The isoxazoles (521), together with the oxadiazoles (522), result from the addition of arenenitrile oxides to the enamine PhC(NH$_2$)=CHCN.[295] Photolysis of the bis-isoxazoline (523) gives, *inter alia*, 3,5-diphenylisoxazole.[296]

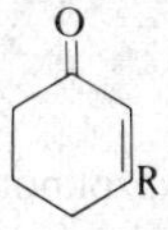

(517) 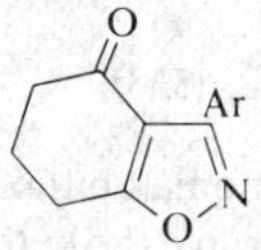    (518) 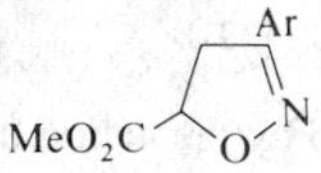    (519) 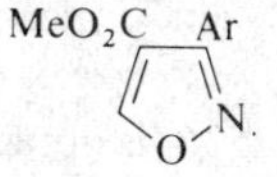    (520)

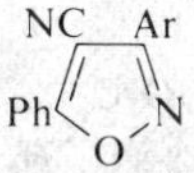

(521) 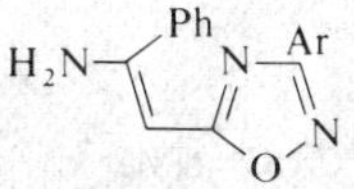    (522) 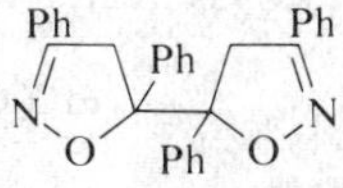    (523)

5-Hydrazino-isoxazoles (524; R$^1$, R$^2$ = Me, Ph, *etc.*) rearrange on heating to produce mixtures of 1-amino-pyrazoles (525), 4-amino-pyrazolin-5-ones (526), and triazinones (527) *via* the suggested diradicals (528).[297] The isoxazolium salt

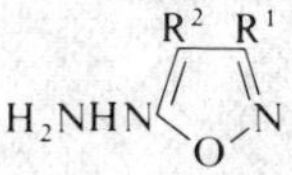

(524) 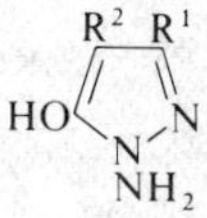    (525) 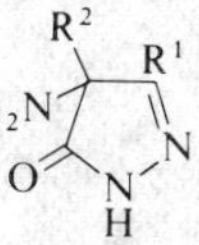    (526)

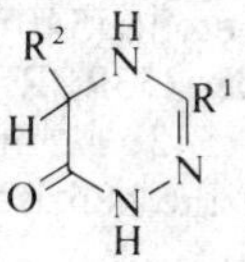

(527) 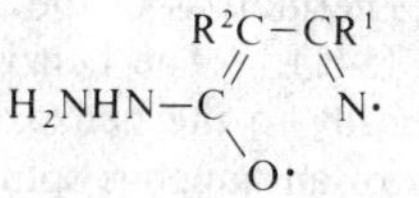    (528)

(529) is cleaved by potassium methoxide to compound (531), presumably by way of the intermediate (530).[298] The kinetics of the base-induced decomposition of the isoxazoles (532; R$^1$, R$^2$ = H or Ar) indicate concerted proton abstraction and fission of the oxygen–nitrogen bond to form the cyano-enolates (533).[299] The

[293] A. A. Akhrem, F. A. Lakhvich, V. A. Khripach, and A. G. Pozdeev, *Synthesis*, 1978, 43.
[294] K. N. Houk, Y.-M. Chang, R. W. Strozier, and P. Caramella, *Heterocycles*, 1977, **7**, 793.
[295] A. Corsaro, U. Chiacchio, and G. Purrello, *J.C.S. Perkin I*, 1977, 2154.
[296] T. Mukai, H. Saiki, T. Miyashi, and Y. Ikegami, *Heterocycles*, 1977, **6**, 1599.
[297] G. Adembri, A. Camparini, F. Ponticelli, and P. Tedeschi, *J.C.S. Perkin I*, 1977, 971.
[298] C. Kashima, N. Mukai, E. F. DeRose, Y. Tsuda, and Y. Omote, *Heterocycles*, 1977, **7**, 241.
[299] A. De Munno, V. Bertini, and F. Lucchesini, *J.C.S. Perkin II*, 1977, 1121.

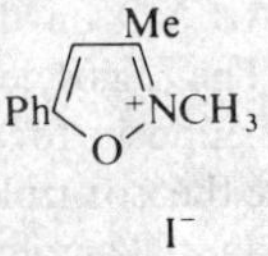

Me
Ph⟶O⁺NCH₃
I⁻
(529)

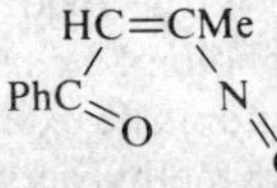

HC=CMe
PhC=O    N⟍CH₂
(530)

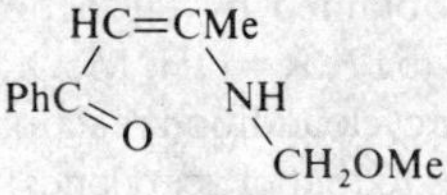

HC=CMe
PhC=O    NH⟍CH₂OMe
(531)

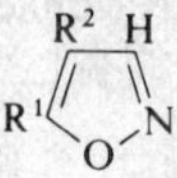

R²  H
R¹⟶O—N
(532)

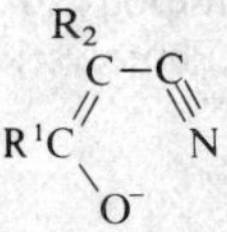

R₂
  C—C
R¹C      N
   O⁻
(533)

diamide (536) is obtained by the action of ammonia on the isoxazolinone (534); the reaction is initiated by proton abstraction to give the keten imine (535).[300] An analogous keten imine can be isolated as its potassium salt (538) when the isoxazolinone (537) is treated with potassium t-butoxide.[301]

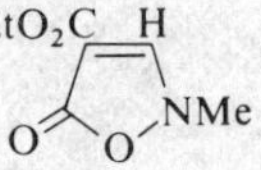

EtO₂C  H
O⟶O—NMe
(534)

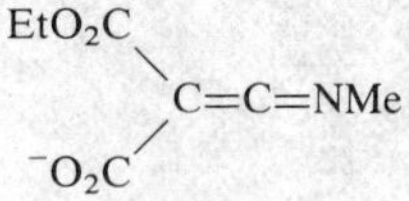

EtO₂C
       C=C=NMe
⁻O₂C
(535)

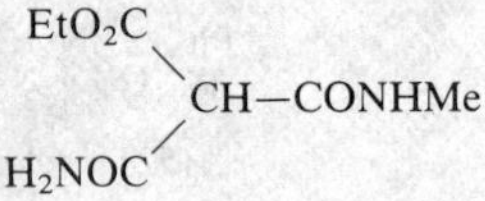

EtO₂C
       CH—CONHMe
H₂NOC
(536)

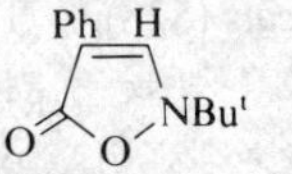

Ph  H
O⟶O—NBuᵗ
(537)

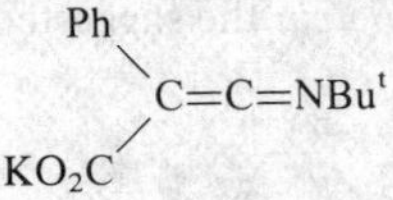

Ph
     C=C=NBuᵗ
KO₂C
(538)

**Benzisoxazoles.**—The reaction of salicylaldehyde oxime with thionyl chloride in pyridine to yield 1,2-benzisoxazole (539; R = H) constitutes a new synthesis of this ring system.[302] Thermolysis of the azido-ketone (540) affords the linearly annelated anthranil (541).[303] The benzisoxazoles (539; R = H, Me, or Ph) rearrange photochemically to the benzoxazoles (543) *via* intermediate azirines (542).[304] Irradiation of an aqueous solution of the benzisoxazolium salt (544) gives the phenolic ketone (545).[305]

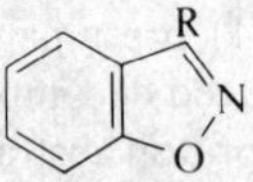

R
⟶N
O
(539)

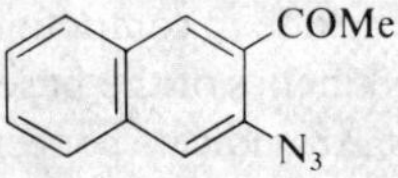

COMe
N₃
(540)

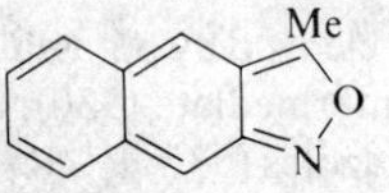

Me
O
N
(541)

[300] R. Pepino, R. Bossio, V. Parrini, and E. Belgodere, *Gazzetta*, 1976, **106**, 1135.
[301] D. J. Woodman, W. H. Campbell, and E. F. DeBose, *Heterocycles*, 1977, **7**, 247.
[302] U. R. Kalkote and D. D. Goswami, *Austral. J. Chem.*, 1977, **30**, 1847.
[303] W. Friedrichsen and P. Kaschner, *Annalen*, 1977, 1959.
[304] K. H. Grellmann and E. Tauer, *J. Photochem.*, 1977, **6**, 365.
[305] N. F. Haley, *J. Org. Chem.*, 1977, **42**, 3929.

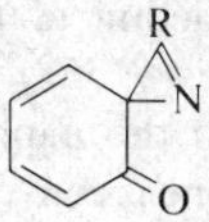

(542) 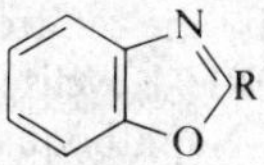　　　　　(543) 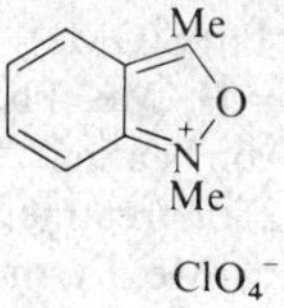　　　　　(544) 　　　　　(545)

**Oxazoles.**—Recent developments in the chemistry of oxazolinones have been reviewed.[306] Tosylalkyl isocyanides TosCHRNC (R = alkyl) react with aromatic aldehydes to form oxazoles (546).[151] The action of hydrogen chloride on a mixture of benzoyl cyanide and an aromatic aldehyde results in chloro-oxazoles (547).[307] Tungsten(VI) chloride catalyses the formation of 2-methyl-5-phenyl-oxazole (548) from acetonitrile and diazoacetophenone.[308] Thermolysis of the

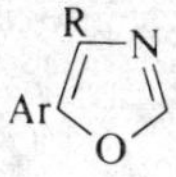

(546) 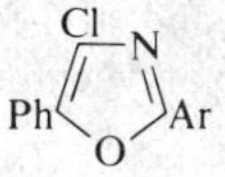　　　　　　(547) 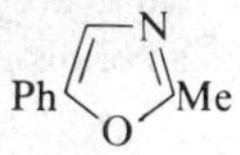　　　　　　(548)

phthalimido-aziridines (549; R = Me, Ph, or OMe) affords mixtures of the oxazolines (550) and oxazoles (551).[309] In the Pomeranz–Fritsch synthesis of isoquinolines from the acetals ArCH=NCH$_2$CH(OMe)$_2$, oxazoles (552) are formed as by-products.[310]

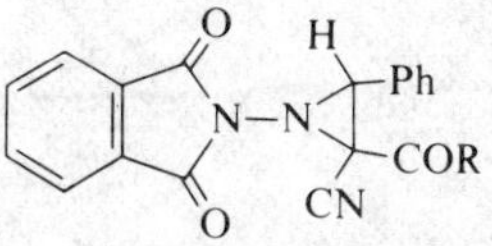

(549) 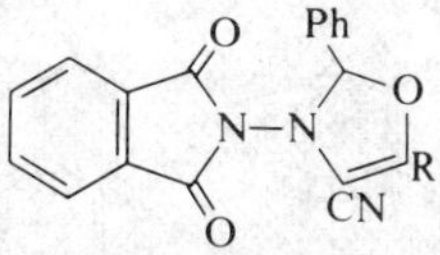　　　　　　　(550) 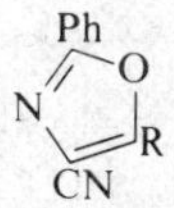　　　　　　　(551) 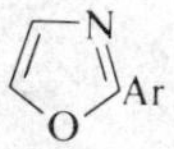　　　　　(552)

Dye-sensitized photo-oxidation of the ether (553) gives a mixture of the peroxide (554) and the triamide (555).[311] The Cornforth rearrangement (556) → (557) has been studied theoretically; MINDO/3 calculations for the cases (556; R$^1$ = H, R$^2$ = NH$_2$, R$^3$ = OMe) and (556; R$^1$ = CH=CH$_2$, R$^2$ = NH$_2$, R$^3$ = OMe) indicate an unusual type of intermediate, which cannot be represented by a classical formula.[312]

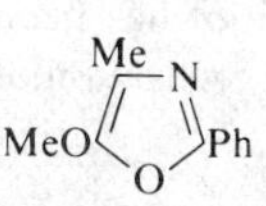

(553) 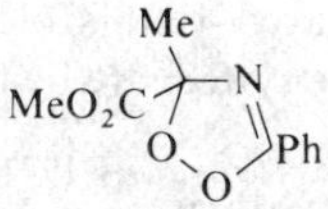　　　　　(554) 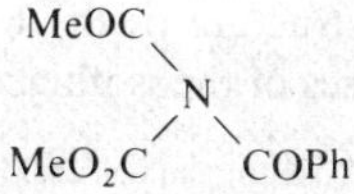　　　　　(555) 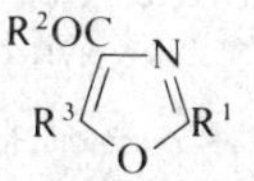　　　　　(556) 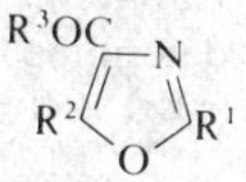　　　　　(557)

[306] R. Filler and Y. S. Rao, *Adv. Heterocyclic Chem.*, 1977, **21**, 175.
[307] M. Davis, R. Lakhan, and B. Ternai, *J. Heterocyclic Chem.*, 1977, **14**, 317.
[308] K. Kitakani, T. Hiyama, and H. Nozaki, *Bull. Chem. Soc. Japan*, 1977, **50**, 1647.
[309] H. Person, K. Luanglath, M. Baudru, and A. Foucaud, *Bull. Soc. chim. France*, 1976, 1989.
[310] E. V. Brown, *J. Org. Chem.*, 1977, **42**, 3208.
[311] (a) M. L. Graziano, M. R. Iesce, A. Carotenuto, and R. Scarpati, *J. Heterocyclic Chem.*, 1977, **14**, 261; (b) *ibid.*, p. 1215.
[312] M. J. S. Dewar and I. J. Turchi, *J.C.S. Perkin II*, 1977, 724.

**Benzoxazoles.**—Benzoxazoles (559) are obtained by the action of the thio-phosphoryl compound $PhPS(OMe)_2$ on the nitrones (558).[313] The naphthaquinones (560; R = H, Me, Ph, or $NH_2$) cyclise to the naphtho[2,3-*d*]oxazole-4,9-diones (561) on heating.[314] 1,2-Naphthoquinone is converted into the condensed oxazole (562) on treatment with ethyl azidoformate.[315] Acenaph-tho[1,2-*d*]oxazoles (564; R = Me, Et, or Ph) are formed when 2-diazoacenaph-thenone (563) is irradiated in the presence of cyanides RCN.[316] Phenanthro-oxazoles result when 2-aryl-4,5-diphenyl-oxazoles are irradiated in the presence of oxygen or iodine;[317] the *p*-nitrophenyl derivative is obtained by the action of *p*-nitrobenzyl chloride on the monoimine or the mono-oxime of phenan-thraquinone.[318]

2-Chloro-3-ethylbenzoxazolium salts (565) are remarkably versatile reagents; the fluoroborate converts $\alpha$-hydroxy-carboxylic acids into ketones, *e.g.* benzilic acid into benzophenone,[319] methyl thiocarbamates RNHCSOMe into iso-cyanates RNCO,[320] and alkyl- and aryl-formamides into isocyanides;[321] the chloride transforms alcohols into alkyl chlorides.[322]

(558)                    (559)                         (560)

(561)                    (562)

(563)                    (564)                         (565)

**Isothiazoles.**—The $X$-ray structure of the hydroxy-isothiazole (566) has been determined.[323] New syntheses of the isothiazole system are by the cyclization of

313  R. Nagase, T. Kawashima, M. Yoshifuji, and N. Inamoto, *Heterocycles*, 1977, **8**, 243.
314  A. S. Hammam and A.-M. Osman, *J. prakt. Chem.*, 1977, **319**, 254.
315  M. S. Chauhan, D. M. McKinnon, and R. G. Cooke, *Canad. J. Chem.*, 1977, **55**, 2363.
316  O. Tsuge and M. Koga, *Heterocycles*, 1977, **6**, 411.
317  V. N. R. Pillai and M. Ravindram, *Indian. J. Chem.*, 1977, **15B**, 1043.
318  W. I. Awad, Z. S. Salih, and Z. H. Aiube, *J.C.S. Perkin I*, 1977, 1280.
319  T. Mukaiyama and Y. Echigo, *Chem. Letters*, 1978, 49.
320  Y. Echigo, Y. Watanabe, and T. Mukaiyama, *Chem. Letters*, 1977, 1345.
321  Y. Echigo, Y. Watanabe, and T. Mukaiyama, *Chem. Letters*, 1977, 697.
322  T. Mukaiyama, S. Shoda, and Y. Watanabe, *Chem. Letters*, 1977, 383.
323  J. L. McVicars, M. F. Mackay, and M. Davis, *J.C.S. Perkin II*, 1977, 1332.

the dithiocarboxylate $PhCH(CN)CS_2Na$ to the thiolate (567)[324] and the formation of the cycloalkano-derivative (569) by treatment of the thiocyanate (568) with ammonia.[325] The benzo[c]isothiazole (571) is produced by thermolysis of the thiosulphinylimine (570).[326] *o*-t-Butylthio-benzaldehyde oxime cyclises to the benzo[b]isothiazole (572) in polyphosphoric acid; the isothiazolobenzisothiazole (573) has been obtained from the corresponding di-t-butylthio-dioxime.[327]

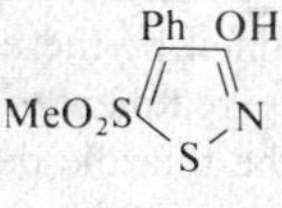

(566) 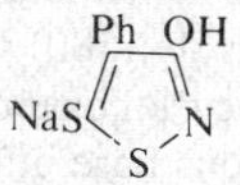      (567) 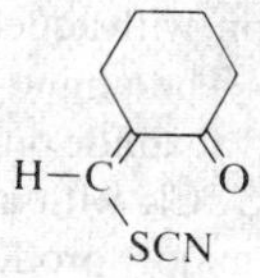      (568) 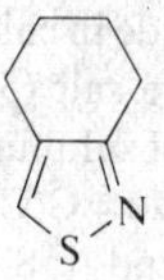      (569)

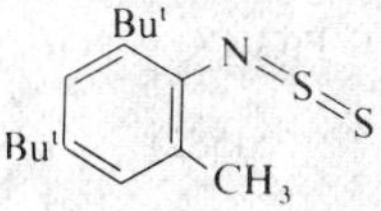

(570) 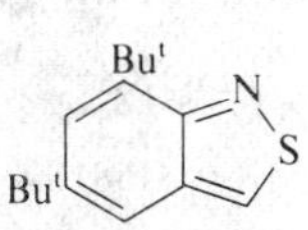      (571) 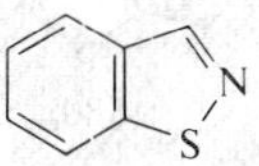      (572) 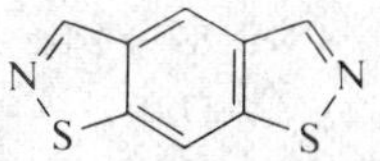      (573)

Benzoylation of compound (574) gives the unstable benzoyloxy-benzisothiazole (575), which rapidly loses sulphur to form the benzoxazinone (576).[328]

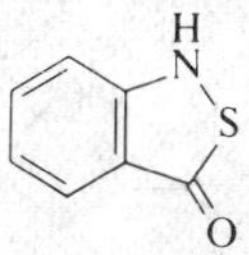

(574) 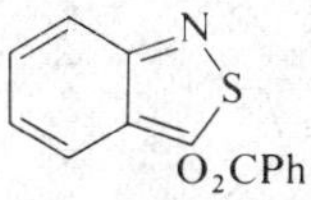      (575) 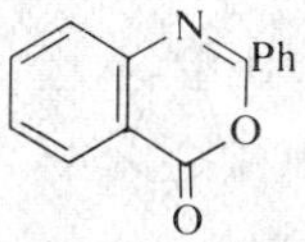      (576)

A general reaction of *N*-substituted isothiazolium salts with carbanions is the formation of dihydrothiophens by initial attack at the sulphur atom, followed by ring-opening and recyclisation; an example is shown in Scheme 4.[329] Flash vacuum pyrolysis transforms the benzisothiazole *SS*-dioxide (577) into 2-phenylbenzoxazole (578).[330]

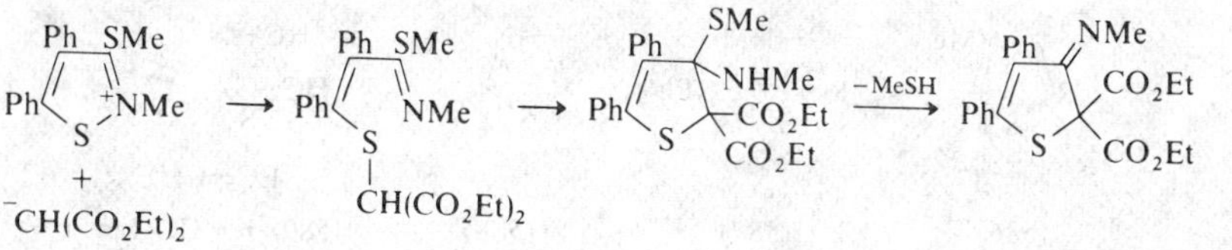

**Scheme 4**

[324] M. Davis, M. C. Dereani, J. L. McVicars, and I. J. Morris, *Austral. J. Chem.*, 1977, **30**, 1815.
[325] B. Schulze, S. Herre, R. Brämer, C. Laux, and M. Mühlstädt, *J. prakt. Chem.*, 1977, **319**, 305.
[326] Y. Inagaki, R. Okazaki, and N. Inamoto, *Tetrahedron Letters*, 1977, 293.
[327] O. Meth-Cohn and B. Tarnowski, *Synthesis*, 1978, 58.
[328] M. Davis and S. P. Pogany, *J. Heterocyclic Chem.*, 1977, **14**, 267.
[329] D. M. McKinnon, M. E. R. Hassan, and M. Chauhan, *Canad. J. Chem.*, 1977, **55**, 1123.
[330] R. A. Abramovitch and S. Wake, *J.C.S. Chem. Comm.*, 1977, 673.

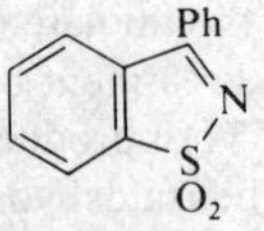

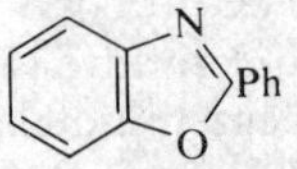

(577)                    (578)

**Thiazoles.**—Toluene-*p*-sulphonylmethyl isocyanide reacts with carbon disulphide in chloroform, with aqueous alkali, under phase-transfer conditions, to yield the salt (579).[331] The amino-thiazole (580) is obtained by the combined action of sulphur and cyanamide on thiocyclohexanone.[332] Treatment of the ester $H_2NCMe=C(CO_2Et)SCCl_3$ with aqueous base affords the thiazole derivatives (581) and (582) as major products.[333] Potassium cyanimidothiocarbonates

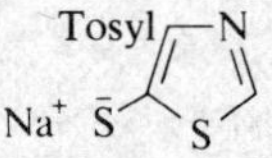

(579) 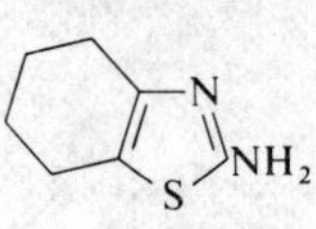              (580) 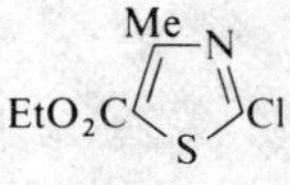              (581) 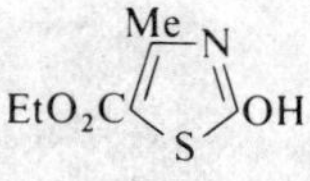              (582)

$NCN=C(SR)S^- K^+$ (R = alkyl) react with chloroacetamide to give the amino-thiazoles (583).[334] Irradiation of 2-phenyl-2-thiazoline (584) in acetonitrile leads to a mixture of the corresponding thiazole, benzonitrile, and ethylene sulphide.[335] Heating the dithiobiuret $Me_2NCSNHCSNH_2$ with phenacyl bromide affords the 2,5'-bithiazole (585) in excellent yield.[336]

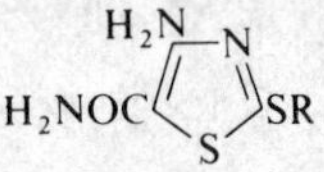

(583) 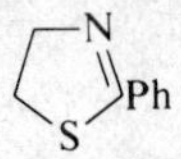                 (584) 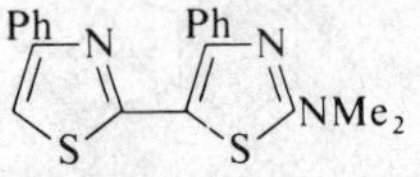                 (585)

*N*-Methylthiazolium salts catalyse the crossed condensation of dissimilar aldehydes to give, in most cases, both possible acyloins.[337] Treatment of the thiazolium salts (586; R = Me or Ph) with benzaldehyde under basic conditions gives rise to the radicals (587).[338] The 1:2 adduct of thiazolium *N*-imine (588) to dimethyl acetylenedicarboxylate has been shown to be the pyrazole (589).[339]

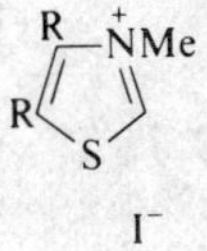

(586) 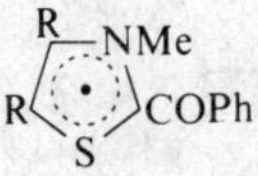          (587) 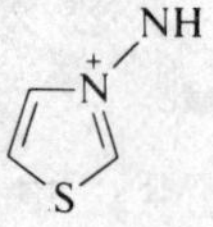          (588) 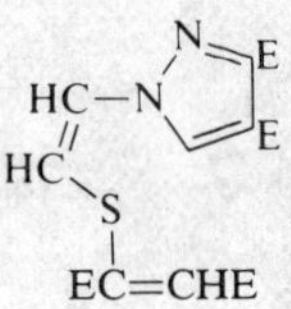          (589) E = CO₂Me

[331] A. M. Van Leusen and J. Wildeman, *Synthesis*, 1977, 501.
[332] Yu. A. Sharanin, K. Ya. Lopatinskaya, and L. G. Sharanina, Deposited Document, 1975, VINITI 2560, available from BLLD (*Chem. Abs.*, 1977, **87**, 151 157).
[333] R. K. Howe, *J. Org. Chem.*, 1977, **42**, 3230.
[334] D. Wobig, *Annalen*, 1977, 400.
[335] N. Suzuki, K. Kuroyanagi, and Y. Izawa, *Chem. and Ind.*, 1977, 313.
[336] Y. Yamamoto, R. Yoda, and K. Sekine, *Chem. Letters*, 1977, 1299.
[337] H. Stetter and G. Dämbkes, *Synthesis*, 1977, 403.
[338] A. I. Vovk, A. F. Babicheva, V. D. Pokhodenko, and A. A. Yasnikov, *Dopovidi Akad. Nauk Ukrain. R.S.R.*, Ser. B. 1977, 907 (*Chem. Abs.*, 1978, **88**, 50 711).
[339] K. T. Potts and D. R. Choudhury, *J. Org. Chem.*, 1977, **42**, 1648.

**Benzothiazoles.**—The kinetics of deprotonation at the *C*-methyl group in the benzothiazolium salt (590; X = S)[340,341] and in the benzoselenazolium (590; X = Se)[341] and benzimidazolium[341] analogues have been determined; the rates follow the order Se > S > NMe.[341] *N*-Acyl-*o*-methylthio-anilines (591; R = alkyl or Ar) are converted into benzothiazoles (592) by the action of phosphonitrile

(590)          (591)          (592)

dichloride.[342] The dichlorobenzothiazole (594) is the main product of the reaction of the amino-alcohol (593) with thionyl chloride.[343] *o,o'*-Diaminodiphenyl disulphide gives 2-arylamino-benzothiazoles (592; R = NHAc) on treatment with aryl isothiocyanates.[344] Methyl derivatives of aromatic compounds react with aniline and sulphur in hexamethylphosphoric triamide to yield the benzothiazoles (592; R = Ph, Ar, or 2- or 5-pyridyl); the naphthothiazoles (595) are similarly

(593)          (594)          (595)

obtained from 1-naphthylamine.[345] Azamonomethinecyanines (597; R = alkyl or Ph) are produced in high yields by successive treatment of the nitroso-imines (596) with Grignard reagents and perchloric acid.[346] The action of alkyl-

(596)          (597)

magnesium bromides on nitro-benzothiazoles, except 4-nitrobenzothiazole, affords alkylnitroso-compounds regiospecifically; thus 5-nitrobenzothiazole (598) reacts with butylmagnesium bromide to give 4-butyl-5-nitrosobenzo-thiazole (599), 6-nitrobenzothiazole yields 7-butyl-6-nitrosobenzothiazole, and 7-nitrobenzothiazole affords a mixture of the 4- and 6-butyl-7-nitroso-deriva-tives; in contrast, 4-nitrobenzothiazole forms 7-butyl-4-nitrobenzothiazole.[347]

[340] J. A. Zoltewicz and L. S. Helmick, *J. Org. Chem.*, 1978, **43**, 1718.
[341] M. Bologa, A. Barabas, V. I. Denes, and F. Chiraleu, *J. Labelled Compounds Radiopharm.*, 1977, **13**, 11 (*Chem. Abs.*, 1977, **87**, 38 623).
[342] G. Rosini and A. Medici, *Synthesis* 1977, 892.
[343] G. Berti, G. Catelani, B. Lombardi, and L. Monti, *Gazzetta*, 1977, **107**, 175.
[344] E. Vinkler and F. Klivenyi, *Acta Chim. Acad. Sci. Hung.*, 1977, **94**, 35 (*Chem. Abs.*, 1978, **88**, 170 021).
[345] J. Perregaard and S. O. Lawesson, *Acta Chem. Scand. (B)*, 1977, **31**, 203.
[346] K. Akiba, K. Ishikawa, and N. Inamoto, *Bull. Chem. Soc. Japan*, 1978, **51**, 535.
[347] G. Bartoli, R. Leardini, M. Lelli, and G. Rosini, *J.C.S. Perkin I*, 1977, 884.

Azulene breaks the sulphur–sulphur bond in di(benzothiazol-2-yl) disulphide to give the azulenyl thioether (600).[348] The azidobenzothiazolium salt (601) transfers a diazo-group to phenols; with 2-naphthol, for instance, the diazo-ketone (602) is obtained.[349]

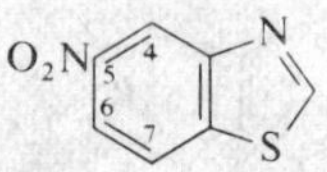

(598) 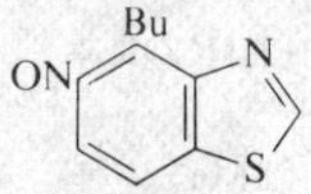                    (599) 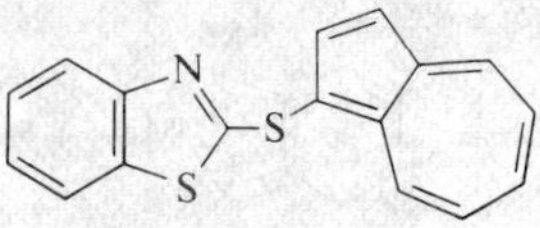                         (600)

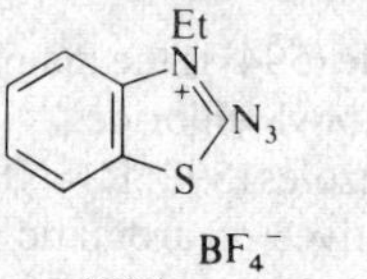

BF$_4^-$

(601) 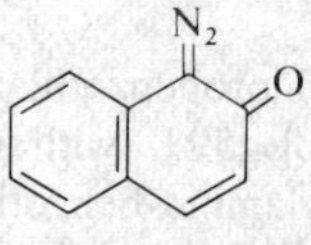                      (602)

## 6  Systems containing Three Heteroatoms

**Triazoles and Benzotriazoles.**—1,2,3-*Triazoles*. The $X$-ray structure of the amino-triazole ester (603) has been reported.[350] A new general synthesis of 1-alkyl-*vic*-triazoles is exemplified by the reaction of the lithiated nitrosoamine LiCH$_2$NMeNO with benzonitrile to yield (604).[351] The bis-toluene-$p$-sulphonyl-hydrazone of benzil cyclises to the triazole (605) in in the presence of mercury(II)

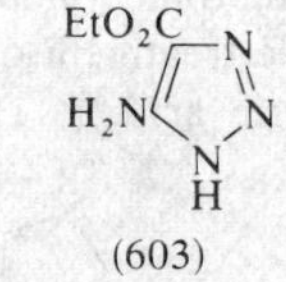

(603) 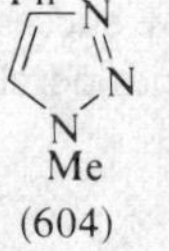              (604) 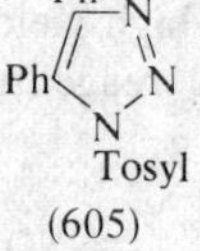            (605)

acetate.[352] The nucleophilic acetylene Me$_2$NC≡CSMe adds $p$-nitrophenyl azide to give the triazole (606), as expected; arenesulphonyl azides, on the other hand, form the imines (607).[353] The reaction of 3-azido-cyclopropenes with acetylenes

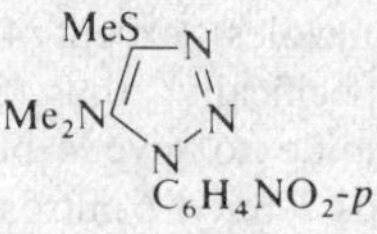

(606) 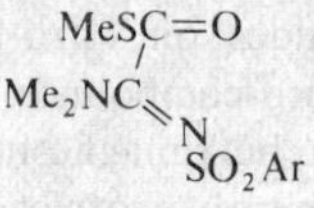                           (607)

affords mixtures of 2-cyclopropenyl-1,2,3-triazoles and cyclopropenyl-azirines, and rearrangement of the cyclopropene ring is observed for each type of product;

[348] Yu. N. Porshnev, V. I. Erikhov, and M. I. Cherkashin, *Khim. geterotsikl. Soedinenii*, 1977, 1278.
[349] H. Balli, V. Müller, and A. Sezen-Gezgin, *Helv. Chim. Acta*, 1978, **61**, 104.
[350] L. Parkanyi, A. Kalman, G. Argay, and J. Schawartz, *Acta Cryst.*, 1977, **B33**, 3102.
[351] D. Seebach, D. Enders, R. Dach, and R. Pieter, *Chem. Ber.*, 1977, **110**, 1879.
[352] R. N. Butler and A. B. Hanahoe, *J.C.S. Chem. Comm.*, 1977, 622.
[353] D. Frank, G. Himbert, and M. Regitz, *Chem. Ber.*, 1978, **111**, 183.

thus the azide (608) and 1-diethylaminopropyne yield not only the triazole (609) and the azirine (610), but also compounds (611) and (612).[354] Thermolysis of the diazouracil derivative (613) affords the triazole (614).[355]

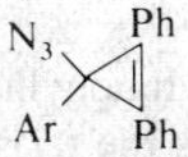

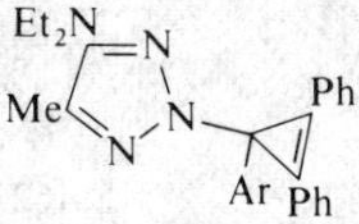

(609)

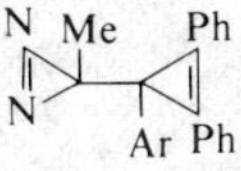

(610)

(608)

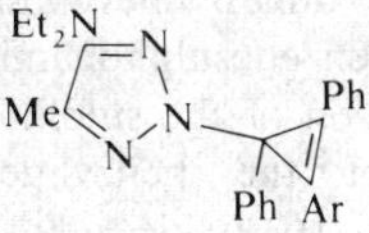

(611)

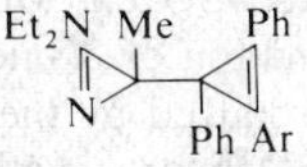

(612)

Ar = $p$-MeC$_6$H$_4$

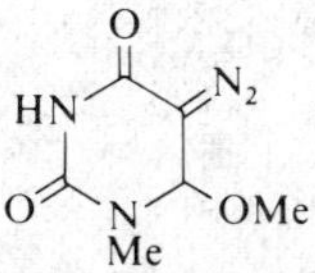

(613)

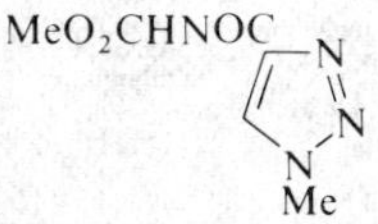

(614)

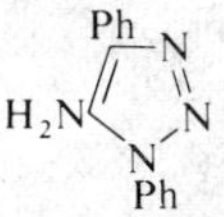

(615)

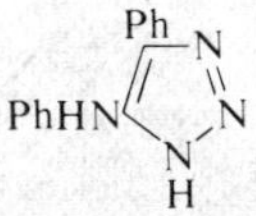

(616)

The amino-triazole (615) undergoes a reversible photo-Dimroth rearrangement to the anilino-derivative (616); the equilibrium mixture contains more of the former isomer. In the thermal reaction, on the other hand, the second isomer is favoured.[356] 1-Acetoxybenzotriazole (617) exists in equilibrium with the *N*-oxide (618).[357] Benzotriazolyl oxide (619), generated by the action of lead dioxide on the corresponding hydroxy-compound, reacts with aromatic compounds ArR to yield carbazolyl oxides (621) *via* o-nitrosophenyl radicals (620).[358]

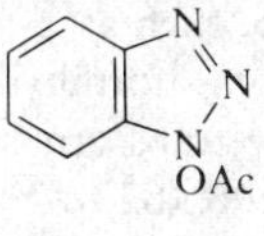

(617)

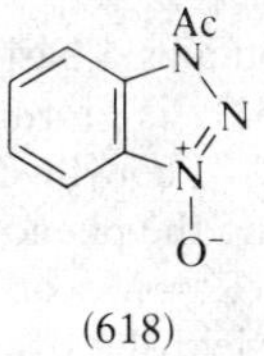

(618)

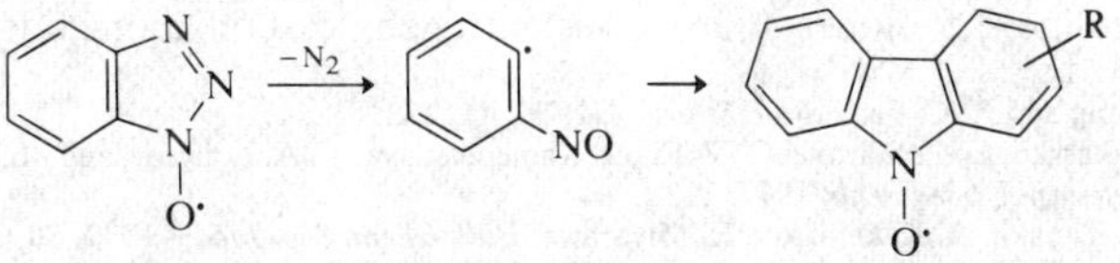

(619)     $\xrightarrow{-N_2}$     (620)     $\rightarrow$     (621)

[354] H. Neunhöffer and H. Ohl, *Chem. Ber.*, 1978, **111**, 299.
[355] T. C. Thurber and L. B. Townsend, *J. Heterocyclic Chem.* 1977, **14**, 647.
[356] Y. Ogata, K. Takagi, and E. Hayashi, *Bull. Chem. Soc. Japan*, 1977, **50**, 2505.
[357] K. Horiki, *Tetrahedron Letters*, 1977, 1901.
[358] H. G. Aurich, G. Bach, K. Hahn, G. Küttner, and W. Weiss, *J. Chem. Res. (S)*, 1977, 122.

*1,2,4-Triazoles.* The structure of 1,2,4-triazole has been established as the 1*H*-form (622) by gas-phase electron diffraction;[359] 3-amino-1,2,4-triazole exists as the 2*H*-tautomer (623), as shown by *X*-ray crystallography.[360]

(622)                    (623)

Condensation of α-chlorobenzylidene benzenesulphonylhydrazone, $PhClC=NNHSO_2Ph$, with *NN'*-dimethylbenzamidine yields the triazole (624) with elimination of *N*-methylbenzenesulphonamide.[361] The triazole derivative (626) is produced by the reaction of the sulphinimide PhNSO with the ester (625).[362] Dehydrogenation of the benzylidene-hydrazidines $PhCH=N-NHCMe=NNHAr$ with mercury(II) oxide yields the cyclic products (627).[363]

(624)                    (625)                    (626)

The action of formic acid on 6-phenylhydrazinouracil (628) causes ring contraction to the triazole derivative (629).[364] Treatment of acyl-hydrazines $R^1CONHNH_2$ with alkyl isothiocyanates $R^2NCS$ gives triazolinethiones (630), if

(627)                    (628)                    (629)                    (630)

one of the reagents contains a basic group, such as 2-morpholinoethyl.[365] The pyrazolyl-triazole. (631) is formed from formhydrazide and α-cyanophenylacetaldehyde.[366] Phenylhydrazine transforms the 4-oxo-1,3-benzoxazinium salt (632) into the phenolic triazole (633).[367]

Urazoles (634; $R^1$, $R^2$ = Me or Ph) suffer ring fission on treatment with benzonitrile oxide to yield the oxadiazolinones (635).[368] Photolysis of 4-aryl-1,2,4-triazoline-3,5-diones (636) gives nitrogen, carbon monoxide, and aryl

[359] J. F. Chiang and K. C. Fu, *J. Mol. Structure*, 1977, **41**, 223.
[360] V. V. Makarskii, G. L. Starova, O. V. Frank-Kamenetskaya, V. A. Lopyrev, and M. G. Voronkov, *Khim. geterotsikl. Soedinenii*, 1977, 1138.
[361] S. Ito, Y. Tanaka, A. Kakehi, and H. Miyazawa, *Bull. Chem. Soc. Japan*, 1977, **50**, 2969.
[362] G. Seitz and T. Kämpchen, *Arch. Pharm.*, 1977, **310**, 269.
[363] M. Takahashi, H. Tan, K. Fukushima, and H. Yamazaki, *Bull. Chem. Soc. Japan*, 1977, **50**, 953.
[364] G. E. Wright, *J. Heterocyclic Chem.*, 1977, **14**, 701.
[365] J. P. Henichart, R. Houssin, and B. Lablanche, *J. Heterocyclic Chem.*, 1977, **14**, 615.
[366] S. A. Lang, Jr., F. M. Lovell, and E. Cohen, *J. Heterocyclic Chem.*, 1977, **14**, 65.
[367] G. N. Dorofeenko, Yu. I. Ryabukhin, S. B. Bulgarevich, V. V. Mezheritskii, and O. Yu. Ryabukhina, *Zhur. org. Khim.*, 1977, **13**, 2549.
[368] G. A. Hoyer and G. Boroschewski, *Arch. Pharm.*, 1977, **310**, 255.

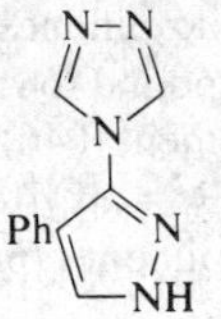

(631) 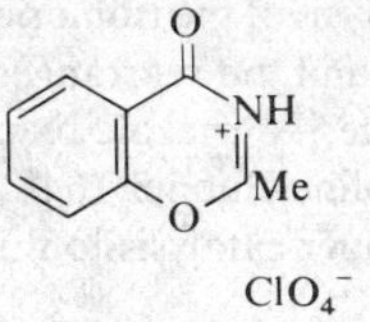        (632) 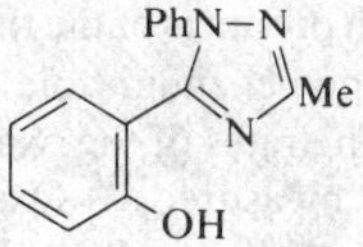        (633)

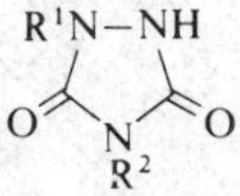

(634) 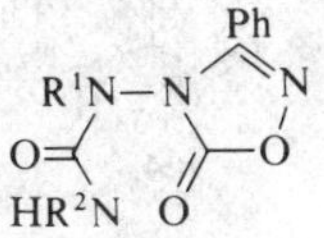        (635)

isocyanates.[369] The phenyl derivative is a potent dienophile; it has been used to trap the isoindene (637)[370] and in studies of the steric course of Diels–Alder additions to 1,2-dihydrophthalic anhydride and related compounds,[371] and bridged [10]annulenes and propellanes.[372] The *N*-methyl analogue forms the single cycloadduct (638) with bullvalene.[373]

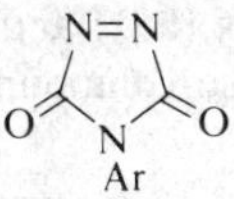

(636) 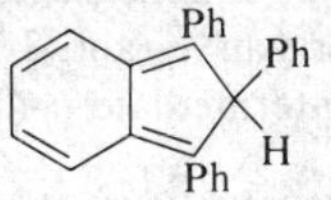        (637) 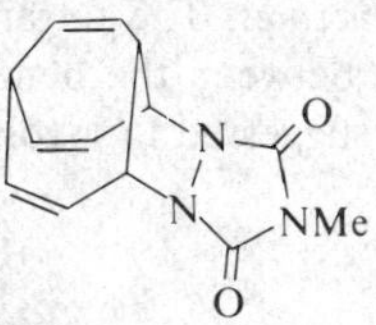        (638)

**Oxadiazoles.**—1,2,4-*Oxadiazoles*. The chemistry of 1,2,4-oxadiazoles has been reviewed.[374] Thermolysis of *N*-aroyl-benzimidoylsulphilimines (639) yields the oxadiazoles (640),[375] which are also formed by treatment of the imino-dihydropropyrimidines (641) with hydroxylamine.[376]

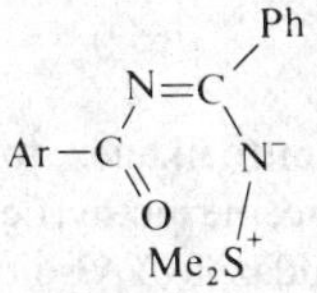

(639) 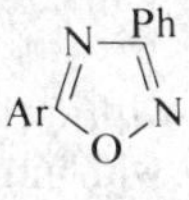        (640) 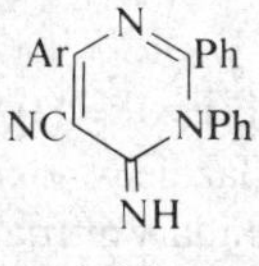        (641)

[369] H. Wamhoff and K. Wald, *Chem. Ber.*, 1977, **110**, 1699.
[370] K. Kamal de Fonseka, C. Manning, J. J. McCullough, and A. J. Yarwood, *J. Amer. Chem. Soc.*, 1977, **99**, 8257.
[371] P. Ashkenazi, D. Ginsburg, G. Scharf, and B. Fuchs, *Tetrahedron*, 1977, **33**, 1345.
[372] (a) P. Ashkenazi, D. Ginsburg, and E. Vogel, *Tetrahedron*, 1977, **33**, 1169; (b) J. Kalo, D. Ginsburg, and E. Vogel, *ibid.*, p. 1177; (c) J. Kalo, E. Vogel, and D. Ginsburg, *ibid.*, p. 1183.
[373] P. Cernuschi, C. De Micheli, and R. Gandolfi, *Tetrahedron Letters*, 1977, 3667.
[374] L. B. Clapp, *Adv. Heterocyclic Chem.*, 1976, **20**, 65.
[375] T. Fuchigami and K. Odo, *Bull. Chem. Soc. Japan*, 1977, **50**, 1793.
[376] S. Robev, *Doklady Bolg. Akad. Nauk*, 1977, **30**, 1031 (*Chem. Abs.*, 1978, **88**, 22 768).

The main products of the thermolysis of the isomeric oxadiazolinones (642) and (644) are the benzimidazole (643) and the rearranged compound (645), respectively.[377] The rates of the oxadiazole → triazole rearrangement (646) → (647) have been measured.[378] Oxadiazoline-thiones (648; $R^1$, $R^2$ = alkyl or Ph) rearrange photochemically or by copper catalysis to thiadiazolinones (649).[379]

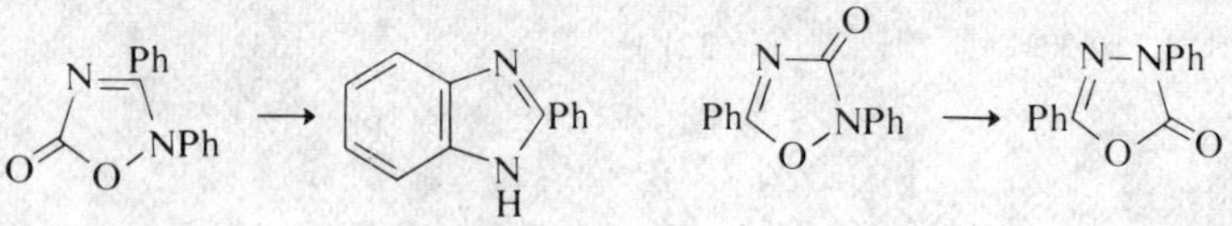

(642)          (643)          (644)          (645)

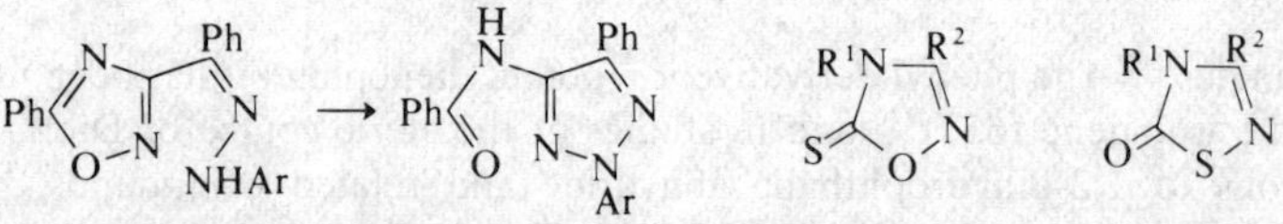

(646)          (647)          (648)          (649)

1,2,5-*Oxadiazoles.* The tetramer of fulminic acid has structure (650).[380] The equilibrium between the benzofurazan 1- and 3-oxides (651) and (653) (R = halogen) is suggested to involve intermediate *o*-dinitroso-compounds (652).[381]

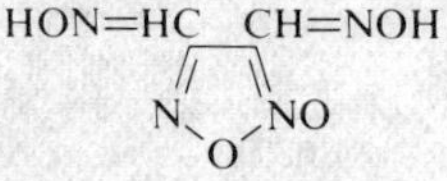

$$HON{=}HC \quad CH{=}NOH$$

(650)

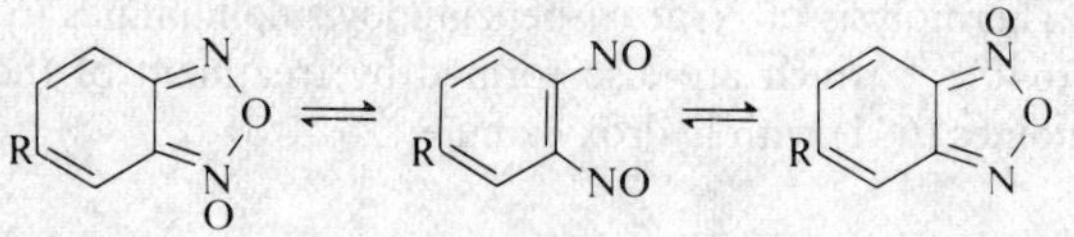

(651)          (652)          (653)

The oxide (651; R = H) reacts with primary nitroalkanes to give *N*-hydroxybenzimidazole *N'*-oxides; *e.g.* with nitroethane, the methyl derivative (654) is formed; secondary nitroalkanes yield 2*H*-benzimidazole *NN'*-dioxides, such as compound (655) from 2-nitropropane.[382] Treatment of the benzofurazan oxide with benzylideneacetone and butylamine yields the quinoxaline *N*-oxide (656), but in the presence of ammonia the corresponding *NN'*-dioxide is obtained.[383]

[377] J. H. Boyer and P. S. Ellis, *J.C.S. Chem. Comm.*, 1977, 489.
[378] D. Spinelli, V. Frenna, A. Corrao, and N. Vivona, *J.C.S. Perkin II*, 1978, 19.
[379] A. Pelter and D. Sumengen, *Tetrahedron Letters*, 1977, 1945.
[380] C. Grundmann, G. W. Nickel, and R. K. Bansal, *Annalen*, 1975, 1029.
[381] S. Uematsu and Y. Akahori, *Chem. and Pharm. Bull. (Japan)*, 1978, **26**, 25.
[382] D. W. S. Latham, O. Meth-Cohn, H. Suschitzky, and J. A. L. Herbert, *J.C.S. Perkin I*, 1977, 470.
[383] G. S. Lewis and A. F. Kluge, *Tetrahedron Letters*, 1977, 2491.

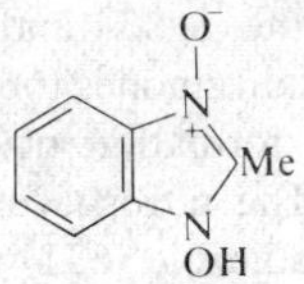

(654) 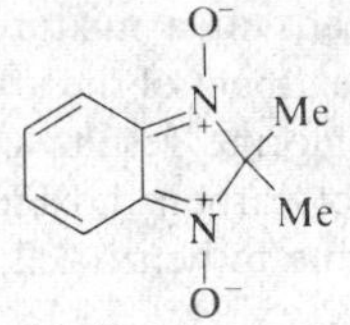          (655) 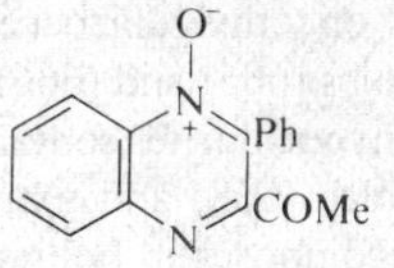                    (656)

*1,3,4-Oxadiazoles.* The iminopyrazoline (657) is transformed into the oxadiazoline (658) on treatment with potassium hydroxide.[384] Irradiation of 2,5-diphenyl-1,3,4-oxadiazole (659) in alcohols ROH gives alkyl benzoates, benzamide, and the diphenyltriazole (660) by the mechanism shown in Scheme 5.[385]

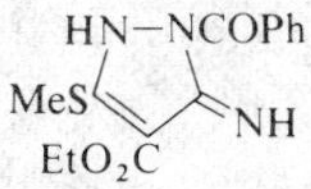

(657) 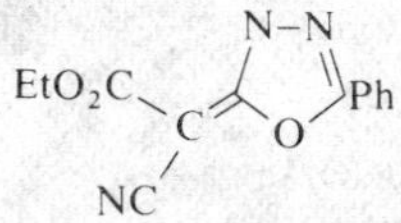                    (658)

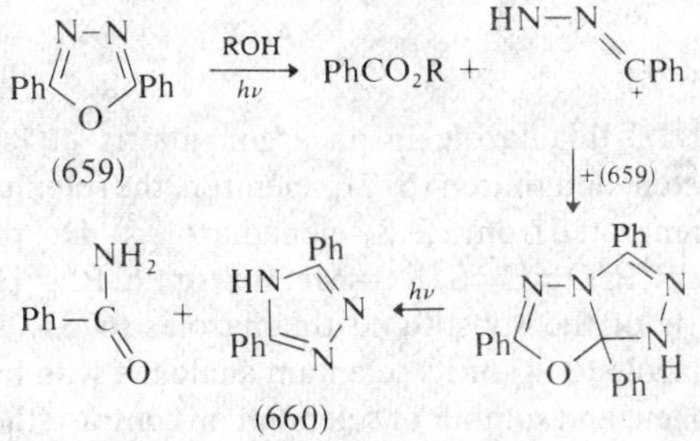

**Scheme 5**

Heating the ethoxy-diazoline (661) with ethyl propiolate yields not only the fission products *p*-nitrobenzonitrile and ethyl *N*-phenylcarbamate, but also ethyl *p*-nitrobenzoate and the pyrazole (662); the last two products arise from the rearranged 1,3-dipolar compound (663).[386]

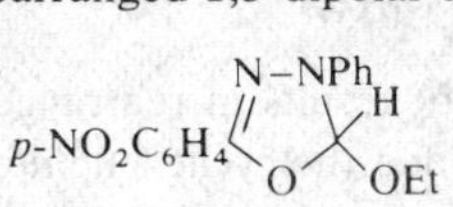

(661) 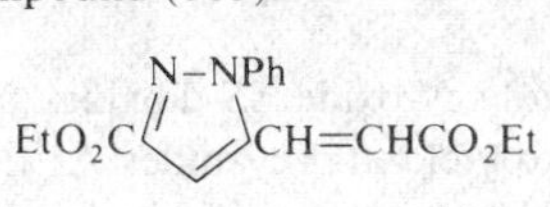                    (662) 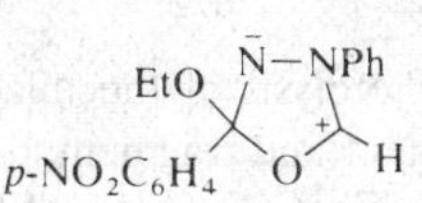                    (663)

**Thiadiazoles.**—*1,2,3-Thiadiazoles.* Syntheses of 5-amino-1,2,3-thiadiazoles (664; $R^1$ = Ar or ArSO$_2$; $R^2$ = H or PhCO) from isothiocyanates $R^1$NCS and diazo-compounds $R^2$CHN$_2$ have been described.[387] The action of thiocarbonyl

[384] K. Peseke, *Z. Chem.*, 1977, **17**, 262.
[385] O. Tsuge, K. Oe, and M. Tashiro, *Chem. Letters*, 1977, 1207.
[386] G. Scherowsky and B. Kundu, *Annalen*, 1977, 1235.
[387] (a) M. Uher, M. Hrobonova, A. Martvon, and J. Lesko, *Chem. Zvesti*, 1976, **30**, 213; (b) M. Uher, A. Rybar, A. Martvon, and J. Lesko, *ibid.*, p. 217; (c) A. Martvon, M. Uher, S. Stankovsky, and J. Sura, *Coll. Czech. Chem. Comm.*, 1977, **42**, 1557.

chloride on ethyl diazoacetate leads to a mixture of the 1,2,3- and 1,3,4-thiadiazoles (665) and (666).[388] Treatment of the aryl-thiadiazinones (667) with t-butyl hypochlorite containing chlorine results in the formation of 4-aryl-thiadiazoles (668) and extrusion of carbon dioxide.[389] The nitrosoacetanilide (669) decomposes in boiling benzene to benzo-1,2,3-thiadiazole (671) *via* the diazonium acetate (670).[390]

(664)            (665)            (666)            (667)            (668)

(669)                    (670)                    (671)

Irradiation of 1,2,3-thiadiazole in an argon matrix at 8 K yields mercap-toacetylene, thioketen, and thiiren (672); seleniren, the selenium analogue of the latter, is similarly generated from 1,2,3-selenadiazole.[391] Deeply coloured, highly reactive thioketens $R^1R^2C=C=S$ ($R^1 = Me$, $Bu^t$, or Ph; $R^2 = H$ or Ph) are formed by flash thermolysis of the substituted thiadiazoles (673).[392,393] Treatment of cyclo-octanothiadiazole (674) or its selenium analogue with butyl-lithium yields cyclo-octyne, nitrogen, and sulphur or selenium; in contrast, the thiophen (675) is produced from cycloheptano-1,2,3-thiadiazole under the same conditions.[394]

(672)            (673)            (674)                    (675)

Photolysis of the thiadiazole *N*-oxide *SS*-dioxides (676) results in rearrange-ment and fragmentation to yield aryl cyanides $Ar^2CN$ and the cyclic sulphites (677).[395] The rates of the base-catalysed Dimroth rearrangement (678) → (679) have been measured.[396]

[388] P. Demaree, M. C. Daria, and J. M. Muchowski, *Canad. J. Chem.*, 1977, **55**, 243.
[389] G. Ege, P. Arnold, G. Jooss, and R. Naronha, *Annalen*, 1977, 791.
[390] J. Brennan, J. I. G. Cadogan, and J. T. Sharp, *J.C.S. Perkin I*, 1977, 1844.
[391] (a) A. Krantz and J. Laureni, *J. Amer. Chem. Soc.*, 1977, **99**, 4842; (b) *Ber. Bunsengesellschaft phys. Chem.*, 1978, **82**, 13.
[392] G. Seybold and C. Heibl, *Chem. Ber.*, 1977, **110**, 1225.
[393] H. Bühl, B. Seitz, and H. Meier, *Tetrahedron*, 1977, **33**, 449.
[394] M. Barth, H. Bühl, and H. Meier, *Chem.-Ztg.*, 1977, **101**, 452.
[395] (a) G. Trickes, U. Plücken, and H. Meier, *Z. Naturforsch.*, 1977, **32b**, 956; (b) G. Trickes, H. P. Braun, and H. Meier, *Annalen*, 1977, 1347.
[396] M. Uher, V. Knoppova, and A. Martvon, *Chem. Zvesti*, 1976, **30**, 514.

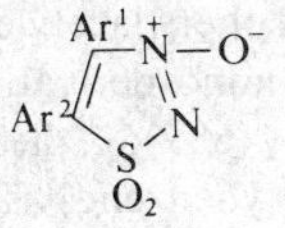

(676) 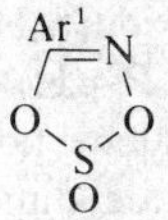    (677) 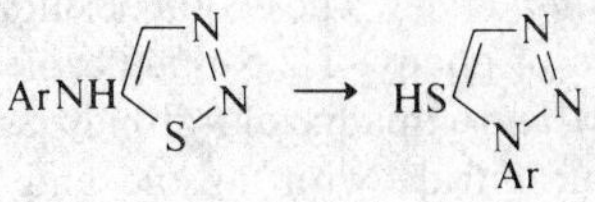    (678)    (679)

**1,2,4-*Thiadiazoles*.** 5-Anilino-1,2,4-thiadiazoles (680)–(682) are formed by rearrangement of heterocyclic thioureas; the reaction rates decrease in the given order.[397]

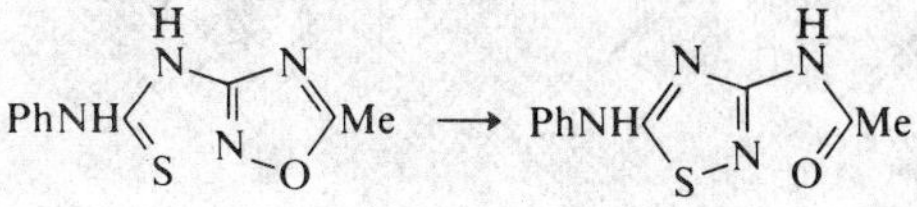

(680)

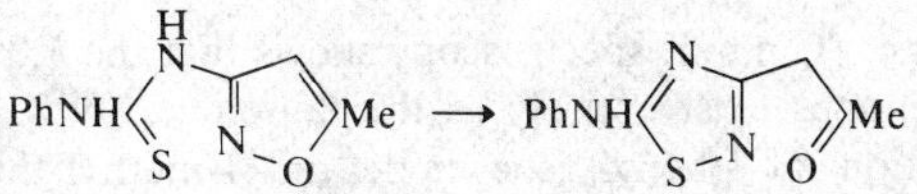

(681)

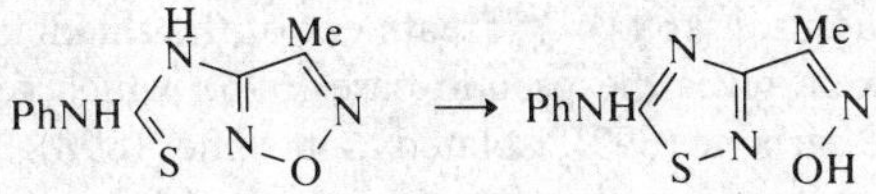

(682)

The action of nucleophilic reagents on the thiadiazolium salts (683) and (685) has been investigated. Hard nucleophiles generally attack a carbon atom, whereas soft nucleophiles react at sulphur. Thus sodium hydroxide reacts with the two salts to yield mainly the benzamidines (684) and (686), respectively, via initial attack at C-5 and subsequent ring opening and loss of sulphur whereas treatment of compound (685) with sodium sulphide, sodium thiosulphate, sodium thiophenate, or sodium borohydride gives the thiobenzoyl derivative (687) by reaction at the sulphur atom (see Scheme 6).[398]

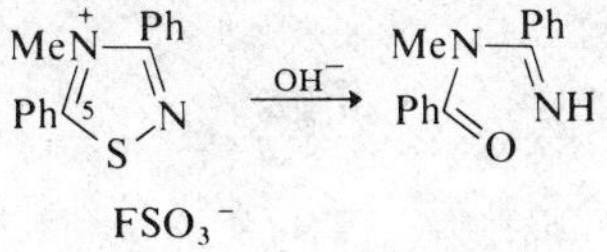

$FSO_3^-$

(683)    (684)

$FSO_3^-$

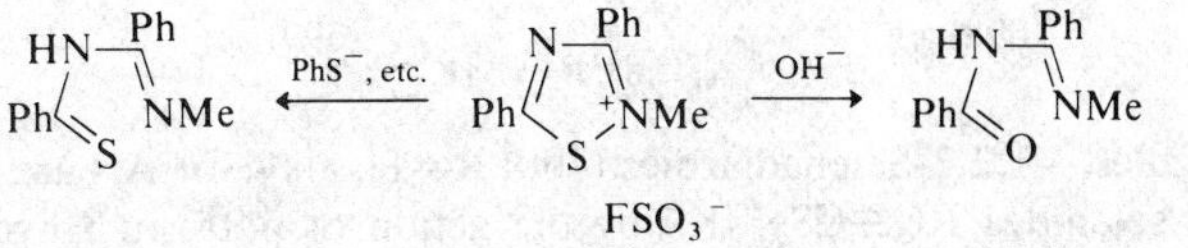

(687)    (685)    (686)

**Scheme 6**

[397] N. Vivona, G. Cusmano, and G. Macaluso, *J.C.S. Perkin I*, 1977, 1616.
[398] S. Crook and P. Sykes, *J.C.S. Perkin I*, 1977, 1791.

1,2,5-*Thiadiazoles*. The action of sulphur nitride, $S_4N_4$, on phenylacetylene gives a mixture of phenyl-1,2,5-thiadiazole (688; R = H), its amino-derivative (688; R = $NH_2$), and a thiadiazolo-dithiatriazapentalene, possibly (689);[399] the reaction of sulphur nitride with anisole and the three dimethoxybenzenes is equally complex, yielding in each case the benzo-bis- and -tris-thiadiazoles (690) and (691).[400]

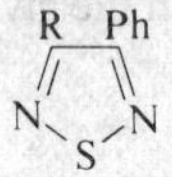

(688) 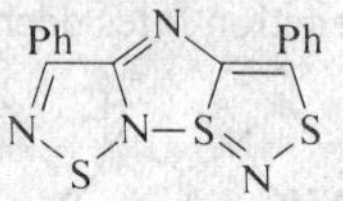                    (689) 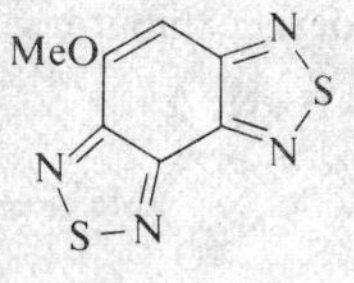                         (690) 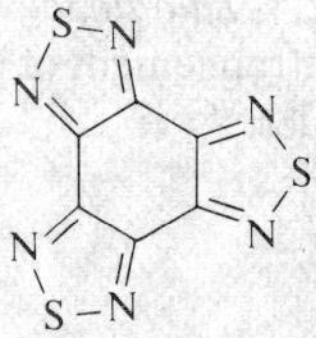                         (691)

1,3,4-*Thiadiazoles*. [13]C n.m.r. spectroscopy shows that the 1,3,4-thiadiazoline-thiones and -selenones exist entirely as the tautomers (692; X = S or Se).[401] Whereas the action of phenyl isocyanide on 4-methyl-thiosemicarbazide, $MeNHCSNHNH_2$, gives only the triazolinethione (693; R = Me), the reaction with 4-phenyl-thiosemicarbazide yields a mixture of the phenyl analogue (693; R = Ph) and the thiadiazole (694).[402] Treatment of the thiadiazolium salt (695) with sodium hydroxide gives the pseudo-base (696), which readily undergoes α-elimination to the carbene (697), isolated as its dimer (698). The pseudo-base reacts with dimethyl acetylenedicarboxylate to form the rearranged adduct (699).[403]

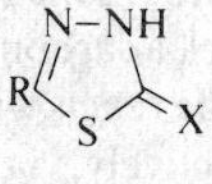

(692) 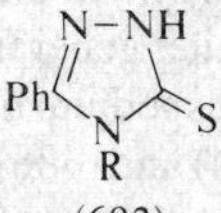                    (693) 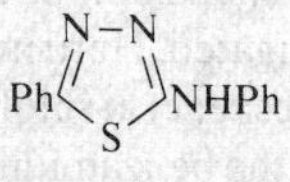                         (694)

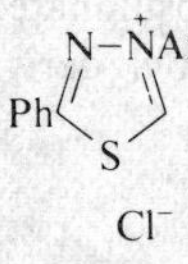

(695) 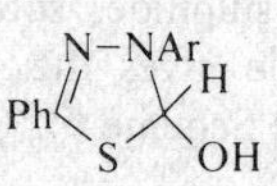                    (696) 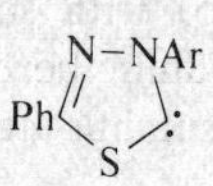                         (697)

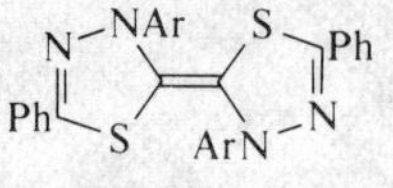

(698) 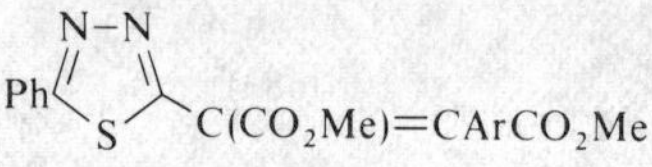                                (699)

Ar = *p*-$NO_2C_6H_4$

**Selenadiazoles.**—1,2,3-Selenadiazoles (700; R = H, alkyl, or Ar) are cleaved to acetylenic selenides $RC{\equiv}CSe^-$ $K^+$ by the action of potassium hydroxide or

[399] M. Tashiro, S. Mataka, and K. Takahashi, *Heterocycles*, 1977, **6**, 933.
[400] S. Mataka, K. Takahashi, and M. Tashiro, *J. Heterocyclic Chem.*, 1977, **14**, 962.
[401] J. R. Bartels-Keith, M. T. Burgess, and J. M. Stevenson, *J. Org. Chem.*, 1977, **42**, 3725.
[402] S. Treppendahl and P. Jakobsen, *Acta Chem. Scand.* (*B*), 1977, **31**, 264.
[403] G. Scherowsky, K. Dünnbier, and G. Hoefle, *Tetrahedron Letters*, 1977, 2095.

ethoxide.[404,405] Photolysis of diphenyl-1,2,4- (701) or -1,2,5-selenadiazole (702) produces the same transient product, identified spectroscopically as benzonitrile selenide, $PhC\equiv\overset{+}{N}-\overset{-}{Se}$.[406]

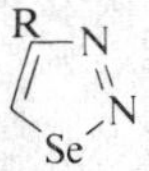

**Dithiazoles.**—1,2,4-Dithiazolium salts (703) are formed from thioamides $ArCSNH_2$ by the action of perchloric acid[407] or bromine.[408] Treatment of benzo-1,2,3-dithiazolium chloride (704) with diethylamine yields compound (705).[409] The *S*-dioxide (707) is formed on heating the azide (706).[410]

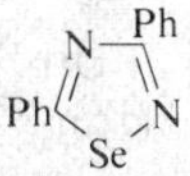

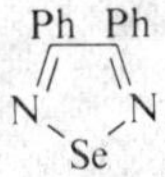

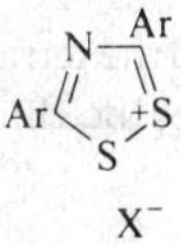

## 7 Systems containing Four Heteroatoms

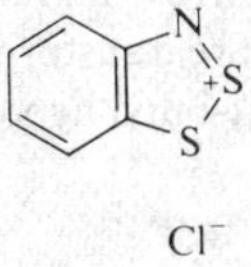

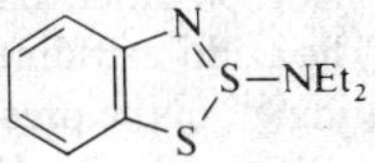

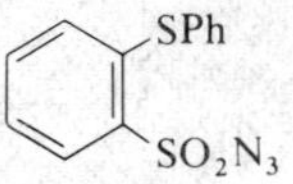

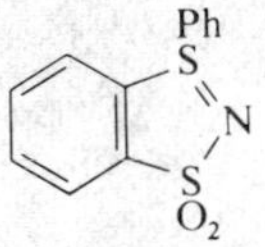

**Tetrazoles.**—Both 1-phenyltetrazole (708) and 2-phenyltetrazole are protonated at N-4.[411] 5-Aryl-4-*p*-dimethylaminophenyl-tetrazoles are formed by the action of hydrazoic acid on *N*-*p*-dimethylaminophenyl-nitrones $ArCH=N(O)-C_6H_4NMe_2$.[412] Tetrazole-5-aldehyde (709), prepared by the action of hydrazoic acid on the acetal $(EtO)_2CHCN$, exists as the dimer (710) in the solid state; it reacts with piperidine to give the betaine (711).[413] The carbodi-imide (712) rearranges to the tetrazole (713) on heating.[414] The formation of the 5-diphenyl-amino-tetrazoles (715) by treatment of the chloroiminium chlorides (714) with

[404] A. Shafiee, I. Lalezari, and F. Savabi, *Synthesis*, 1977, 764.
[405] F. Malek-Yazdi and M. Yalpani, *Synthesis*, 1977, 328.
[406] (*a*) C. L. Pedersen and N. Hacker, *Tetrahedron Letters*, 1977, 3982; (*b*) C. L. Pedersen, N. Harrit, M. Poliakoff, and I. Dunkin, *Acta Chem. Scand.* (*B*), 1977, **31**, 848.
[407] J. Liebscher and H. Hartmann, *Annalen*, 1977, 1005.
[408] S. Leistner, G. Wagner, and M. Ackermann, *Z. Chem.*, 1977, **17**, 223.
[409] Yu. I. Akulin, B. K. Strelets, and L. S. Efros, *Khim. geterotsikl. Soedinenii*, 1977, 849.
[410] R. A. Abramovitch, C. I. Azogu, I. T. McMaster, and D. P. Vanderpool, *J. Org. Chem.*, 1978, **43**, 1218.
[411] A. Könnecke, E. Lippmann, and E. Kleinpeter, *Tetrahedron*, 1977, **33**, 1399.
[412] D. Moderhack, *J. Heterocyclic Chem.*, 1977, **14**, 757.
[413] D. Moderhack, *Chem.-Ztg.*, 1977, **101**, 403.
[414] V. I. Gorbatenko, V. N. Fetyukhin, N. V. Melnichenko, and L. I. Samarai, *Zhur. org. Khim.*, 1977, **13**, 2320.

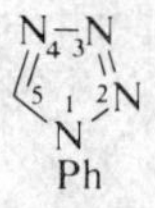

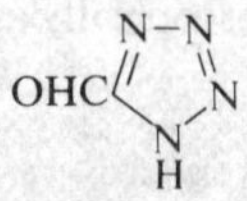

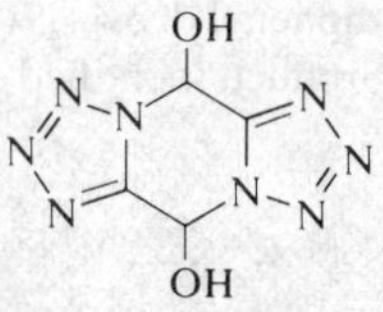

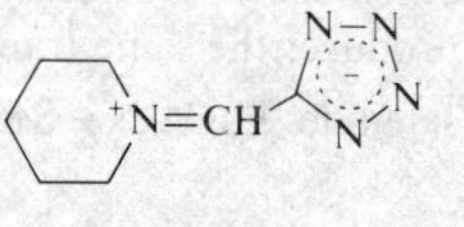

(708)            (709)                    (710)                         (711)

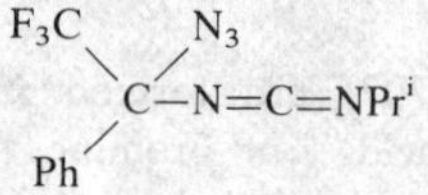

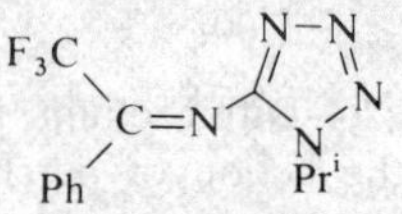

(712)                    (713)

trimethylsilyl azide or other azides involves migration of the aryl group (see Scheme 7).[415] The azido-aziridines (716) are transformed into the tetrazoles (717; R = H) on heating;[416] in the presence of t-butyl hypochlorite, the chloro-derivatives (717; R = Cl) are obtained.[417]

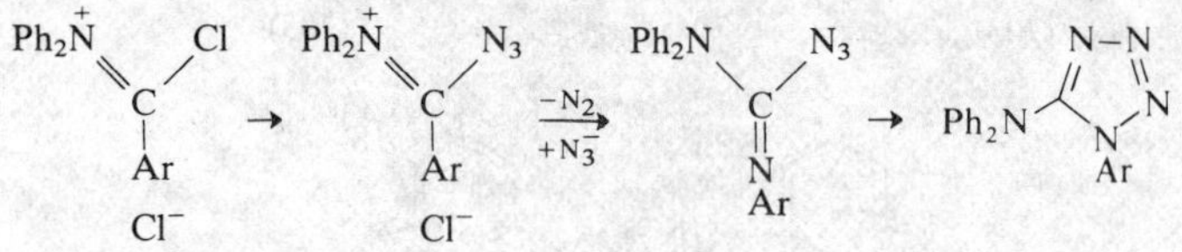

(714)                                                                              (715)

**Scheme 7**

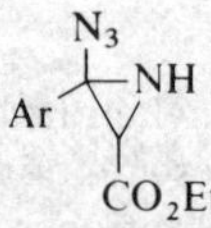

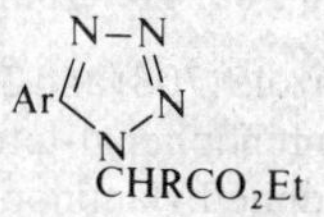

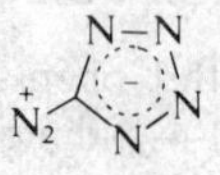

(716)                    (717)                    (718)

Thermolysis of diazotetrazole (718) gives rise to atomic carbon; in the presence of propane, a mixture of but-1-ene, *cis*- and *trans*-but-2-ene, isobutene, and methylcyclopropane, as well as propene, methane, and acetylene, is obtained.[418] Tris(methylimino)methane (720), the first nitrogen analogue of trimethylenemethane, is generated by low-temperature photolysis of the tetrazoline (719).[419] 1-Methoxy-5-phenyltetrazole (721) rearranges to the tetrazole *N*-oxide (722) at 200 °C.[420] The main product of the reaction of diazomethane with the tetrazolyl disulphide (723) is the rearranged thione (724).[421] Irradiation of the

[415] R. Imhof, D. W. Ladner, and J. M. Muchowski, *J. Org. Chem.*, 1977, **42**, 3709.
[416] G. Szeimies, K. Mannhardt, and W. Mickler, *Chem. Ber.*, 1977, **110**, 2922.
[417] G. Szeimies and K. Mannhardt, *Chem. Ber.*, 1977, **110**, 2939.
[418] P. B. Shevlin and S. Kammula, *J. Amer. Chem. Soc.*, 1977, **99**, 2627.
[419] H. Quast, L. Bieber, and W. C. Danen, *J. Amer. Chem. Soc.*, 1978, **100**, 1306.
[420] J. Plenkiewicz, *Tetrahedron Letters*, 1978, 399.
[421] W. Tochtermann, H. Gustmann, and C. Wolff, *Chem. Ber.*, 1978, **111**, 566.

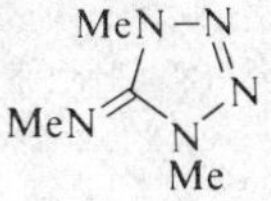

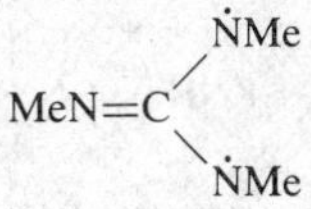

MeN—N
MeN    N
    N
    Me
(719)

                    ṄMe
    MeN=C
                    ṄMe
(720)

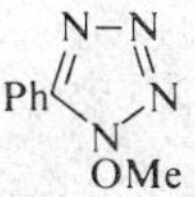

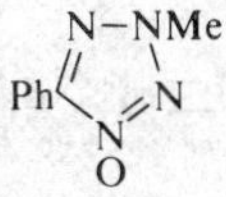

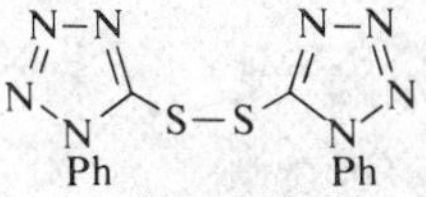

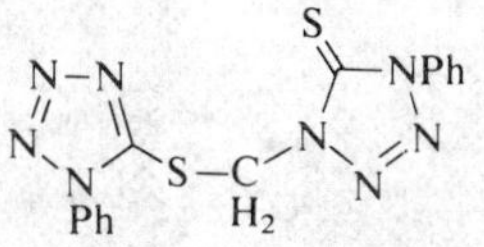

N—N
Ph    N
    N
  OMe
(721)

N—NMe
Ph    N
    N
   O
(722)

N—N
N    N
  S—S
 Ph
N—N
    N    N
       Ph
(723)

N—N
N    N
  S—C
 Ph   H₂
            S
          ∥
N—NPh
    N    N
(724)

tetrazole derivative (725) yields the cyclopentano-pyrazole (727) by intra-molecular 1,3-dipolar cycloaddition to the transient nitrile imine (726); the propargyl ether (728) is similarly transformed into the pyrazolo-benzopyran (729).[422,423]

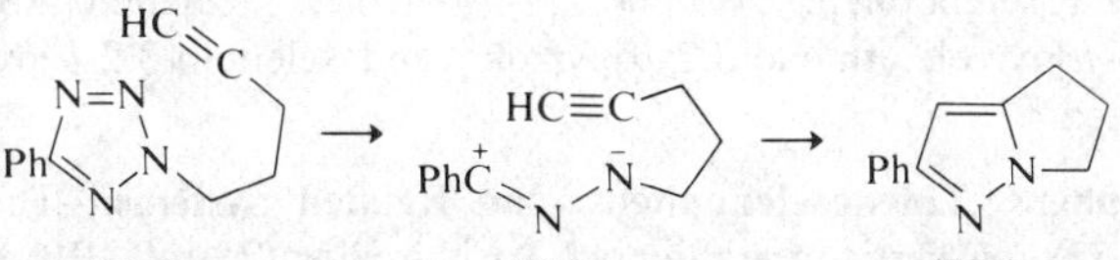

HC≡C
N=N
Ph    N
    N
(725)

→

HC≡C
      +        −
PhC=N—N
(726)

→

Ph    N
       N
(727)

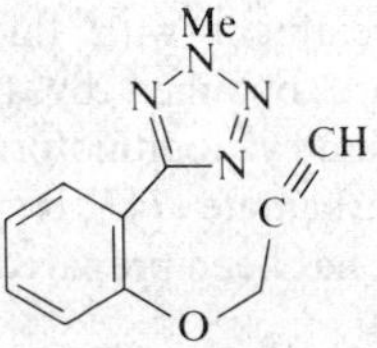

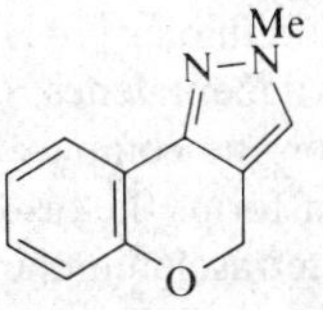

Me
N—N
N    N
      CH
     C
      O
(728)

Me
N—N
         N
     O
(729)

**Miscellaneous Ring Systems.**—The chemistry of 1,2,3,4-thiatriazoles has been reviewed.[424] The thiatriazole derivatives (730) and (731) exist in the tautomeric forms shown.[425] The formation of amino-thiatriazoles (731) from isothiocyanates is more rapid with trimethylsilyl azide than with hydrazoic acid.[426]

The reaction of benzonitrile with 'trichlorocyclotrithiazene' (732) gives 4-phenyl-1,2,3,5-dithiadiazolium chloride (733).[427] Treatment of the sulpho-di-imide $Me_3SiN=S=NSiMe_3$ with ClCOSCl yields 5-oxo-5$H$-1,3$\lambda^4$,2,4-dithiadiazole (734).[428] The action of thionyl chloride on amidoximes (735) affords 1,2,3,5-oxathiadiazole $S$-oxides (736); these compounds decompose to sulphur dioxide and carbodi-imides (737) on heating.[429]

[422] A. Padwa, S. Nahm, and E. Sato, *J. Org. Chem.*, 1978, **43**, 1664.
[423] H. Meier and H. Heimgartner, *Helv. Chim. Acta*, 1977, **60**, 3035.
[424] A. Holm, *Adv. Heterocyclic Chem.*, 1976, **20**, 145.
[425] G. L'Abbé, S. Toppet, A. Willcox, and G. Mathys, *J. Heterocyclic Chem.*, 1977, **14**, 1417.
[426] L. Floch, A. Martvon, M. Uher, J. Lesko, and W. Weiss, *Coll. Czech. Chem. Comm.*, 1977, **42**, 2945.
[427] G. G. Alange, A. J. Banister, B. Bell, and P. W. Millen, *Inorg. Nuclear Chem. Letters*, 1977, **13**, 143.
[428] R. Neidlein, P. Leinberger, A. Gieren, and B. Dederer, *Chem. Ber.*, 1978, **111**, 698.
[429] A. Dondoni, G. Barbaro, and A. Battaglia, *J. Org. Chem.*, 1977, **42**, 3372.

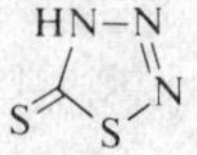

(730) 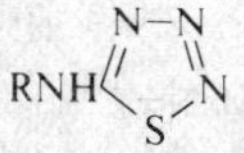            (731)            (732)            (733) Cl⁻

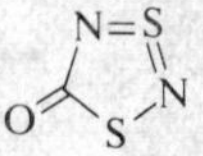

(734) 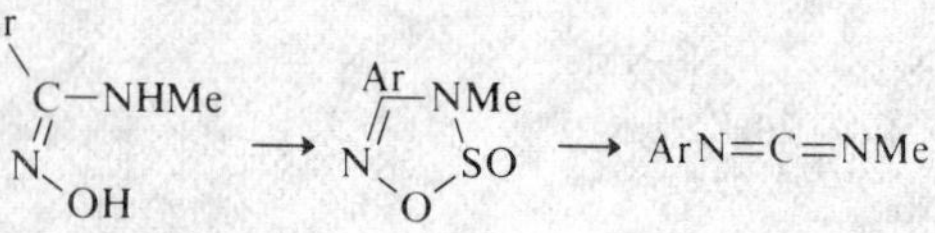            (735)            (736) → ArN=C=NMe  (737)

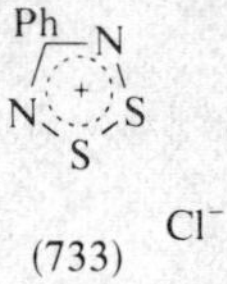

## 8 Compounds containing Two Fused Five-membered Rings

Reviews on heteropentalenes have appeared.[430,431] The photoelectron spectra of thieno[2,3-*b*]selenophen, selenolo[2,3-*b*]selenophen, thieno[3,2-*b*]selenophen, selenolo[3,2-*b*]selenophen, thieno[2,3-*b*]pyrrole, selenolo[2,3-*b*]pyrrole, pyrrolo[2,3-*b*]pyrrole, thieno[3,2-*b*]pyrrole, and selenolo[3,2-*b*]pyrrole have been reported.[432]

**Thienothiophens, Thienoselenophens, and Related Systems.**—The thiophen derivative (738) cyclises to the thieno[2,3-*b*]thiophen (739).[433] Electrochemical reduction of carbon disulphide in acetonitrile gives the di-anion (740), which forms the dithiolodithiole (741) on treatment with thiophosgene.[434] The di(acylimino)tetrathiapentalenes (742) are obtained by the action of acid chlorides RCOCl on the complex of sodium cyanodithioformate with DMF.[435] Perchloric acid transforms the bis-dithiocarbamate (743; $R_2N$ = piperidino) into the salt (744); the tetraselenium analogue has been prepared similarly.[436]

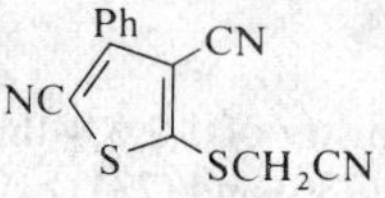

(738) 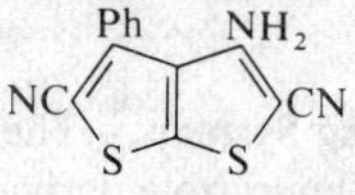            (739)

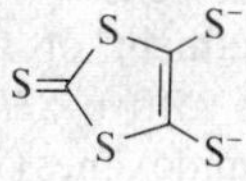

(740) 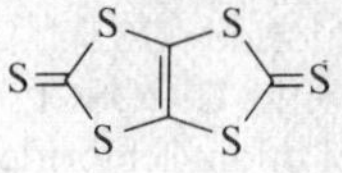            (741) 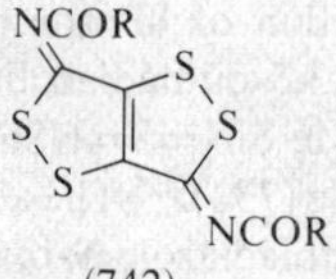            (742)

[430] K. T. Potts, *Chem. Heterocyclic Compounds*, 1977, **30**, 317.
[431] J. P. Paolini, *Chem. Heterocyclic Compounds*, 1977, **30**, 1.
[432] R. Gleiter, M. Kobayashi, J. Spanget-Larsen, S. Gronowitz, A. Konar, and M. Farnier, *J. Org. Chem.*, 1977, **42**, 2230.
[433] M. Augustin, W. D. Rudorf, and U. Schmidt, *Tetrahedron*, 1976, **32**, 3055.
[434] W. P. Krug, A. N. Bloch, and D. O. Cowan, *J.C.S. Chem. Comm.*, 1977, 660.
[435] A. Reiter, P. Hansen, and H. U. Kibbel, *Z. Chem.*, 1977, **17**, 221.
[436] R. R. Schumaker and E. M. Engler, *J. Amer. Chem. Soc.*, 1977, **99**, 5519.

Acetylene reacts with selenium to yield a mixture of about thirty products, one of which is the selenoloselenophen (745).[437] The selenolo[3,4-c]selenophens (747; R = Me or CO$_2$Et), generated by heating the oxides (746), have been trapped as Diels–Alder adducts with N-phenylmaleimide.[438] The related selenolothiophens (748) add N-phenylmaleimide at the thiophen ring to form the condensed selenophens (749); in contrast, dimethyl acetylenedicarboxylate reacts at the selenophen ring, giving the benzo[c]thiophen (751) by extrusion of sulphur from the initial adduct (750).[439]

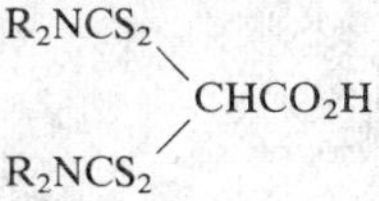

(743)

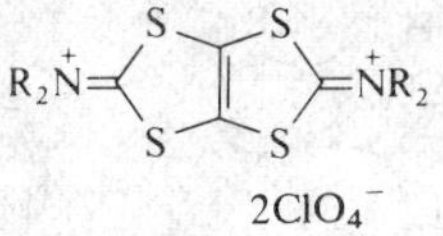

(744)

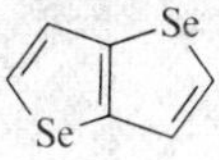

(745)

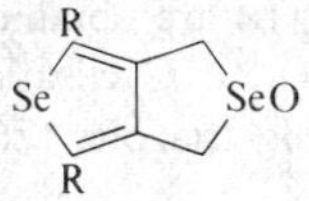

(746)

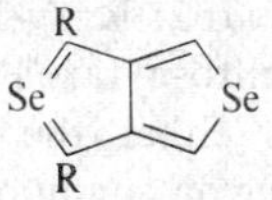

(747)

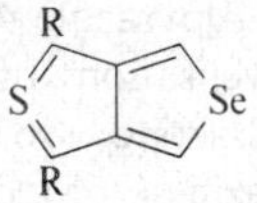

(748)

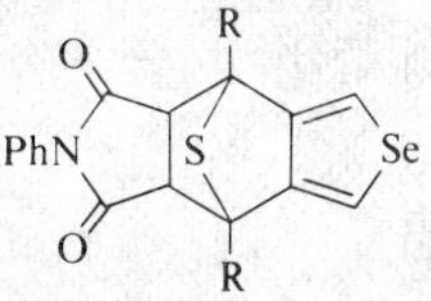

(749)

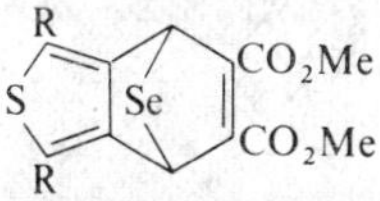

(750)

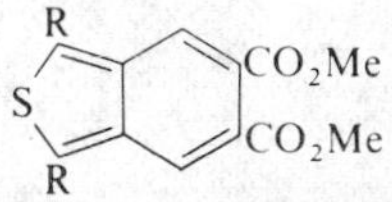

(751)

The X-ray crystal structures of selenolo[2,3-b]benzothiophen (752) and of selenolo[3,2-b]benzothiophen (753) have been determined.[440]

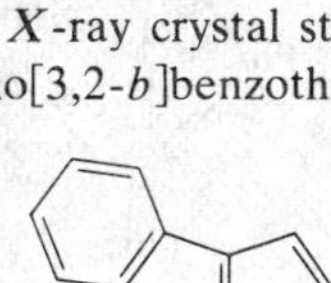

(752)

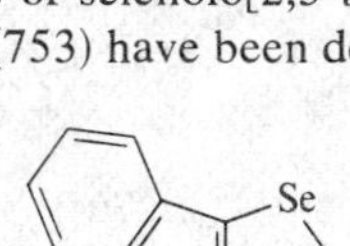

(753)

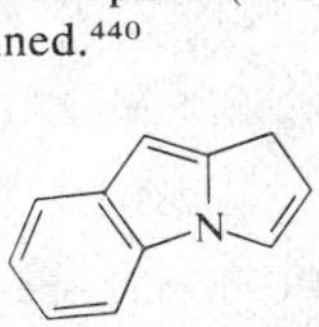

(754)

**Nitrogen Systems.**—A review on the synthesis of pyrrolo[1,2-a]indoles (754) and related systems has appeared.[441] Treatment of azobenzene (755) with 2,6-dimethylphenyl isocyanide (756) in the presence of dicobalt octacarbonyl yields a mixture of the imino-indazoline (757) and the indazolo-indazole (758).[442] Heating 5-amino-1-o-chlorophenyl-pyrazoles (759; R = CN or CO$_2$Et) in DMF in the presence of copper oxide yields the pyrazolo-benzimidazoles (760).[443]

[437] S. Gronowitz, A. Konar, and A. B. Hornfeldt, *Chemica Scripta*, 1976, **10**, 159.
[438] S. Gronowitz and A. Konar, *J.C.S. Chem. Comm.*, 1977, 163.
[439] L. E. Saris and M. P. Cava, *Heterocycles*, 1977, **6**, 1349.
[440] H. Campsteyn, L. Dupont, J. Lamotte, and M. Vermeire, *Cryst. Struct. Commun.*, 1977, **6**, 639.
[441] T. Kametani and K. Takahashi, *Heterocycles*, 1978, **9**, 293.
[442] Y. Yamamoto and H. Yamazaki, *J. Org. Chem.*, 1977, **42**, 4136.
[443] M. A. Khan and V. L. T. Ribeiro, *Heterocycles*, 1977, **6**, 979.

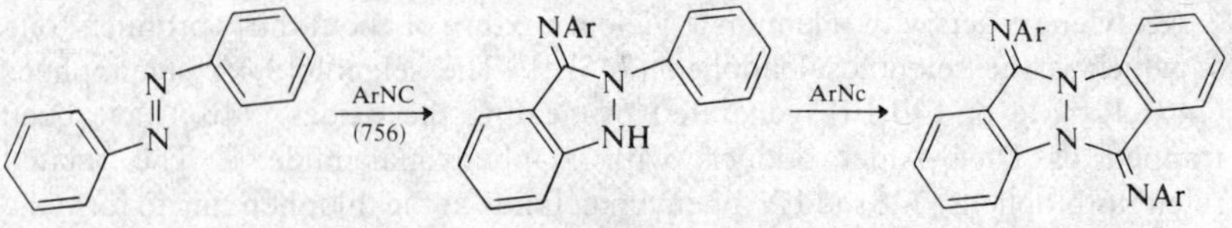

(755)                              (757)                                    (758)

Ar = 2,6-Me$_2$C$_6$H$_3$

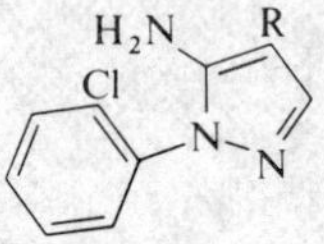

(759) 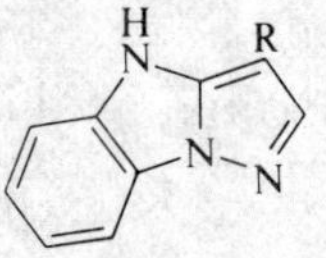                    (760)

Imidazo[4,5-*c*]pyrazole 4-oxides (762; R = *p*-NO$_2$C$_6$H$_4$) are obtained by base-catalysed cyclisation of the nitro-pyrazoles (761).[444] Cyclisation of the pyrazolylhydrazone (763) gives either the pyrazolo-triazole (764) or the pyrazolo-triazine (765), depending on conditions.[445]

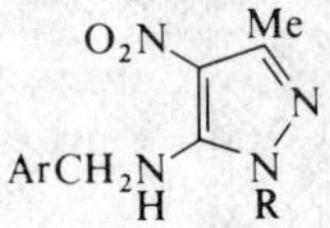

(761) 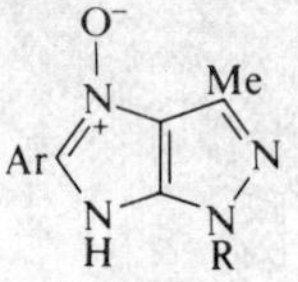                    (762)

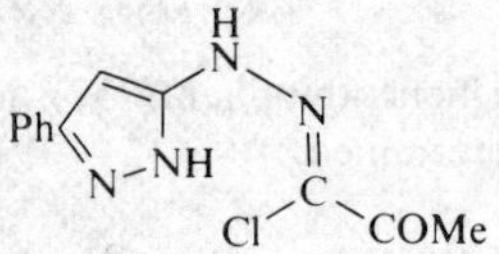

(763) 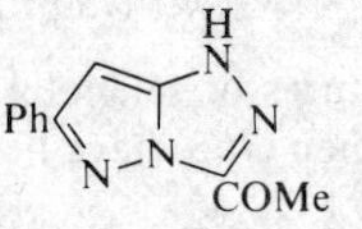                    (764) 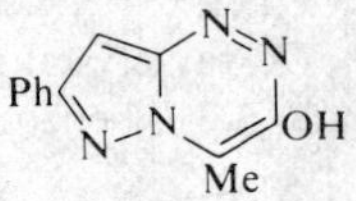                    (765)

The $^{13}$C n.m.r. spectra of numerous triaza- tetra-aza-, and penta-aza-pentalenes, *e.g.* (766)–(768), have been reported.[446]

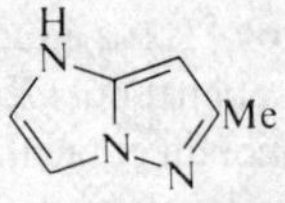

(766) 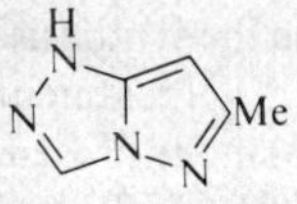          (767) 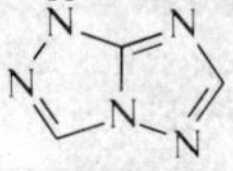          (768)

**Mixed Systems.**—4*H*-Furo[3,2-*b*]indole (769), the parent of a new ring system, has been prepared by heating 2-*o*-azidophenylfuran;[447] 2-*o*-azidophenyl-

[444] M. Lange, R. Quell, H. Lettau, and H. Schubert, *Z. Chem.*, 1977, **17**, 94.
[445] M. H. Elnagdi, M. R. H. Elmoghayar, E. M. Kandeel, and M. K. A. Ibrahim, *J. Heterocyclic Chem.*, 1977, **14**, 227.
[446] R. Faure, E. J. Vincent, R. M. Claramunt, and J. Elguero, *Org. Magn. Resonance*, 1977, **9**, 508.
[447] A. Tanaka, K. Yakushijin, and S. Yoshina, *J. Heterocyclic Chem.*, 1977, **14**, 975.

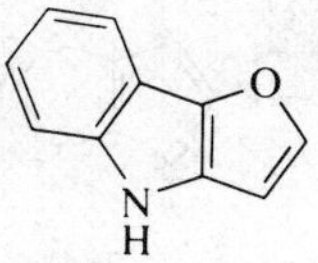

(769) 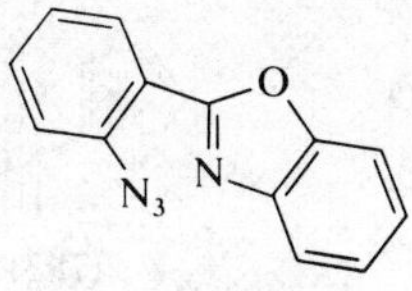          (770) 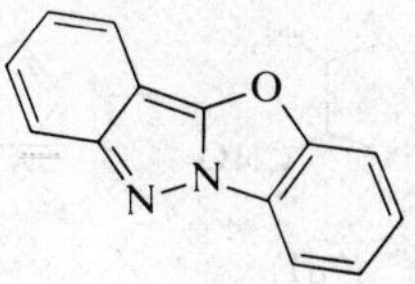          (771)

benzoxazole (770) similarly affords the indazolo-benzoxazole (771).[448] The protected hydrazino-thiophen (772) reacts with ethyl acetoacetate in acetic acid to yield the thieno-pyrrole (773).[449] Sequential treatment of the

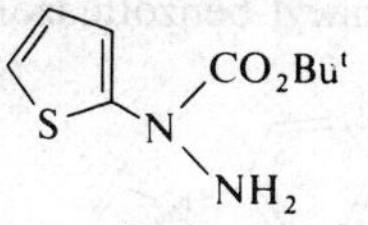

(772) 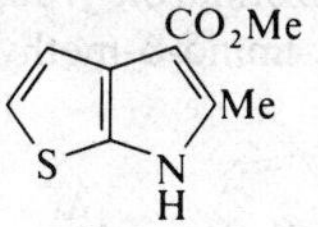          (773)

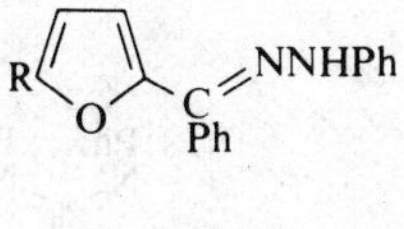

(774) 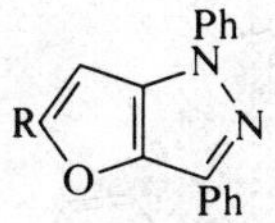          (775) 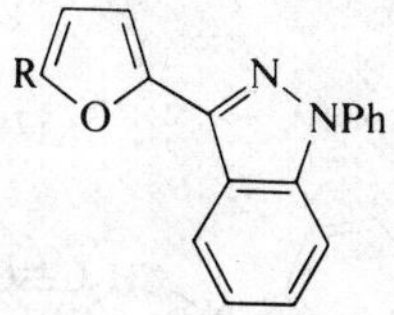          (776)

phenylhydrazones (774) with lead tetra-acetate and boron trifluoride yields either furo[3,2-*c*]pyrazoles (775; R = H, Me, or Br) or furyl-indazoles (776; R = $CO_2Me$ or $NO_2$), depending on the nature of the substituent R.[450] Reductive cyclisation of the nitro-thiophens (777) leads to the thieno[3,2-*b*]pyrroles (778); the thieno-pyrazoles (780) are similarly formed from the anils (779).[451] Furo-,

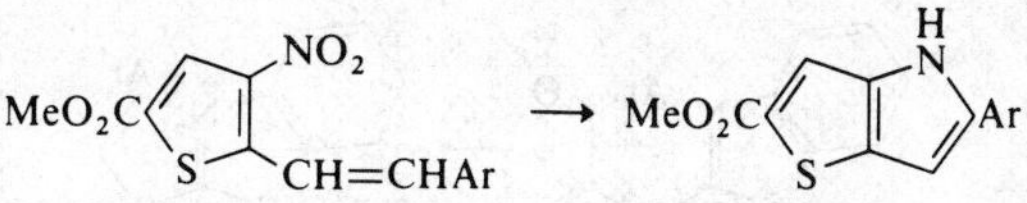

(777)          (778)

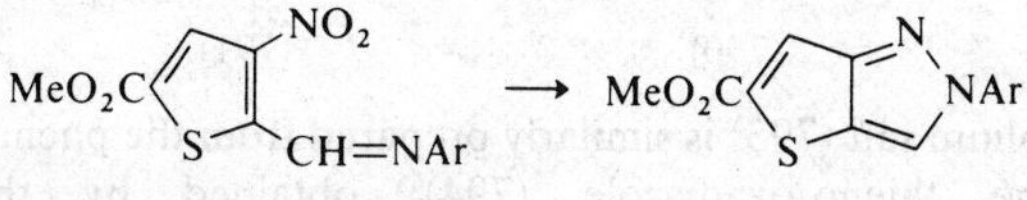

(779)          (780)

thieno-, and selenolo-[3,2-*c*]isoxazoles (782; X = O, S, or Se) result from the thermal decomposition of the corresponding azido-aldehydes (781).[452] The thioacetamide (783) is transformed into the selenolo-thiazole (784) by the action

[448] G. S. Reddy and K. K. Reddy, *Indian J. Chem.*, 1977, **15B**, 84.
[449] D. Binder, G. Habison, and C. R. Noe, *Synthesis*, 1977, 487.
[450] S. Yoshina, A. Tanaka, and S.-C. Kuo, *Yakugaku Zasshi*, 1977, **97**, 955.
[451] V. M. Colburn, B. Iddon, and H. Suschitzky, *J.C.S. Perkin I*, 1977, 2436.
[452] S. Gronowitz, C. Westerlund, and A. B. Hornfeldt, *Chemica Scripta*, 1976, **10**, 165.

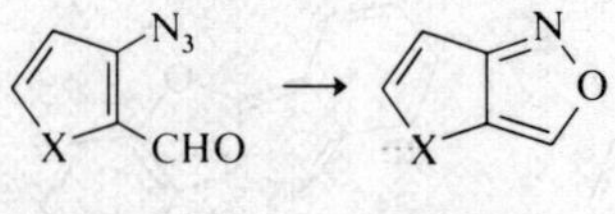

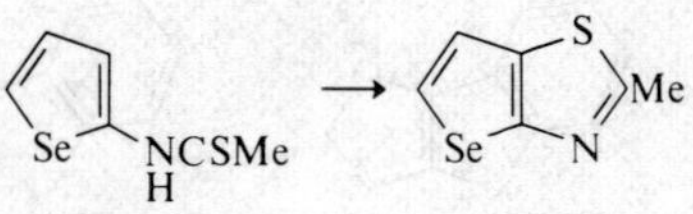

(781)                    (782)                         (783)                    (784)

of alkaline potassium iron(III) cyanide.[453] The thiazolo-benzothiazolium salts
(785) react with 1-morpholinocyclohexene to yield pyrrolo[2,1-*b*]benzothiazoles
(786).[454] Treatment of 2-amino-6-methylbenzothiazole with propargyl bromide
yields the imine (787), which cyclises in the presence of acids or bases to the
imidazo[2,1-*b*]benzothiazole (788).[455] The 2-phenyl derivative of this ring system
is obtained from 2-imino-6-methyl-3-phenacyl-benzothiazoline.[456]

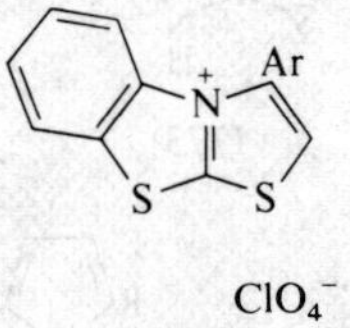

(785) 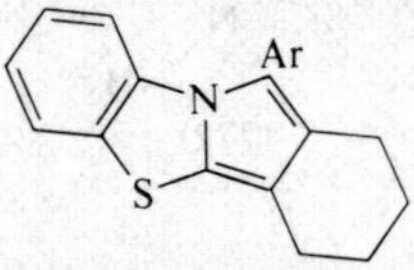                                          (786)

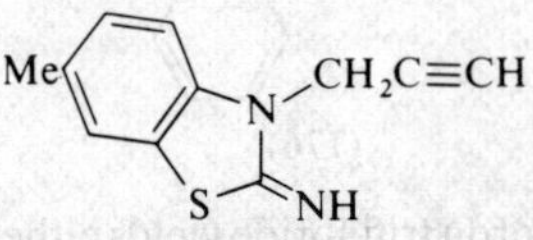

(787) 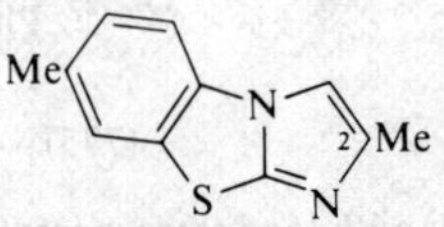                    (788) 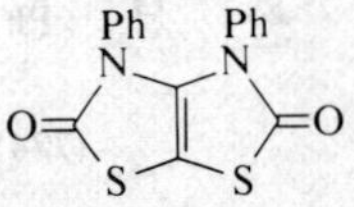                         (789)

The thiazolo-thiazole (789) is formed by the action of ClSCOCl on 1,1-
dianilino-2-nitroethylene.[457] Treatment of the benzothiazoles (790) with
perchloric acid yields the thiazolo[2,3-*b*]benzothiazolium salts (791);[458] the

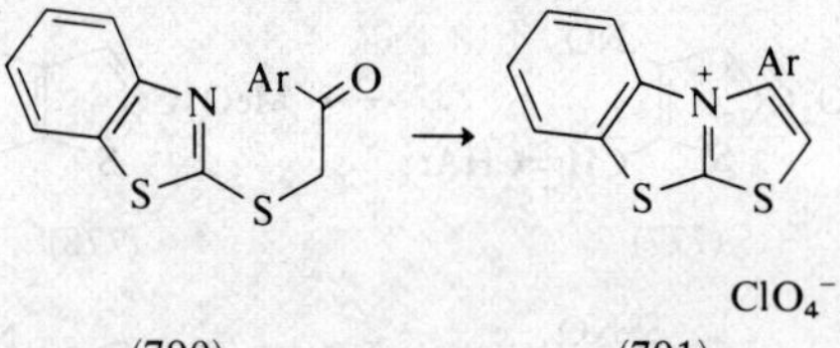

(790)                                          (791)

thiazolotriazolium salt (793) is similarly prepared from the phenacylthiotriazole
(792).[459] The thieno-oxadiazole (794), obtained by the action of
diphosphorus decasulphide on 3,4-dibenzoyl-1,2,5-oxadiazole, reacts with *N*-

[453] P. I. Abramenko and V. G. Zhiryakov, *Khim. geterotsikl Soedinenii*, 1977, 1495.
[454] S. Sawada, T. Miyasaka, and K. Arakawa, *Chem. and Pharm. Bull.* (*Japan*), 1978, **26**, 275.
[455] A. N. Krasovskii, N. P. Grin, I. I. Soroka, P. M. Kochergin, and E. I. Bogatyreva, *Khim. Farm. Zhur.*,
1977, **11**, 25 (*Chem. Abs.*, 1977, **87**, 201 412).
[456] A. Singh, J. Mohan, and H. K. Pujari, *Indian J. Chem.*, 1976, **14B**, 997.
[457] H. Schäfer, B. Bartho, and K. Gewald, *J. prakt. Chem.* 1977, **319**, 149.
[458] S. Sawada, T. Miyasaka, and K. Arakawa, *Chem. and Pharm. Bull.* (*Japan*), 1977, **25**, 3370.
[459] F. S. Babichev, V. A. Kovtunenko, A. K. Tyltin, and I. G. Lelyukh, *Khim. geterotsikl. Soedinenii*,
1977, 1132.

phenylmaleimide to yield a mixture of *endo-* and *exo-*adducts (795).[460] The condensation of phenacyl bromide with 5-amino-1,2,4-thiadiazole affords the imidazo[1,2-*d*]-1,2,4-thiadiazole (796);[461] with 2,5-diamino-1,3,4-thiadiazole, a derivative (797) of the imidazo[2,1-*b*]-1,3,4-thiadiazole system is formed.[462]

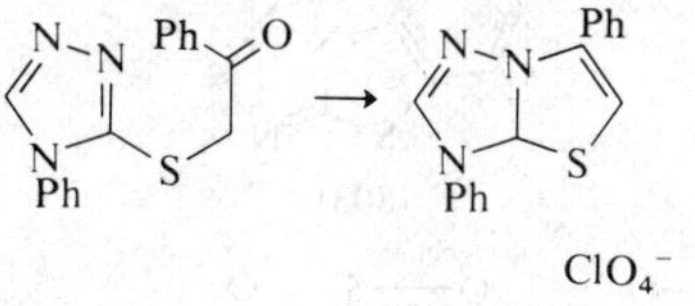

(792)        (793) 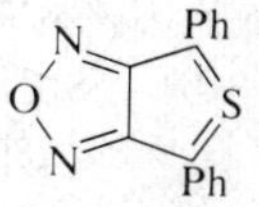        (794) 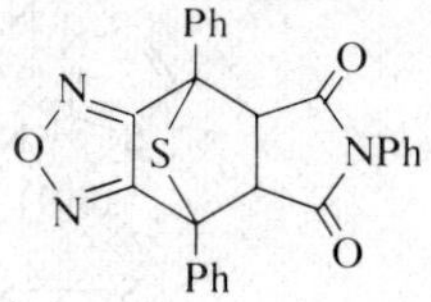        (795)

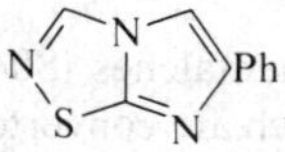

(796) 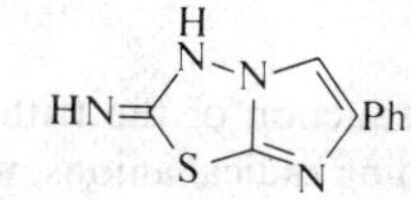        (797)

The reaction of 2-thiocyanatobenzimidazole with imidazole unexpectedly yields the imidazolyl-thiadiazolobenzimidazole (798).[463] The triazolothiadiazolium salt (800) has been prepared by treating the meso-ionic triazolium thiolate (799) with dichloromethyleneaniline, $Cl_2C=NPh$, in the presence of magnesium perchlorate.[464]

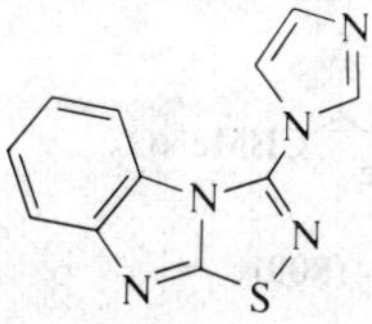

(798) 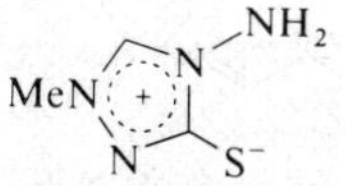        (799) 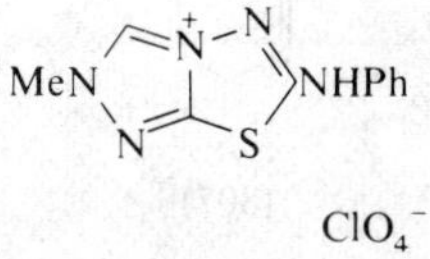        (800)

Azide–tetrazole tautomerism has been investigated by $^1H$ n.m.r. spectroscopy and dipole-moment measurements for the 2-azidothiazole system, (801) ⇌ (802), and for 2-azidobenzothiazole, 2-azidobenzoxazole, and 3-azidoisoxazole.[465] Chlorosulphonation of the tetrazolobenzothiazole (803) with chlorosulphonic acid yields 2-azido-6-chlorosulphonyl-benzothiazole.[466]

**Hypervalent Sulphur Compounds.**—The gas-phase ESCA spectrum of trithiapentalene (804; R = H) has been determined,[467] and the results of *ab initio*

[460] O. Tsuge, T. Takata, and M. Noguchi, *Heterocycles*, 1977, **6**, 1173.

[461] L. Pentimalli, G. Milani, and F. Biavati, *Gazzetta*, 1977, **107**, 1.

[462] H. Paul, A. Sitte, and R. Wessel, *Monatsh.*, 1977, **108**, 665.

[463] R. D. Haugwitz, B. Toeplitz, and J. Z. Gougoutas, *J.C.S. Chem. Comm.*, 1977, 736.

[464] A. Y. Lazaris, S. M. Shmuilovich, and A. N. Egorochkin, *Izvest. Akad. Nauk S.S.S.R., Ser. khim.*, 1977, 1827.

[465] R. Faure, J. P. Galy, E. J. Vincent, J. P. Fayet, P. Mauret, M. C. Vertut, and J. Elguero, *Canad. J. Chem.*, 1977, **55**, 1728.

[466] V. N. Skopenko, L. F. Avramenko, V. Ya. Pochinok, and N. E. Kruglyak, *Ukrain. khim. Zhur.* (*Russ. Edn.*), 1977, **43**, 518 (*Chem. Abs.*, 1977, **87**, 152 084).

[467] L. J. Saethre, S. Svensson, N. Martensson, U. Gelius, P. A. Malmquist, E. Basilier, and K. Siegbahn, *Chem. Phys.*, 1977, **20**, 431.

SCF MO calculations on this compound have been reported.[468] The microwave spectrum of the 1,6-dioxa-analogue (805) shows no evidence for valence tautomerism of the type (805) ⇌ (806).[469]

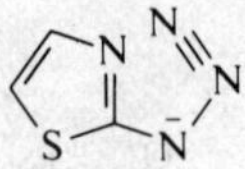

(801) 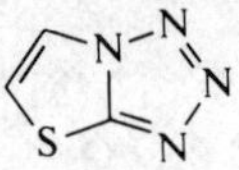          (802) 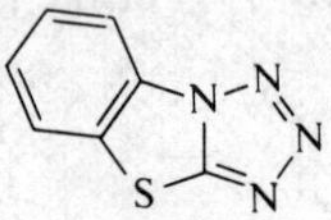          (803)

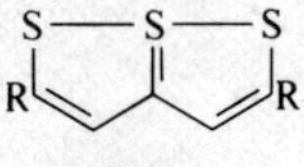

(804) 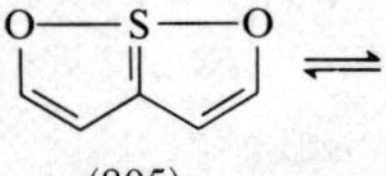          (805) 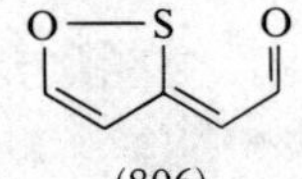          (806)

Electrochemical reduction of the trithiapentalenes (804; R = H, Me, or Ph) yields the corresponding radical anions, which are converted into the thiapyran-thiones (807).[470] 1-Phenyl-6,6a-dithia-1,2-diazapentalene (808) is nitrated mainly in position 3, but the 3,4-dimethyl derivative undergoes 'nitro-dediazoniation' to the dithiole (809). Treatment of compound (808) with nitrous acid yields the phenylazo-oxadithiazapentalene (810);[471] a similar reaction is the transformation of dioxathiapentalene (805) into the oxathiadiazapentalene (811) by the action of nitrobenzene-*p*-diazonium fluoroborate.[472]

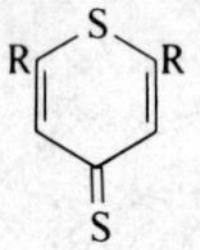

(807) 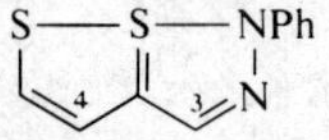          (808) 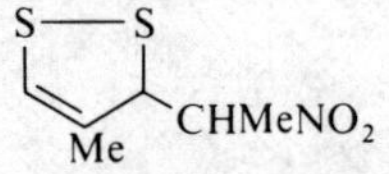          (809)

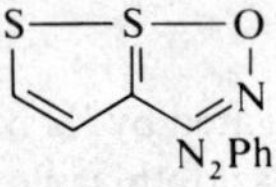

(810) 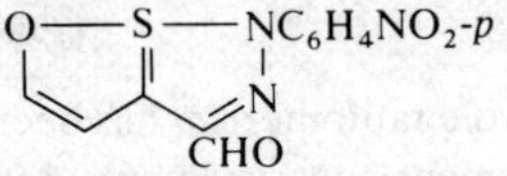          (811)

The formation of the dioxathiadiazapentalenes (813) from aminothiatriazole (812) and aroyl chlorides was discovered independently in two laboratories.[473,474] The hydrochloride of the iminothiadiazoline (814) reacts with benzoyl chloride,

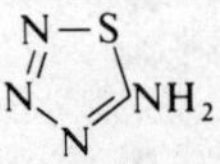

(812) 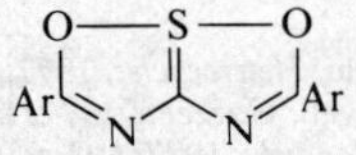          (813) 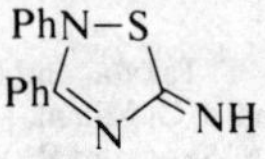          (814)

[468] S. C. Nyburg, G. Theodorakopoulos, and I. G. Csizmadia, *Theor. Chim. Acta.*, 1977, **45**, 21.
[469] T. Pedersen, S. V. Skaarup, and C. T. Pedersen, *Acta Chem. Scand.* (*B*), 1977, **31**, 711.
[470] F. Gerson, R. Gleiter, and H. Ohya-Nishiguchi, *Helv. Chim. Acta*, 1977, **60**, 1220.
[471] R. M. Christie and D. H. Reid, *J.C.S. Perkin I*, 1977, 848.
[472] D. H. Reid and R. G. Webster, *J.C.S. Perkin I*, 1977, 854.
[473] R. J. S. Beer and I. Hart, *J.C.S. Chem. Comm.*, 1977, 143.
[474] G. L'Abbé, G. Verhelst, and G. Vermeulen, *Angew. Chem. Internat. Edn.*, 1977, **16**, 403.

*N*-phenylbenzimidoyl chloride (PhCCl=NPh), and phenyl isothiocyanate to yield the heterothiapentalenes (815)–(817), respectively.[474]

(815)  (816)  (817)

## 9 Meso-ionic Compounds

The synthesis of 3-alkylthio-5-phenyl-1,2-dithiolium 4-olates (818; R = Salkyl) from the dithio-esters PhCOCH(OMe)CS$_2$alkyl has been described.[475] The carbon–carbon distances in the diphenyl analogue (818; R = Ph) indicate that there is some aromatic character;[476] *X*-ray crystallography of the triketone ArCOCOCOAr (Ar = *p*-BrC$_6$H$_4$) gives no evidence for the meso-ionic structure (819).[476]

(818)  (819)  (820)

2-*p*-Nitrophenyl-4-phenyl-oxazolin-5-one exists as the meso-ionic tautomer (820), as shown by photoelectron spectroscopy.[477] The münchnones (821; Ar$^1$ = Ph, Ar$^2$ = *p*-MeC$_6$H$_4$) and (821; Ar$^1$ = *p*-MeC$_6$H$_4$, Ar$^2$ = Ph) add carbon disulphide with high regioselectivity to form the cycloadducts (822), which lose carbon dioxide to yield meso-ionic thiazolium thiolates (823).[478] The

(821)  (822)  (823)

diphenylmünchnone (821; Ar$^1$ = Ar$^2$ = Ph) reacts with the Diels–Alder adduct (824) of cyclopentadiene to *p*-benzoquinone to give the acid (826); this is formed by opening of the lactone ring of the initial product (825). On heating, the acid fragments to cyclopentadiene, carbon dioxide, hydrogen, and the isoindolo-quinone (827).[479] Diphenylmünchnone functions as the keten (828) in its reactions with the azirine (829) to form the oxazoline derivative (830),[480] and with *o*-benzoquinone to give compound (831).[481] Treatment of *S*-benzoyl-*α*-

[475] D. Barillier, P. Rioult, and J. Vialle, *Bull. Soc. chim. France*, 1977, 659.
[476] J. R. Cannon, K. T. Potts, C. L. Raston, A. F. Sierakowski, and A. H. White, *Austral. J. Chem.*, 1978, **31**, 297.
[477] P. Espinasse, G. Kille, A. Kalt, and G. Nanse, *Tetrahedron Letters*, 1977, 4587.
[478] R. Huisgen and T. Schmidt, *Annalen*, 1978, 29.
[479] (*a*) W. Friedrichsen and W. D. Schröer, *Tetrahedron Letters*, 1977, 1603; (*b*) J. Lukac, H. Heimgartner, and H. Schmid, *Chimia* (*Switz.*), 1977, **31**, 138.
[480] J. Lukac, J. H. Bieri, and H. Heimgartner, *Helv. Chim. Acta*, 1977, **60**, 1657.
[481] W. Friedrichsen and I. Schwarz, *Tetrahedron Letters*, 1977, 3581.

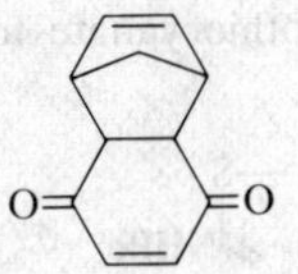

(824) 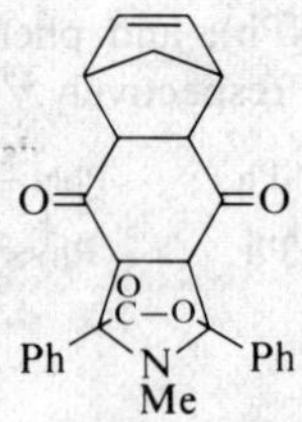            (825) 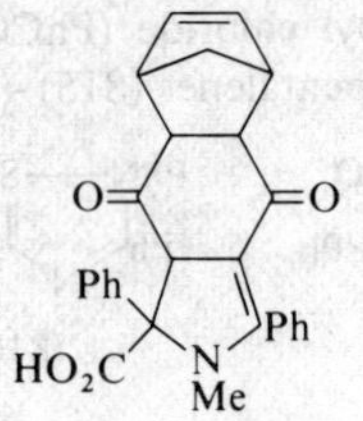            (826) 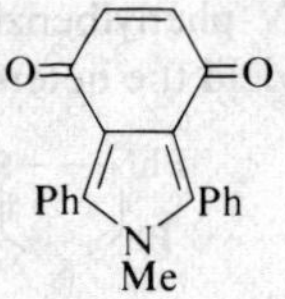            (827)

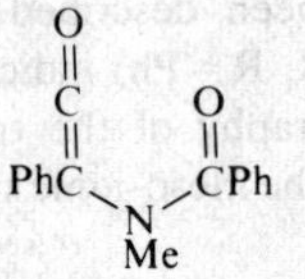

(828) 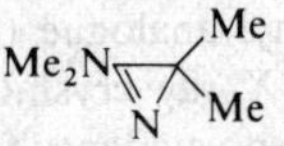            (829) 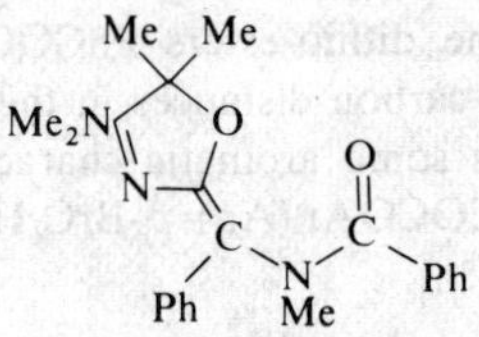            (830)

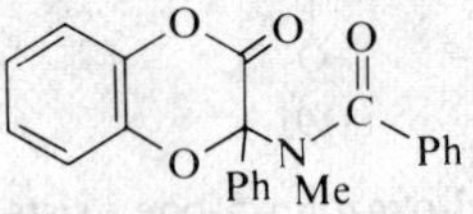

(831)

mercaptophenylacetic acid (832) with dimethyl acetylenedicarboxylate in the presence of trifluoroacetic anhydride yields the thiophen derivative (835), which is formed by way of the oxathiolium olate (833) and the cycloadduct (834).[482]

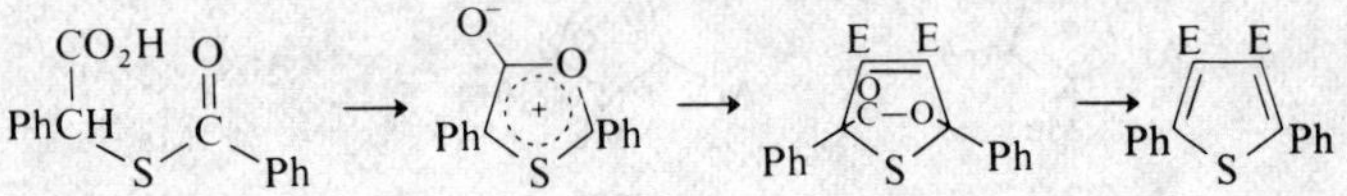

(832)            (833)            (834)            (835)

E = CO₂Me

E = $CO_2Me$

Irradiation of the thiazolium 4-olate (836) in methanol affords a mixture of the rearranged thiazolin-2-one (837) and the β-lactam (838).[483]

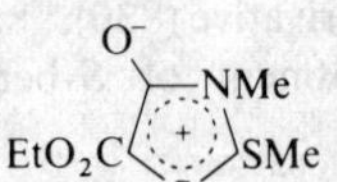

(836) 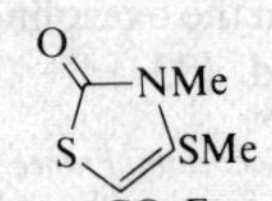            (837) 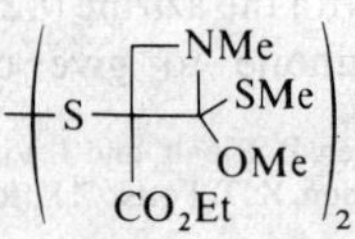            (838)

[482] K. Masuda, J. Adachi, and K. Nomura, *Chem. and Pharm. Bull. (Japan)*, 1977, **25**, 1471.
[483] D. H. R. Barton, E. Buschmann, J. Häusler, C. W. Holzapfel, T. Sheradsky, and D. A. Taylor, *J.C.S. Perkin I*, 1977, 1107.

Photolysis of 3,4-diphenylsydnone (839) generates the nitrile imine (840), which reacts with tetraphenylcyclopentadienone to yield the *spiro*-adduct (841);[484] in the case of *N-o*-allylphenyl-*C*-phenylsydnone (842), the corresponding nitrile imine (843) forms the intramolecular cycloadduct (844).[485]

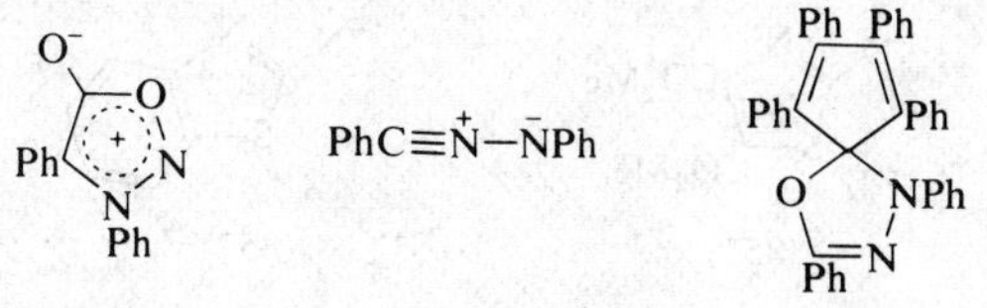

(839)          (840)          (841)

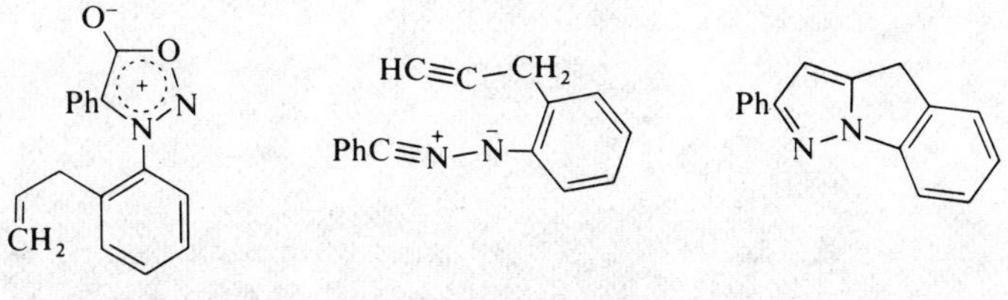

(842)          (843)          (844)

Full details of the thermal isomerization of dehydrodithizone (845) to 1,3-diphenyltetrazolium 5-thiolate (846) and of the reaction of dehydrodithizone with pentacarbonyliron to yield the phenylazo-thiadiazolinone (847) have appeared.[486] The cycloadduct (848) of dehydrodithizone to tetraphenylcyclopentadienone undergoes a spontaneous ring-opening reaction to afford a mixture of the stable ylide (849) and the phenylazo-compound (850);[487]

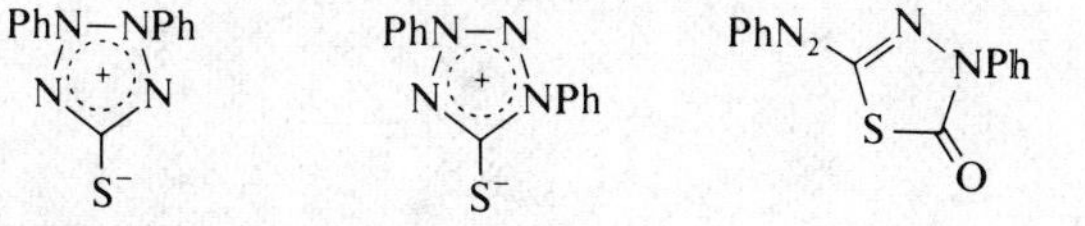

(845)          (846)          (847)

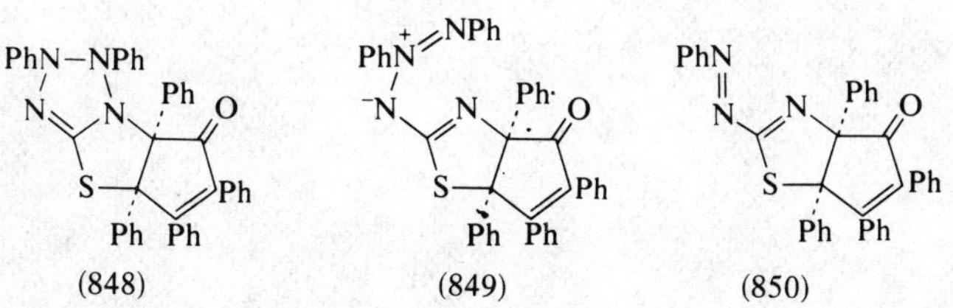

(848)          (849)          (850)

[484] S. K. Kar, *Indian J. Chem.*, 1977, **15B**, 184.
[485] H. Meier, H. Heimgartner, and H. Schmid, *Helv. Chim. Acta*, 1977, **60**, 1087.
[486] P. N. Preston and K. Turnbull, *J.C.S. Perkin I*, 1977, 1229; see these Reports, Vol. 6, p. 80.
[487] G. V. Boyd, T. Norris, and P. F. Lindley, *J.C.S. Perkin I*, 1977, 965.

dimethyl acetylenedicarboxylate and dehydrodithizone similarly yield the ylide (851), and benzyne forms 2-phenylazo-benzothiazole and azobenzene by loss of phenylnitrene from the intermediate (852).[488]

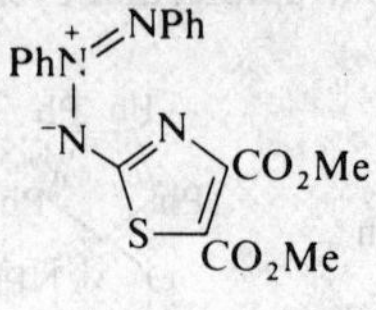

(851) 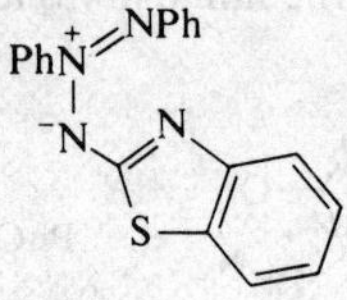                                        (852)

[488] G. V. Boyd, T. Norris, P. F. Lindley, and M. M. Mahmoud, *J.C.S. Perkin I*, 1977, 1612.

# 3
# Six-membered Homocyclic Compounds

BY A. W. SOMERVILLE

## 1 Introduction

The material in this chapter has been organised into ring-type and functional-group order so that required information may be rapidly obtained. Monocyclic systems and their alkyl derivatives are reviewed first, followed by related compounds, in order of their functional groups. This is followed by polycyclic systems (non-fused and fused). Derivatives of these are covered under the appropriate parent compound heading, appearing in the same order as that in the earlier sections. Later sections include more specialised systems such as helicenes and cyclophanes. Synthesis and reactivity are discussed collectively under the appropriate system heading.

## 2 Monocyclic Compounds

**Benzene and its Alkylated Derivatives.**—Interest in direct synthesis of six-membered carbocyclic compounds, lacking over the years, has been stimulated by the current interest in cycloaddition reactions. A useful synthesis of 1,2,4-triphenylbenzene (1) by the Diels–Alder reaction of 2,3,5-triphenylcyclopentadienone (2) with dimethyl acetylenedicarboxylate has been reported[1] (Scheme 1).

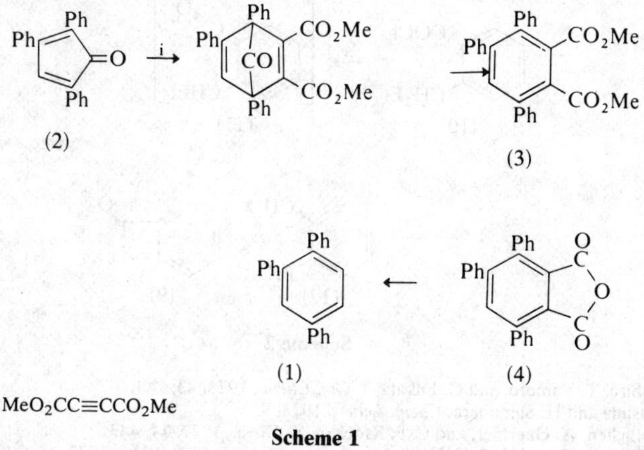

Reagents: i, MeO$_2$CC≡CCO$_2$Me

**Scheme 1**

[1] M. Ballester, J. Castaner, and J. Morell, *Anales de Quim.*, 1977, **73**, 439.

91

Friedel–Crafts alkylations have been reported by a number of groups. The t-butylation of *p*-xylene by 2-t-butyl- and 2,6-di-t-butyl-*p*-cresols, using a modified AlCl₃–MeNO₂ procedure,[2] has led to a new preparation of 2-t-butyl-*p*-xylene. Such reactions can lead to a mixture of expected and unexpected products, the latter arising from carbo-cation rearrangements. The reaction of benzene with functionally substituted 2-chloropropanes RCH₂CHClMe (R = OH or OMe)[3] gave, unusually, phenylpropane in addition to 1-substituted 2-phenylpropane, 1,1-diphenylpropane, 1,2-diphenylpropane, and 1,1-diphenylpropene.

Alkylation of benzene and toluene with cyclohexane[4] is promoted by the addition of 2-methylpropane and generation of the t-butylium ion. The expected cyclohexylbenzene is accompanied by the rearrangement products (5) and (6), from benzene and the methylcyclopentylium ion, and products resulting from coupling of this ion either with itself or with the cyclohexylium ion.

The fused system 3,4-bis(trimethylsilyl)benzocyclobutene (7) has been obtained[5] by cyclotrimerization of the alkyne 1-trimethylsilylhexa-1,5-diyne in the presence of dicarbonyl(cyclopentadienyl)cobalt and converted (by Bu$^n$Li) into 1,2-dihydrocyclobuta[*c*]benzyne, which was trapped as the adduct (8).

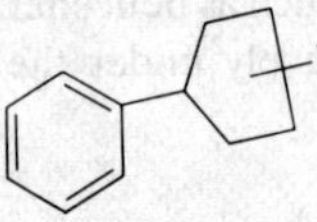

(5) 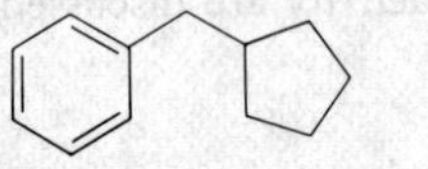                    (6) 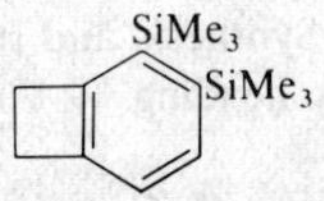                    (7) 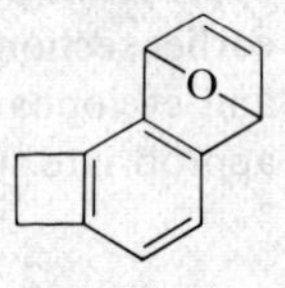                    (8)

The benzocyclobutenone system (9) has been prepared by gas-phase pyrolysis of *ortho*-alkyl-substituted aromatic acyl chlorides (10).[6] Scheme 2 illustrates how elimination of hydrogen chloride results in the *ortho*-quinonoid keten (11) and hence (9). If the alkyl side-chain bears β-H atoms, (11) is stabilized by 1,5-hydrogen shift, giving the aldehyde (12).

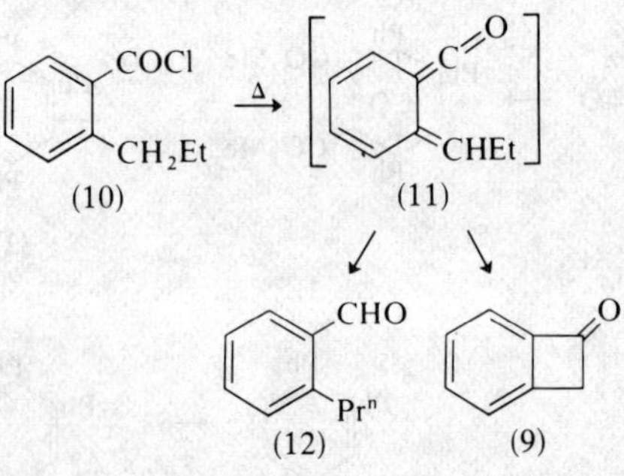

**Scheme 2**

[2] M. Tashiro, T. Yamoto, and G. Fukata, *J. Org. Chem.*, 1978, **43**, 743.
[3] H. Matsuda and H. Shinohara, *Chem. Letters*, 1978, 95.
[4] R. Miethchen, A. Gaertner, and C. F. Kroeger, *Z. Chem.*, 1977, **17**, 443.
[5] R. L. Hillard, III, and K. P. C. Vollhardt, *Angew. Chem. Internat. Edn.*, 1977, **16**, 399.
[6] P. Schiess and M. Heitzmann, *Angew. Chem. Internat. Edn.*, 1977, **16**, 469.

A reaction sequence widely used for ring closure and the formation of C—C bonds is illustrated in Scheme 3. In this instance[7] the product is the angular di(cyclobuteno)benzene (13), but this method has been widely applied to the synthesis of cyclophanes, as will be reported later.

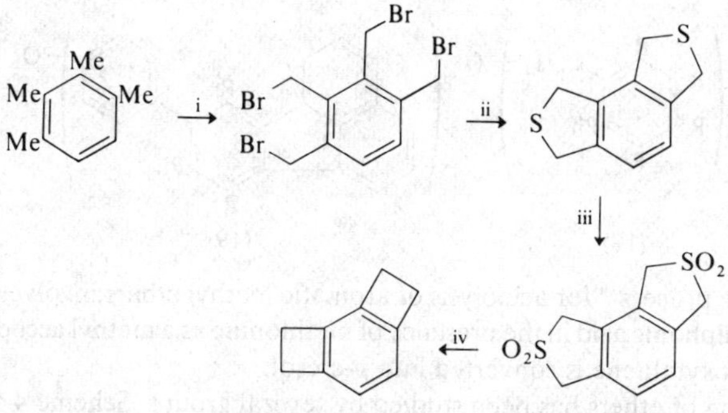

(13)

Reagents: i, Br$_2$, $h\nu$; ii, Na$_2$S; iii, MeCO$_3$H; iv, heat, 280 °C

**Scheme 3**

Reviews have appeared on aspects of benzenoid reactivity. These include reports on the Hofmann–Loeffler–Freytag reaction on benzene rings,[8] valence isomers of benzene,[9] and the organic photochemistry of benzene.[10]

Salicylamide complexes (preferably with LiAlH$_4$) have been reported[11] to be successful aids in the hydrogenation of benzene, toluene, and xylenes to the corresponding naphthenes. Photocyclisations of stilbenes and related compounds[12] in the presence of $\pi$-acceptors (*e.g.* tetracyanoethene, chloranil) have led to good yields of the corresponding polycyclic aromatic hydrocarbons. A dihydro-compound such as (14) is formed in the absence of the $\pi$-acceptor, and it can be oxidized to (15).

(14)                          (15)

**Ethers.**—A new dienophile, bis-(4,4'-diphenylethynyl) ether (16), has been made[13] and utilized in a convenient preparation of a new series of bis-ethers (18) and (19) by cycloaddition reactions with cyclopentadienones (17).

[7] E. Giovannini and H. Vuilleumier, *Helv. Chim. Acta.* 1977, **60**, 1452.
[8] R. Oda, *Kagaku* (*Kyoto*), 1977, **32**, 78 (*Chem. Abs.*, 1977, **87**, 84 633).
[9] A. H. Schmidt, *Chem. Unserer Zeit*, 1977, **11**, 118 (*Chem. Abs.*, 1977, **87**, 166 941).
[10] D. Bryce-Smith and A. Gilbert, *Tetrahedron*, 1977, **33**, 2459.
[11] P. Patnaik and S. Sarkar, *Tetrahedron Letters*, 1977, 2531.
[12] J. Bendig, M. Beyermann, and D. Kreysig, *Tetrahedron Letters*, 1977, 3659.
[13] K. Ghosh and A. J. Bhattacharya, *Indian J. Chem., sect. B.* 1977, **15**, 672.

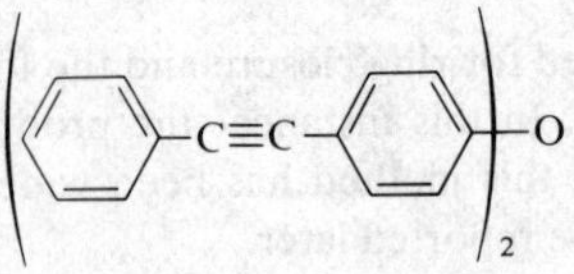

(16) 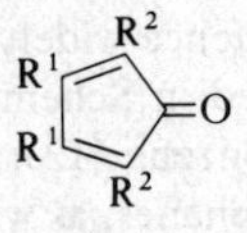                                                    (17)

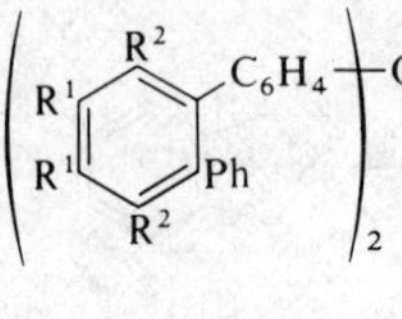

(18) 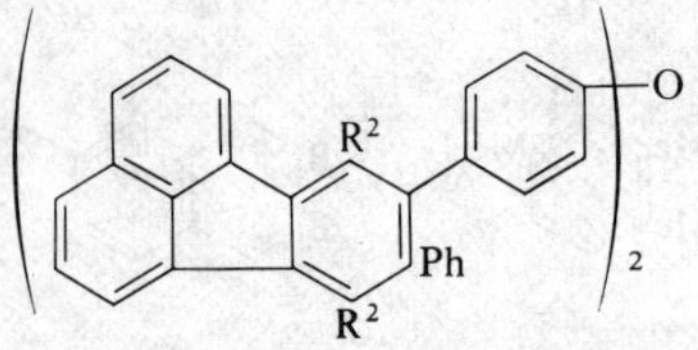                                                    (19)

A simple process[14] for acidolysis of aromatic methyl ethers involves the use of methanesulphonic acid in the presence of methionine as a methyl acceptor. In this way, methoxytoluene is converted into *p*-cresol.

Oxidation of ethers has been studied by several groups. Scheme 4 shows how the anodic oxidation[15] of 4-methoxyphenyl 3-methoxybenzoate (20; R = *m*-MeOC$_6$H$_4$CO) gives the *ortho*-coupled biphenyls (21) and the *ortho*- and *para*-quinones (22) formed by further oxidation of (21).

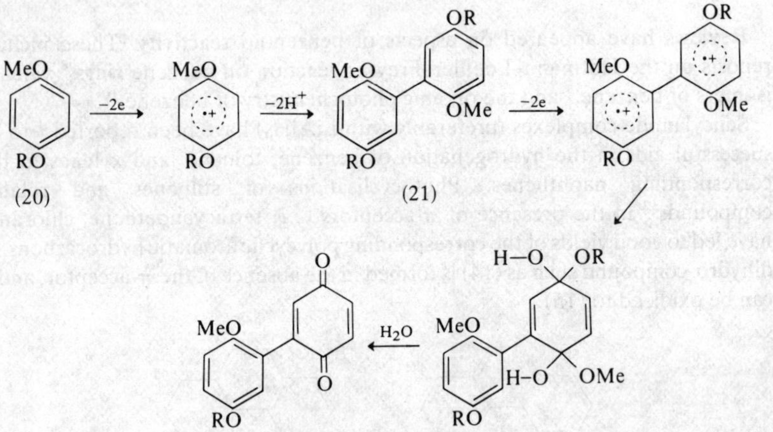

R = *m*-MeOC$_6$H$_4$CO-

**Scheme 4**

The oxidation[16] of the benzaldehyde acetals (23) with 2,3-dichloro-5,6-dicyanobenzoquinone gave mixtures of the corresponding aldehydes, the addition products (24), and 2,3-dichloro-5,6-dicyanodihydroquinone.

[14] N. Fujii, H. Irie, and H. Yajima, *J.C.S. Perkin I*, 1977, 2288.
[15] M. Sainsbury and J. Wyatt, *J.C.S. Perkin I*, 1977, 1750.
[16] J. Iwamura, N. Iwamoto, and N. Hirao, *Nippon Kagaku Kaishi*, 1977, **7**, 1009.

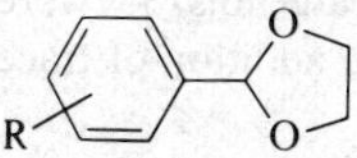

(23) 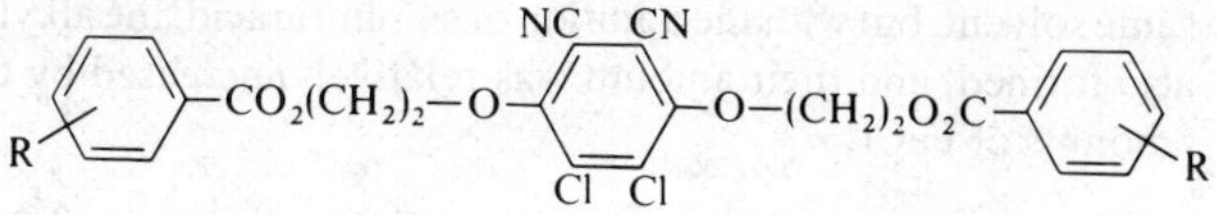          (24)

R = H, *o*-, *m*-, or *p*-OMe, Cl, or Me

**Alcohols and Phenols.**—The discovery of a convenient preparation of 3-methoxy-4-isopropylbenzyl alcohol has resulted from a systematic study[17] concerned with the preferential oxidation of methyl groups attached to aromatic rings. This showed that the oxidation of aromatic compounds of low ionisation potential with $Pb(OAc)_4$ can lead to selective oxidation of Me in preference to $Pr^i$. *m*-Methoxy-*p*-cymene (25), for example, can be converted into 4-isopropyl-3-methoxybenzyl acetate (26).

Several reports have appeared on the synthesis of specialized phenols. Most of the possible isomers of deuteriated phenol (27) have been obtained[18] by reductive dehalogenation of the corresponding bromo-phenols with Raney alloys in an alkaline $D_2O$ solution.

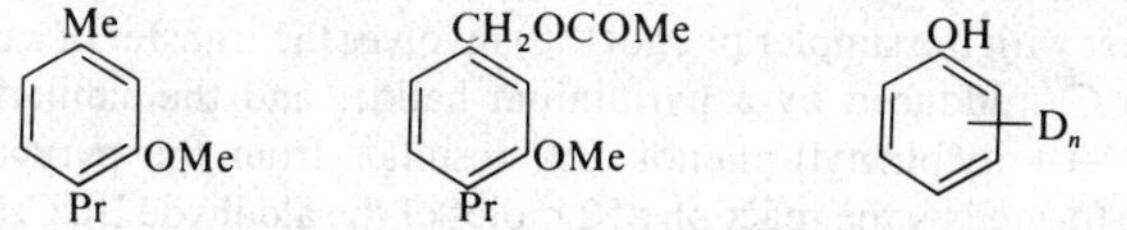

(25)          (26)          (27)

3-(t-Butyl)-4-methoxyphenol (28) has been prepared[19] selectively from 3-t-butylhydroquinone (29) by the use of pivaloyl chloride as a blocking group (Scheme 5).

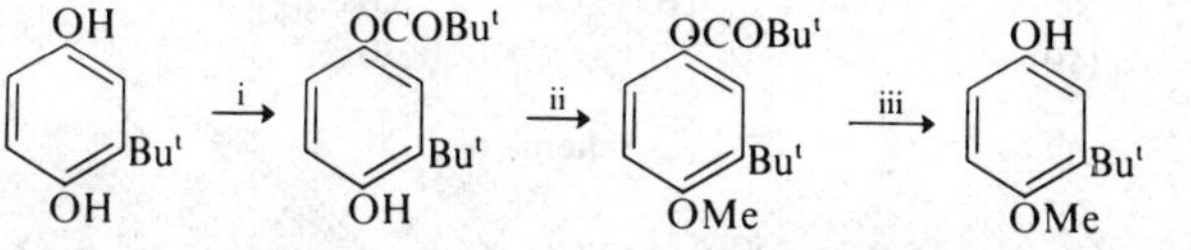

(29)          (28)

Reagents: i, Bu$^t$COCl; ii, Me$_2$SO$_4$; iii, KOH

**Scheme 5**

5-Acyl-guaiacols (30), together with the 4-isomers (31), have been obtained as mixtures[20] from the Fries rearrangement of the guaiacol esters (32).

A further study of aromatization and enolization[21] of alkylcyclohexanones has been reported. In AcOH as solvent (33) gave only the 2-alkyl-phenols (34). In the

[17] V. V. Dhekne and A. S. Rao, *Synth. Comm.*, 1978, **8**, 135.
[18] M. Tashiro, A. Iwasaki, and G. Fukata, *J. Org. Chem.*, 1978, **43**, 196.
[19] L. K. T. Lam and K. Farhat, *Org. Prep. Proced. Internat.*, 1978, **10**, 79 (*Chem. Abs.*, 1978, **88**, 190 294).
[20] M. Robert, *Bull. Soc. chim. France*, 1977, 901.
[21] M. Kablaoui, *Div. Petrol. Chem., Amer. Chem. Soc. Preprint*, 1975, **20**, 635 (*Chem. Abs.* 1977, **87**, 134 211).

same solvent, but with the addition of sulphuric acid, the alkyl-catechols (35) were also formed, and their amount was relatively increased by the addition of trace amounts of $FeCl_3$.

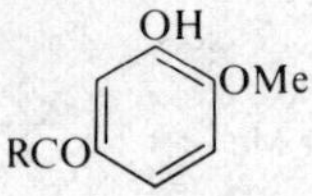

(30) 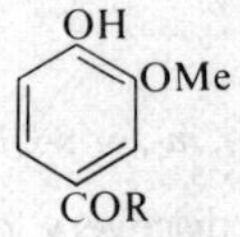                                 (31) 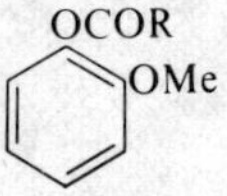                                 (32)

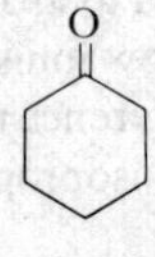

(33) 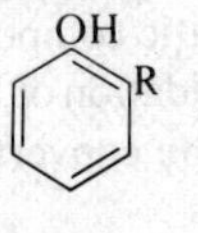                  (34) 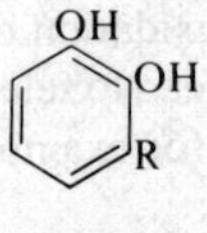                  (35) 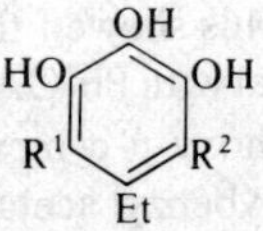          (36) $R^1 = R^2 = Me$
                                                          (37) $R^1 = H$, $R^2 = Me$
                                                          (38) $R^1 = R^2 = H$

An interesting formation of highly substituted trihydroxybenzenes (36) and (37) together with the simpler phenol (38) involves the transfer of a methyl group from O to C,[22] induced by a pyridinium halide, and the similarly congested 2,4,6-tris-(4-t-butylphenyl)-phenol (39) results[23] from the pyrylium salt (40) (Scheme 6) formed by the reaction of 2 moles of the aldehyde $Bu^tC_6H_4CHO$ with 1 of the ketone $Bu^tC_6H_4COMe$.

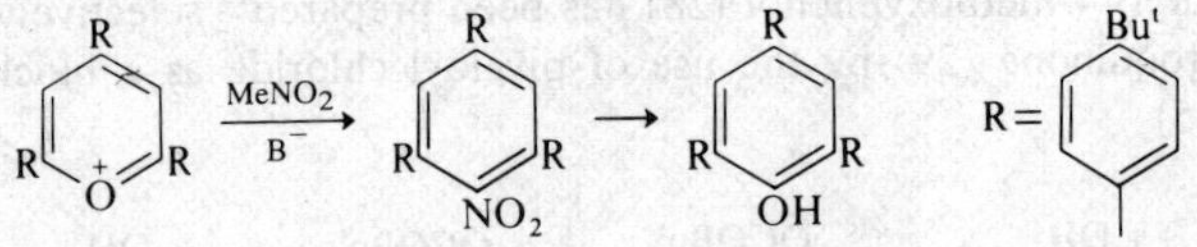

(40)                                 (39)                $R = $

**Scheme 6**

Alkylenedioxy-phenols such as (41) have been prepared[24] by treatment of benzenetriols (42) with $\alpha\omega$-dichloroalkanes $Cl(CH_2)_nCl$ ($n = 1$—3) in DMSO, and o- and p-nitrosophenols (43)—(46) (as the monoxime of the quinonoid forms) are formed[25] by Claisen cyclocondensation of the oximes (47) with ketones (48).

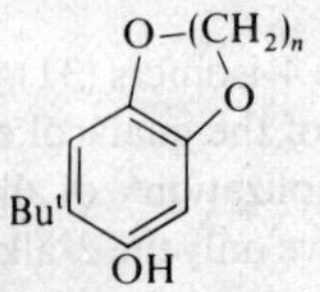

(41) 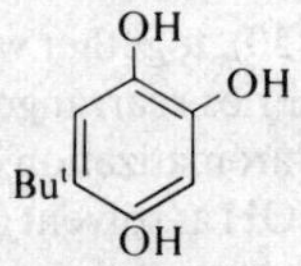                                 (42)

[22] J. P. Bachelet, P. Demerseman, and R. Royer, *Tetrahedron Letters*, 1977, 4407.
[23] K. Dimroth, W. Tuencher, and H. Ketsch, *Chem. Ber.*, 1978, **111**, 264.
[24] H. Fujita and M. Yamashita, *Nippon Kagaku Kaishi*, 1977, **6**, 925.
[25] E. Yu Belyaev, L. M. Gornostaev, A. P. Es'ki, M. S. Tovbis, G. A. Suboch, and A. V. El'tsov, *J. Org. Chem. (U.S.S.R.)*, 1977, **13**, 2149.

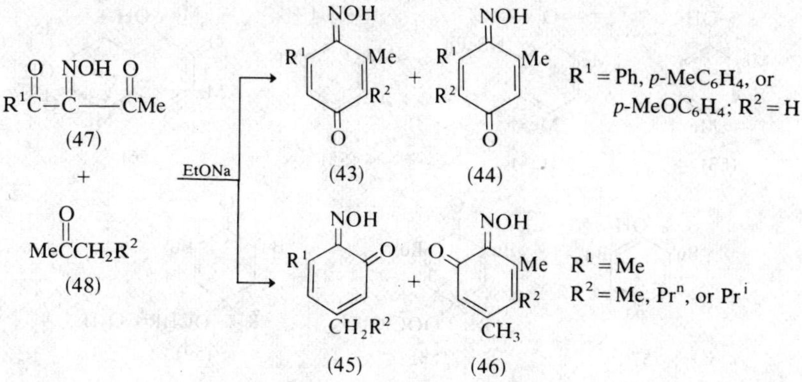

The methylthiomethyl group has been used[26] for the protection of phenols since it is stable under conditions in which primary alcohol ethers are hydrolysed, allowing selective removal of the phenolic protecting group.

In a continuing study of the oxidation of phenols,[27] the formation of a mixture of the spiro-acetal (49) and the phenol (52) from 2-methyl-4-t-butylphenol has been described. The spiro-acetal arises from cross-coupling of the radicals (50) and (51) followed by further oxidation of the trimeric species (52), as shown in Scheme 7.

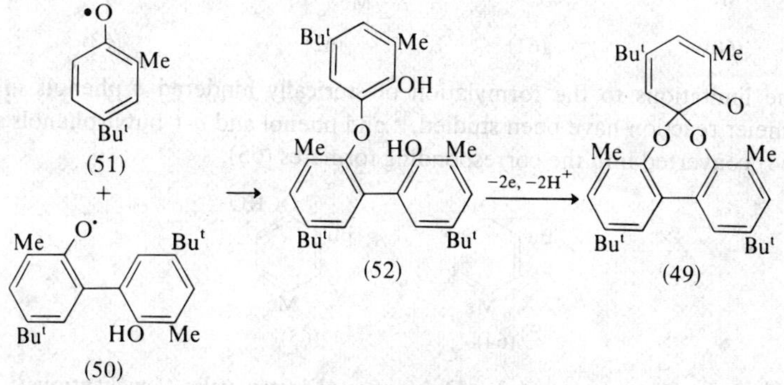

**Scheme 7**

Diphenylseleninic anhydride, $(PhSeO)_2O$, has been used for the oxidation of phenols. The relatively new phenylselenimine (54) has been formed[28] from (53) in the presence of hexamethyldisilazane, while the phenolate ion derived from 2,6-xylenol (55) resulted[29] in the o-hydroxydienone dimer (56).

Base-catalysed oxygenation of the phenol (57) followed by Schotten–Baumann acylation yields the p-peroxy-quinol esters (58), which undergo a novel base-catalysed rearrangement[30] with $KOBu^t$ to give (59).

[26] R. A. Holton and R. G. Davis, *Tetrahedron Letters*, 1977, 533.
[27] D. T. Dalgleish, N. P. Forrest, D. C. Nonhebel, and P. L. Pauson, *J.C.S. Perkin I*, 1977, 584.
[28] D. H. R. Barton, A. G. Brewster, S. V. Ley, and M. N. Rosenfeld, *J.C.S. Chem. Comm.*, 1977, 147.
[29] D. H. R. Barton, S. V. Ley, P. D. Magnus, and M. N. Rosenfeld, *J.C.S. Perkin I*, 1977, 567.
[30] A. Nishinaga, K. Nakamura, K. Yoshida, and T. Matsuura, *Chem. Letters*, 1977, 303.

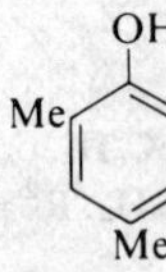

OH
Me
Me
(53)

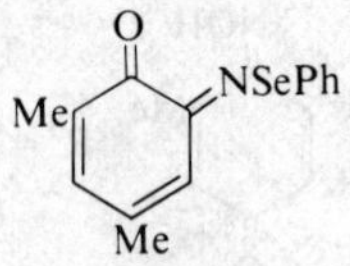

O
Me       NSePh
Me
(54)

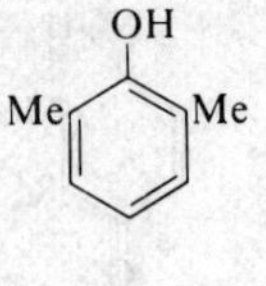

OH
Me       Me
(55)

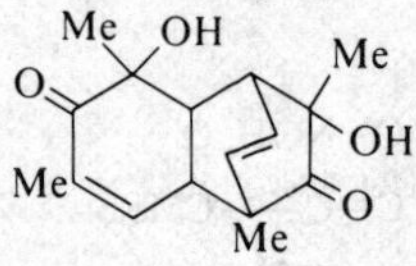

Me   OH        Me
O                  OH
Me                 O
      Me
(56)

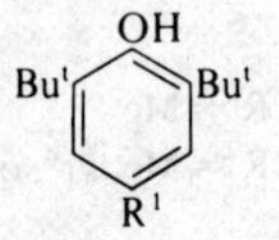

OH
Bu$^t$      Bu$^t$
R$^1$
(57)

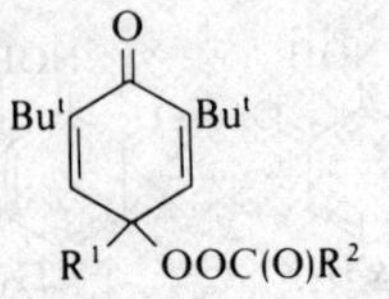

O
Bu$^t$      Bu$^t$
R$^1$   OOC(O)R$^2$
(58)

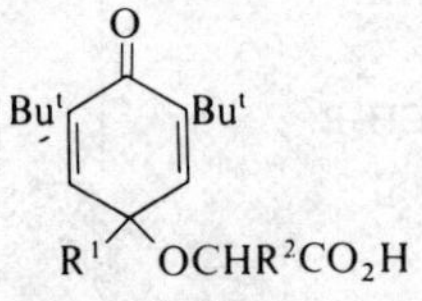

O
Bu$^t$      Bu$^t$
R$^1$   OCHR$^2$CO$_2$H
(59)

Studies on the bromination[31] of 3,4-dimethyl-phenols have revealed abnormal reaction paths. The compounds (60) and (61) yield the tribromo-dienone (62), which rearranges autocatalytically (and possibly homocatalytically) to 3-bromomethyl-2,6-dibromo-4-methylphenol (63).

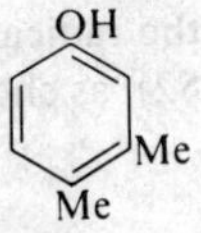

OH
Me
Me
(60)

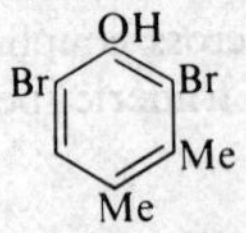

OH
Br       Br
Me
Me
(61)

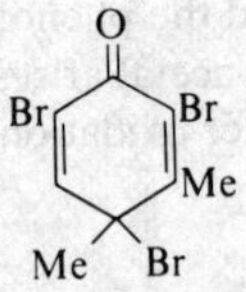

O
Br       Br
Me
Me   Br
(62)

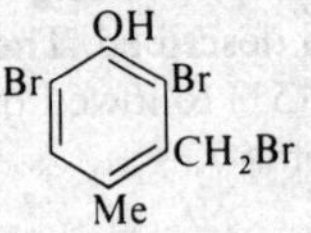

OH
Br       Br
CH$_2$Br
Me
(63)

The limitations to the formylation of sterically hindered *o*-phenols in the Vilsmeier reaction have been studied,[32] and phenol and *o*-t-butyl-phenols such as (64) converted into the corresponding formates (65).

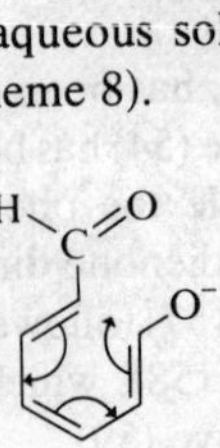

OH
Bu$^t$
Me
(64)

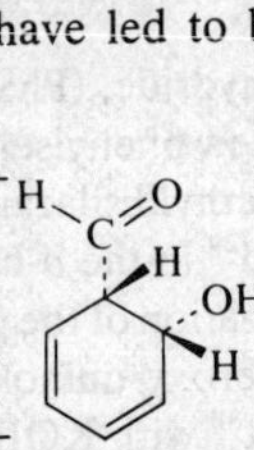

OCHO
Bu$^t$
Me
(65)

**Aldehydes, Ketones, and Quinones.**—Electrocyclic cyclisations[33] of heptamethines (66) in aqueous solution have led to benzaldehyde and other aromatic aldehydes (Scheme 8).

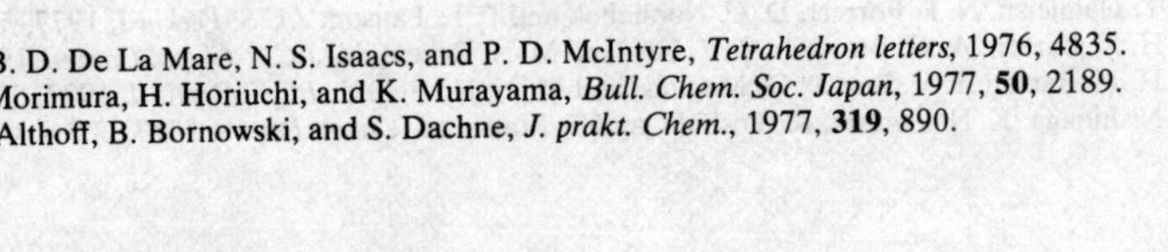

H   O$^-$
  C
O
(66)

H   O
  C
    O$^-$

H$^+$

H   O
  C
H
    OH
H

$-H_2O$

H   O
  C

**Scheme 8**

[31] P. B. D. De La Mare, N. S. Isaacs, and P. D. McIntyre, *Tetrahedron letters*, 1976, 4835.
[32] S. Morimura, H. Horiuchi, and K. Murayama, *Bull. Chem. Soc. Japan*, 1977, **50**, 2189.
[33] H. Althoff, B. Bornowski, and S. Dachne, *J. prakt. Chem.*, 1977, **319**, 890.

The preparation of [*formyl*-$^2$H]aldehydes such as (67) is not readily achieved by simple exchange reactions, but exchange has been accomplished[34] with several aromatic aldehydes, using benzoin-type conditions (Scheme 9).

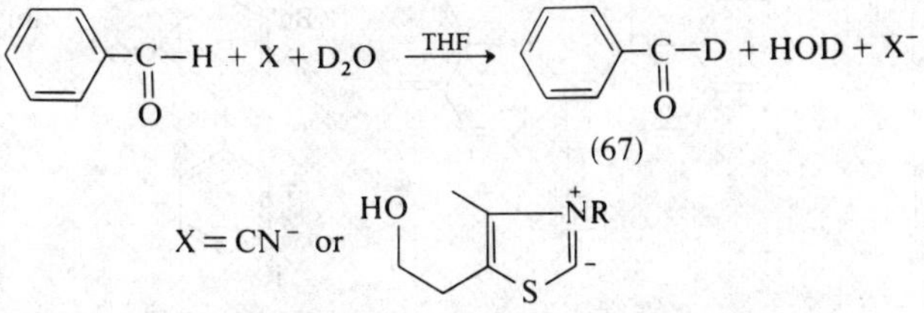

(67)

$$X = CN^- \text{ or}$$

(R = Et or PhCH$_2$)

**Scheme 9**

Metal oxides have been employed[35] to catalyse the acylation of aromatics, leading to dimethylbenzophenones and 4-substituted benzophenones. The reaction of t-butylbenzene with benzoyl chloride in the presence of small amounts of ferric oxide leads to 4-t-butylbenzophenone. Substituted benzophenones have also been obtained[36] by hydrolysis of the dilithium salts (68) formed by the reaction of lithium carboxylates with organolithium derivatives, and benzophenone-2'-methoxy-2-sulphinic acids (69) have been successfully prepared[37] by displacement of the sulphone linkage in thioxanthen-9-one 10,10-dioxides (70)

$$ArCO_2Li + PhLi \rightarrow PhCAr(O^- Li^+)_2 \xrightarrow{H^+} PhCOAr$$

(68)

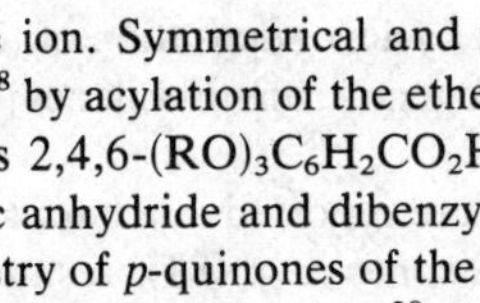

(69) 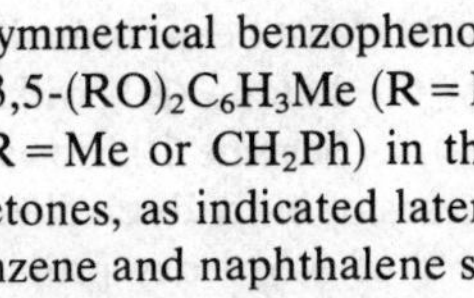                    (70)

by methoxide ion. Symmetrical and unsymmetrical benzophenones have also been formed[38] by acylation of the ethers 3,5-(RO)$_2$C$_6$H$_3$Me (R = Me or PhCH$_2$) with the acids 2,4,6-(RO)$_3$C$_6$H$_2$CO$_2$H (R = Me or CH$_2$Ph) in the presence of trifluoroacetic anhydride and dibenzyl ketones, as indicated later (see ref. 58).

The chemistry of *p*-quinones of the benzene and naphthalene series has been the subject of a recent review.[39] A synthesis of *o*-quinones has also been reported[40] in which the phenols (71) undergo oxidation with elemental oxygen in

[34] T. Chancellor, M. Quill, D. E. Bergbreiter, and N. Newcomb, *J. Org. Chem.*, 1978, **43**, 1245.

[35] J. O. Morley, *J.C.S. Perkin II*, 1977, 601.

[36] P. Hodge, G. M. Perry, and P. Yates, *J.C.S. Perkin I*, 1977, 680.

[37] O. F. Bennett, G. Saluti, and F. X. Quinn, *Synth. Comm.*, 1977, **7**, 33.

[38] E. G. Sundholm, *Tetrahedron*, 1977, **33**, 991.

[39] 'Methods in Organic Chemistry' (Houben-Weyl) Vol. 7/3a, ed. E. Mueller, George Thieme, Stuttgart, 1977.

[40] A. Nishinaga, K. Nishizawa, H. Tomita, and T. Matsuura, *Synthesis*, 1977, 270.

the presence of the bis-(3-salicylideneaminopropyl)aminecobalt(II) complex (72)
to hydroperoxides (73), which decompose to the *o*-benzoquinones (74).

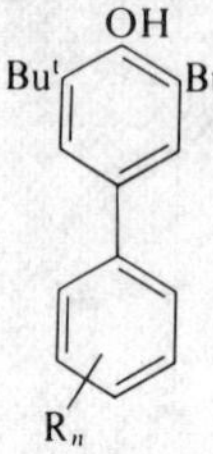

(71) 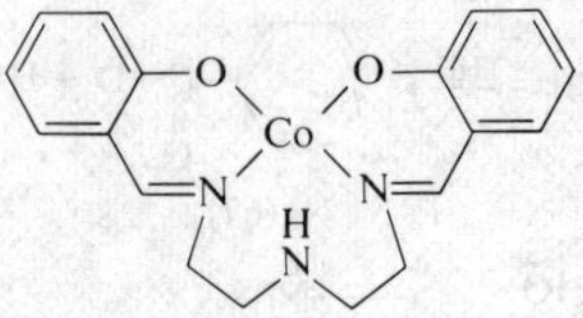                          (72) 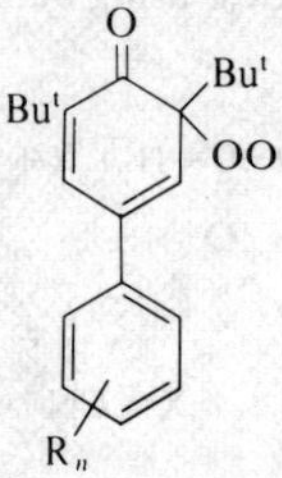                     (73) 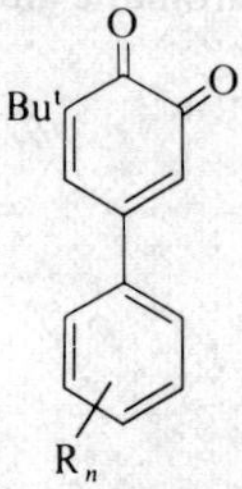              (74)

Reactions of aldehydes and their derivatives have been the subject of several
reports.   Benzene-1,2-dicarboxaldehyde    condenses[41]   with    2,3-dihydro-
naphthazarone to give the *p*-quinone (75), and a similar reaction occurs
between diketones such as $PhCOCH_2CH_2COPh$ and naphthalene-1,4-diol. 4-
Nitrobenzaldehyde undergoes disproportionation[42] in the presence of cyanide
ions and methanol, resulting in methyl 4-nitrobenzoate as the main product. The
acid-catalysed rearrangement[43] of arylhydrazones (76) derived from aromatic
aldehydes and diaryl ketones leads to amino-biphenyls (77).

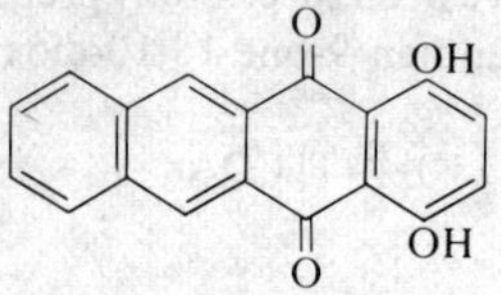

(75) 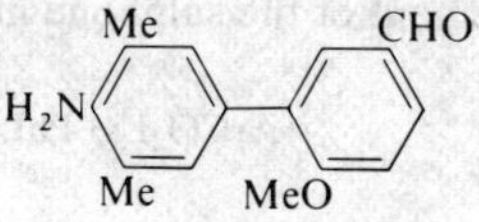                                      (77)

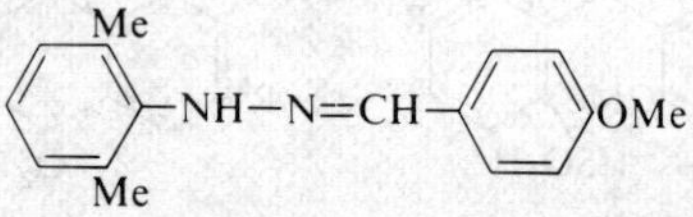

(76)

Terphenyls (78) have been formed[44] by cyclocondensation of (79) [formed by
the reaction of the ketone (80) with halogeno-trimethylsilanes] with a second
molecule of (80) as Scheme 10 shows.

Other syntheses of bi- and ter-phenyls are reported in a later section.

Chlorination[45] of *o*-hydroxy-propiophenones, *e.g.* (81), results in the cyclo-
hexadienones (82), which are hydrolysed to the expected chloro-compound (83).

Photolysis[46] of 2,5-dimethyl α-chloroacetophenone (84) yields 6-methylindan-
1-one (85).

[41]  B. Serpaud and Y. Lepage, *Bull. Soc. chim. France*, 1977, 539.
[42]  J. C. Trisler and S. K. McKinney, *Tetrahedron Letters*, 1977, 3125.
[43]  R. Fusco and F. Sannicolo, *Tetrahedron Letters*, 1977, 3163.
[44]  A. H. Schmidt and M. Russ, *Chem. Ztg.*, 1978, **102**, 26.
[45]  A. Ahmad and G. Hussain, *Pakistan J. Sci. Ind. Res.*, 1976, **19**, 101.
[46]  W. R. Bergmark, *J.C.S. Chem. Comm.*, 1978, 61.

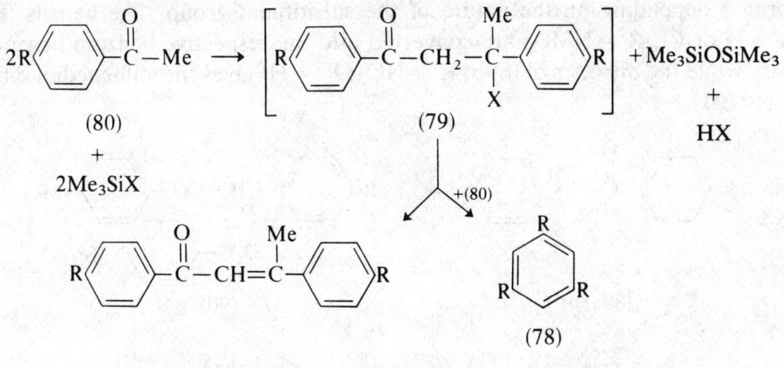

(80)
+
2Me₃SiX

R = Ph, p-MeC₆H₄, p-ClC₆H₄, or p-FC₆H₄

**Scheme 10**

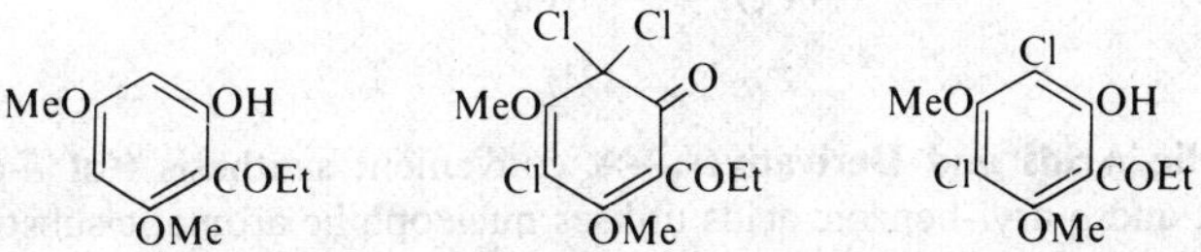

(81)     (82)     (83)

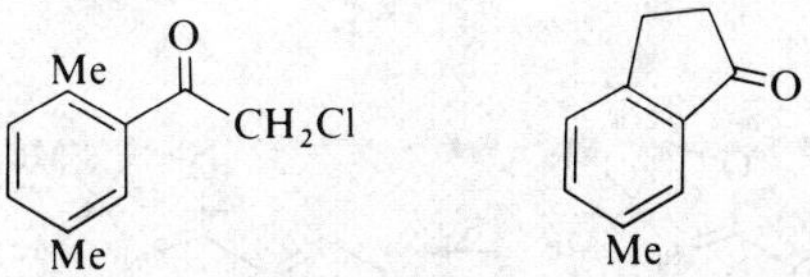

(84)     (85)

Studies[47] on the thermal decomposition of 2-azidosulphonyl-benzophenone (86) showed that the cyclization product (87), a rearrangement product (88), and orthanilic acid are formed.

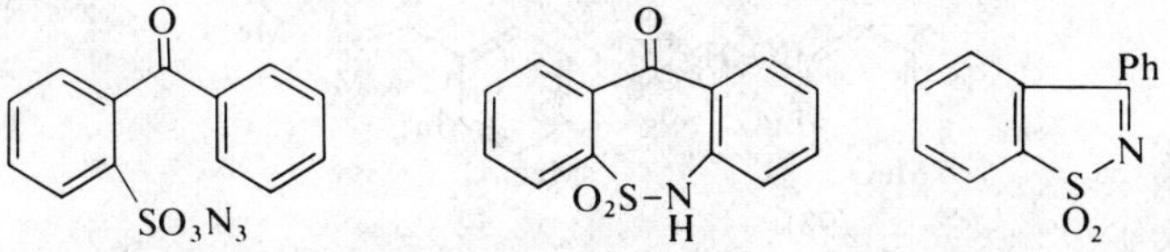

(86)     (87)     (88)

Symmetrical diketones RC(O)C(O)R (R = Ph, 4-MeC₆H₄, 4-MeOC₆H₄, or furyl) have been obtained[48] in good yield by oxidation of the corresponding acyloins with alkaline $K_3Fe(CN)_6$, and a detailed investigation[49] has been carried out on the reaction of unsymmetrical benzils with CN⁻ in DMSO. The latter was initiated in an attempt to clarify some ambiguities in the literature, the products

[47] R. A. Abramovitch and D. P. Vanderpool, *J.C.S. Chem. Comm.*, 1977, 18.
[48] R. A. El-Zaru and A. A. Jarrar, *Chem and Ind.*, 1977, 741.
[49] A. Kawasaki and Y. Ogata, *J. Org. Chem.*, 1977, **42**, 2506.

formed depending on the nature of the substituent group. The benzils (89a; $R^1$ = H or Cl; $R^2$ = NMe$_2$) are converted into the respective benzoin benzoates (90), while the nitrobenzil (89b; $R^1$ = NO$_2$; $R^2$ = H) gives the stilbenediol dibenzoate (91).

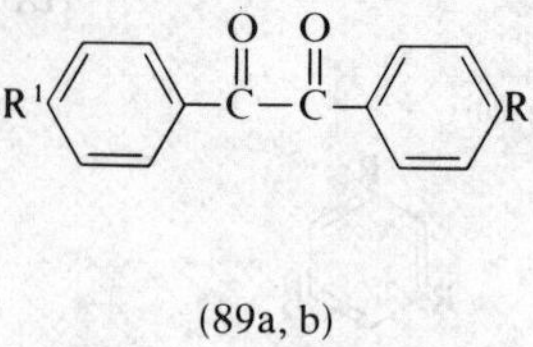

(89a, b) 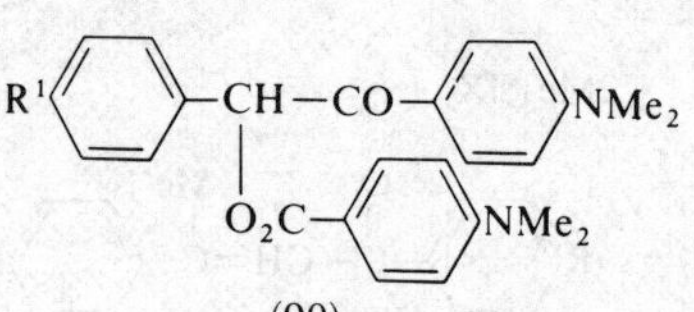                                                    (90)

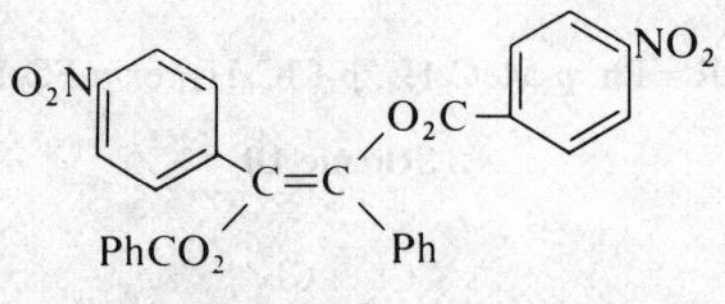

(91)

**Carboxylic Acids and Derivatives.**—A convenient synthesis[50] of *o*-alkyl-, *o*-alkenyl-, and *o*-aryl-benzoic acids utilises nucleophilic aromatic substitution of *o*-methoxyaryl-oxazolines (Scheme 11). The oxazolines (92) are converted into the acids (93) by organometallic replacement of the methoxy-group, and

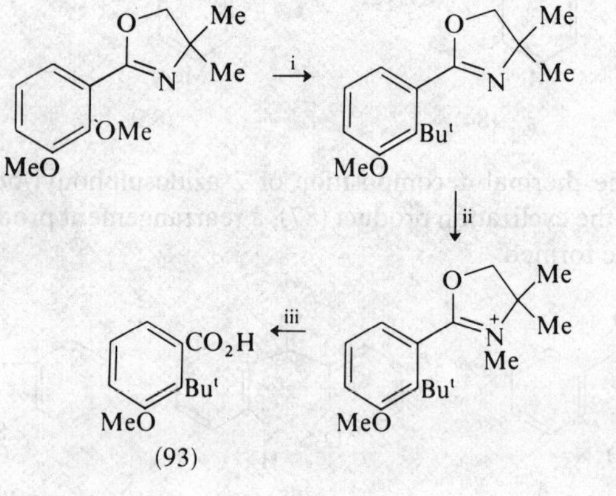

(93)

Reagents: i, Bu$^t$Li or Bu$^t$MgBr; ii, MeI; iii OH$^-$

**Scheme 11**

several ring-deuteriated anthranilates, *e.g.* (94) and (95), have been formed by photolysis of suitably deuteriated methyl 2-pyridylacetates. The labelling patterns have been interpreted[51] in terms of a biradical intermediate involving

[50] A. I. Meyers, R. Gabel, and E. D. Mihelich, *J. Org. Chem.*, 1978, **43**, 1372.
[51] K. Tagaki and Y. Ogata, *J.C.S. Perkin II*, 1977, 1980.

scission of the N-1—C-6 bond of the pyridine ring and bond formation between C-4 or C-6 and C-7 (Scheme 12); see also p. 150.

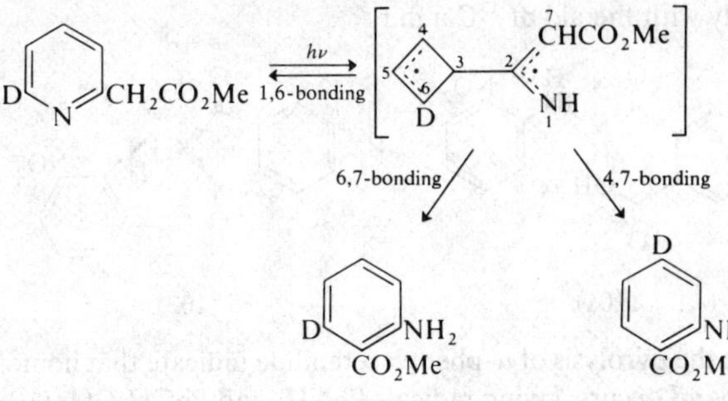

**Scheme 12**

Phenoxides react with nitro, fluoro-, or chloro-substituted phthalic anhydrides (96) to give substituted phthalic anhydrides (97). The success of the reaction[52] is dependent on the conditions and the leaving group, and, despite the formation of by-products, can lead to product yields of up to 87% when the leaving group is F.

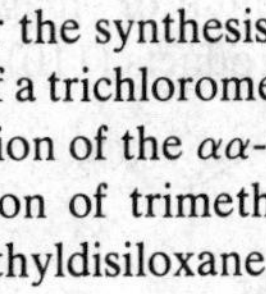

A mild method[53] for the synthesis of acid chlorides has been described which involves the reaction of a trichloromethyl-arene (98) with an organosilicon oxide, resulting in the formation of the $\alpha\alpha$-dichlorobenzyloxytrimethylsilane (99). This undergoes $\beta$-elimination of trimethylchlorosilane (Scheme 13), which can be hydrolysed to hexamethyldisiloxane and re-used.

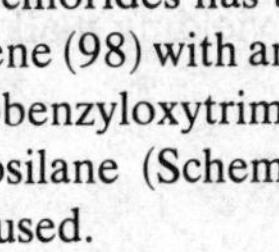

$$XC_6H_4CCl_3 + (Me_3Si)_3O \rightarrow [XC_6H_4CCl_2OSiMe_3] \xrightarrow{\beta\text{-elimination}} XC_6H_4COCl + Me_3SiCl$$

(98) X = Hal          (99)

**Scheme 13**

An improvement in the synthesis of phenacyl esters,[54] avoiding the frequently accompanying hydrolysis, utilises the reaction of phenacyl bromide with carboxylic acids in the presence of KF at ambient temperature. Alkyl pyrocarbonates such as (100) may be prepared[55] from esters of chloroformic acid (Scheme 14).

$$PhO_2CCl + NaOCO_2Me \rightarrow PhO_2C-O-CO_2Me$$

(100)

**Scheme 14**

[52] F. J. Williams, H. M. Relles, P. E. Donahue and J. S. Manello, *J. Org. Chem.*, 1977, **42**, 3425.
[53] T. Nakano, K. Ohkawa, H. Matsumoto, and Y. Magai, *J.C.S. Chem. Comm.*, 1977, 808.
[54] J. H. Clark and J. M. Miller, *Tetrahedron Letters*, 1977, 599.
[55] B. Razden and H. Schlude, *Chem. Ber.*, 1978, **111**, 803.

The exclusive product of cyclization of a 1,2-disubstituted acyl hydrazide (101) is the *N*-substituted phthalimide (102), the structure of which was assigned unambiguously with the aid of $^{13}$C n.m.r.[56]

(101)          (102)

Studies[57] on the pyrolysis of α-phenylacetanilide indicate that homolysis of the amide C—N bond occurs, giving radicals PhNH· and PhCH$_2$CO· (103). Loss of CO occurs from (103), and the products that were isolated are *o*- and *p*-amino-diphenylmethanes (104). Similar pyrolysis in the presence of *o*-toluidine or isoquinoline results in products such as (105).

(104)          (105)

**Nitro- and Nitroso-derivatives and Hydroxylamines.**—The reduction of β-nitro-styrenes (106) to the nitrophenylethanes (107) and subsequent base-catalysed condensation with aromatic aldehydes Ar'CHO gives rise[58] to the substituted nitro-styrene (108), which on further reduction leads to the dibenzyl ketone (109). Nitrobenzene reacts with organo-lithium or -magnesium derivatives, as shown in Scheme 15, yielding *p*-alkylated products (110).[59]

(106)          (107)          (108)          (109)

A number of novel methods have been described for the reduction of nitro- to amino-groups. A search[60] for more active H donors for transfer hydrogenation resulted in the application of formic, phosphinic, and phosphorous acids, or their salts, in the presence of catalysts such as Pd, Pt, or Rh. In this way *p*-fluoronitro-benzene is reduced to *p*-fluoroaniline in high yield, while nitro-compounds possessing halogen other than fluorine are not affected. Unusual reductions[61] of nitro-compounds, azo-compounds, and related systems include the highly specific simple distillation of the compound in liquid paraffin. This process, which can be used in the presence of carbonyl, ester, or nitrile groups, and for the reduction of

[56] G. E. Martin and J. W. Munson, *J. Heterocyclic Chem.*, 1977, **14**, 349.
[57] M. Z. A. Badr, M. M. Aly, and S. S. Salem, *Tetrahedron*, 1977, **33**, 3155.
[58] E. McDonald and R. T. Martin, *Tetrahedron Letters*, 1977, 1317.
[59] F. Kienzle, *Helv. Chim. Acta*, 1978, **61**, 449.
[60] I. D. Entwistle, A. E. Jackson, R. A. W. Johnstone, and R. P. Telford, *J.C.S. Perkin I*, 1977, 443.
[61] L. Bin Din, J. M. Lindley, and O. Meth-Cohn, *Synthesis*, 1978, 23.

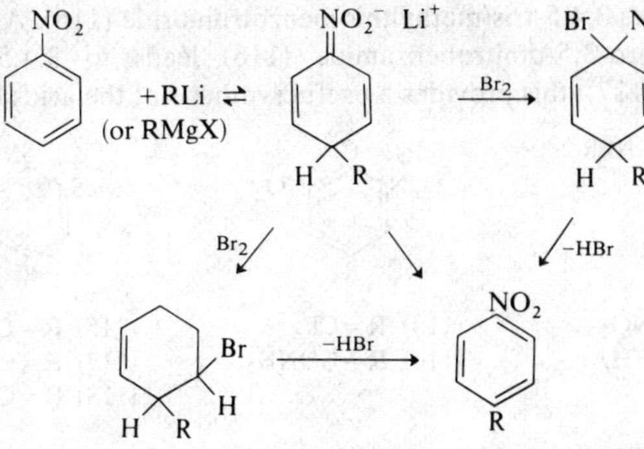

(110) R = Me or Bu$^n$

**Scheme 15**

nitro-heterocycles without reducing the hetero-ring, proceeds by way of a concerted hydrogen transfer, illustrated in Scheme 16 for the azo-reduction.

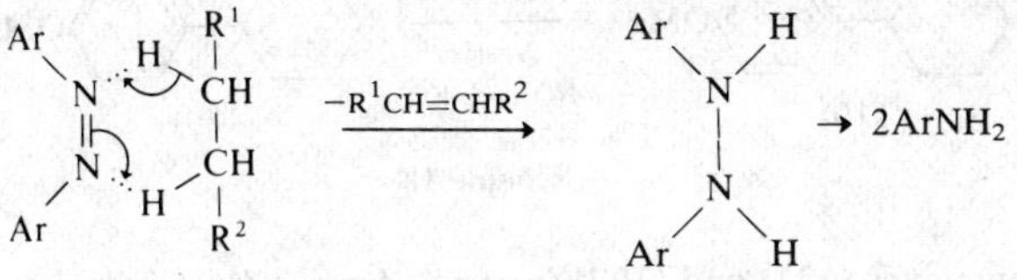

**Scheme 16**

Amines have also been formed[62] from nitro-compounds by the use of an alkaline solution of thiourea dioxide (111); see Scheme 17.

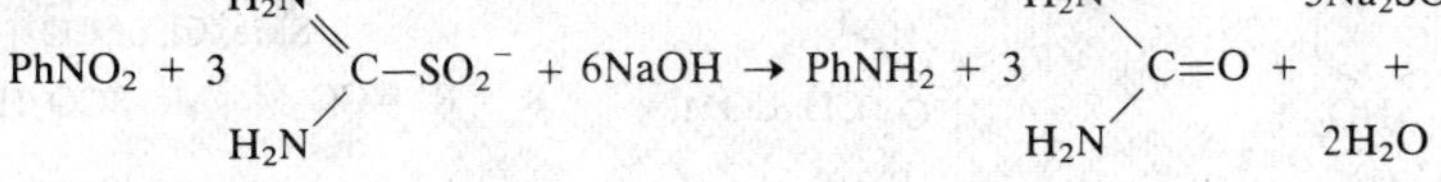

$$PhNO_2 + 3 \quad \underset{H_2N}{\overset{H_2N^+}{\diagdown}} C-SO_2^- + 6NaOH \rightarrow PhNH_2 + 3 \quad \underset{H_2N}{\overset{H_2N}{\diagdown}} C=O + \quad \underset{+}{\overset{3Na_2SO_3}{}} + 2H_2O$$

(111)

**Scheme 17**

The direct conversion of nitro-arenes into anilides has recently been brought about, making use of the reducing ability of low-valent molybdenum.[63] Molybdenum(II) species are conveniently prepared by heating $Mo(CO)_6$ with organic acids, and their mild nature makes possible the reduction of nitro-groups in compounds containing an alkene, such as (112), to the corresponding anilide, *i.e.* (113), in the presence of ethanoic acid.

Displacement of a nitro-group by the methanethiolate anion has been investigated. Treatment[64] of 4-chloro-3,5-dinitrobenzotrifluoride (114) with

[62] K. Nakagawa, S. Mineo, S. Kawamura, and K. Minami, *Yakugaku Zasshi*, 1977, **97**, 1253.
[63] Tse-Lok Ho, *J. Org. Chem.*, 1977, **42**, 3755.
[64] J. R. Beck and J. A. Yahner, *J. Org. Chem.*, 1978, **43**, 2048.

methanethiol gives 3,4,5-tris(methylthio)benzotrifluoride (115). A similar reaction[65] of 4-chloro-3,5-dinitrobenzamide (116) leads to 3,4,5-tris(methylthio)benzamide (117); this provides a useful synthesis of the acids (118).

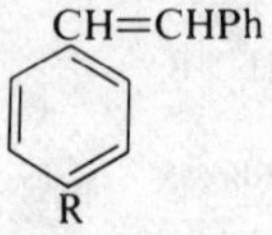

(112) R = NO₂      (114) R = CF₃      (115) R = CF₃
(113) R = NHAc      (116) R = CONH₂      (117) R = CONH₂
                                               (118) R = CO₂H

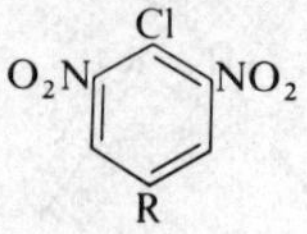

Photo-excited aromatic nitro-compounds have been used in the selective hydroxylation of electron-rich aromatic substituents at the benzylic position.[66] Scheme 18 shows how, in this way, *p*-cyclohexylanisole (119) is converted into (120).

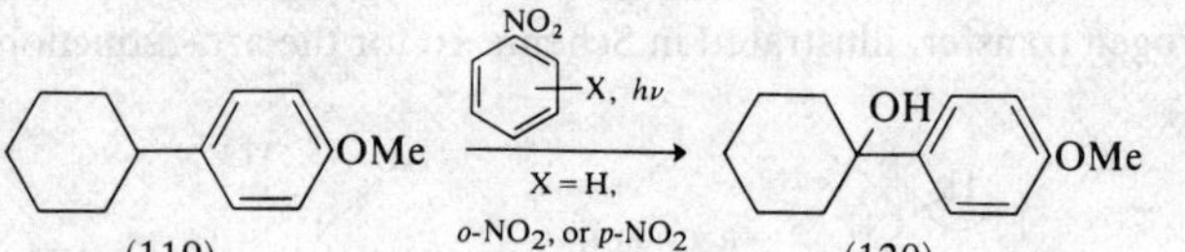

**Scheme 18**

The σ-complexes (121) and (122) formed[67] from 2,4,6-trinitrobenzenes with potassium methoxide and acetone respectively have been studied, together with those formed[68] from 2,4- and 3,5-dinitrobenzenes with acetone, *i.e.* (123)—(126).

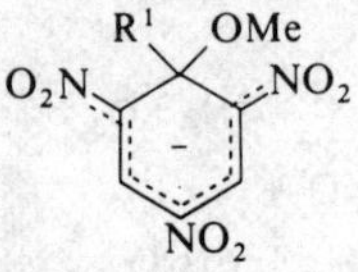

$R^1 = H$; $R^2 = $ Me, OMe, $CO_2Me$,
                SMe, Cl, or $CO_2H$

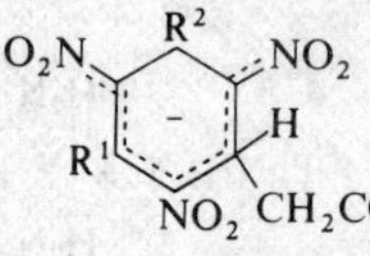

$R^1 = R^2 = CO_2Me$, Me, $SCO_2H$,
            or $CO_2^{-}$

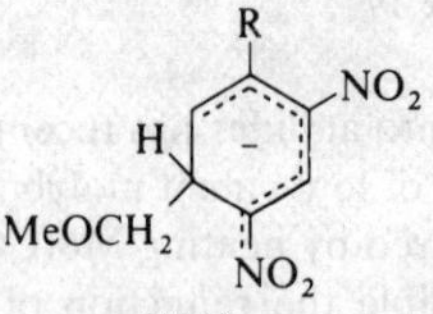

(121)          (122)

(123) 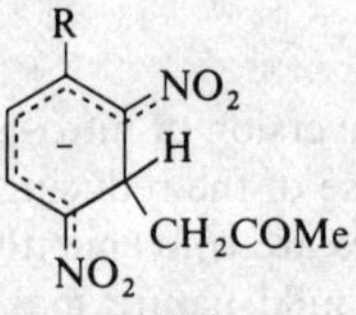          (124)

R = OMe, $CO_2Me$, or SCN

[65] J. R. Beck and J. A. Yahner, *J. Org. Chem.*, 1978, **43**, 2052.
[66] J. Libman, *J.C.S. Chem. Comm.*, 1977, 868.
[67] A. Ya. Kaminskii, S. S. Gitis, L. I. Khabarova, V. Sh. Gollibehik, and E. G. Kaminskaya, *Tezisy Vses. Simp. Org. Sint. Benzoidnye Aromat. Soedin, 1st*, 1974, 12 (*Chem. Abs.*, 1977, **87**, 39 042).
[68] A. Ya. Kaminskii, S. S. Gitis, E. A. Bronshtein, and E. E. Gol'teuzen, *Tezisy Vses. Simp. Org. Sint. Benzoidnye Aromat. Soedin. 1st*, 1974, 14 (*Chem. Abs.*, 1977, **87**, 39 054).

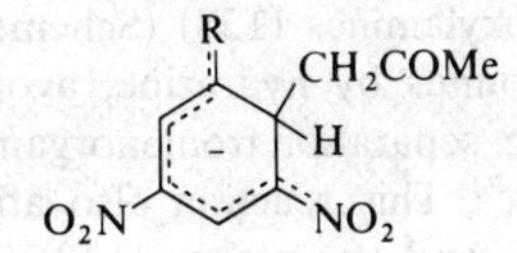

(125) 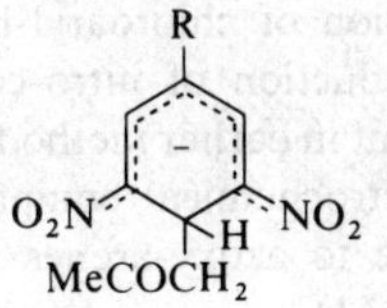    (126)

R = Me, OMe, CO$_2$Et, or CF$_3$

The deoxygenation of certain sterically hindered aromatic compounds has been studied, contrasting results being obtained.[69] 2,5-Di-t-butylnitrosobenzene with TEP gives a complex mixture containing 2,5-di-t-butylaniline (127), 2,5,2',5'-tetra-t-butyl-azobenzene (128), *N*-(2,5-di-t-butylphenyl)-α-t-butyl-α-2-(5-t-butylpyridyl) nitrone (129), and diethyl *N*-(2,5-di-t-butylphenyl)-phosphoramidate (130), while the similar reaction of 2,4,6-tri-t-butylnitrosobenzene yielded mainly 3,3-dimethyl-5,7-di-t-butyl-dihydroindole (131) together with small amounts of 2,4,6-tri-t-butylaniline (132) and 2-(2-methylallyl)-4,6-di-t-butylaniline (133).

ArNH$_2$    ArN=NAr    ArN=C    ArN—P(OEt)$_2$    Ar =

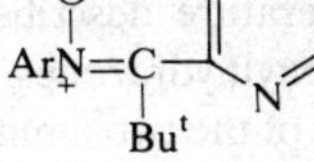

(127)    (128)    (129)    (130)

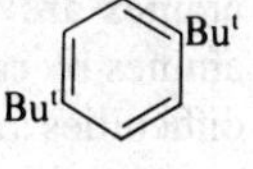

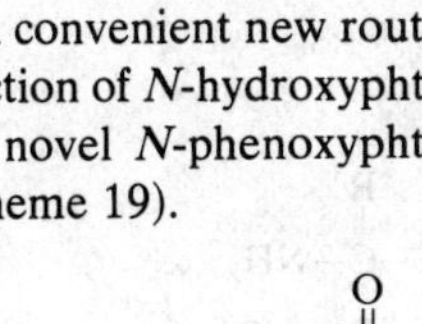

(131) 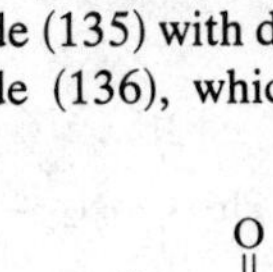    (132) 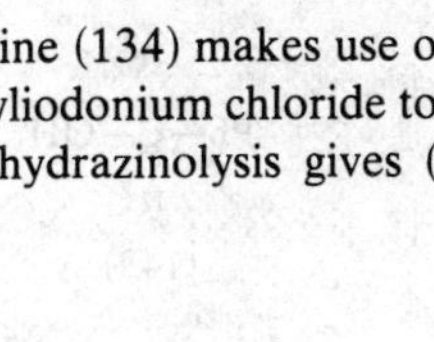    (133)

A convenient new route[70] to *O*-phenylhydroxylamine (134) makes use of the reaction of *N*-hydroxyphthalimide (135) with diphenyliodonium chloride to give the novel *N*-phenoxyphthalimide (136), which on hydrazinolysis gives (134) (Scheme 19).

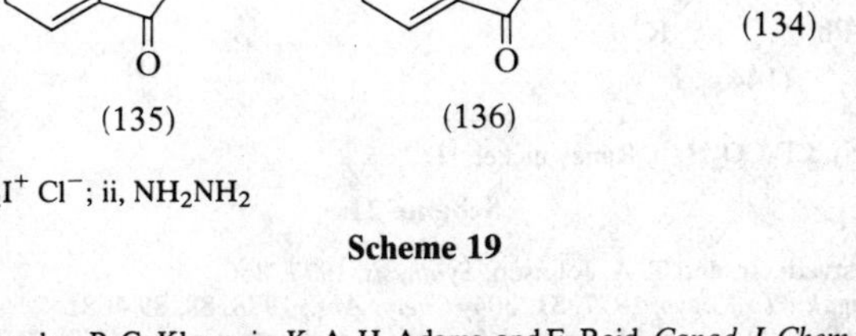

(135)    (136)    (134)

Reagents: i, Ph$_2$I$^+$ Cl$^-$; ii, NH$_2$NH$_2$

**Scheme 19**

[69] L. Ross, C. Barclay, P. G. Khazanie, K. A. H. Adams, and E. Reid, *Canad. J. Chem.*, 1977, **55**, 3273.
[70] J. I. G. Cadogan and A. G. Rowley, *Synth. Comm.*, 1977, **7**, 365.

The preparation of chloroaryl-hydroxylamines (138) (Scheme 20), using a Pd-catalysed reduction of nitro-compounds by hydrazine, avoids the disadvantages inherent in earlier methods, *viz.* separation from inorganic materials or the use of electrochemical equipment.[71] This reaction also affords a useful high-yield route to azoxy-arenes (139) and azo-arenes (140) by way of the nitroso-arene (141).

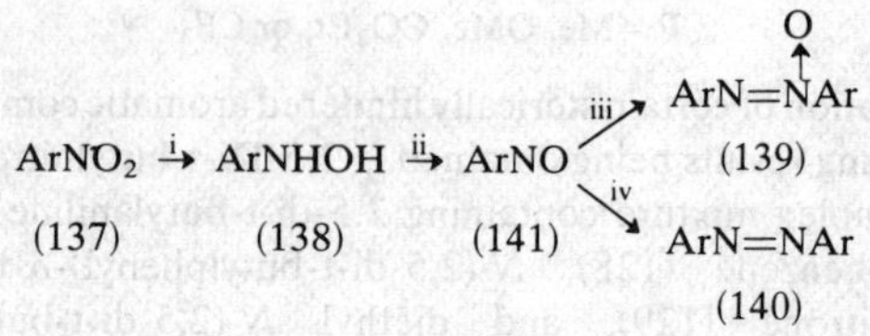

$$
\begin{array}{c}
& & & \overset{O}{\underset{\uparrow}{}} \\
& & \xrightarrow{\text{iii}} & \text{ArN=NAr} \\
& & & (139) \\
\text{ArNO}_2 \xrightarrow{\text{i}} \text{ArNHOH} \xrightarrow{\text{ii}} \text{ArNO} & & \\
(137) & (138) & (141) & \xrightarrow{\text{iv}} \text{ArN=NAr} \\
& & & (140)
\end{array}
$$

Reagents: i, $NH_2NH_2$, Pd/C, THF, EtOH; ii, $I_2$ or $FeCl_3$; iii, ArNHOH; iv, $ArNH_2$

**Scheme 20**

**Amines and Related Compounds.**—Recent advances in the amination of aromatic compounds have been reviewed,[72] but certain specific syntheses of aromatic amines are worthy of mention. The literature describes attempts to prepare amines by catalytic hydrogenation of phenylhydrazones. This has suffered from difficulties arising from self-condensation of the two amines formed, but a recent report[73] describes how this can be overcome by performing the reactions in the presence of aqueous hydrochloric acid and a suitable solvent, resulting in distribution of the hydrogenolysis products between two phases. In this way *m*-aminobenzylamine has been obtained from the phenylhydrazone of *m*-nitrobenzaldehyde in high yield.

An improved synthesis (Scheme 21) of $\alpha\alpha$-disubstituted benzylamines (142) has also been reported[74] which utilises the reaction of tertiary alcohols (143) or alkenes (144) with sodium azide followed by reduction with Raney nickel.

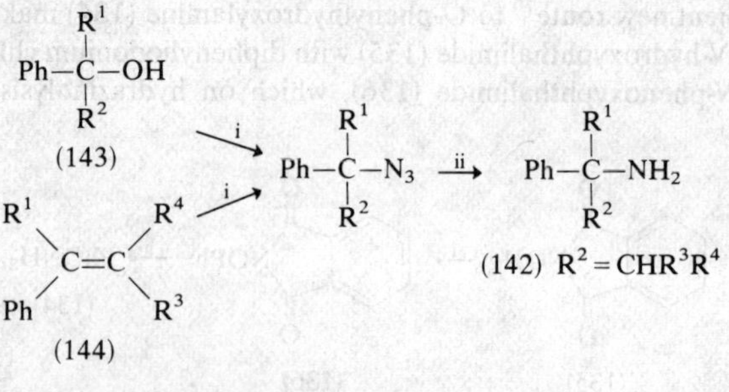

$$
\begin{array}{c}
\overset{R^1}{\underset{|}{}} \\
\text{Ph—C—OH} \\
\overset{|}{R^2} \\
(143)
\end{array}
\xrightarrow{\text{i}}
\begin{array}{c}
\overset{R^1}{\underset{|}{}} \\
\text{Ph—C—N}_3 \\
\overset{|}{R^2}
\end{array}
\xrightarrow{\text{ii}}
\begin{array}{c}
\overset{R^1}{\underset{|}{}} \\
\text{Ph—C—NH}_2 \\
\overset{|}{R^2} \\
(142) \quad R^2 = CHR^3R^4
\end{array}
$$

$$
\begin{array}{c}
\overset{R^1}{\underset{Ph}{}}\text{C=C}\overset{R^4}{\underset{R^3}{}} \\
(144)
\end{array}
\xrightarrow{\text{i}}
$$

Reagents: i, $NaN_3$, $CF_3CO_2H$; ii, Raney nickel, $H_2$

**Scheme 21**

[71] C. S. Rondestvedt, Jr. and T. A. Johnson, *Synthesis*, 1977, 850.
[72] K. Nara, *Kagaku To Kogyo*, 1977, **51**, 304 (*Chem. Abs.*, 1978, **88**, 89 408).
[73] A. A. Siddiqui, N. H. Khan, M. Ali, and A. R. Kidwai, *Synth. Comm.*, 1977, **7**, 71.
[74] D. Balderman and A. Kalir, *Synthesis*, 1978, 24.

A new route to nucleophilically substituted *o*-phenylenediamines (145), which are potentially powerful in heterocyclic synthesis, has been introduced.[75] This involves the reaction of isobenzimidazole-2-spirocyclohexane (146) (obtained in high yield from *o*-phenylenediamine and cyclohexanone) with secondary amines in ethanol to give the monosubstituted isobenzimidazole (147), or in DMSO or sulpholan to give the disubstituted compound (148), as shown in Scheme 22. The reaction is essentially nucleophilic addition followed by oxidation of the resultant dihydro-isobenzimidazole (149) by an unchanged molecule of (146).

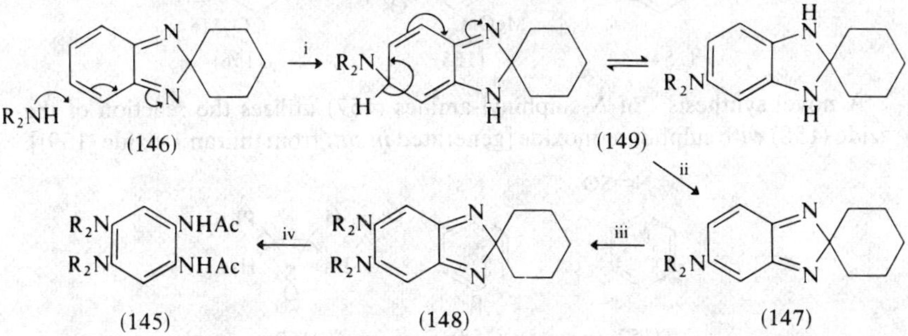

Reagents: i, EtOH, room temperature; ii, [O]; iii, piperidine, DMSO, 100 °C; iv, Pd/C, Ac$_2$O, AcOH

**Scheme 22**

A simple synthesis of substituted cyano-anilines has been described.[76] Malononitrile reacts with C$_{4-6}$ cycloalkanones and aldehydes RCHO to give (150), which with alkali is converted into (151).

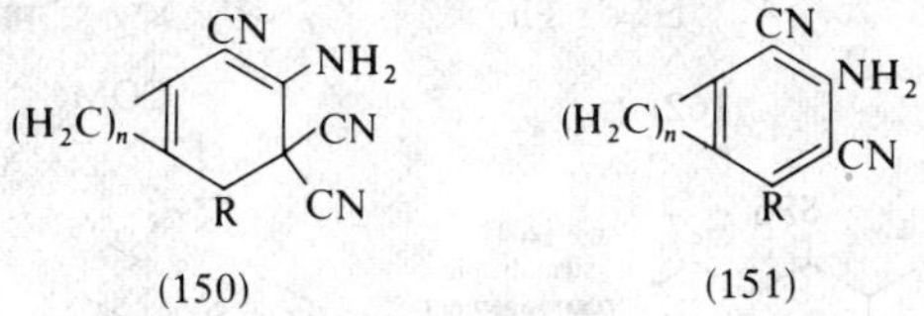

R = Bu$^n$, indolyl, or R'C$_6$H$_4$; *n* = 3, 4, or 5

Tertiary aromatic amines (152) have been formed[77] by trimerisation of ynamines (153) in the presence of transition-metal complexes (Scheme 23), the actual product depending on the nature of the catalyst.

*NN*-Diethyl-2,6-dimethylaniline (154), a highly hindered tertiary amine, has been considered.[87] This is illustrated by the decomposition o *N*-nitrosoacetanilide (169) and benzenediazonium acetate by two competing routes to give converted into (154) in four steps.

[75] A. M. Jefferson and H. Suschitzky, *J.C.S. Chem. Comm.*, 1977, 189.
[76] Yu. A. Sharinin, V. P. Marshtupa, and L. G. Sharanina, *Tezisy Vses. Simp. Org. Sint. Benzoidnye Aromat. Soedin. 1st*, 1974, 70 (*Chem. Abs.*, 1977, **87**, 514).
[77] J. Ficini, J. D'Angelo, and S. Falou, *Tetrahedron Letters*, 1977, 1645.
[78] T. A. Bryson and D. M. Donelson, *J. Org. Chem.*, 1977, **42**, 2930.

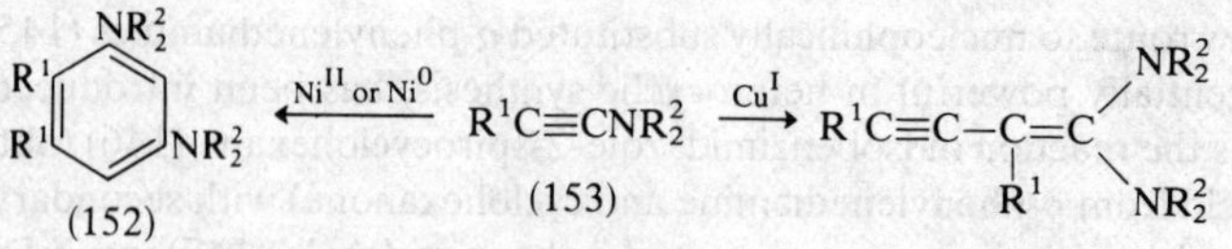


(152) (153)

**Scheme 23**

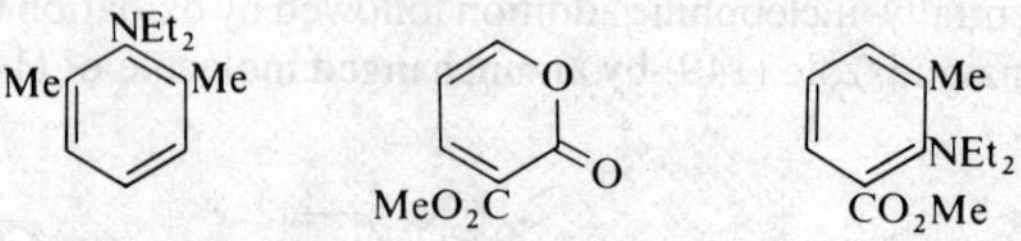


(154) (155) (156)

A novel synthesis[79] of *N*-sulphinyl-amines (157) utilizes the reaction of the azide (158) with sulphur monoxide [generated *in situ* from thiiran 1-oxide (159)].

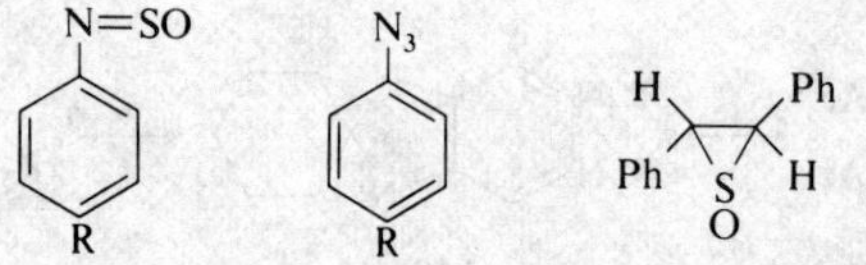


(157) (158) (159)

Alkylation of amines has received attention from several groups. Some 2-(1-thiophenoxybutyl)-acetanilides (160) have been obtained[80] by [2,3]-sigmatropic rearrangement of azasulphonium ylides (161), formed from the sulphimine (162) by reaction with acetic anhydride (Scheme 24).

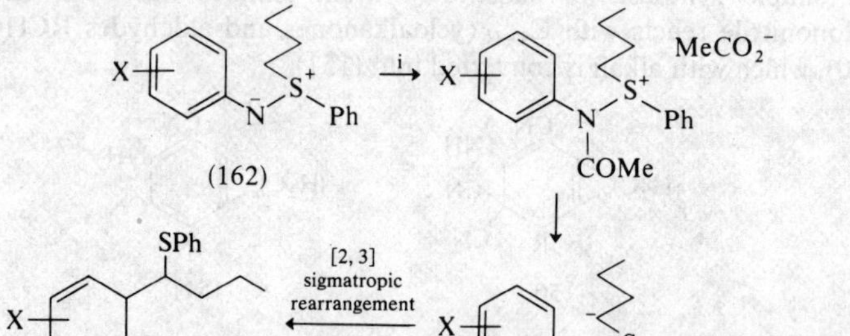


(162) (161) (160)

Reagents: i, $(MeCO)_2O$, 0 °C

**Scheme 24**

[79] B. F. Bonini, G. Maccagnani, and G. Mozzanti, *Tetrahedron Letters*, 1977, 1185.
[80] P. G. Gassman and P. L. Parton, *Tetrahedron Letters*, 1977, 2055.

Experimental conditions have been developed[81] for the alkylation of primary and secondary amines with primary and secondary alcohols in the presence of $Al(OBu^t)_3$ and Raney nickel.

Thermolysis studies of arylamines[82] have shown that homolytic fission of the N—Ar bond occurs, giving aryl and amino radicals, the former undergoing dimerisation to the corresponding biaryl while the amino radical abstracts H, forming ammonia.

De-amination reactions have been the subject of several reports. A general method[83] utilises reductive de-amination of $RNH_2$ to hydrocarbons with hydroxylamine-$O$-sulphonic acid in the presence of aqueous alkali, and similar rapid conversion[84] of amines into hydrocarbons occurs by the use of alkyl nitrites in DMF (Scheme 25).

$$ArNH_2 + RONO \rightarrow ArN{=}NOR + H_2O$$

$$ArN{=}NOR \rightarrow ArN_2{}^{\cdot} + RO^{\cdot}$$

$$ArN_2{}^{\cdot} \rightarrow Ar^{\cdot} + N_2$$

$$Ar^{\cdot} + SolH \rightarrow ArH + Sol^{\cdot}$$

**Scheme 25**

Conversion of the C—N bond in dinosylsulphonimide (163) into a C—O bond in 3,5-dinitrobenzoate esters (164) occurs by reaction[85] with the lithium salts (165) in DMF.

$$RN(SO_2C_6H_4NO_2{-}4)_2$$

(163)

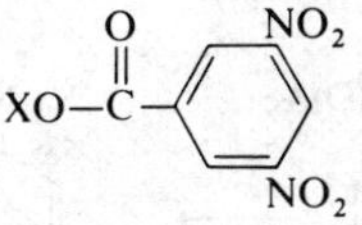

(164) X = R
(165) X = Li

Several reports have appeared on the reactions of nitroso-amines. $N$-nitrosodibenzylamine (166) reacts[86] with phenacyl bromide in the presence of silver hexafluoroantimonate to give the $N$-nitrene (167), which in ether as solvent results in bibenzyl and (168), while the major product in benzene is diphenylmethane.

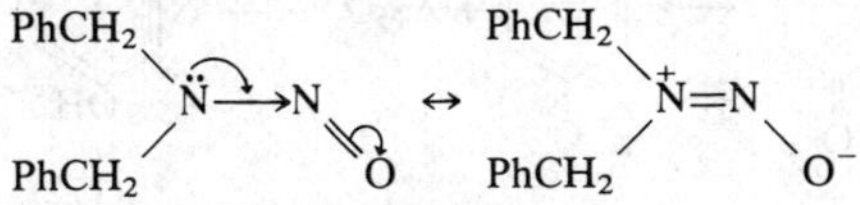

(166)

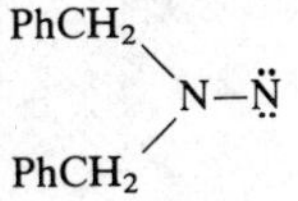

(167)

$$PhCH{=}NN(CH_2Ph)_2$$

(168)

[81] M. Botta, F. De Angelis, and R. Nicoletti, *Synthesis*, 1977, 722.
[82] M. Z. A. Badr, M. M. Aly, and A. E. Abdel-Rahman, *Indian J. Chem., Sect. B*. 1977, **15**, 381.
[83] G. A. Doldouras and J. Kollonitsch, *J. Amer. Chem. Soc.*, 1978, **100**, 341.
[84] M. P. Doyle, J. F. Dellaria, Jr., B. Siegfried, and S. W. Bishop, *J. Org. Chem.*, 1977, **42**, 2494.
[85] V. A. Curtis, H. S. Schwartz, A. F. Hartman, R. M. Pick, L. W. Kolar, and R. J. Baumgarten, *Tetrahedron Letters*, 1977, 1969.
[86] K. Nishiyama and J. P. Anselme, *J. Org. Chem.*, 1978, **43**, 2045.

Acyl-aryl-nitrosamines decompose in benzene by a radical and an ionic path, and the relative efficiency of a series of alkenes in promoting the ionic path has been considered.[87] This is illustrated by the decomposition of *N*-nitrosoacetanilide (169) and benzenediazonium acetate by two competing routes to give benzyne (by acetate-induced elimination), which can be trapped as the furan adduct (170), and phenyl radicals by a chain reaction (Scheme 26).

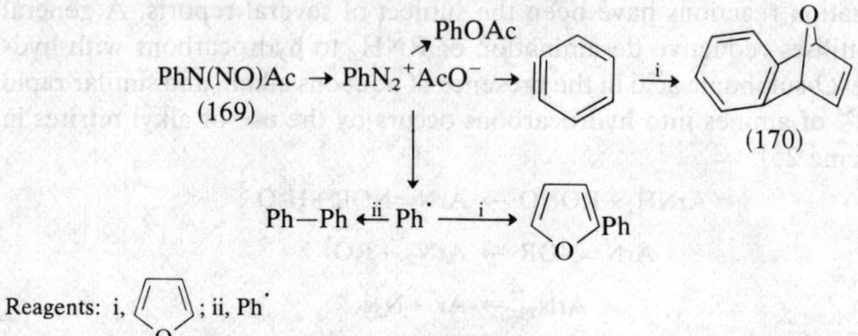

Reagents: i, furan; ii, Ph·

**Scheme 26**

1,1-Diphenylethene increases the yield of ionic products by a factor of 10, and further use[88] of this compound has been made in converting 4-chloro-*N*-nitrosoacetanilide (171) into 4-hydroxybenzenediazonium chloride (172) (as shown in Scheme 27) and acetic anhydride.

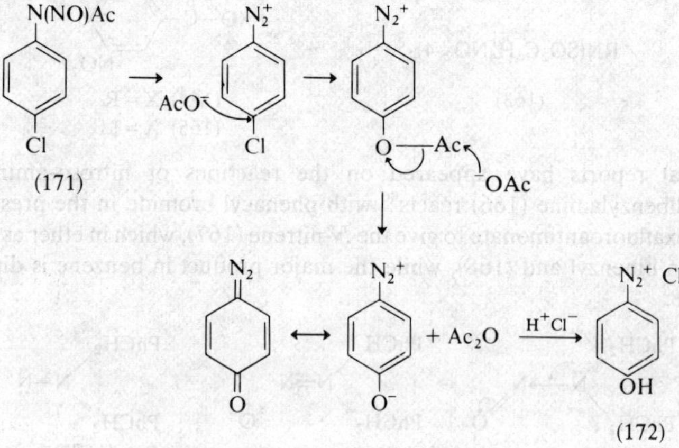

**Scheme 27**

The reaction route shown in Scheme 27 is supported by the 1,2-diphenyl-ethene-promoted reaction of (173) in xylene, which gives 2-cyclo-pentadienylidene-1,3-benzodioxide (174), formed by the reaction of 2-diazo-cyclohexadienone (175) with the carbene 2-oxocyclohexa-3,5-dienylidene that is generated on decomposition of (175).

[87] J. Brennan, J. I. G. Cadogan, and J. T. Sharp, *J. Chem. Res. (S)*, 1977, 107.
[88] J. Brennan, J. I. G. Cadogan, and J. T. Sharp, *J.C.S. Perkin I*, 1977, 1844.

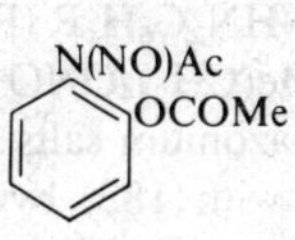

(173)

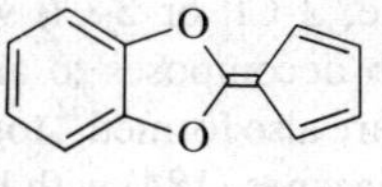

(174)

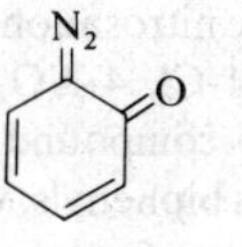

(175)

**Azo- and Diazo-compounds, Hydrazines, Azides, and Related Compounds.—** Reference has been made earlier[71] to a useful route to azo- and azoxy-compounds. Azo-compounds, *e.g.* (176), have also been made[89] by the reduction of nitrobenzene in benzene, using the cobalt carbonyl $Co_2(CO)_8$.

Diazonium compounds have also been widely investigated. The bromolysis[90] of 4,4′-diamino-octachloroazobenzene (177) with $Br_2$ in the presence of $I_2$ gives 4-amino-2,3,5,6-tetrachlorobenzenediazonium dibromo-iodide (178) and 4-bromo-2,3,5,6-tetrachloroaniline (179).

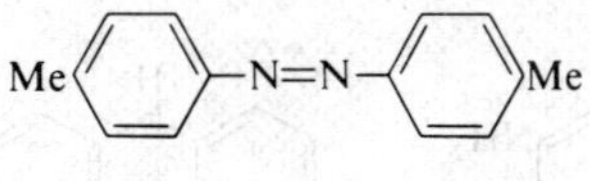

(176)

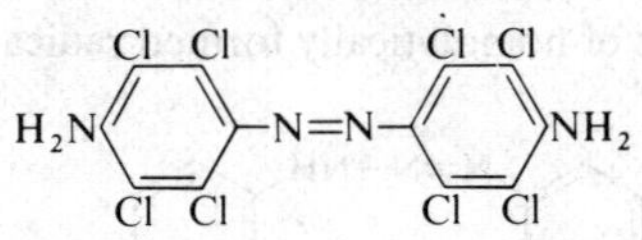

(177)

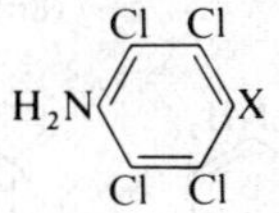

(178) $X = N_2^+ IBr_2^-$
(179) $X = Br$

A high-yield conversion[91] of arenediazonium fluoroborates into the corresponding arylhydrazines occurs in methylene chloride, using benzeneselenol, PhSeH, while the reaction[92] of diazonium salts with Janovsky σ-complexes (180) formed from *m*-dinitrobenzene and $K^+$ $MeCOCH_2^-$ gives (181) and (182) by displacement of the nitro-group with the diazonium residue (Scheme 28).

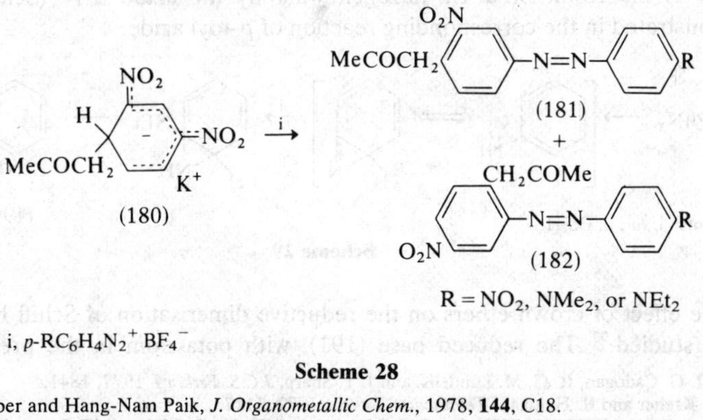

$R = NO_2$, $NMe_2$, or $NEt_2$

Reagents: i, $p$-$RC_6H_4N_2^+ BF_4^-$

**Scheme 28**

[89] H. Alper and Hang-Nam Paik, *J. Organometallic Chem.*, 1978, **144**, C18.
[90] M. Ballester, J. Riera, and J. Garcia-Oricain, *Anales de Quim.*, 1976, **72**, 892.
[91] F. G. James, M. J. Perkins, O. Porta, and B. V. Smith, *J.C.S. Chem. Comm.*, 1977, 131.
[92] A. Ya. Kaminskii, S. S. Gitis, I. L. Bagal, Yu. D. Grudtsyn, T. P. Ikher, L. V. Illarionova, and N. D. Stepanov, *J. Org. Chem.* (*U.S.S.R.*), 1977, **13**, 733.

Aprotic nitrosation of 1,3-diaryl-triazenes $RC_6H_4NHN_2C_6H_4R$ (R = H, 4-Me, 4-OMe, 4-Cl, 4-$NO_2$, 2-Me, 2-Cl, or 3-Cl) with $Me(CH_2)_4ONO$ yields[93] the N-nitroso-compound, which decomposes to arenediazonium salts and radicals and yields biphenyls, which are also formed[94] together with (183), by electrophilic substitution of sulphonyl-triazenes (184) with benzene and $AlCl_3$.

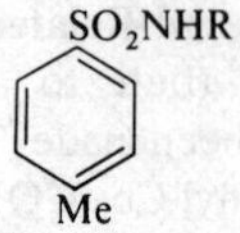

(183) 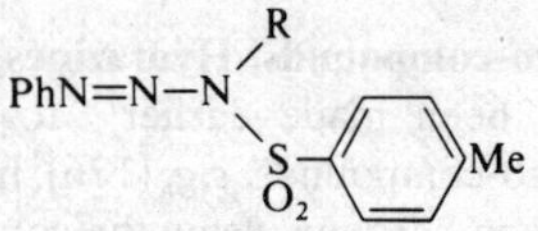                                                                (184)

R = H or Me

The triazenes (185), on photolysis in non-aromatic solvents, give rise[95] to a variety of products (186)—(189) that are consistent with a cage-recombination process of homolytically formed radicals.

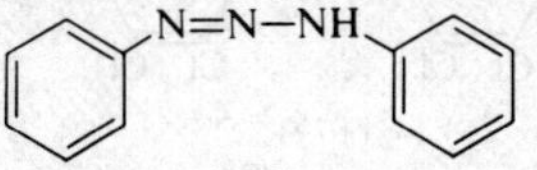

(185) 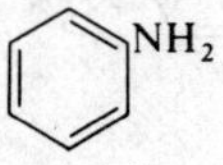                                    (186) 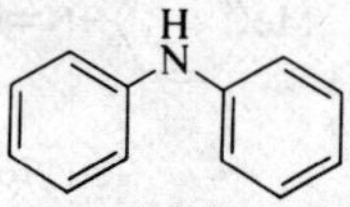                                    (187)

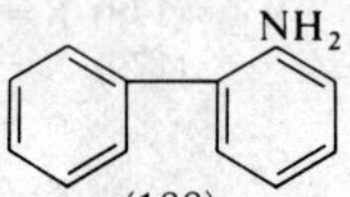

(188) 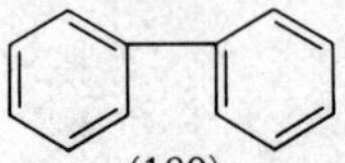                                    (189)

Radical reactions feature in a number of related systems. The photochemical decomposition[96] of aromatic azides in the presence of ethanethiol results not in the usual ring expansion to 2-substituted-3H-azepines but in the formation of thioethoxy-compounds. Phenyl azide gives 2-ethylthio-1-aminobenzene (190), which is the result of a 1,2-nitrogen walk by the azide α-N (Scheme 29) demonstrated in the corresponding reaction of p-toyl azide.

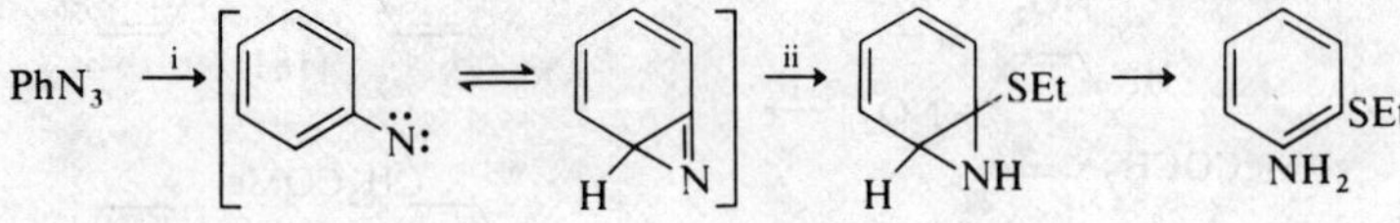

Reagents: i, hν; ii, EtSH                                                            (190)

**Scheme 29**

The effect of crown ethers on the reductive dimerisation of Schiff bases has been studied.[97] The reduced base (191), with potassium in the presence of

[93] J. I. G. Cadogan, R. G. M. Landells, and J. T. Sharp, *J.C.S. Perkin I*, 1977, 1841.
[94] R. Kreher and R. Halpaap, *Tetrahedron Letters*, 1977, 3147.
[95] M. Julliard, M. Scelles, A. Guillmonat, G. Vernin, and J. Metzger, *Tetrahedron Letters*, 1977, 375.
[96] S. E. Carroll, B. Nay, E. F. V. Scriven, H. Suschitzky, and D. R. Thomas, *Tetrahedron Letters*, 1977, 3175.
[97] J. G. Smith and Ying-Luen Chun, *Tetrahedron Letters*, 1978, 413.

18-crown-6 ether, is stabilised through complexation of the metal ion in solution, enabling it to accept a second electron and to form the dianion (192), which undergoes alkylation with 1,2-dibromopropane to give (193).

$$[PhCH=NPh]^{\cdot} K^+ \qquad [PhCH=NPh]^{2-} 2K^+ \qquad \begin{array}{c} PhCHNHPh \\ | \\ CH_2CH=CH_2 \end{array}$$

$$\text{(191)} \qquad\qquad\qquad \text{(192)} \qquad\qquad\qquad \text{(193)}$$

A new route to diazo-iminyl radicals (194), which fragment to produce an arenenitrile and aryl radicals, has been described[98] which utilises the treatment of (195) with thiosulphate.

$$Ar^1N=N-CAr^2 \rightarrow Ar^2CN + Ar^1N=N^{\cdot} \rightarrow Ar^1 \xrightarrow{PhH} PhAr^1$$
$$\underset{\overset{\|}{N}}{\overset{}{\phantom{.}}}$$
$$\text{(194)} \qquad \overset{\cdot}{N}$$

$$Ar^1N=NCAr^2=NOCH_2CO_2H$$
$$\text{(195)}$$

**Sulphur and Selenium Compounds.**—Nitroaryl-sulphonic acids (196) have been made[99] by oxidative treatment of aminoaryl-sulphonic acids (197) with peracetic acid, while the useful acyl thioacyl sulphides (198)[100] have been formed by the reaction of Na or piperidinium dithiocarboxylates (199) with acetyl chloride.

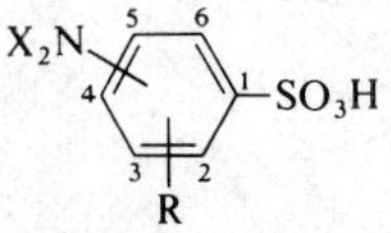

(196) X = O
(197) X = H

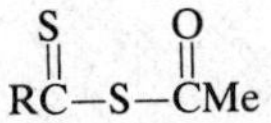

(198) R = Ph or $p$-ClC$_6$H$_4$

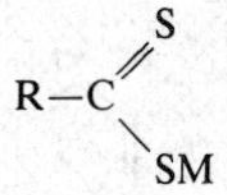

(199) M = Na or C$_5$H$_{12}$N

R = 2-, 3-, or 4-Me; NX$_2$ is in 3-, 4-, or 5-position

The bis-(2-nitrophenyl) disulphide (200) undergoes a photochemically induced reaction with aromatic amines, with homolytic fission of the S—S bond,[101] yielding (201).

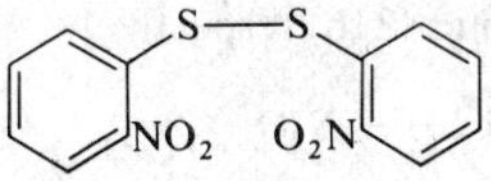

(200)

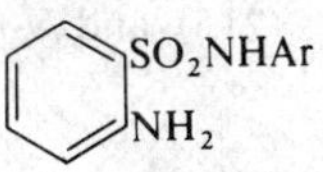

(201)

Arenesulphenyl thiocyanates (202) act as electrophiles[102] towards sulphonium ylides (203) to give the intermediate (204). The SCN$^-$ ion attacks at the methyl group, the R group, or the S—C(SAr) bond to give (205), (206), or (207) respectively (Scheme 30).

[98] A. R. Forrester, M. Gill, E. M. Johansson, C. J. Meyer, and R. Thomson, *Tetrahedron Letters*, 1977, 3601.
[99] E. E. Gilbert, *Synthesis*, 1977, 315.
[100] S. Kato, K. Sugino, M. Yamada, T. Katada, and M. Mizuta, *Angew. Chem. Internat. Edn.*, 1977, **16**, 879.
[101] V. N. R. Pillai, *Chem. and Ind.*, 1977, 665.
[102] H. Matsuyama, H. Minato, and M. Kobayashi, *Bull. Chem. Soc. Japan*, 1978, **51**, 575.

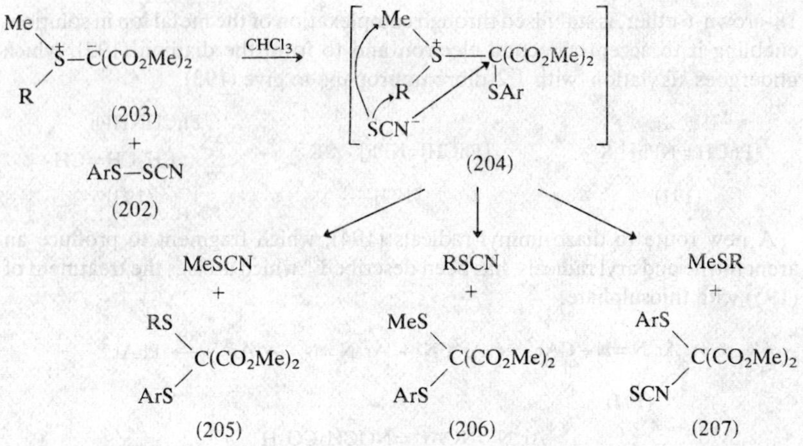

**Scheme 30**

Substitution reactions of thiophenoxides (208) with nitro- and halogeno-phthalimide derivatives (209)[103] and with nitro- and halogen-substituted phthalic anhydrides (210)[104] have been studied. These have given rise to a series of previously unknown thioether imides (211) and a wide range of thioether-substituted anhydrides (212).

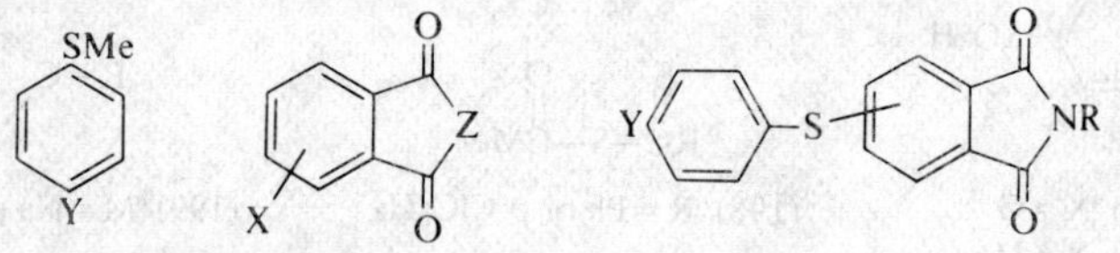

(208)          (209) Z = NR          (211) Z = NR
               (210) Z = O           (212) Z = O

Thiols have been activated[105] by conversion into thionitrites (213) with di-nitrogen tetroxide. The latter react readily with other thiols, sulphinic acids, and amines as shown in Scheme 31, resulting in unsymmetrical disulphides (214), thiosulphonates (215), and N-nitrosamines (216) respectively.

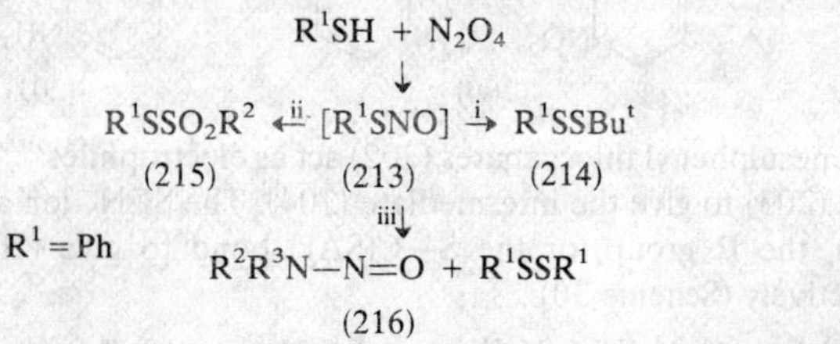

$$R^1SH + N_2O_4$$
$$\downarrow$$
$$R^1SSO_2R^2 \xleftarrow{ii} [R^1SNO] \xrightarrow{i} R^1SSBu^t$$
$$(215) \qquad (213) \qquad (214)$$

$$R^1 = Ph \qquad \downarrow iii$$
$$R^2R^3N-N=O + R^1SSR^1$$
$$(216)$$

Reagents: i, Bu$^t$SH; ii, R$^2$SO$_2$H; iii, R$^2$R$^3$NH

**Scheme 31**

[103] F. J. Williams and P. E. Donahue, *J. Org. Chem.*, 1978, **43**, 250.
[104] F. J. Williams and P. E. Donahue, *J. Org. Chem.*, 1978, **43**, 255.
[105] S. Oae, P. Fukushima, and Y. H. Kim, *J.C.S. Chem. Comm.*, 1977, 407.

The chemistry of the S—N bond in sulphenamide enolate equivalents[106] and in benzenesulphenanilides[107] has been examined. The alkylation of the enolate equivalents (217) derived from the hitherto unavailable (218) by reaction with lithium di-isopropylamide gives (219) (Scheme 32).

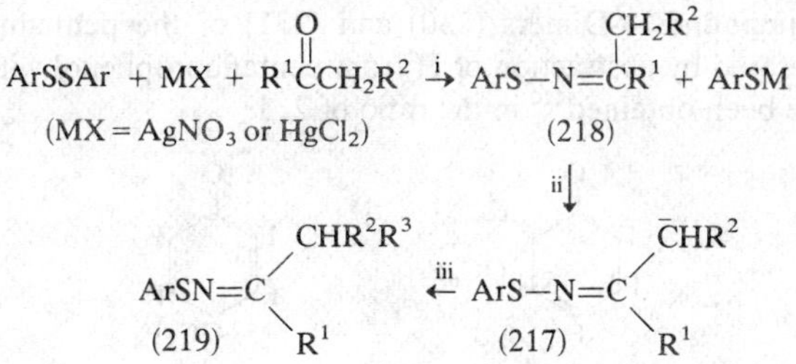

$$ArSSAr + MX + R^1\overset{\overset{O}{\|}}{C}CH_2R^2 \overset{i}{\rightarrow} ArS-N{=}CR^1 + ArSM$$
$$\overset{CH_2R^2}{|}$$

$$(MX = AgNO_3 \text{ or } HgCl_2) \qquad\qquad (218)$$

$$ii\downarrow$$

$$ArSN{=}C\overset{CHR^2R^3}{\underset{R^1}{\diagdown}} \overset{iii}{\leftarrow} ArS-N{=}C\overset{\bar{C}HR^2}{\underset{R^1}{\diagdown}}$$

$$(219) \qquad\qquad (217)$$

Reagents: i, NH$_3$; ii, LDA; iii, R$^3$X

**Scheme 32**

*NN*-Bis-*p*-methoxyphenyl-benzenesulphenamide (220) in liquid SO$_2$ gives the cation radical (221), which undergoes oxidative cleavage to (222) and (223) and formation of (224) and (225) in high yield, together with small amounts of (226) and (227).

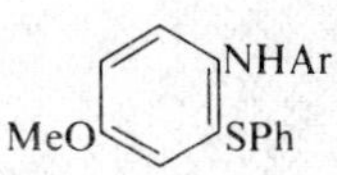

PhS—NAr$_2$      [PhS—NAr$_2$]$^{\cdot}$     PhS$^+$     Ar$_2$N$^{\cdot}$

(220)          (221)          (222)         (223)

Ar$_2$NH      PhS$_2$Ph      (226) 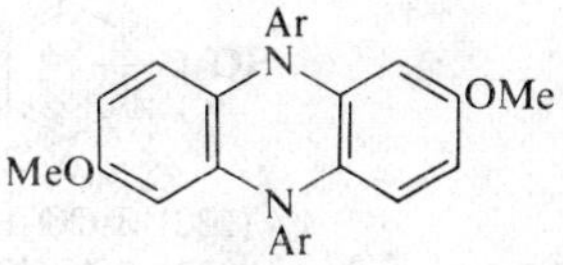      (227)

(224)        (225)

Ar = C$_6$H$_4$OMe-*p*

Arylation reactions of aryldiazosulphones, leading mainly to 4-RC$_6$H$_4$Ph, have been studied.[108] The sulphone (228), with benzene in the presence of AlCl$_3$, eliminates toluene-*p*-sulphinic acid and N$_2$ to give the 4-substituted biphenyl (229). Aryl and biphenylyl 4-tolylsulphoxides are formed as major products in further reactions of the sulphinic acid.

$$4\text{-}RC_6H_4N{=}NSO_2C_6H_4Me \qquad\qquad 4\text{-}RC_6H_4R$$

$$(228) \qquad\qquad\qquad\qquad\qquad (229)$$

Methods for the formation of aromatic selenonium compounds have proved troublesome, but a recent report[109] describes their formation in one step by the reaction of Grignard reagents with selenium oxychloride (Scheme 33).

[106] F. A. Davis and P. A. Mancinelli, *J. Org. Chem.*, 1978, **43**, 1797.
[107] T. Ando, M. Nojima, and N. Tokura, *J.C.S. Perkin I*, 1977, 227.
[108] R. Kreher and R. Halpaap, *Z. Naturforsch.*, 1977, **32b**, 1325 (*Chem. Abs.*, 1978, **88**, 62 088).
[109] Y. Ishi, Y. Iwama, and M. Ogawa, *Synth. Comm.*, 1978, **8**, 93.

$$3ArMgX + SeOCl_2 \rightarrow Ar_3SeOMgX + 2MgXCl$$

$$Ar_3SeOMgX + HY \rightarrow Ar_3Se^+ X^- \text{ (or } Y^-) + MgOHX \text{ (or Y)}$$

**Scheme 33**

**Halogeno-compounds.**—Dimers (230) and (231) of the pentafluorophenoxyl radical, generated by abstraction of H from pentafluorophenol with lead tetra-acetate, have been obtained[110] in the ratio of 2:1.

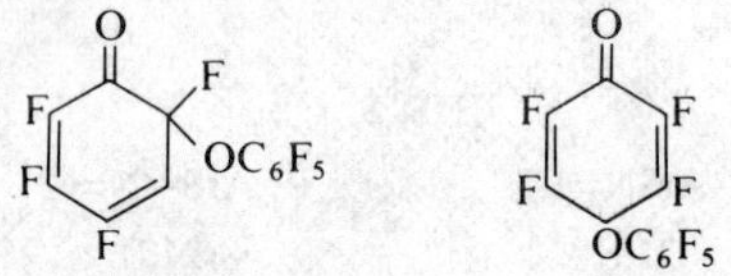

$$\begin{array}{ccc} (230) & : & (231) \\ 2 & : & 1 \end{array}$$

New routes to aromatic fluoro-compounds include fluorination[111] of (232), (233), and (234) with $SF_4$ in anhydrous HF, giving good yields of the corresponding poly(trifluoromethyl)benzenes, and the reaction[112] of hexafluorobenzene with $Me_3SiN=PX_3$, giving $N$-pentafluorophenylphosphinimine, $C_6F_5N=PX_3$ (X = Me, Ph, or $NMe_2$).

Nucleophilic displacement[113] of fluorine or chlorine from fluoro- (or chloro-)nitro-benzophenones (235) takes place with a range of heterocycles. Thus (235) with pyrrole gives (236) and with 2-methylimidazole gives (237).

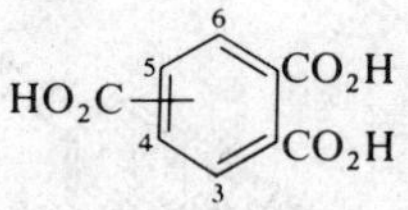

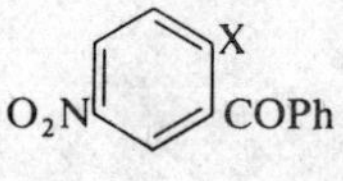

(232) 3-$CO_2H$
(233) 3,4-$(CO_2H)_2$
(234) 4,5-$(CO_2H)_2$

(235) X = F or Cl

(236) X = —N⟨pyrrole⟩

(237) X = —N⟨2-methylimidazole⟩

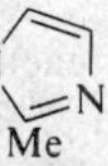

Scheme 34 shows how chlorine is also displaced[114] from chloronitrobenzenes (238) by the desoxybenzoin enolate anion, giving (239). The $p$-Cl is displaced more selectively than $o$-Cl, and this is preparatively useful, while displacement of $o$-Cl is frequently accompanied by the formation of considerable amounts of coloured by-products.

[110] L. S. Kobrina, V. N. Kovtonyuk, and G. G. Yakobson, *J. Org. Chem. (U.S.S.R.)* 1977, **13**, 1331.
[111] L. M. Yagupol'skii, N. V. Kondratenko, L. A. Alekseeva, V. M. Betous, V. N. Boika, V. G. Lukmanov, V. I. Popov, V. P. Sambur, and G. M. Schupak, *Ref. Dokl. Soobshch-Mendeleevsk, S'ezd. Obshch. Prikl. Khim. 11th*, ed. V. N. Rozinskaya, 'Nauka', Moscow, 1975, Vol. 2, p. 73 (*Chem. Abs.*, 1978, **88**, 190 255).
[112] D. Dahmann and H. Rose, *Chem.-Ztg.*, 1977, **101**, 401.
[113] N. W. Gilman, B. C. Helland, G. R. Walsh, and R. I. Fryer, *J. Heterocyclic Chem.*, 1977, **14**, 1157.
[114] M. Jawdosiuk and W. Wilczynski, *Roczniki Chem.*, 1977, **51**, 595.

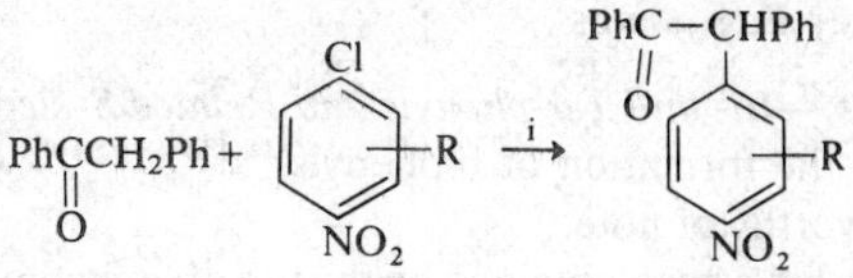

PhCCH$_2$Ph + 〈Cl ─R〉 —$\xrightarrow{i}$→ 〈 ─R〉

(238)         (239)

Reagents: i, KOH in DMSO or NaH in DMF

**Scheme 34**

The inert radical 4-(triphenylphosphoranylideneamino)tetradecachloro-triphenylmethyl (240) is formed[115] by chlorination of the 4-amino-tetradeca-chlorotriphenylmethyl radical (241) to perchloro-fuchsinimine (242) followed by condensation with triphenylphosphine. Its spectrum has been recorded. Detailed studies have also been carried out[116] on the pentachlorophenyl radical, obtained in a variety of ways.

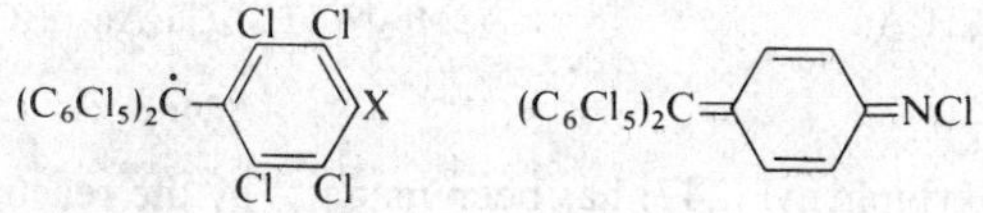

(240) X = N=PPh$_3$
(241) X = NH$_2$    (242)

An interesting synthesis[117] of trichloroethyl-benzenes provides the first example of homolytic displacement by an alkyl group at a saturated carbon atom. In this way benzylbis(dimethylglyoximato)pyridine cobalt, with BrCCl$_3$ in chloroform, leads to PhCH$_2$CCl$_3$ and PhCH$_2$Br.

Periodo-arenes have been prepared[118] by halogenodemercuration of permercurated arenes with I$_3^-$ ions (Scheme 35). Bromo- and iodo-arenes have also been formed,[119] in excellent yields, by an improved Sandmeyer conversion of stable arenediazonium tetrafluoroborates with potassium acetate in the presence of 18-crown-6 and subsequent reaction with bromo-trichloromethane.

$$\text{PhX} + 5\text{Hg(OTf)}_2 \rightarrow \text{C}_6(\text{HgOTf})_5\text{X} + 5\text{TfOH}$$

$$\text{XC}_6\text{H}_4\text{CO}_2\text{H} + 5\text{Hg(OTf)}_2 \rightarrow \text{C}_6(\text{HgOTf})_5\text{X} + 5\text{TfOH} + \text{CO}_2$$

$$p\text{-MeC}_6\text{H}_4\text{SO}_2\text{Na} + 5\text{Hg(OTf)}_2 \rightarrow \text{MeC}_6(\text{HgOTf})_5 + 4\text{TfOH} + \text{SO}_2 + \text{NaOTf}$$

$$\text{C}_6(\text{HgOTf})_5\text{X} + 5\text{I}_3^- \rightarrow \text{C}_6\text{I}_5\text{X} + 5\text{HgI}_2 + 5\text{TfO}^-$$

$$(\text{Tf} = \text{CF}_3\text{CO})$$

**Scheme 35**

[115] M. Ballester, J. Riera, and C. Rovira, *Anales de Quim.*, 1976, **72**, 489.
[116] R. Bolton, P. E. Mitchell, and G. H. Williams, *J. Chem. Res. (S)* 1977, 223.
[117] T. Funabiki, B. Dass Gupta, and M. D. Johnson, *J.C.S. Chem. Comm.*, 1977, 653.
[118] G. B. Deacon and G. J. Farquharson, *Austral. J. Chem.*, 1977, **30**, 170.
[119] S. H. Korzeniowski and G. W. Gokel, *Tetrahedron Letters*, 1977, 3519.

### 3 Polycyclic Systems

**Non-fused Systems.**—*Bi- and Ter-phenyls and Related Systems.* Earlier sections contain reports of the formation of biphenyls,[15,43,93—95,108] while several other methods are also worthy of note.

Simple biphenyls have been obtained by the use of organometallic compounds. The reaction[120] of the aryl-copper (243) with copper trifluoromethanesulphonate results in excellent yields of biphenyls (Scheme 36) by way of the precursor complex (244).

$$(ArCu)_n + (CF_3SO_3)Cu \rightarrow Ar_2Cu_3(CF_3SO_3) \rightarrow Ar{-}Ar$$

(243)                                        (244)

**Scheme 36**

The reaction of copper or gold complexes such as (245) with copper or silver trifluoromethanesulphonates gives hexanuclear cluster compounds such as (246).

$Me_2NC_6H_4Au$                      $(2Me_2NC_6H_4)_4Cu_4Au_2(F_3CSO_3)_2$

(245)                                        (246)

4,4′-Dimethoxybiphenyl (247) has been made[121] by the reaction with Raney nickel of an organotellurium complex (248) obtained by heating anisole with tellurium chloride.

(247)                                        (248)

A novel phenol–benzene coupling reaction,[122] giving rise to 2- and 4-hydroxy-biphenyls, (249) and (250), occurs in the acid-catalysed reaction of *N*-aryl-hydroxylamines (251) with benzene (Scheme 37).

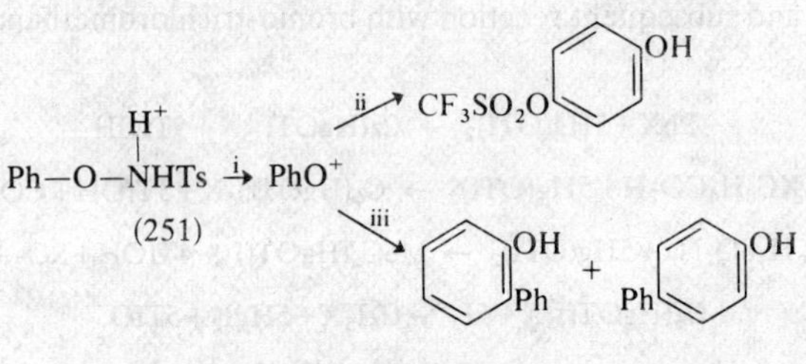

(249)            (250)

Reagents: i, TFA, PhH, TFSA, 1 hour at room temperature; ii, $CF_3SO_2O^-$; iii, PhH

**Scheme 37**

[120] A. Van Koten, J. T. B. H. Jastrzebski, and J. G. Noltes, *J.C.S. Chem. Comm.*, 1977, 203.
[121] J. Bergman, R. Carlson, and B. Sjoberg, *Org. Synth.*, 1977, **57**, 18.
[122] Y. Endo, K. Shudo, and T. Okamoto, *J. Amer. Chem. Soc.*, 1977, **99**, 7721.

The oxidation[123] of 3,3'-di-t-butyl-5,5'-ditritylbiphenyl-2,2'-diol (252) by addition of a solution in benzene to an aqueous solution of $K_3Fe(CN)_6$ and KOH results in the diphenoquinone (253), which isomerises in iso-octane in the presence of MeOH or EtOH to the spiro-substituted benzoxetes (254).

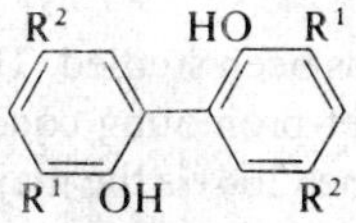

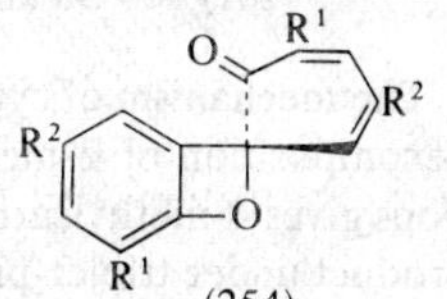

$$R^1 = Ph_3C, R^2 = Bu^t$$

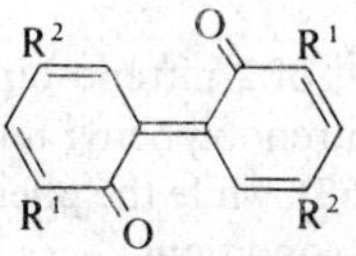

Methanolysis of the diphenic anhydrides (257) yields the isomeric mono-esters (255) and (256).[124]

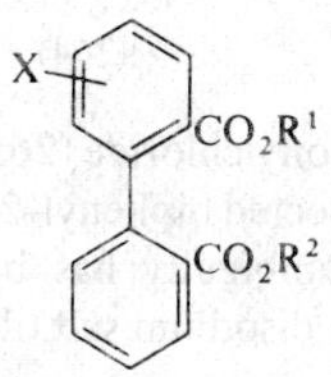

(255) $R^1 = H, R^2 = Me$
(256) $R^1 = Me, R^2 = H$     (257)

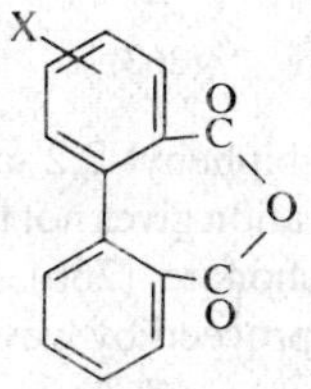

Several reports describe the formation of dimethylamino-biphenyls. The photochemical reaction[125] of p-dimethylaminotoluidine in the presence of an electron acceptor such as tetracyanoethene results in (258), while phenylation[126] of NN-dimethylaniline N-oxide (259) with benzene in the presence of trifluoromethanesulphonic acid gives mainly 4(-NN-dimethylamino)-biphenyl (260) together with small amounts of the 2-isomer.

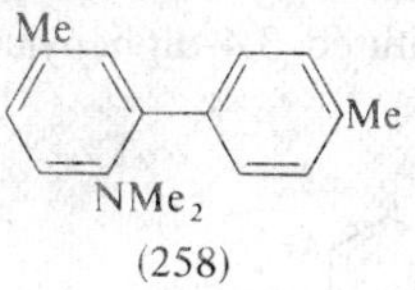

(258) 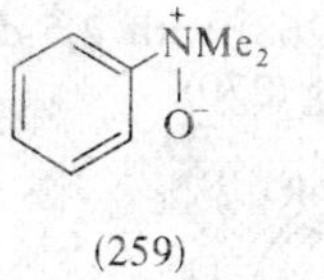      (259) 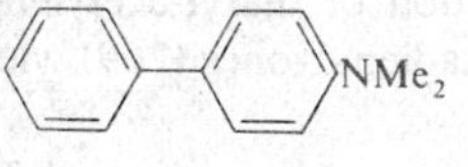      (260)

Earlier methods of preparation of 4-alkyl-4'-cyanobiphenyls, which have found applications as liquid crystals, have led only to low yields, a problem which has recently been overcome with a newly described preparation[127] of 4-alkyl-4'-bromobiphenyls (261) and their conversion into the cyano-compounds. The series of compounds (261) are formed by decomposition of the hydrazones (262) with KOBu$^t$.

[123] H. D. Becker and K. Gustafsson, *Tetrahedron Letters*, 1976, 4883.
[124] S. Kobayashi and M. Kihara, *Yakugaku Zasshi*, 1977, **97**, 901.
[125] M. Ohashi, S. Suwa, and K. Tsujimoto, *J.C.S. Chem. Comm.*, 1977, 348.
[126] K. Shudo, T. Ohta, Y. Endo, and T. Okamoto, *Tetrahedron Letters*, 1977, 105.
[127] M. Jawdosiuk and I. Kmiotek-Skarzynska, *Roczniki Chem.* 1977, **51**, 2023.

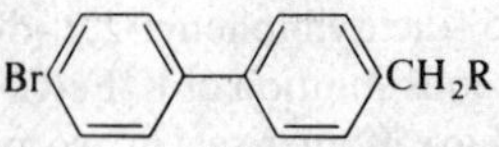

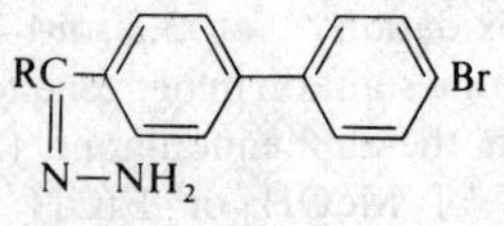

(261) R = Bu$^n$, n-C$_6$H$_{13}$, or n-C$_7$H$_{15}$                    (262)

The mechanism of cyclisation[128] of 2-nitreno-biphenyls has been studied. The decomposition of 2-methyl-1'-nitrenobiphenyl under singlet-promoting conditions gives 4-methylcarbazole (263), while the phenanthridine (264) is the major product under triplet-promoting conditions.

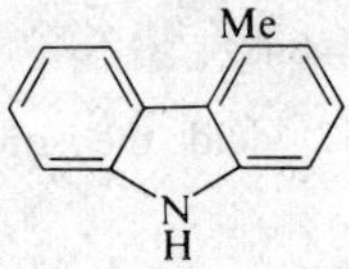

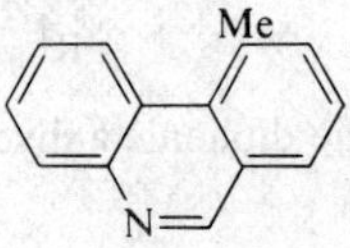

(263)                                    (264)

The reduction of biphenyl-2,2'-disulphonylchloride (265) with sodium sulphite followed by acidification gives not the expected biphenyl-2,2'-disulphinic acid but the cyclic thiosulphonate (266). This ambiguity has been resolved and the reaction shown to proceed by way of the disodium salt of the acid (267).[129]

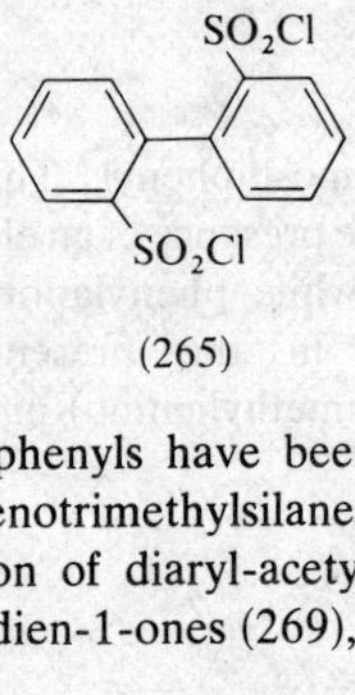

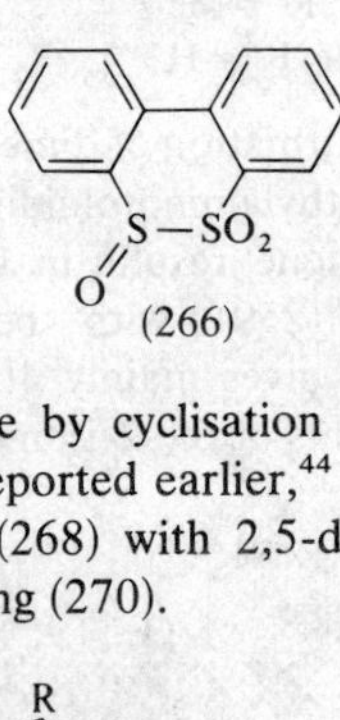

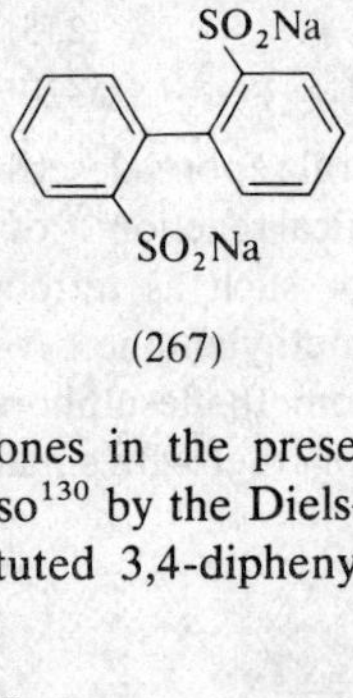

(265)                     (266)                     (267)

Terphenyls have been made by cyclisation of ketones in the presence of halogenotrimethylsilanes, as reported earlier,[44] and also[130] by the Diels–Alder reaction of diaryl-acetylenes (268) with 2,5-disubstituted 3,4-diphenylcyclopentadien-1-ones (269), yielding (270).

ArC≡CAr

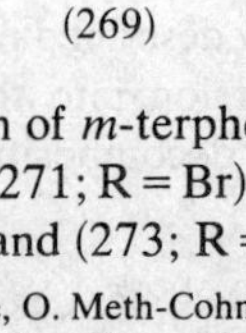

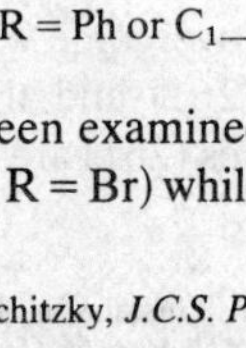

(268)              (269)              (270) R = Ph or C$_{1-3}$ alkyl

Electrophilic substitution of *m*-terphenyl has been examined.[131,132] Bromination results in a mixture of (271; R = Br) and (272; R = Br) while iodination led to (271; R = I), (272; R = I), and (273; R = I).

[128] J. M. Lindley, I. M. McRobbie, O. Meth-Cohn, and H. Suschitzky, *J.C.S. Perkin I*, 1977, 2194.
[129] M. M. Chau and J. L. Kice, *J. Org. Chem.*, 1977, **42**, 3265.
[130] G. Rabilloud and B. Sillion, *Bull. Soc. chim. France*, 1977, 276.
[131] K. Ghosh and A. J. Bhattacharya, *Indian J. Chem.*, Sect. B, 1977, **15**, 678.
[132] G. Rabilloud, B. Masson, B. Sillion, N. Platzer, and J. J. Basselier, *Bull. Soc. chim. France*, 1977, 281.

(271)　　　　　　　(272)　　　　　　　　　　(273)

A mixture of isomeric perfluoro-dimethyldiphenyltetrahydrobiphenyls, such as (274), has been formed[133] by the reaction of pentafluorobenzoyl peroxide with octafluorotoluene. Macrocyclic hydrocarbons (275) have resulted[134] from the sulphides (276) by oxidation to the sulphones followed by pyrolysis.

(274)　　　　　　　　(275)　　　　　　　　(276)

**Fused Systems.**—*Naphthalene and Derivatives.* A new method of aromatization of partially hydrogenated aromatic hydrocarbons overcomes the difficulty of the competing Diels–Alder reaction of reactive arenes. The method is based on deprotonation–hydride elimination[135] in which potassium fencholate (277) serves as the base and fenchone (278) as the hydride acceptor, catalytic amounts only of fencholate being required since it is regenerated in the aromatisation step. In this way, the conversion of 1,2-dihydronaphthalene into naphthalene and of 9,10-dihydroanthracene into anthracene proceeds almost quantitatively.

Hexasubstituted naphthalenes such as (279) have been prepared[136] in high yields by the cycloaddition of benzynes to arylcyclopentadienones (280).

(277)　　　　　　(278)　　　　　　(279)　　　　　　(280)

The crystalline structure of a hemi-Dewar naphthalene (281) has been determined.[137] An unusual Diels–Alder reaction[138] of 2-vinylnaphthalene with benzyne gives not the expected dihydrobenzphenanthrene (282; R = H), but (282; R = Ph) is obtained instead, which can be aromatized. The structure was established unambiguously, using α-deuterio-2-vinylnaphthalene (283).

The first examples of thermal insertion of a C-isocyano carbon into a C—C or C—H bond have been found in the conversion[139] of the isocyano-naphthalene

[133] L. S. Kobrina, V. L. Salenko, and G. G. Yakobson, *J. Org. Chem. (U.S.S.R.)*, 1977, **13**, 1565.
[134] J. Gruetze and F. Voegtle, *Chem. Ber.*, 1977, **110**, 1978.
[135] M. T. Reetz and F. Eibach, *Angew. Chem. Internat. Edn.*, 1978, **17**, 278.
[136] K. Ghosh and A. J. Bhattacharya, *Indian J. Chem., Sect. B*, 1976, **14**, 793.
[137] R. W. Franck, R. Gruska, and J. G. White, *Tetrahedron Letters*, 1977, 509.
[138] Y. Ittah, I. Shakak, J. Blum, and J. Klein, *Synthesis*, 1977, 678.
[139] J. H. Boyer, and J. R. Patel, *J.C.S. Chem. Comm.*, 1977, 855.

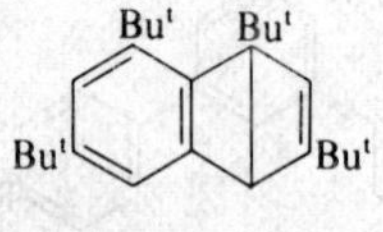

(281) 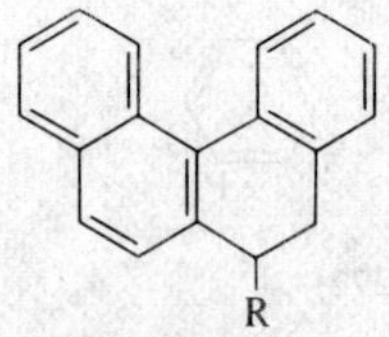                            (282) 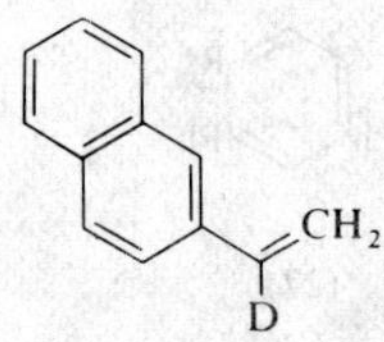                            (283)

(284) into the benzo-cycloheptindole (285) and the benzophenanthridine (286), and in the conversion of isocyanophenylnaphthalene (287) into the benzophenanthridine (288).

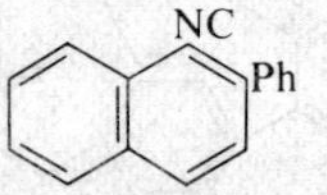

(284) 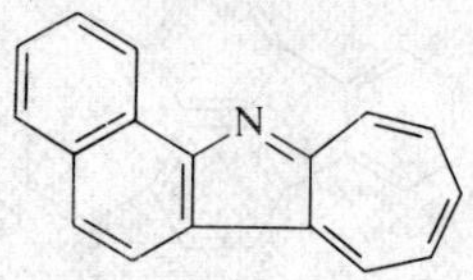                            (285) 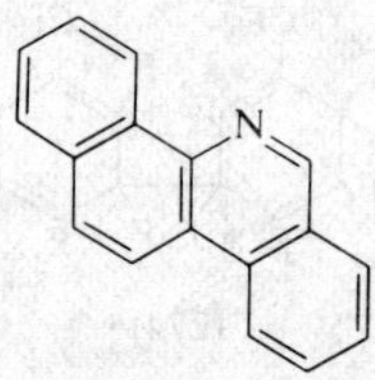                            (286)

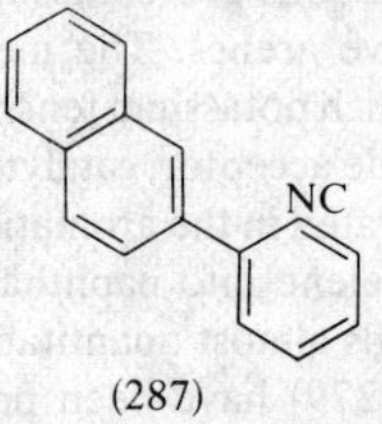

(287) 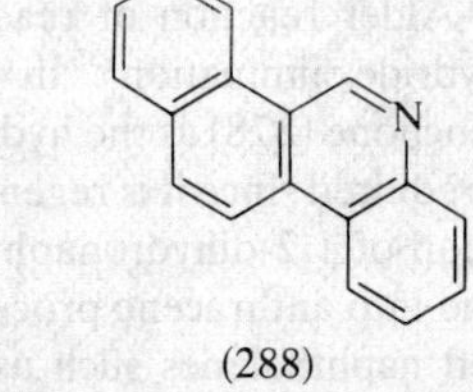                            (288)

Strained molecules based on naphthalene have been described.[140] Lithiation of 1,8-di-iodonaphthalene and 5,6-dibromoacenaphthene and the subsequent cycloaddition reaction with acenaphthenequinone and pyracenequinone gives diols, which with HF yield the acenaphth[1,2-*a*]acenaphthalenes (289), (290), and (291).

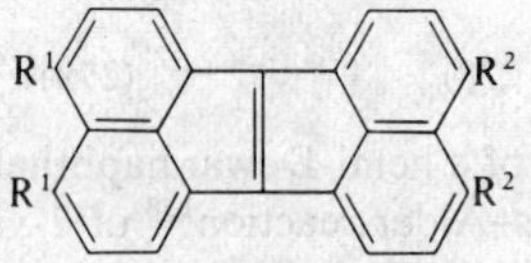

(289)  R$^1$ = R$^2$ = H
(290)  R$^1$ = H, R$^2$R$^2$ = CH$_2$CH$_2$
(291)  R$^1$R$^1$ = R$^2$R$^2$ = CH$_2$CH$_2$

New synthetic methods[141] have been described for the preparation of 1-hydroxy-2,3-disubstituted naphthalenes (292) and 1,4-dihydroxy-2,3-disubstituted naphthalenes, as illustrated in Scheme 38. 1*H*-2-Benzofuran-1-one-3-(phenylsulphone) (294; Ar = Ph, R$^3$ = H) may be similarly converted into

[140] R. H. Mitchell, T. Fyles, and L. M. Ralph, *Canad. J. Chem.*, 1977, **55**, 1480.
[141] F. M. Hauser and R. P. Rhee, *J. Org. Chem.*, 1978, **43**, 178.

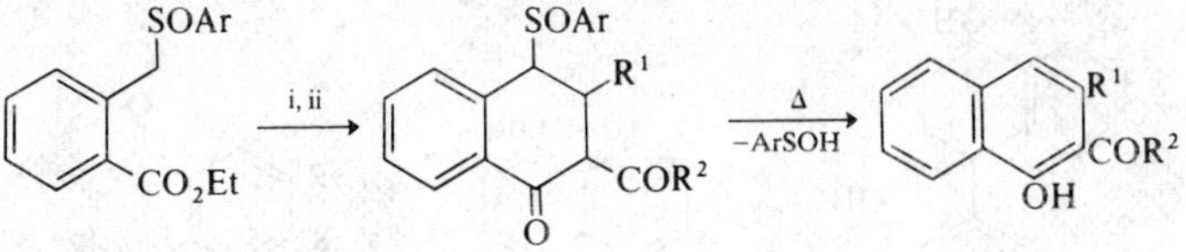

(293)                                                  (292)

Reagents: i, LiNPr$_2^i$, THF, at $-78$ °C; ii, R$^1$CH=CHCOR$^2$ [R$^1$ = H, Me, or CH$_2$SMe; R$^2$ = Me or OEt; or R$^1$R$^2$ = (CH$_2$)$_3$]

**Scheme 38**

1,4-dihydroxy-naphthalenes, which, because they are unstable in air, are converted into dimethyl ethers (295; R$^3$ = H).

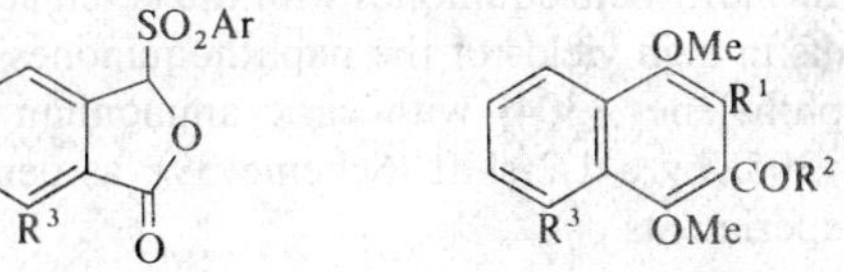

(294)                                         (295)

Sterically hindered 1-t-butyl- and 1-t-pentyl-2-naphthols (296) have been reported[142] for the first time, and their autoxidation to the unstable hydroperoxides (297) has been followed. The rate of autoxidation and the stability of the hydroperoxides depend on the bulk of the alkyl group.

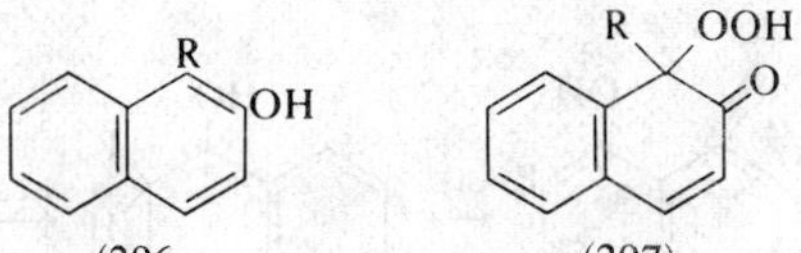

(296                                          (297)

Photochemical dimerization[143] of 2-alkoxy-naphthalenes gives mixtures of dienes (298), (299), and (300).

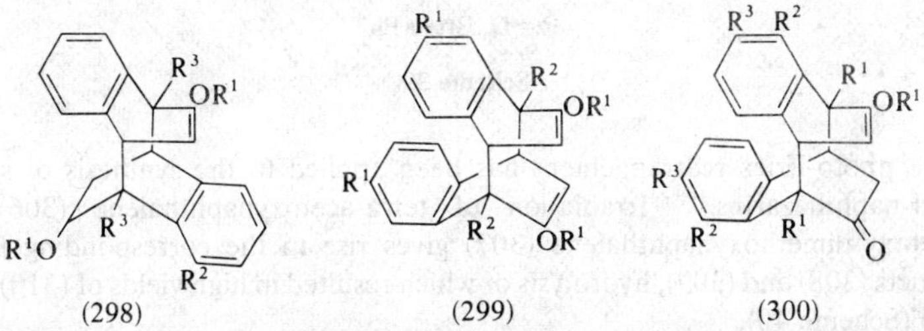

(298)                                        (299)                                     (300)

The irradiation of 1-naphthoxide anion in the presence of methoxide also gives a photodimer (301), by nucleophilic photosubstitution.[144] 1-Alkoxynaphthalene undergoes thiocyanation by the copper(II) thiocyanate method to give (302), the

[142] J. Carnduff and P. A. Brady, *J. Chem. Res. (S)*, 1977, 235.
[143] T. Teitei, D. Wells, T. H. Spurling, and W. H. F. Sasse, *Austral. J. Chem.*, 1978, **31**, 85.
[144] T. Kitamura, T. Imagawa, and M. Kawanisi, *J.C.S. Chem. Comm.*, 1977, 81.

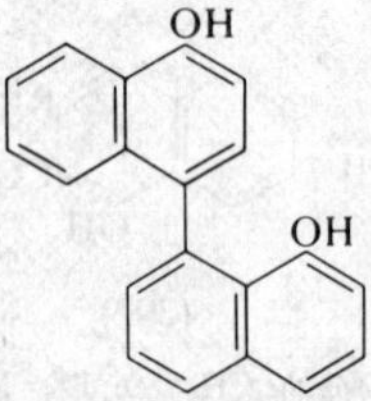

(301) 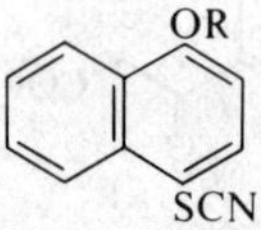                        (302) 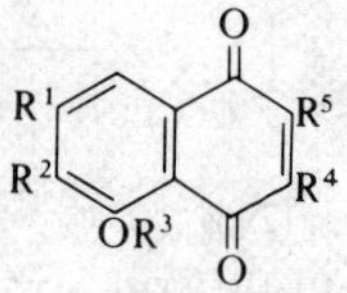                        (303)

techniques employed[145] permitting a higher reaction temperature than that usually possible.

The synthesis and reactions of naphthoquinones have received attention. The reaction[146] of 2,6-dichloro-benzoquinones with the keten acetal $CH_2=C(OEt)_2$ in acetic acid results in high yields of the naphthoquinones (303), while oxidation[147] of the naphthalenes (304) with ceric ammonium sulphate gives the naphthoquinones (305) by a 1,2-shift (Scheme 39), as demonstrated by deuterium-labelling experiments.

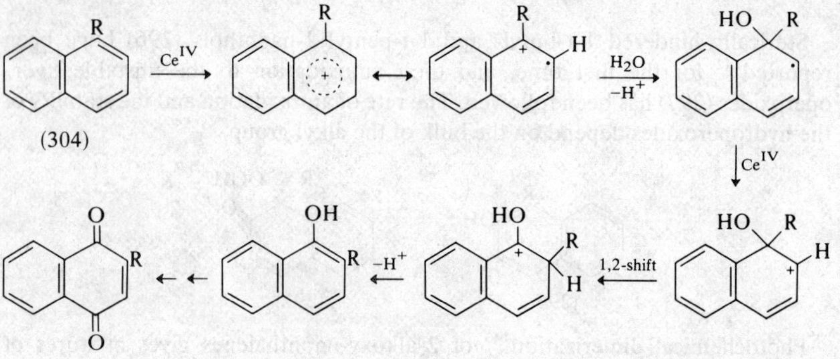

R = D, Br, or Ph

**Scheme 39**

The photo-Fries rearrangement has been applied to the synthesis of some acetyl-naphthazarins.[148] Irradiation of tetra-acetoxynaphthalene (306) or diacetoxy-dimethoxynaphthalene (307) gives rise to the corresponding Fries products (308) and (309), hydrolysis of which resulted in high yields of (310) and (311) (Scheme 40).

The reaction of 2-hydroxy-1,4-naphthoquinone (312) with methyl vinyl ketone has been studied.[149] The product isolated in the presence of secondary amines

[145] K. Fujiki, *Bull. Chem. Soc. Japan*, 1977, **50**, 3065.
[146] J. L. Grandmaison and P. Brassard, *Tetrahedron*, 1977, **33**, 2047.
[147] M Periasamy and V. M. Bhatt, *Tetrahedron Letters*, 1977, 2357.
[148] E. Escobar, F. Farina, R. Martinez-Utrilla, and M. C. Paredes, *J. Chem. Res. (S)*, 1977, 266.
[149] H. Rafart, T. Valderrama, and J. C. Vega, *Anales de Quim.*, 1976, **72**, 804.

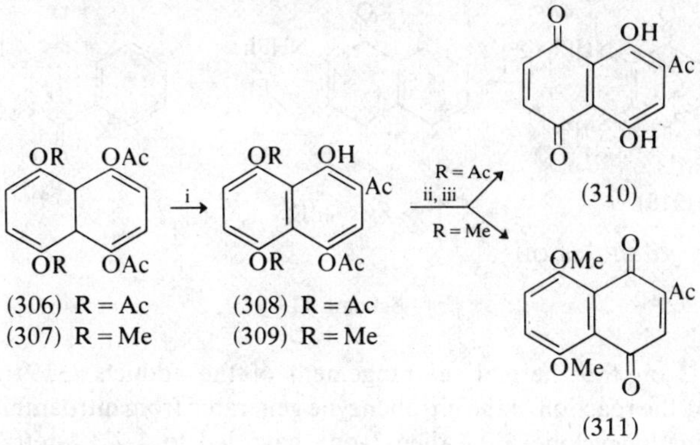

Reagents: i, $h\nu$; ii, H⁺; iii, FeCl₃

**Scheme 40**

such as piperidine, *viz.* 2-piperidino-9,10-anthraquinone (313), is believed to be formed by way of the intermediate 2-hydroxy-3-(3-oxobutyl)-1,4-naphthoquinone (314), which with the secondary amine gives the piperidino-9,10-phenanthraquinone (315) as well as (313).

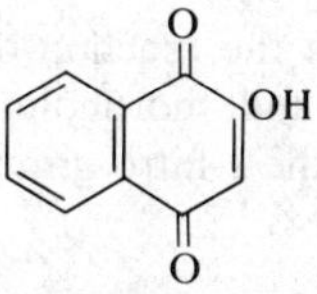

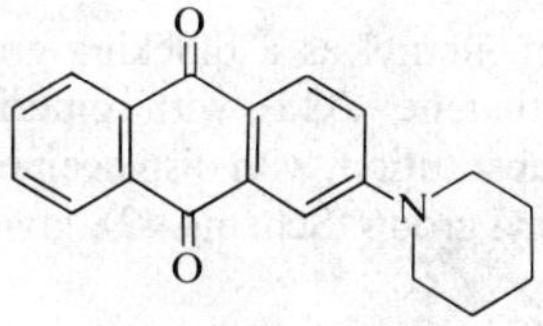

Stable *o-* and *p*-naphthoquinone methides leading to (316) have been obtained[150] by a two-stage process, starting with 6-methoxy-1,3,8-tri-hydroxynaphthalene, and studies[151] on the stability conferred on ylides by the phenylimino-group (PhN=) have led to the synthesis shown in Scheme 41 for the formation, in high yield, of the ylide (317) from 2-anilino-3-methylthio-1,4-naphthoquinone (318).

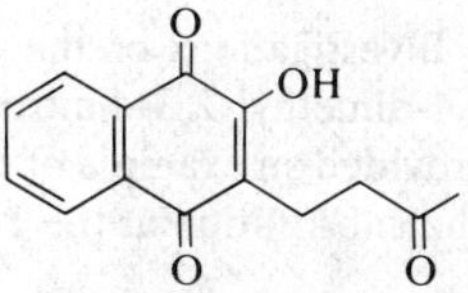

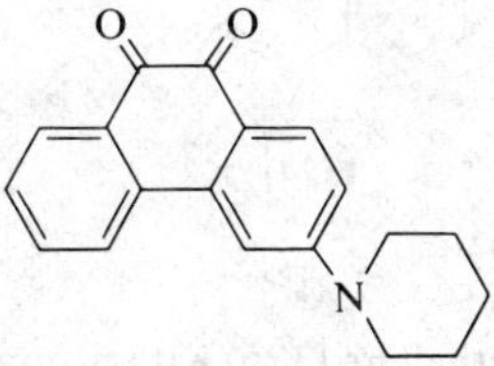

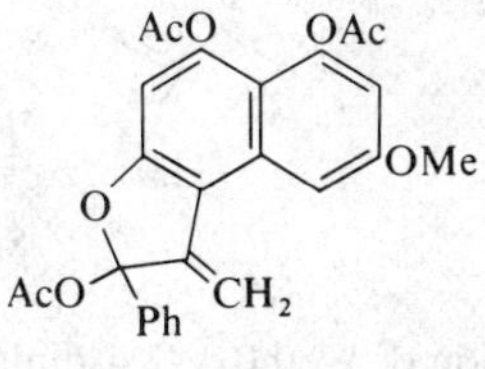

[150] A. T. Henriques, J. A. Rabi, P. M. Baker, and K. S. Brown, Jr., *Synthesis*, 1977, 713.
[151] J. W. Marsico, Jr., G. O. Morton, and L. Goldman, *J. Org. Chem.*, 1977, **42**, 2164.

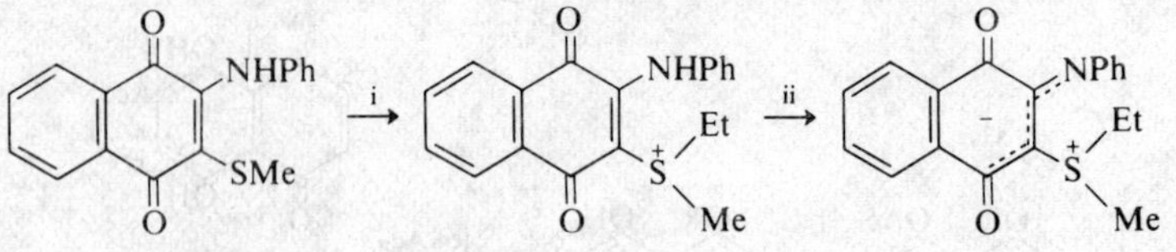

(318)                                    (317)

BF$_4^-$

Reagents: i, Et$_3$O$^+$ BF$_4^-$; ii, OH$^-$

**Scheme 41**

Studies[152] on the thermal rearrangement of the adducts (319) and (320) prepared by the reaction of the nitrobenzyne generated from nitroanthranilic acid with hexamethylcyclohexa-2,4-dien-1-one have led to 1,2,3,4-tetramethyl-5-nitronaphthalene (321) and 1,2,3,4-tetramethyl-6-nitronaphthalene (322).

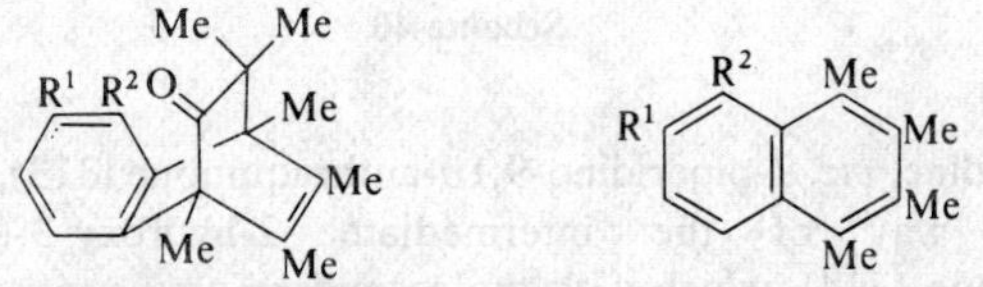

(319) R$^1$ = NO$_2$, R$^2$ = H          (321) R$^1$ = NO$_2$, R$^2$ = H
(320) R$^1$ = H, R$^2$ = NO$_2$          (322) .R$^1$ = H, R$^2$ = NO$_2$

Investigations of the use of methyl as a blocking group in the reaction of 1,4-dimethyl-2,3-dinitronaphthalene (323) with piperidine and morpholine provided an example of *tele*-substitution, with displacement of the 2-nitro-group and substitution at the 1-methyl group (Scheme 42), giving (324).[153]

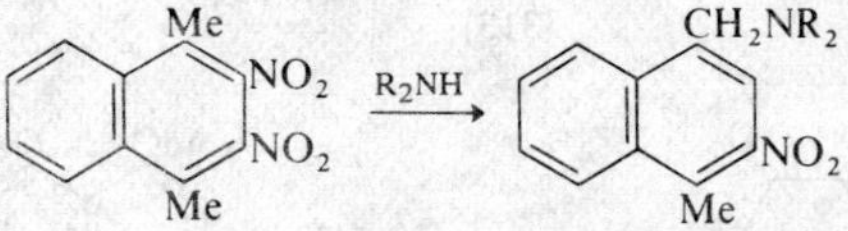

(323)                                    (324)

**Scheme 42**

The reaction of 1-ethoxy-2,4-dinitronaphthalene (325) with n-butylamine has been followed by high-resolution flow n.m.r. spectroscopy,[154] which has made possible the characterisation of the transient σ-complex (326) that is formed initially from the nucleophilic substitution.

Naphthalene-1,4-imines with bridgehead substituents (327) have been made[155] by cycloaddition of benzyne or tetrachlorobenzyne with the appropriate pyrrole. Photochemical reactions of naphthonitriles have been investigated, with

[152] A. Oku and A. Matsui, *Bull. Chem. Soc. Japan*, 1978, **51**, 331.
[153] G. Guanti, S. Thea, M. Novi, and C. Dell'Erba, *Tetrahedron Letters*, 1977, 1429.
[154] C. A. Fyfe, A. Koll, S. W. H. Dasnji, C. D. Malkiewich, and P. A. Toole, *Canad. J. Chem.*, 1977, **55**, 1468.
[155] J. M. Vernon, M. Ahmed, and J. M. Moran, *J. C. S. Perkin I*, 1977, 1084.

(325)          (326)          (327)

interesting results. The irradiation of 2-naphthonitrile and *cis*-MeOCH=CHMe in benzene gave[156] a mixture of (328) and (329) in a 7:3 ratio, while the similar reaction[157] of 1- and 2-naphthonitriles with $Me_2C=CMe_2$ resulted in (330) and (331) respectively.

(328)          (329)          (330) $R^1 = CN, R^2 = H$
                              (331) $R^1 = H, R^2 = CN$

Studies on naphthalenes fused to alicyclic rings have been reported. Dehydrochlorination of the tetrahydro-cyclopropa[*b*]napthalene (332) with KOBu$^t$ gave[158] a high yield of the *gem*-dichlorocyclopropa[*b*]naphthalene (333) while 1,2,3,4,7,8,9,10-octahydrodicyclohepta[*de, ij*]naphthalene and 2,7-dimethylpyrene (335) are obtained[159] by the reaction sequence shown in Scheme 43.

*Anthracene and Related Compounds.* The formation of anthracene by elimination of hydride from 9,10-dihydroanthracene has been reported in the preceding section.[135] An improved high-yield synthesis[160] of 7,12-dimethylbenz-[*a*]anthracene (336), a compound of biological interest, utilizes the reaction of 7,12-benz[*a*]anthraquinone with MeLi. Subsequent reaction with dry HCl in ethyl acetate produces 7-(chloromethyl)-12-methylbenz[*a*]anthracene, reduction of which gives (336).

9-Methoxy-7,12-dimethylbenz[*a*]anthracene (337) has been made[161] from 5-methoxy-3-methyl-3-(2-naphthyl)phthalide (338), obtained by hydrolysis of the reaction product of 2-(2-lithio-4-methoxyphenyl)-4,4'-dimethyl-2-oxazoline with methyl 2-naphthyl ketone.

Electrochemical reduction[162] of anthracene in DMF and $Ac_2O$ with $Bu_4^tN^+ I^-$ present gave mainly (339) together with small amounts of *cis*- and *trans*-9,10-diacetyl-9,10-dihydroanthracene, while the reductive coupling[163] of the electrolytically generated anthracene anion radical with 1,3-dihalogenopropane has been studied, this coupling resulting in, for example, cyclopentanodihydroanthracene (340).

[156] K. Mizuno, C. Pac, and H. Sakurai, *J. Org. Chem.*, 1977, **42**, 3315.
[157] J. J. McCullough, R. C. Miller, and Wei-Sai Wu, *Canad. J. Chem.*, 1977, **55**, 2909.
[158] A. R. Browne and B. Halton, *J.C.S. Perkin I*, 1977, 1177.
[159] D. E. Laycock, R. J. Wain, and R. H. Wightman, *Canad. J. Chem.*, 1977, **55**, 21.
[160] M. S. Newman and V. Sankaran, *Tetrahedron Letters*, 1977, 2067.
[161] M. S. Newman and S. Kumar, *J. Org. Chem.*, 1978, **43**, 370.
[162] H. Lund, *Acta Chem. Scand. (B)*, 1977, **31**, 424.
[163] E. Hobolth and H. Lund, *Acta Chem. Scand. (B)*, 1977, **31**, 395.

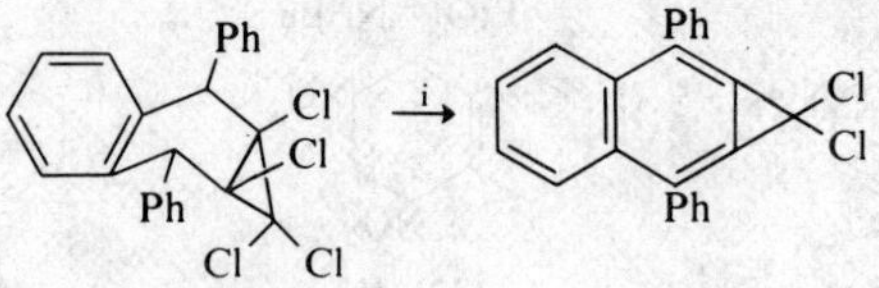

(332)            (333)

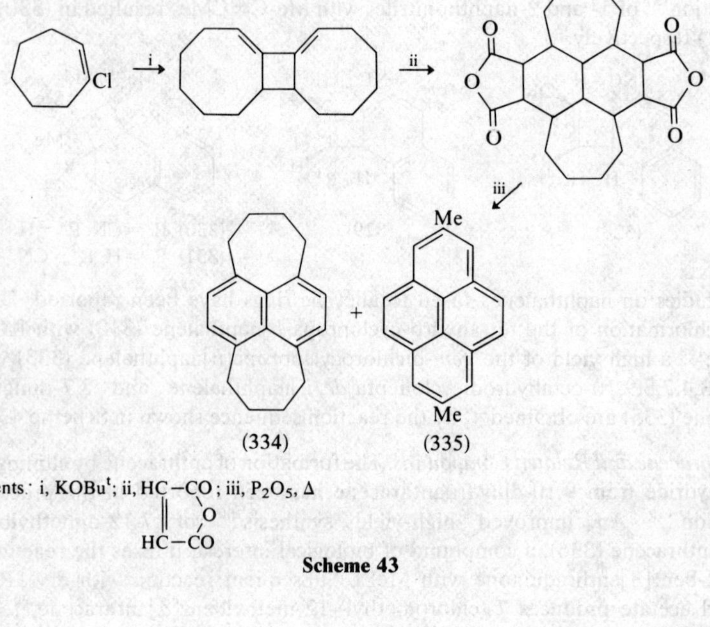

(334)      +      (335)

Reagents: i, KOBu$^t$; ii, HC—CO ; iii, P$_2$O$_5$, Δ
                                        O
                                    HC—CO

**Scheme 43**

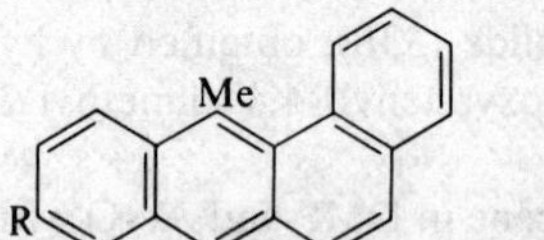

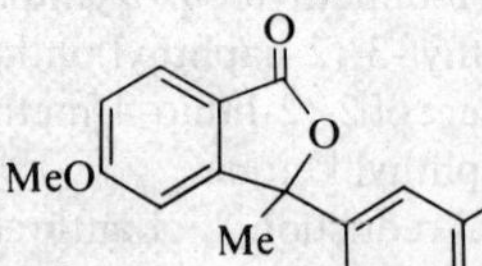

(336) R = H                   (338)
(337) R = OMe

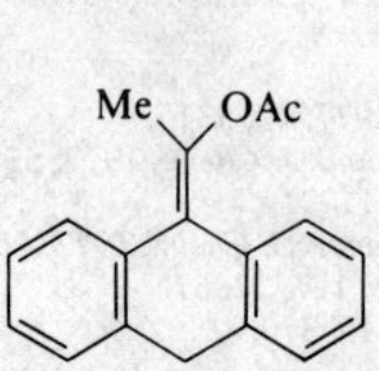

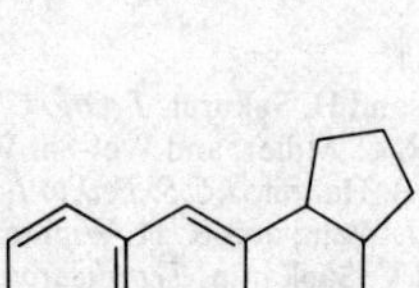

(339)                      (340)

The reaction of chlorocarbene with the benz[*de*]anthracenyl anion has also been studied;[164] it produces 4,5-benzocyclohepta[1,2,3-*de*]naphthalene (341) together with the isomer (342). A useful synthetic entry[165] into the 9-position of anthracene has been effected, using 1-cyano-1-methylethyl radical, and leading to (343).

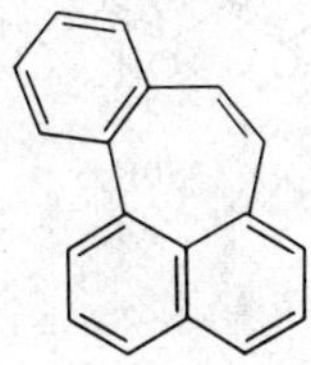

(341) 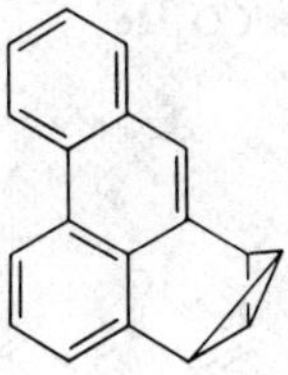                    (342) 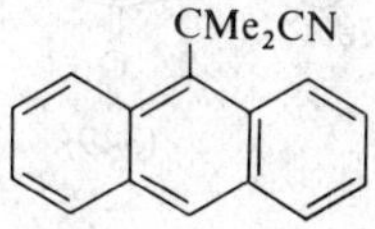                    (343)

A series of reactions starting with 4,7-dimethylisatin results[166] in decamethyl-anthracene, treatment of which with $CH_2Cl_2$ and TFA gives the tautomeric form (344).

Irradiation of decamethylanthracene (in $C_6H_6$ or $Et_2O$) gives the 9,10-Dewar isomer (345),[166] while studies of steric interactions in a similar molecule, *i.e.* 1,4,5,8,9-pentamethylanthracene, showed[167] that there is loss of aromaticity, the reaction of this compound with KOBu$^t$ in DMSO leading to a mixture of (346; R = H) and (346; R = OH).

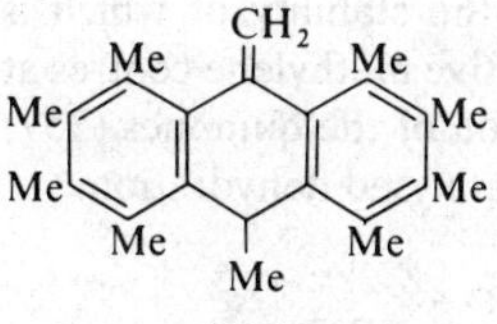

(344) 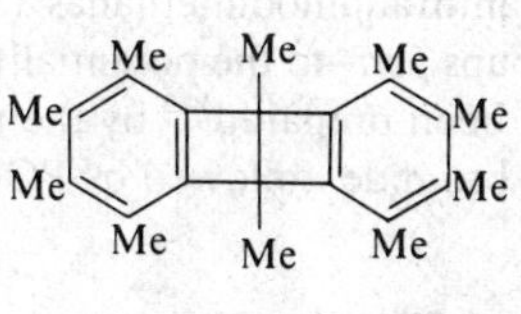                    (345) 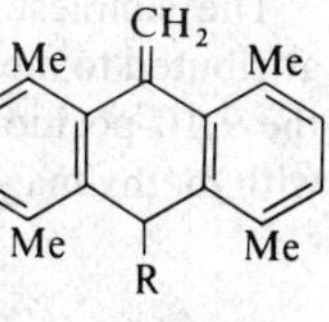                    (346)

Photo-oxygenation[168] of 7,12-dimethylbenz[*a*]anthracene gave the bicyclic acetal (347), while 9-methyl-anthracenes with a free 10-position undergo substitution when they react with a phenyl radical, leading to 9-methylene-10-phenyl-9,10-dihydroanthracene (348).[169]

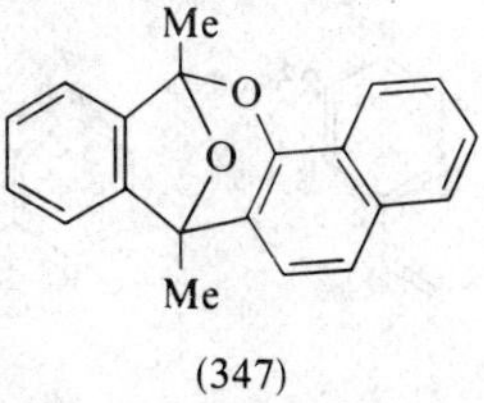

(347) 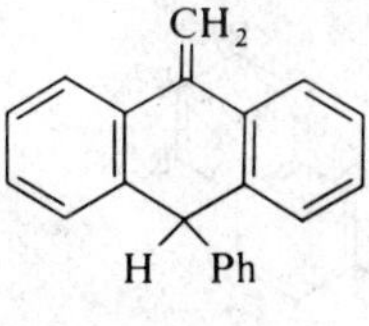                    (348)

[164] R. M. Pagni, M. Burnett, and A. C. Hazell, *Tetrahedron Letters*, 1977, 163.
[165] T. Mitsuhashi, S. Otsuka, and M. Oki, *Tetrahedron Letters*, 1977, 2441.
[166] H. Hart and B. Ruge, *Tetrahedron Letters*, 1977, 3143.
[167] H. Hart and H. Wachi, *J.C.S. Chem. Comm.*, 1977, 409.
[168] M. K. Lagani, W. A. Austin, and R. E. Davies, *Tetrahedron Letters*, 1977, 2467.
[169] F. M. Gromarty, R. Henriquez, and D. C. Nonhebel, *J. Chem. Res. (S)*, 1977, 309.

New syntheses[170] of anthrols and anthraquinones utilise the benzophenone carbanion (349), the reaction of which with KOBu$^t$ in DMF gives (350) and (351).

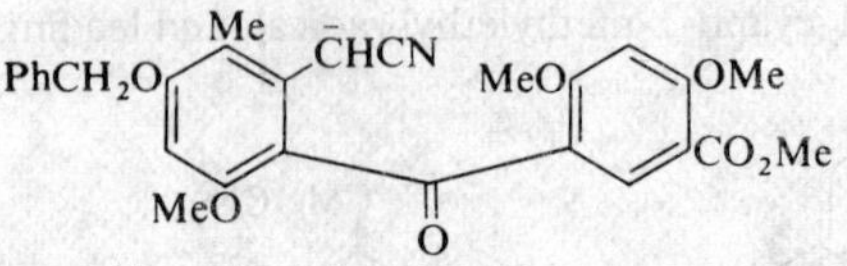

(349) 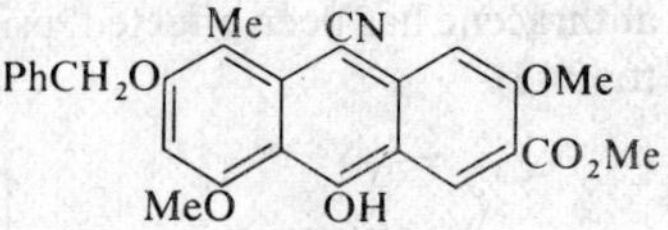                                                                 (350)

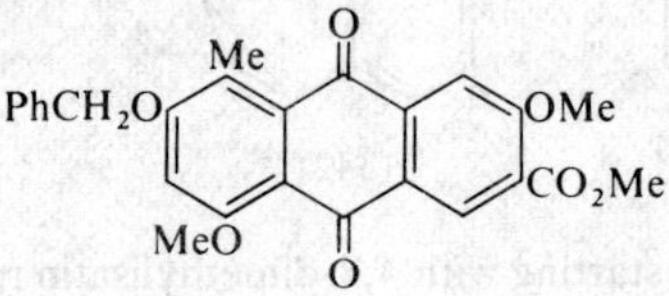

(351)

The allyloxy-anthraquinone (352) has been converted into the aldehyde (353) by a thermally induced rearrangement[171] followed by migration of a double bond, ozonolysis, and methylation. 7,12-Benz[*a*]anthraquinones (354) have been synthesized by the Diels–Alder reaction[172] of 1,4-phenanthraquinones (355) with buta-1,3-diene.

The simplest stable anthraquinodimethanes (356), the stability of which is attributed to methyl groups *peri*- to the potentially reactive methylene centres at the 9,10-position, have been prepared[173] by the reaction of the quinones (357) with methylmagnesium bromide, followed by POCl$_3$-catalysed dehydration.

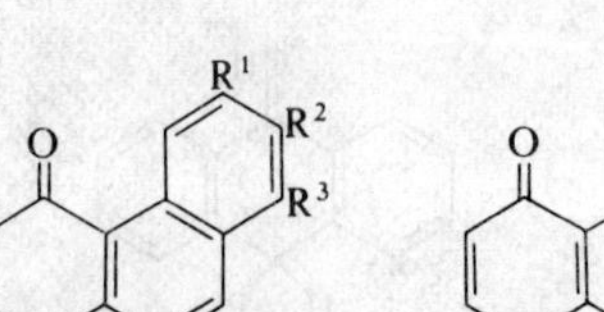

(352) 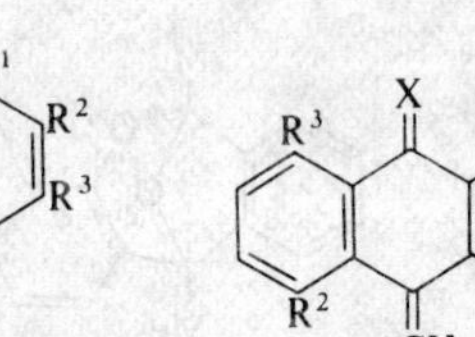                                                                 (353)

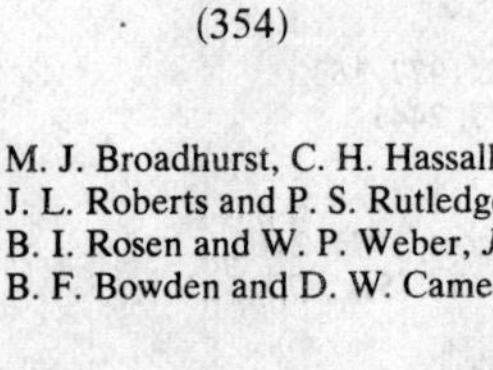

(354) 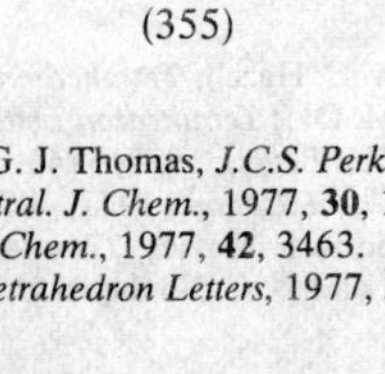                          (355) 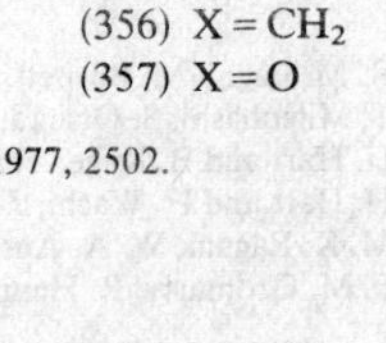                          (356) X = CH$_2$
                                                             (357) X = O

[170] M. J. Broadhurst, C. H. Hassall, and G. J. Thomas, *J.C.S. Perkin I*, 1977, 2502.
[171] J. L. Roberts and P. S. Rutledge, *Austral. J. Chem.*, 1977, **30**, 1743.
[172] B. I. Rosen and W. P. Weber, *J. Org. Chem.*, 1977, **42**, 3463.
[173] B. F. Bowden and D. W. Cameron, *Tetrahedron Letters*, 1977, 383.

A general synthesis (see Scheme 44) of 1-, 2-, 3-, and 4-substituted benz[*a*]anthracene-7,12-diones (358) has been described.[174] It uses the reaction of ring-substituted styrenes (359) with 1,4-naphthoquinone in the presence of chloranil.

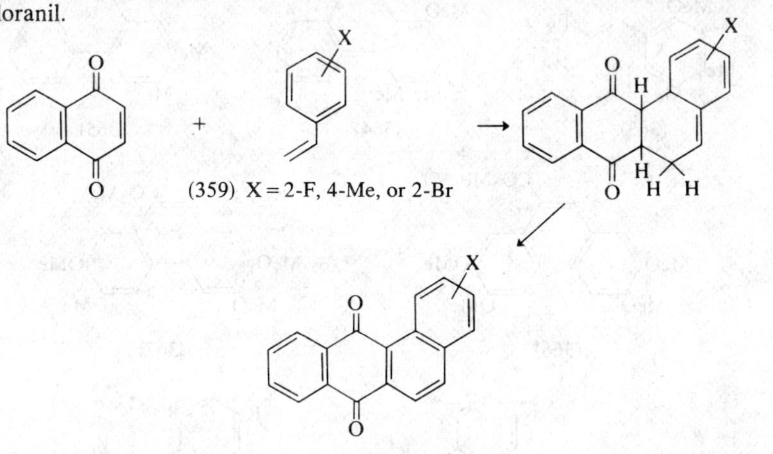

(359) X = 2-F, 4-Me, or 2-Br

(358)

**Scheme 44**

The Diels–Alder reaction[175] of 9,10-dichloro-anthracenes with 1,1-dichloro-ethene gives a mixture of (360) and (361), and an interesting high-temperature reaction[176] of hydrogen sulphide with the 9-halogeno-derivatives of polynuclear aromatic hydrocarbons led to the corresponding sulphide $R_2S$, *e.g.* (362).

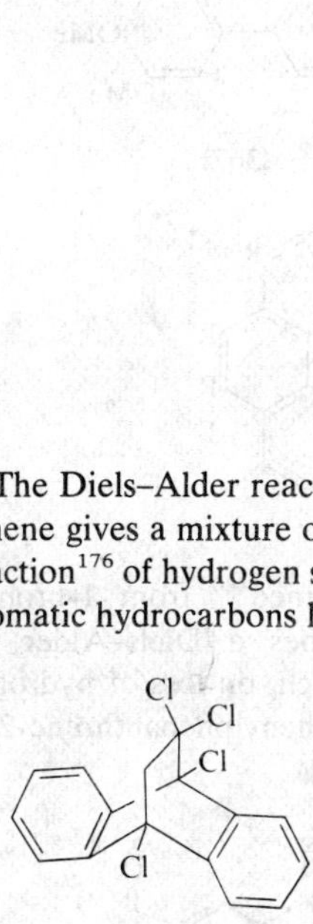

(360)

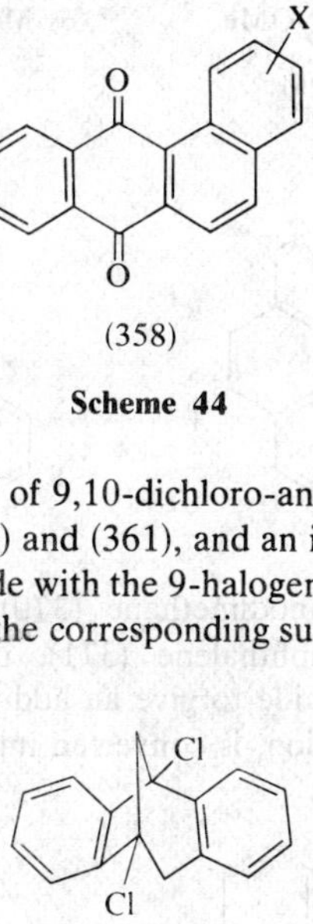

(361)

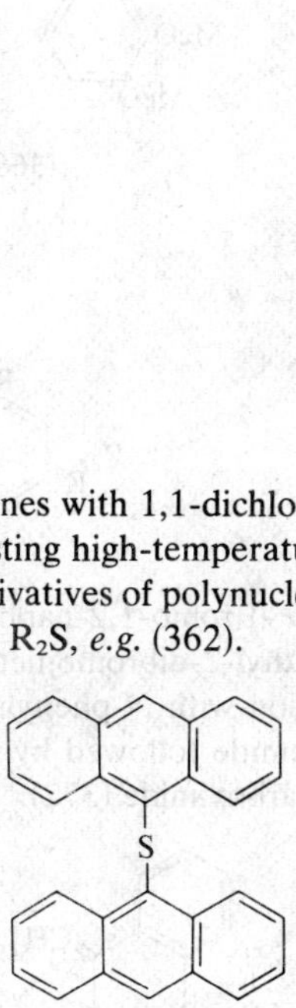

(362)

*Phenanthrene and Derivatives.* Activity in the field of phenanthrene synthesis has been considerable. Photodehydrocyclization of stilbenes has been mentioned earlier,[12] and a similar reaction[177] of 2-methoxy-4,5-dimethylstilbene (363) leads to 1-methoxy-3,4-dimethylphenanthrene (364) and 2,3-dimethylphenanthrene (365).

Intramolecular coupling of the related 1,2-diaryl-ethenes (366) to give (367) has been brought about[178] by using vanadium oxyfluoride in TFA and $CH_2Cl_2$,

[174] W. B. Manning, J. E. Tomaszewski, G. M. Muschik, and R. I. Sato, *J. Org. Chem.*, 1977, **42**, 3465.
[175] T. Miettinen, *Acta Chem. Scand.* (*B*), 1977, **31**, 761.
[176] M. G. Voronkov, E. N. Deryagina, and G. M. Ivanova, *J. Org. Chem.* (*U.S.S.R.*), 1977, **13**, 2391.
[177] E. N. Marvell, J. K. Reed, W. Gaenzler, and H. Tong, *J. Org. Chem.* 1977, **42**, 3783.
[178] A. J. Liepa and R. E. Summons, *J.C.S. Chem. Comm.*, 1977, 826.

and the photocyclization[179] of 2-styryl-naphthalenes (368) results similarly in alkyl-substituted benzo[c]phenanthrenes (369).

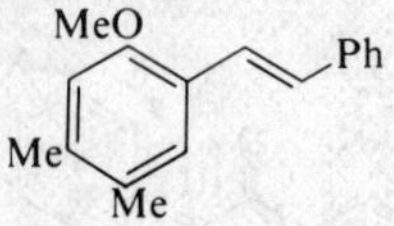

(363) 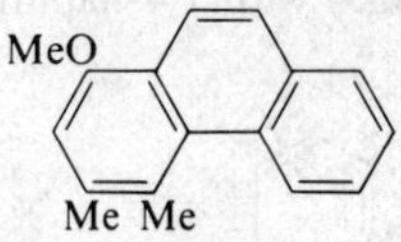                          (364) 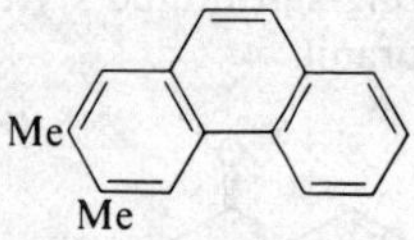                          (365)

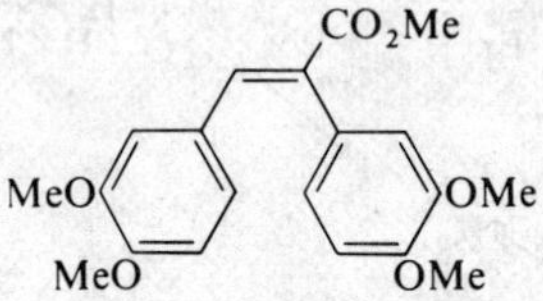

(366) 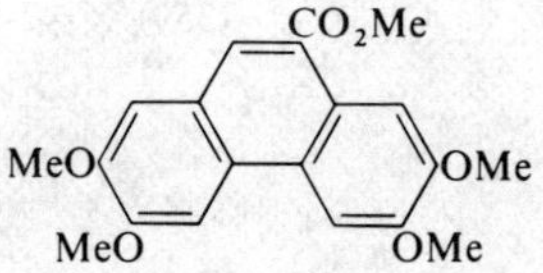                                        (367)

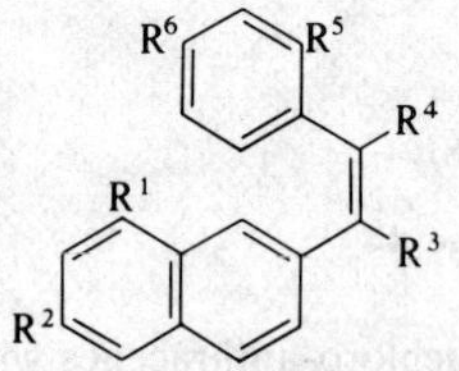

(368) 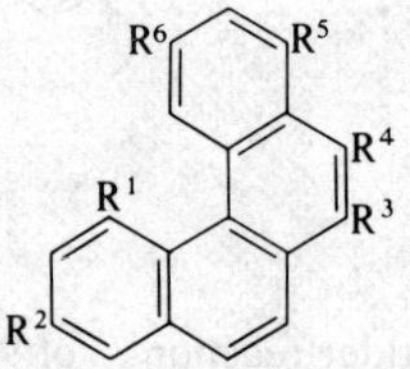                              (369)

$\alpha'$-Bromo-1,2-naphthoquinodimethane (370), obtained[180] from 1-bromo-methyl-2-dibromomethyl-naphthalene (371), undergoes a Diels–Alder reaction with N-phenylmaleimide to give an adduct which, on loss of hydrogen bromide followed by oxidation, is converted into N-phenylphenanthrene-2,3-dicarboximide (372).

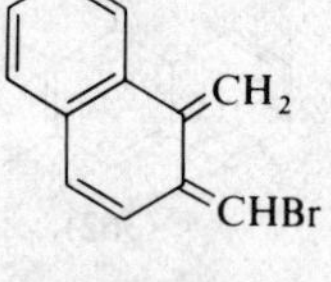

(370) 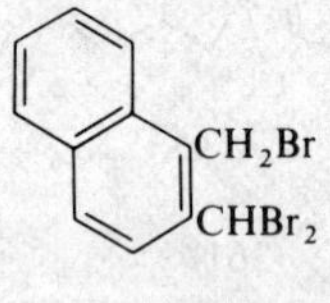                              (371) 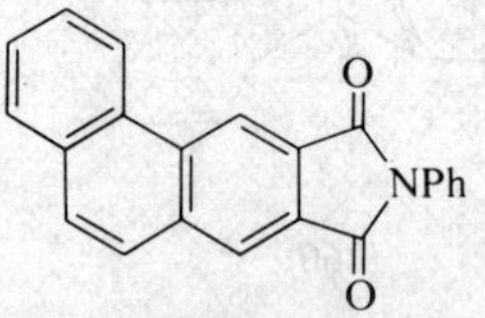                              (372)

Another approach to the synthesis of phenanthroid compounds that is of general applicability utilises the condensation[181] of protected p-quinones such as the monoketal (373) with phenethyl carbanions, resulting in the p-quinol (374), which cyclizes to the phenanthrene (375).

The hydrophenanthrene (376) has been obtained[182] by standard reactions of the adduct (377) formed by Diels–Alder reaction of the benzyne (378) with the

[179] D. L. Nagel, R. Kupper, K. Antonson, and L. Wallcave, *J. Org. Chem.*, 1977, **42**, 3626.
[180] G. W. Gribble, E. J. Holubowitch, and M. C. Venuti, *Tetrahedron Letters*, 1977, 2857.
[181] D. A. Evans, P. A. Cain, and R. Y. Wong. *J. Amer. Chem. Soc.*, 1977, **99**, 7083.
[182] W. Tochtermann, G. Frey, and H. A. Klein, *Annalen*, 1977, 2018.

tetrahydrobenzo[*b*]furan (379), and a similar reaction[183] of 4,5-dimethoxybenzyne (380) (generated from 4,5-dibromoveratrole and Bu$^n$Li) with the tetrahydrobenzo[*b*]furanol (381) gives the hexahydro-epoxyphenanthrenol (382), which may be converted into the phenanthrene (383).

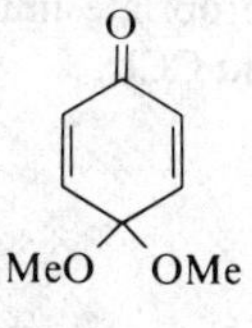

(373) 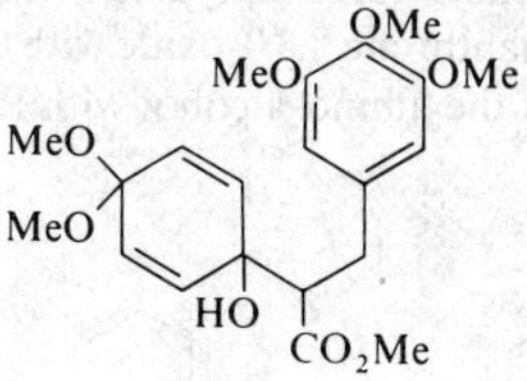    (374) 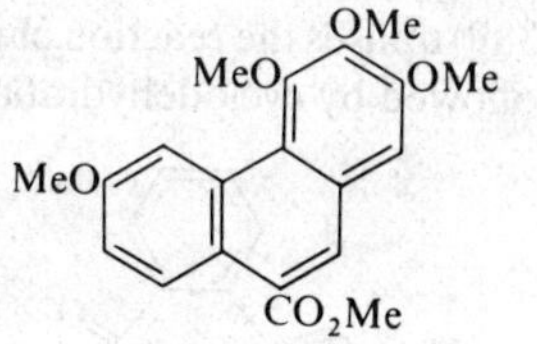    (375)

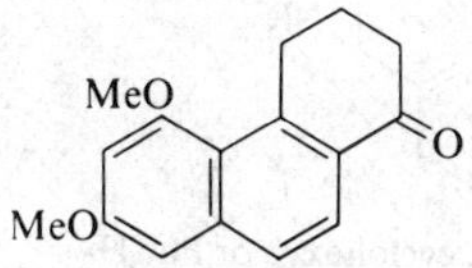

(376) 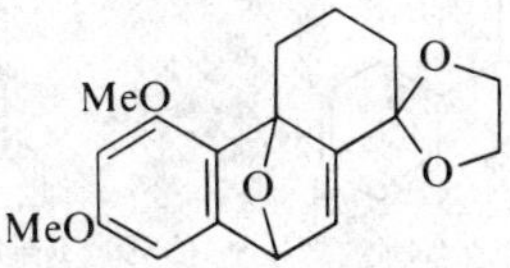    (377) 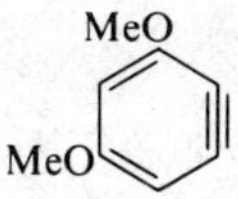    (378)

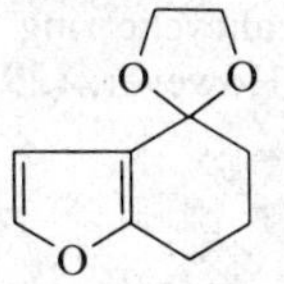

(379) 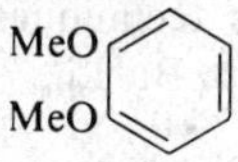    (380) 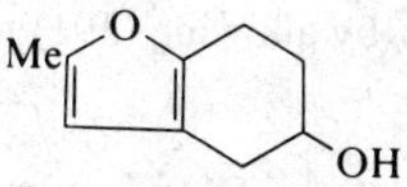    (381)

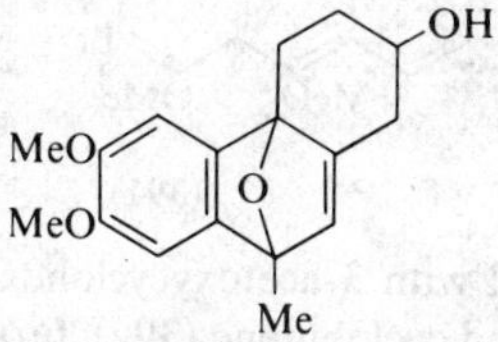

(382) 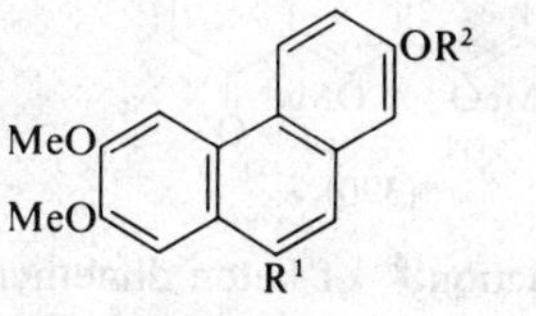    (383)

The mechanism of the photoconversion[184] of α-phenylcinnamic esters, *e.g.* (384), in a protic solvent, into 9,10-dihydrophenanthrenes has been studied. The primary step is the concerted photo-induced electrocyclisation of (384) to (385), which undergoes proton exchange with the solvent to give (386). The final product (387) results from a homolytic H abstraction–addition reaction.

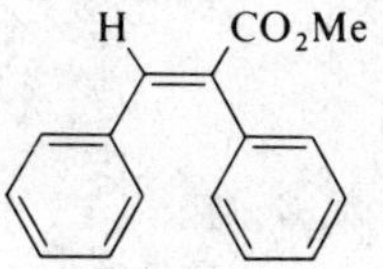

(384) 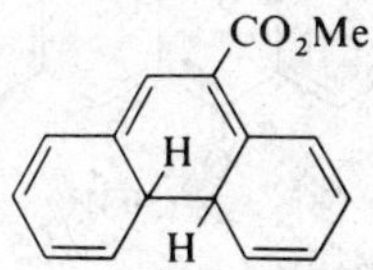    (385) 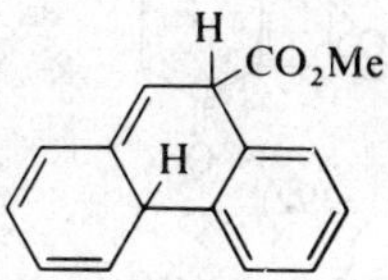    (386) 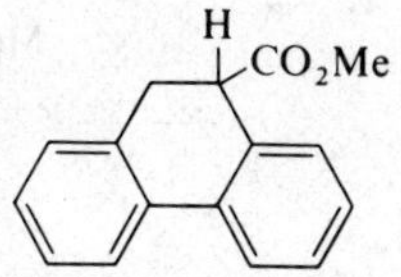    (387)

[183] W. Tochtermann, R. Strickler, H. A. Klein, and E. Biegi, *Chem. Ber.*, 1977, **110**, 2456.
[184] P. H. G. Ophet Veld and W. H. Laarhoven, *J. Amer. Chem. Soc.*, 1977, **99**, 7221.

Interest has been shown in phenanthraquinone monoxime. The reactions of this compound have been reviewed[185] and particular attention has been paid to its rearrangement to (388).

Stable *N*-alkyl-azirino-arenes have been reported.[186] Thus a general route to 1-butyl-, 1-cyclohexyl-, and 1-benzyl-1a,9b-dihydrophenanthra[9,10-*b*]azirines (389) utilises the reaction of phenanthrene 9,10-oxide with the appropriate amine followed by cyclodehydration of the amino-alcohol with PPh$_3$ and CCl$_4$.

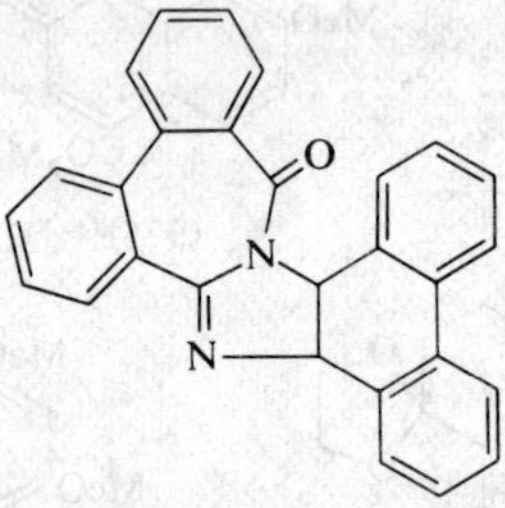

(388) 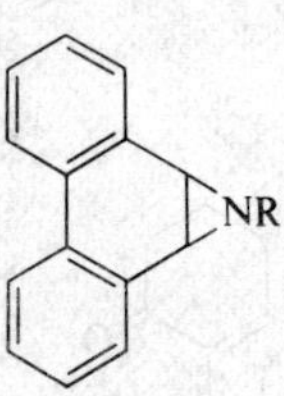                    (389)   R = Bu$^n$, cyclohexyl, or PhCH$_2$

*Polycyclic Systems.* Several reports have appeared on the synthesis of naphthacenes. The introduction of a functionality into the alicyclic ring cannot be effected by conventional Friedel–Crafts techniques. However, (390) can be obtained[187] by allowing (391) to react with Bu$^n$Li.

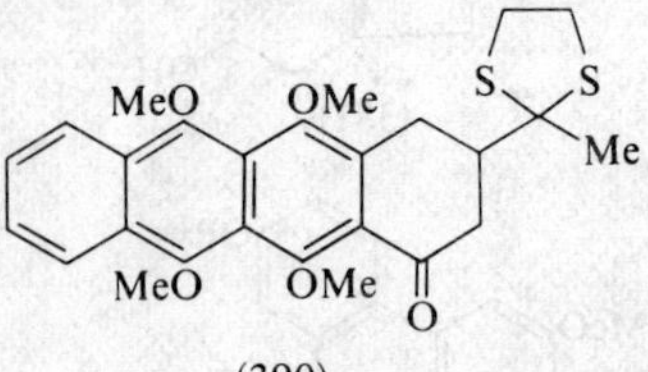

(390) 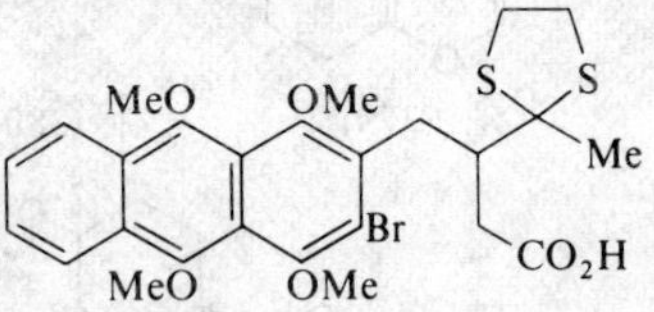                              (391)

The reaction[188] of keten dimethylacetal with 3-acetoxycyclohex-2-en-1-one gives rise to the stable, easily handled, fused cyclobutene (392), further reaction of which, followed by oxidation, resulted in the naphthacenetrione (393). The naphthacenes (394) originate[189] by dimerization of the polyphenyl-allenes R$^1$CCl=C=CR$^2$Ph (R$^1$ = Ph, R$^2$ = Cl; R$^1$ = Cl, R$^2$ = Ph), by way of bis-allyl diradicals, the dihydronaphthalenes (395), and then the naphthocyclobutenes (396).

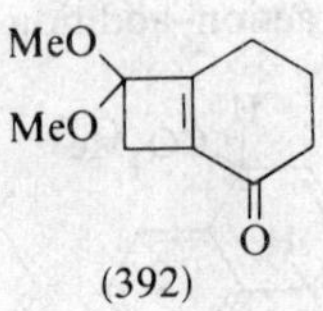

(392) 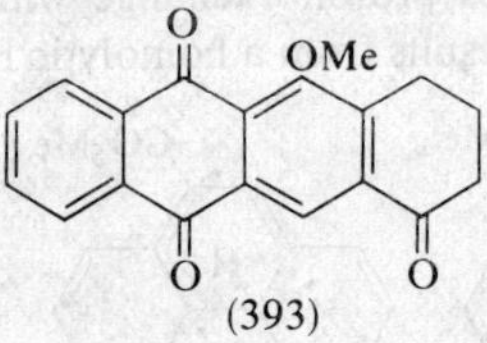                              (393)

[185] S. Ranganathan and C. S. Panda, *Heterocycles*, 1977, **7**, 529.
[186] Y. Ittah, I. Shahak, and J. Blum, *J. Org. Chem.*, 1978, **43**, 397.
[187] R. J. Boatman, B. J. Whitlock, and H. W. Whitlock, Jr., *J. Amer. Chem. Soc.*, 1977, **99**, 4822.
[188] R. K. Boeckman, Jr., M. H. Delton, T. Nagasaka, and T. Watanabe *J. Org. Chem.*, 1977, **42**, 2946.
[189] J. Rigaudy and J. Cardevielle, *Tetrahedron*, 1977, **33**, 767.

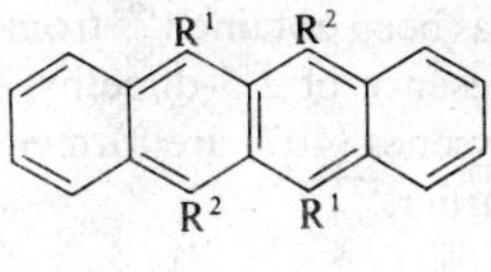

(394) 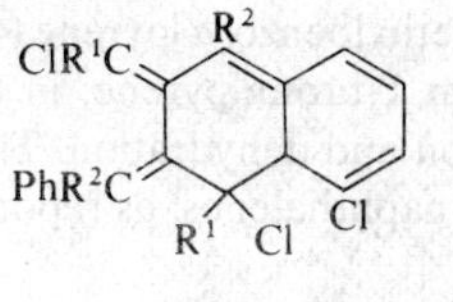          (395) 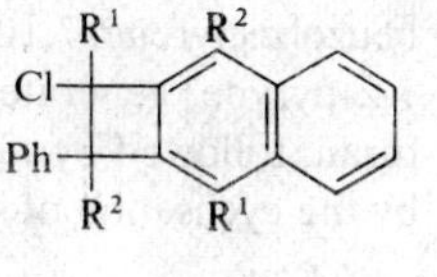          (396)

Pyrene and 4-methyl- and 4,5-dimethyl-pyrene have been prepared from 4*H*-cyclopenta[*def*]phenanthrenes by a reaction sequence[190] involving carboxylation and then conversion, through the ester, into the carbinol, which undergoes Wagner–Meerwein rearrangement. 5,12-Diphenylnaphthacene undergoes photochemical oxidation,[191] giving a mixture of endoperoxides (397) and (398); the major product, (397), was subsequently converted into the diol (399).

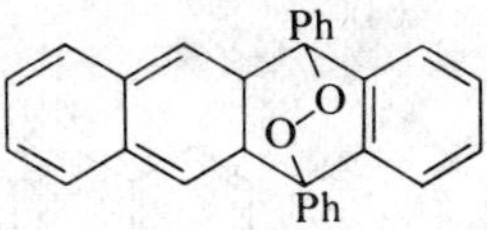

(397) 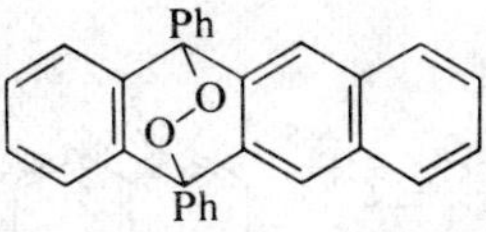          (398) 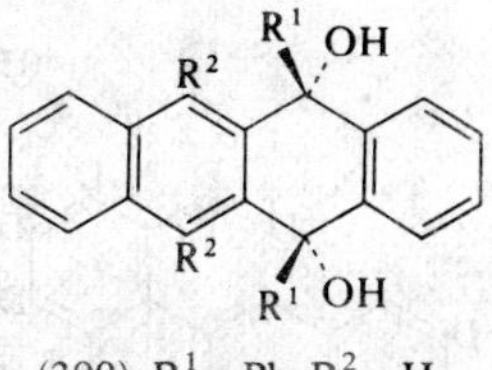          (399) R$^1$ = Ph, R$^2$ = H;
                                     R$^1$ = H, R$^2$ = Ph

Some interesting compounds have been obtained in the anthrone series. The reaction of 10-methyleneanthrone with acid catalysts in HOAc gives[192] anthrone-10-spiro-1'-cyclobutane-2'-spiro-10'-anthrone (400), further treatment of which

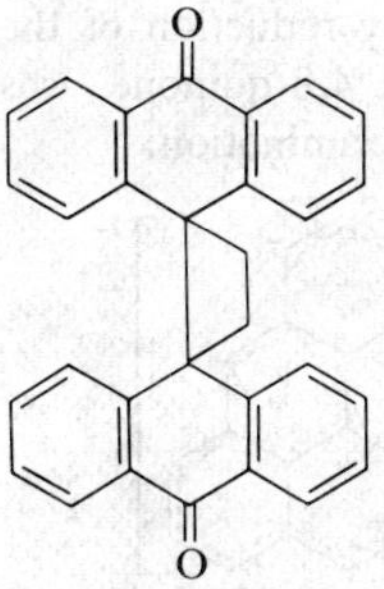

(400) 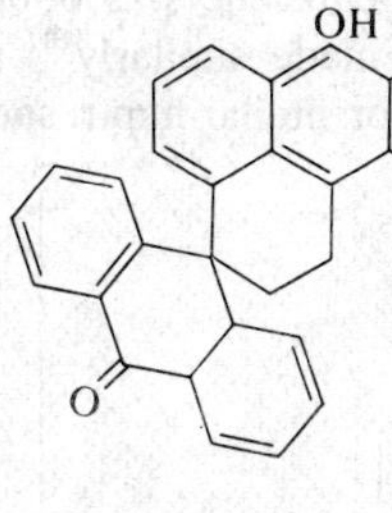          (401) 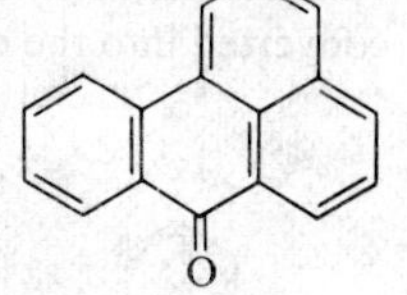          (402)

leads to (401). Benzanthrone (402), on heating with a mixture of Cu, ZnCl$_2$, and NaCl, produces[193] violanthrene B (403) and isoviolanthrene B (404).

Perdeuteriated polycyclic arenes have been employed to test hypotheses of chemical carcinogenesis, and in this context perdeuteriobenzo[*a*]pyrene (405) has been prepared[194] by the reaction of deuteriobenzene, Br$_2$, and AlBr$_3$ in a sealed ampoule, the reaction product being similarly treated with

[190] T. Kimuna, M. Minabe, and K. Suzuki, *J. Org. Chem.*, 1978, **43**, 1247.
[191] J. Rigaudy and D. Sparfel, *Bull. Soc. chim. France*, 1977, 742.
[192] K. Hirakawa and S. Nakogawa, *J. Org. Chem.*, 1978, **43**, 1804.
[193] J. Aoki, M. Takekawa, S. Fujisawa, and S. Iwashima, *Bull. Chem. Soc. Japan*, 1977, **50**, 1017.
[194] J. C. Seibles, D. M. Bollinger, and M. Orchin, *Angew. Chem. Internat. Edn.*, 1977, **16**, 656.

benzo[*a*]pyrene. 7,10-Dimethylbenzo[*a*]pyrene (406) has been obtained[195] from
1,2-pyryne, generated from 1-bromopyrene, in the presence of 2,5-dimethyl-
furan, followed by reduction and dehydration. The chrysenes (407) are formed
by the cyclisation of styryl-naphthalenes, as reported earlier.[177]

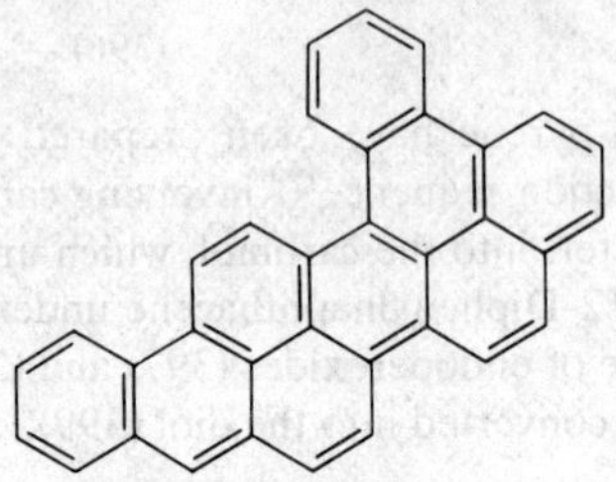

(403) 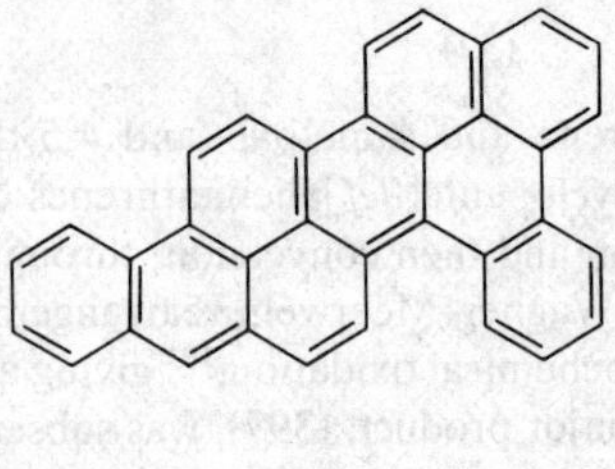                                                                (404)

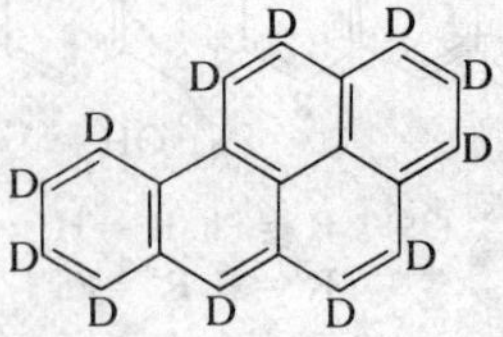

(405) 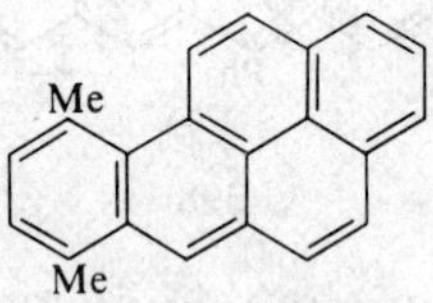                                     (406) 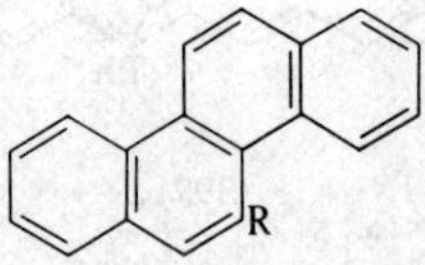                        (407) R = Me or Et

The dibenzo[*dc, jk*]pentacenyl cation (408) and the corresponding dianion
(409) have both been prepared[196] from the corresponding hydrocarbon, and their
n.m.r. spectra interpreted. The pentacene was obtained by reduction of the
7,9-quinone, and a triangulene, made similarly[197] from a 4,8-quinone, was
converted into the dianion (410) for similar n.m.r. spectral examination.

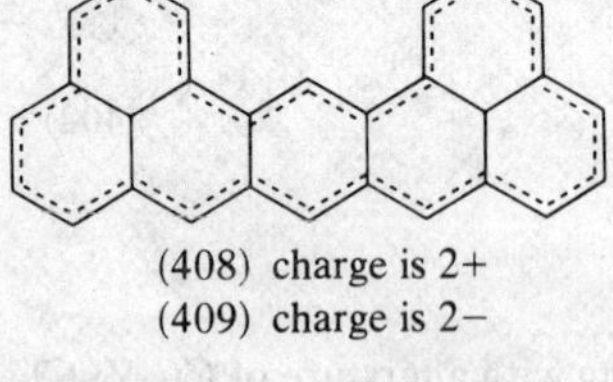

(408) charge is 2+                                     (410)
(409) charge is 2−

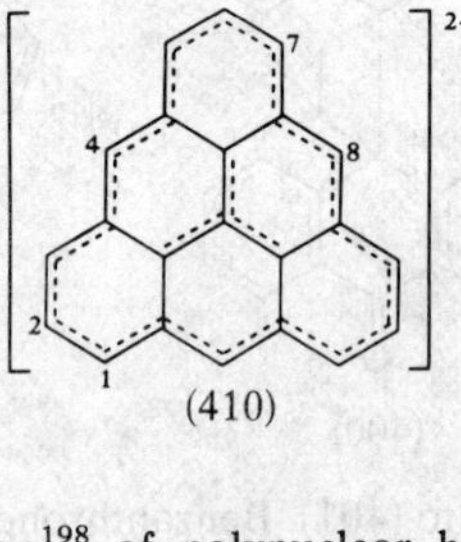

Studies on the mild regiospecific hydrogenation[198] of polynuclear hydrocarbons over Pd/C has led to the conversion of (411) into (412). The spironaph-
thalene (413)[199] and 5,5′-disubstituted 2,2′-spirobi-indane (414)[200] have been
described.

195  M. S. Newman and S. Kumar, *J. Org. Chem.*, 1977, **42**, 3284.
196  O. Hara, K. Yamamoto, and I. Murata, *Tetrahedron Letters* 1977, 2431.
197  O. Hara, K. Tanaka, K. Yamamoto, T. Nakazawa, and I. Mureta, *Tetrahedron Letters*, 1977, 2435.
198  P. P. Fu and R. G. Harvey, *Tetrahedron Letters*, 1977, 415.
199  D N. Chatterjee and S. R. Chakraborty, *J. Indian Chem. Soc.*, 1976, **53**, 812.
200  H. Neudeck and K. Schloegl, *Chem. Ber.*, 1977, **110**, 2624.

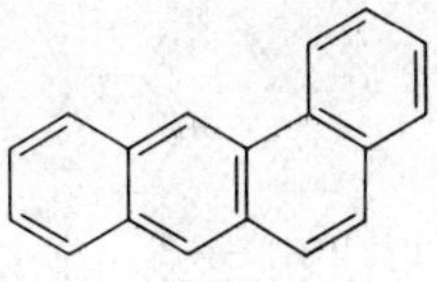

(411) 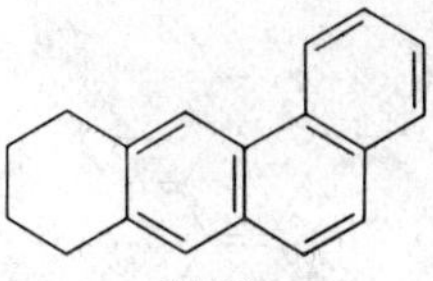　　　　　　　　　(412)

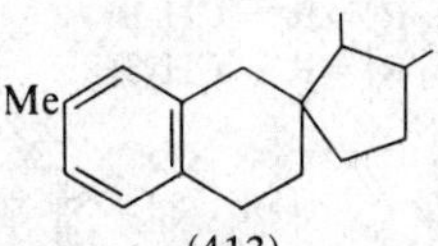

(413) 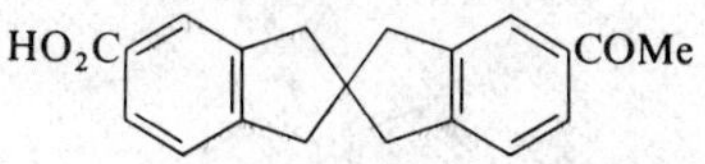　　　　　　　　　(414)

*Helicenes.* A systematic investigation of chemically induced asymmetric photosyntheses of helicenes has led to the preparation[201] of hexahelicenes (415) from the 1,2-diaryl-ethenes (416); the bridged helicene (417) has been obtained[202] from the heptahelicene diester (418) by standard reactions.

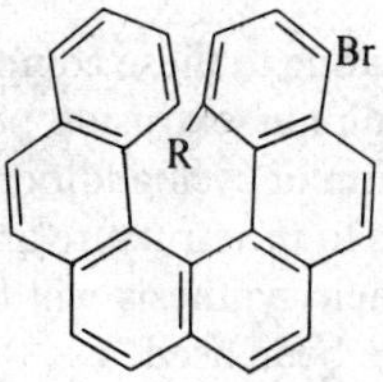

(415) 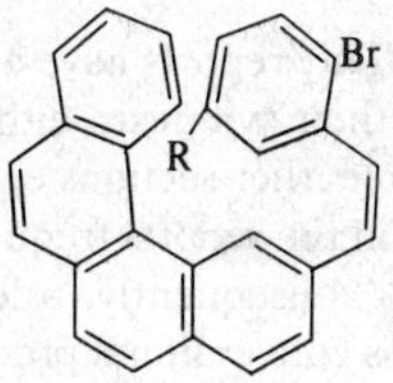　　　　　　　　　(416)

$R = CO_2CH(Me)Bu^t$

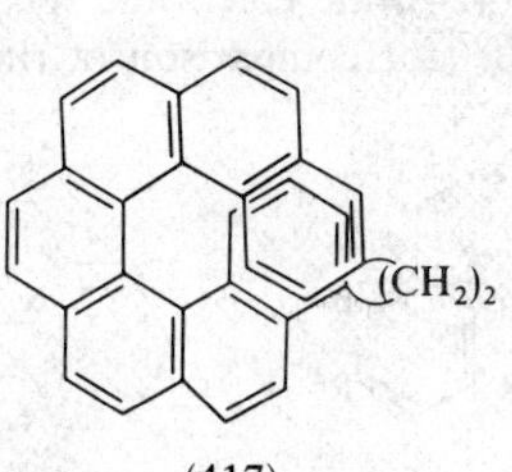

(417) 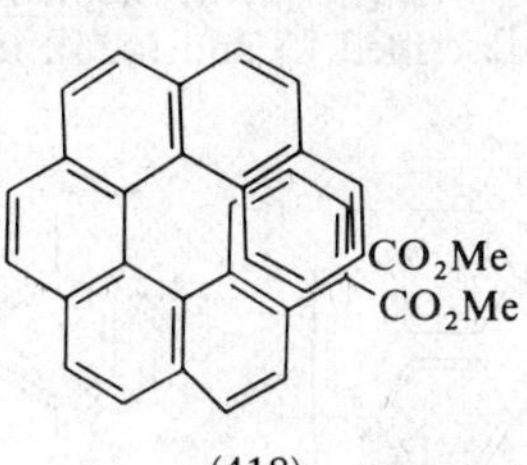　　　　　　　　　(418)

The first representative of a new type of helical structure (419), having features of both helicenes and cyclophanes, has been formed[203] by pyrolysis of a sulphone made from the bis(bromomethyl)-quaterphenyl (420a) and the corresponding dithiol (420b). An intramolecular carbene-insertion reaction involving the helicene skeleton has been reported[204] in which 1-formyl-[6]helicene toluene-*p*-sulphonylhydrazone (421) is converted into (422), the structure of which has been determined unambiguously by $^1H$ and $^{13}C$ n.m.r. spectroscopy.

[201] Y. Cochez, R. H. Martin, and J. Jespers, *Israel J. Chem.*, 1977, **15**, 29.
[202] M. Joly, N. Defay, R. H. Martin, J. P. Declerq, and G. Germain, *Helv. Chim. Acta*, 1977, **60**, 537.
[203] F. Voegtle, M. Atzmueller, W. Wehner, and J. Gruetze, *Angew. Chem. Internat. Edn.*, 1977, **16**, 325.
[204] J. Jespers, N. Defay, and R. H. Martin, *Tetrahedron*, 1977, **33**, 2142.

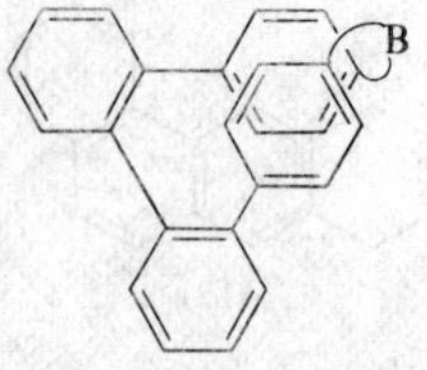

(419)  Ḃ = CH$_2$S(CH$_2$)$_2$SCH$_2$

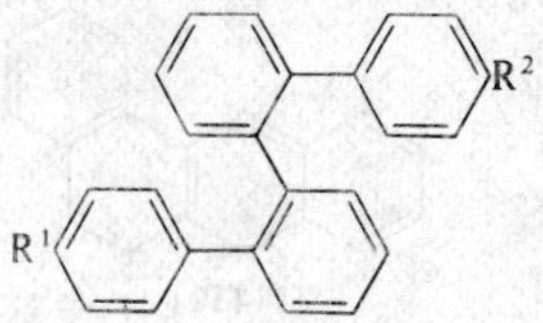

(420)  a; R$^1$ = R$^2$ = CH$_2$Br
       b; R$^1$ = R$^2$ = CH$_2$SH

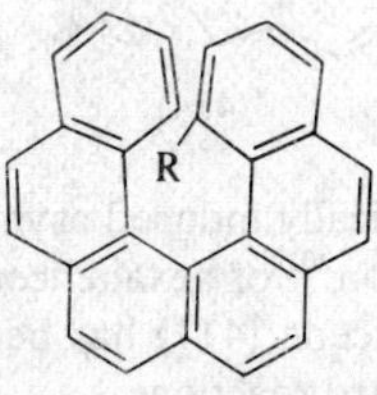

(421)  R = CH=NNHO$_2$SC$_6$H$_4$Me-4

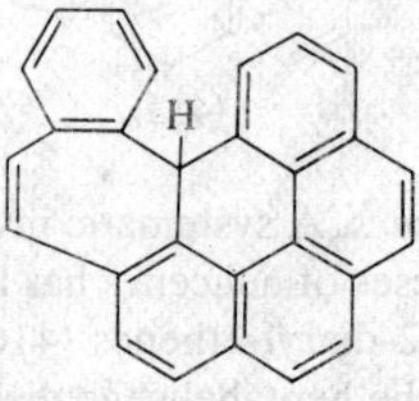

(422)

*Cyclophanes.* Many reports have appeared relating to these compounds. Most of them describe their synthesis, either by the sulphur-extrusion process reported several times in earlier sections or by some form of cycloaddition reaction.

The novelty in the reports frequently relates to the structure and properties of the compound; consequently, reference to their synthesis will be restricted to cases where less conventional procedures have been used.

Among those obtained by C—C bond formation resulting from sulphur extrusion is the compound (423), in which the benzene rings are perpendicular to each other,[205] the [3.3]paracyclophanes (424), the electronic spectra of which have been described,[206] and (425), used[207] for comparison of ring strain and

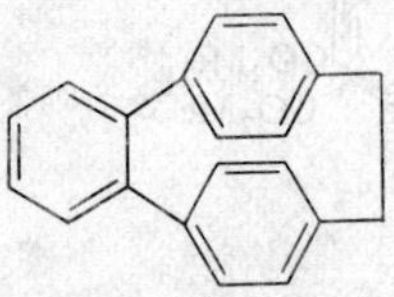

(423)

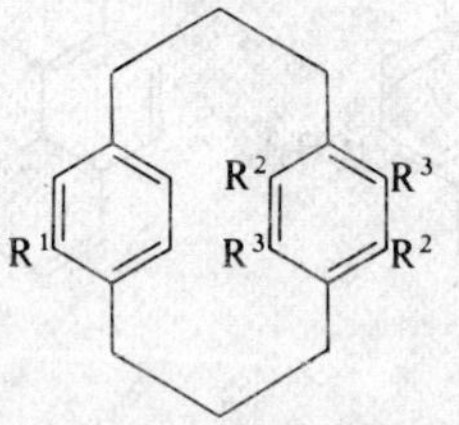

(424)  R$^1$ = CN; R$^2$, R$^3$ = H, OMe

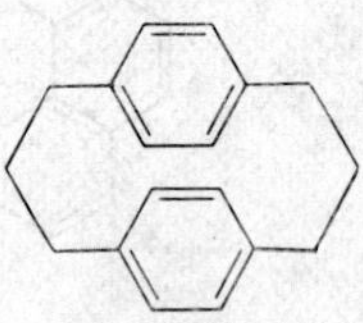

(425)

transannular effects, lying between the [2.2]paracyclophanes, where such effects are pronounced, and the [4.4]paracyclophanes, where they are absent. The [3.3]metabenzophane (426)[208] and [3.3]cyclophane (427)[209] have been prepared similarly.

[205]  N. Jacobson and V. Bockelheide, *Angew. Chem. Internat. Edn.*, 1978, **17**, 46.
[206]  M. W. Haenel, A. Flatow, V. Toglieber, and H. A. Staab, *Tetrahedron Letters*, 1977, 1733.
[207]  D. T. Longone, S. H. Kusefoglu, and J. A. Gladysz, *J. Org. Chem.*, 1977, **42**, 2787.
[208]  L. Rossa and F. Voegtle, *J. Chem. Res. (S)*, 1977, 264.
[209]  T. Otsubo, M. Kitosawa, and S. Misumi, *Chem. Letters*, 1977, 977.

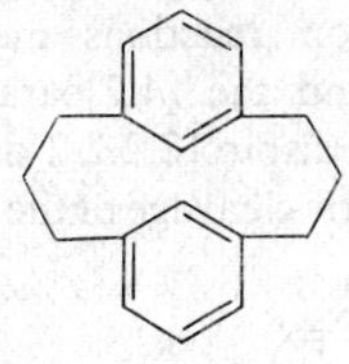

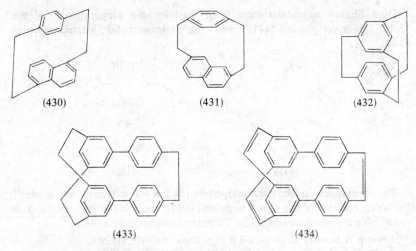

(426) 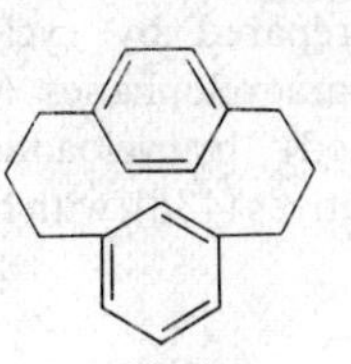          (427)

An attempted preparation[210] of [7]circulene failed, but led to the formation of [2.2](3,6)phenanthro(2,7)naphthaleno-1,11-diene (428), and the synthesis of the similar naphthalenophanediene (429) has been reported,[211] it being formed in high yield.

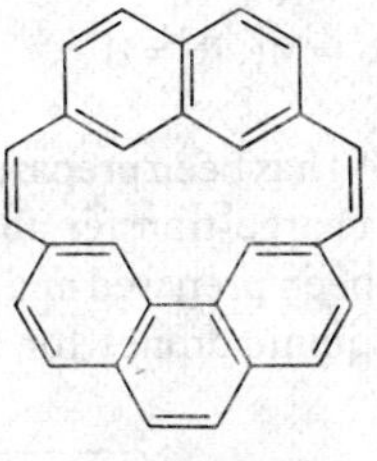

(428) 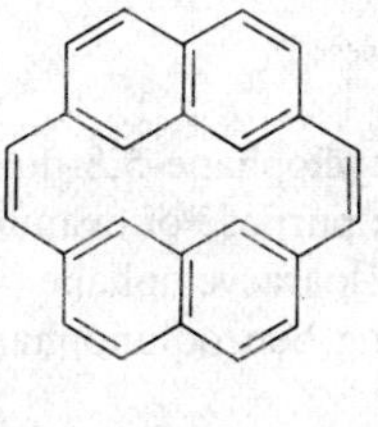          (429)

The [2]naphthaleno[2]paracyclophanes (430) and (431) have been synthesized[212] with a view to observing strain and deformation of the naphthalene unit. [2.2.2](1,2,4)(1,3,5)Cyclophane (432) was made[213] for examination of its strained and skew structure, and the novel triply clamped biphenylophane hydrocarbons (433) and (434) have been prepared for similar steric purposes.[214]

(430)          (431)          (432)

(433)          (434)

[210] P. J. Jessup and J. A. Reiss, *Austral. J. Chem.*, 1977. **30**, 851.
[211] P. J. Jessup and J. A. Reiss, *Austral. J. Chem.*, 1977, **30**, 843.
[212] M. W. Haenel, *Tetrahedron Letters*, 1977, 4191.
[213] M. Nagazaki, K. Yamamoto, and Y. Miura, *J.C.S. Chem. Comm.* 1977, 206.
[214] F. Voegtle and G. Steinhagen, *Chem. Ber.*, 1978, **111**, 205.

Cyclophanes prepared by cycloaddition reactions include[215] the tetra-substituted [2.2]paracyclophanes (435) and the [4.2]paracyclophanes (436), these being obtained[216] by cycloaddition of dispiro[2.2.2.2]deca-4,9-diene (437) with styrene derivatives (438), with homolytic cleavage of the cyclopropane rings.

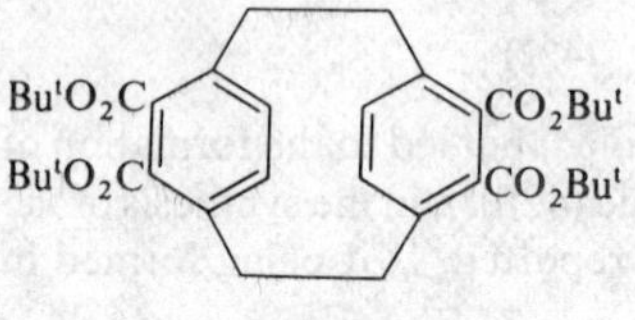

(435) 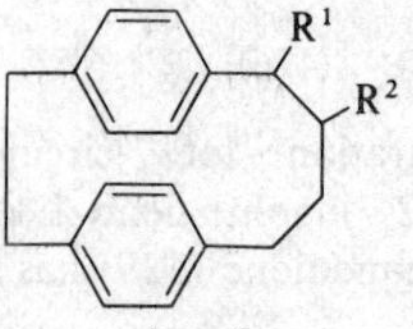          (436)  $R^1 = R^2 = H$,  
                      $R^1 = Me, R^2 = H$

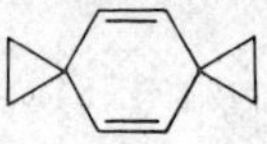

(437)

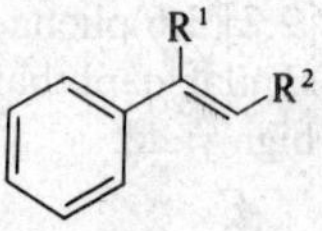

(438)

The [3.3]paracyclophane-5,8-quinone (439) has been prepared[217] by a series of reactions for the purpose of examining the charge-transfer absorption, and the four-layered [2.2]paracyclophane (440) has been prepared and converted[218] into the corresponding benzoquinophanes and quinhydrones for similar purposes.

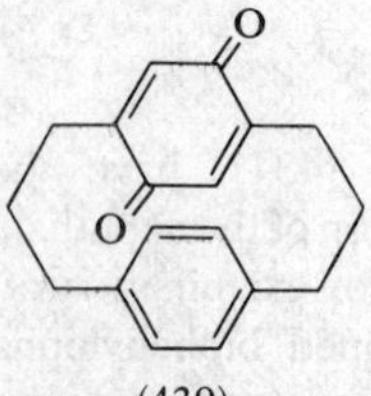

(439) 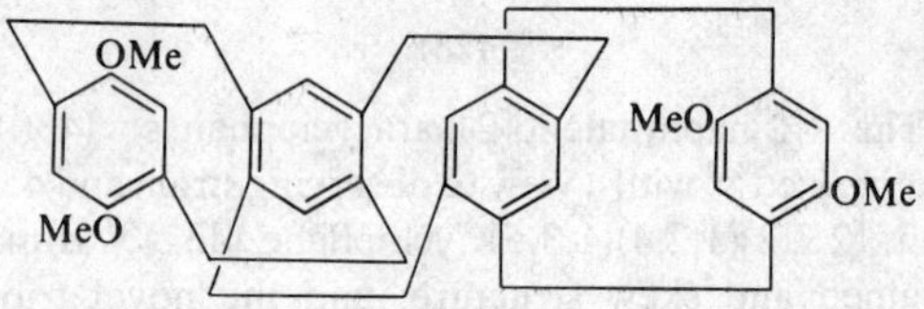                    (440)

High-dilution techniques have been used for the preparation[219] of the [2.1.2.1]paracyclophane (441) and the tetramethyl[3.3]paracyclophanes (442).[220]

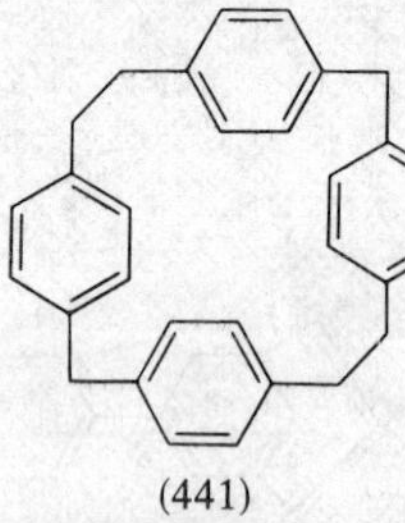

(441) 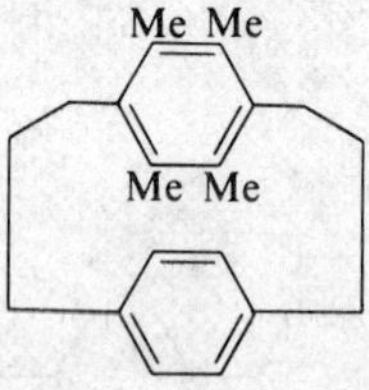                    (442)

The novel short-bridged paracyclophanes (443; $n = 7$ or 8) have been made[211] by Wittig reaction of the spiro-compound (444), and a bis-Wittig reaction was used[222] for the synthesis of the [2.0.2.0]metacyclophanediene (445).

[215] I. Boehm, H. Hermann, K. Menke, and H. Hopf, *Chem. Ber.*, 1978, **111**, 523.
[216] T. Shibata, T. Tsuji, and S. Nishida, *Bull. Chem. Soc. Japan*, 1977, **50**, 2039.
[217] T. Shinmyozu, T. Inazu, and T. Yoshino, *Chem. Letters*, 1977, 1347.
[218] H. A. Staab, U. Zapf, and A. Gurkel, *Angew. Chem. Internat. Edn.*, 1977, **16**, 801.
[219] C. Sergheraert, P. Marcinal and E. Guingnet, *Tetrahedron Letters*, 1977, 2879.
[220] T. Synmyozu, K. Kumagae, T. Inazu, and T. Yoshino, *Chem. Letters*, 1977, 43.
[221] J. W. Van Straten, W. H. DeWolf, and F. Bickelhaupt, *Rec. Trav. Chim.* 1977, **96**, 88.
[222] B. Thulin and O. Wennerstrom, *Tetrahedron Letters*, 1977, 929.

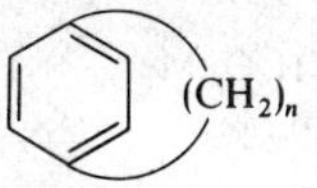

(443) 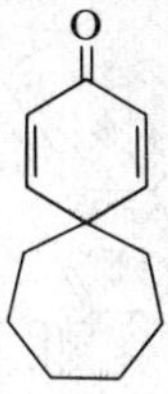  (444) 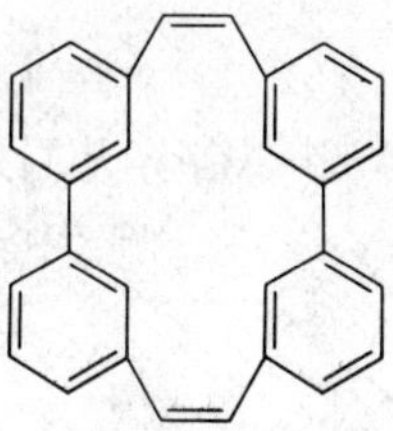  (445)

A new cage molecule $[2_6](1,4)_3(1,3,5)_2$-bicyclophanehexaene (446) has resulted[223] from a multi-stage reaction sequence, and the [10]paracyclophanes (447) from a mixture of two inside–inside [10.2.2]hexadecadienes (448) and (449) obtained[224] by cycloaddition of methyl propiolate with cyclodecene.

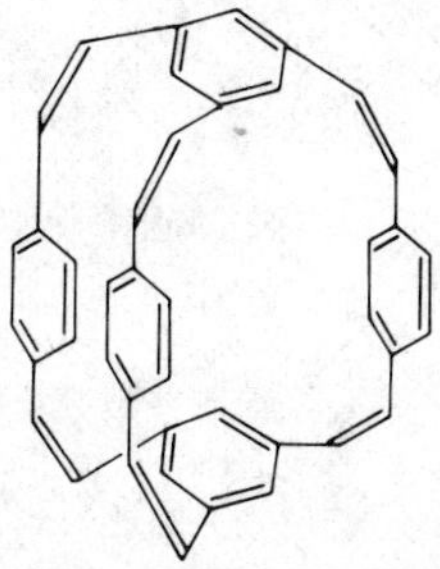

(448)

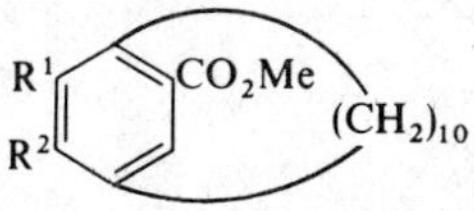

(447)  $R^1 = H$, $R^2 = CO_2Me$
  $R^1 = CO_2Me$, $R^2 = H$

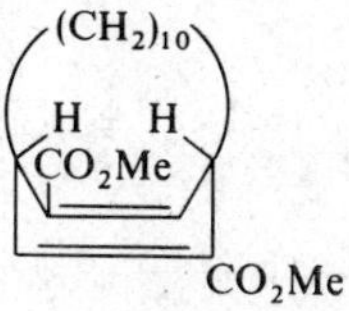

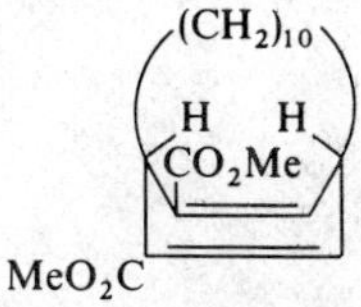

(449)

(446)

The [2.2](2,7)pyrenophane (450) has been formed[225] in six stages from 5,3-dimethyl[2.2]metacyclophane, and the hexaoxa[6.6.6](1,3,5)cyclophane (451) made[226] in the same number of steps from phloroglucinol

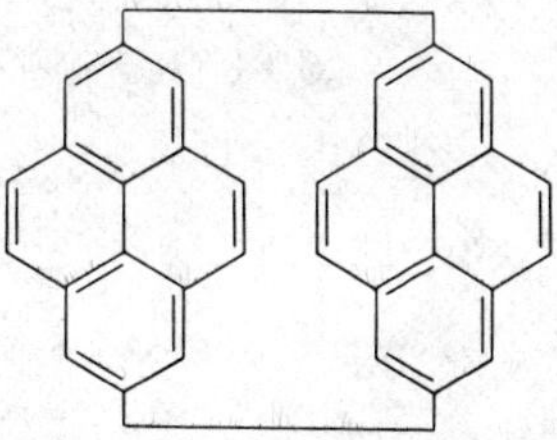

(450) 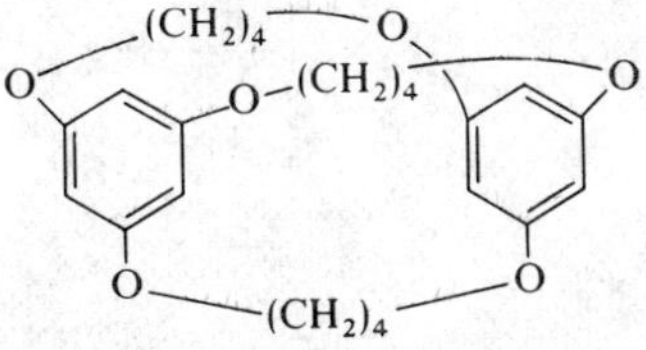  (451)

Octopus molecules in the cycloveratrylene series, such as (452), capable of adopting cavity-containing conformations, have been prepared.[227] A multi-stage

[223] H. E. Hogberg, B. Thulin, and O. Wennerstrom, *Tetrahedron Letters*, 1977, 931.
[224] B. B. Snider and N. J. Hrib, *Tetrahedron Letters*, 1977, 1725.
[225] H. Irngartinger, R. G. H. Kirrstetter, C. Krieger, H. Rodewald, and H. A. Staab, *Tetrahedron Letters*, 1977, 1425.
[226] W. D. Curtis, J. F. Stoddart, and G. H. Jones, *J.C.S. Perkin I*, 1977, 785.
[227] J. A. Hyatt, *J. Org. Chem.*, 1978, **43**, 1808.

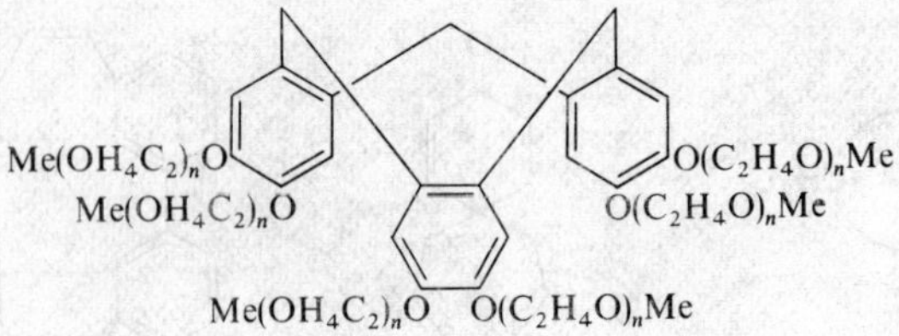

$Me(OH_4C_2)_nO$
$Me(OH_4C_2)_nO$
$O(C_2H_4O)_nMe$
$O(C_2H_4O)_nMe$

$Me(OH_4C_2)_nO$   $O(C_2H_4O)_nMe$

(452)

synthesis[228] has led to the paracycloannulenophane (453) and the naphthalenoannulenophane (454).

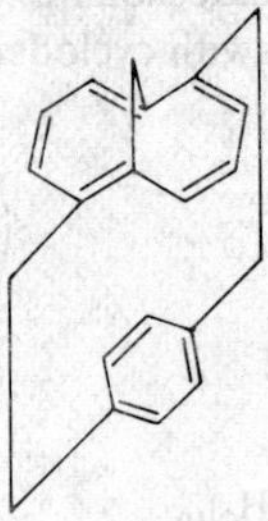

(453)

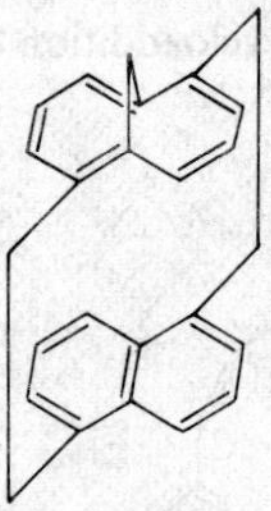

(454)

[228] M. Matsumoto, T. Otsubo, Y. Sakata, and S. Misumi, *Tetrahedron Letters*, 1977, 4425.

# 4

# Six-membered Heterocyclic Systems

BY R. K. SMALLEY

## 1 Systems containing One Heteroatom, and their Benzo-derivatives

**Pyridines.**—*Synthesis.* Vinylamidinium salts, *e.g.* [Me$_2$N$\overset{+}{=}$CHC(R$^1$)$=$CHNMe$_2$] ClO$_4^-$, condense with 3-amino-crotonates and 3-aminocrotononitrile in the presence of sodium methoxide to give 3,5-di-substituted-2-methyl-pyridines (2; R$^1$ = H, OMe, Me, or Ph; R$^2$ = CO$_2$alkyl or CN) in 52–95% yield.[1] Electro-cyclic ring-closure of the azahexatriene intermediate (1) is accompanied by loss of dimethylamine. Ethyl 3-ethoxymethylene-2,4-dioxovalerate [EtOCH= C(Ac)COCO$_2$Et] is advocated as a good precursor for the synthesis of sub-stituted nicotinamide derivatives.[2] For example, with β-aminocrotonamide in hot ethanol, nicotinamide (3; R = CONH$_2$) is produced in 36% yield. Similarly, condensation with β-aminocrotonate yields the diester (3; R = CO$_2$Et) (58%).

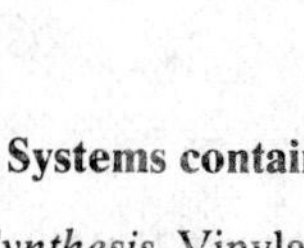

(1) 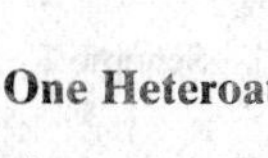     (2) 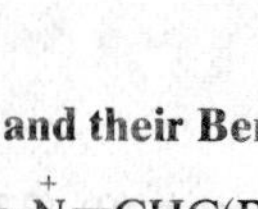     (3)

The versatility of 2*H*-azirines in the synthesis of larger ring heterocycles has been demonstrated further by the thermal rearrangement of their 2-allyl deriva-tives (4) to pyridines and indoles.[3] The allyl-azirine (4; R$^1$ = H, R$^2$ = Ph, R$^3$ = Me), on thermolysis in toluene at 180 °C, gives a minor amount (31%) of the 3-azabicyclo[3.1.0]hex-2-ene (5), as illustrated in Scheme 1, together with 3-allyl-2-methylindole (58%), the latter product being formed by direct vinylnitrene attack at the adjacent phenyl ring. Bicyclohexene (5), on further heating, rear-ranges to 2-methyl-3-phenylpyridine (6; R$^1$ = H).

In contrast, the allylic ester (4; R$^1$ = CO$_2$Me, R$^2$ = Ph, R$^3$ = Me), under similar conditions, yields the pyridine ester (6; R$^1$ = CO$_2$Me) (70%) directly along with the indole derivative (7) (30%). The isomeric allylic ester (4; R$^1$ = CO$_2$Me, R$^2$ = Me, R$^3$ = Ph), however, gives a mixture of pyridine ester (8) (47%) and

[1] C. Jutz, H. G. Lobëring, and K.-H. Trinkl. *Synthesis*, 1977, 326.
[2] T. Kurihara and T. Uno, *Heterocycles*, 1977, **6**, 547.
[3] A. Padwa and P. H. J. Carlsen, *Tetrahedron Letters*, 1978, 433.

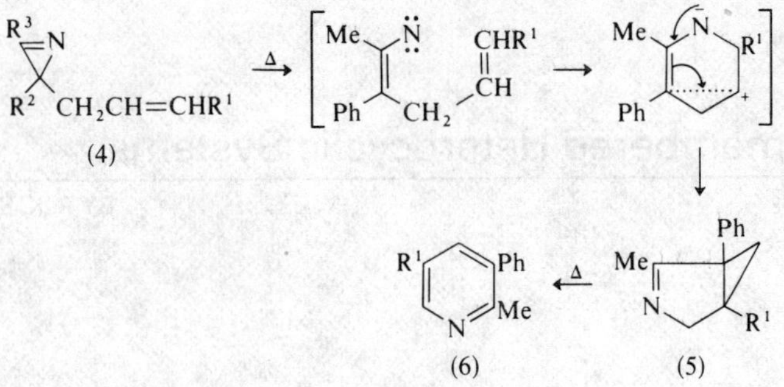

**Scheme 1**

pyrrole-acetic ester (9) (37%). Reasonable, but as yet tentative, mechanistic rationales have been advanced to explain these varied reactions.

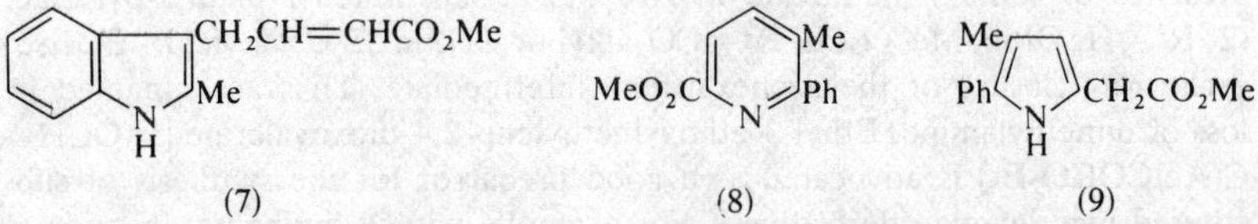

(7)                                  (8)                                  (9)

A remarkably simple two-stage synthesis of 2-aryl-pyridines from cyclo-pentadienes has been reported.[4] The 4a,7a-dihydrocyclopenta[e][1,2]oxazines (10), available from cyclopentadiene and $\alpha$-halogeno-ketoximes, at 200 °C undergo successive electrocyclic ring-opening and ring-closure to form the acyl-dihydropyridines (11), which aromatize by loss of acetaldehyde as illustrated in Scheme 2.

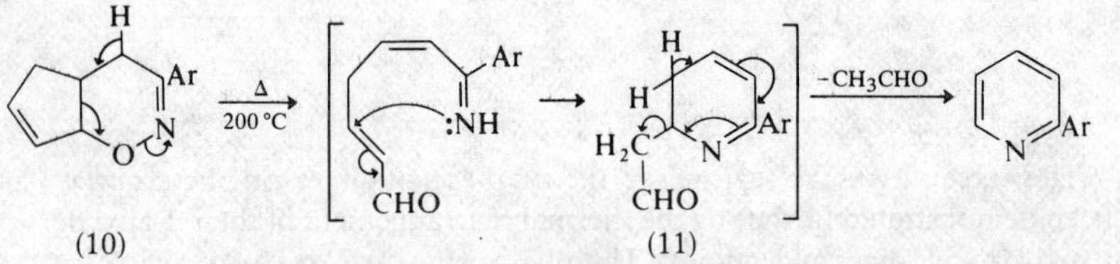

(10)                                              (11)

**Scheme 2**

A new one-stage synthesis of annelated pyridines, involving the cobalt-catalysed co-oligomerization of dialkynes $HC{\equiv}C(CH_2)_nC{\equiv}CH$ ($n$ = 3, 4, or 5) with nitriles RCN (R = alkyl or aryl), has been announced.[5] Under high dilution and in the presence of cyclopentadienylcobalt dicarbonyl, co-oligomerization to pyridines of type (12) takes place in a remarkably selective manner, yields of 43—81% being realised. The process is particularly useful for the preparation of the otherwise not easily accessible 5,6,7,8-tetrahydroquinolines. An interesting extension of the reaction uses ethyl cyanoacetate in place of the nitrile RCN,

[4] R. Faragher and T. L. Gilchrist, *J.C.S. Chem. Comm.*, 1977, 252.
[5] A. Naiman and K. P. C. Vollhardt, *Angew. Chem. Internat. Edn.*, 1977, **16**, 708.

to give 7,8-annelated 2-amino-4-oxo-4$H$-quinolizine-1-carboxylates (13). Another new one-step reaction is the conversion of furoic acid derivatives into 2-amino-3-hydroxy-pyridines.[6] For example, 2-amino-3-hydroxypyridine is obtained in 55% yield by heating furoic acid amide with ammonium chloride and ammonia at 240 °C. 2-Amino-3-cyano-pyridines (15) are available from methoxymethylenemalononitrile [MeOCH=C(CN)$_2$] and enamines.[7] Initial nucleophilic displacement of the methoxy-group by the enamine $\beta$-carbon centre yields dieneamine (14), which on treatment with ammonia cyclizes to the cyano-pyridine (15; R$^1$ = alkyl, R$^2$ = H or alkyl) in variable yield (37—91%).

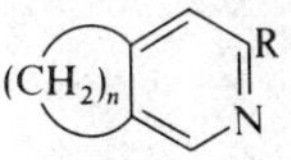

(12) 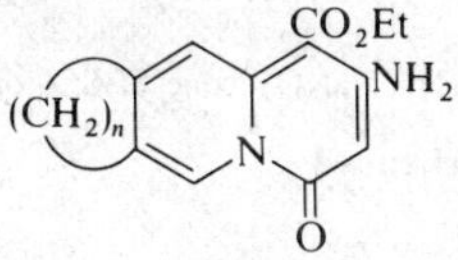  (13) 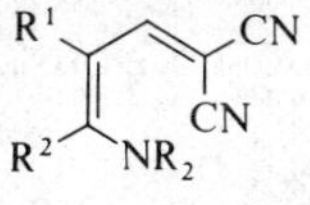  (14) 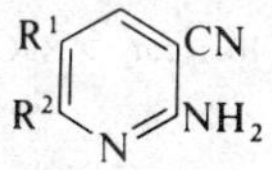  (15)

The use of acrolein in the technical gas-phase synthesis of pyridine derivatives has been reviewed,[8] as has the formation of pyridines, pyrimidines, pyrido-pyrimidines, and pyrazolopyrimidines by ring-closure of $\beta$-enol- and $\beta$-enamino-carbonyl compounds.[9]

Two new syntheses of 2-pyridones have appeared. The first, involving conden-sation of ethyl 2-cyanoacrylates, *e.g.* ArCH=C(CN)CO$_2$Et, with an aryl or alkyl methyl ketone in the presence of ammonium acetate, furnishes 4-aryl-3-cyano-5-substituted and -5,6-disubstituted 2-pyridones in moderate yields (21—53%).[10] The second method, which is particularly useful for 6-alkyl-2-pyridones (17; R$^1$ = alkyl or aryl, R$^2$ = alkyl), involves thermal rearrangement of the pseudo-urea (16), obtained by addition of secondary propargylic alcohols, *e.g.* R$^1$CH$_2$CH(OH)C≡CR$^2$, to 1-cyanopyrrolidine, in boiling xylene[11] (Scheme 3).

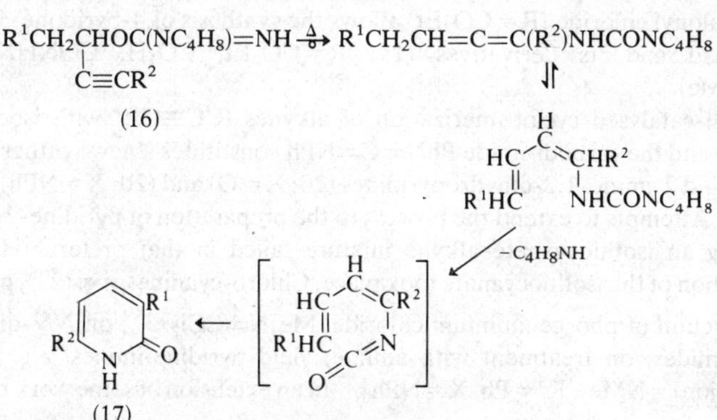

**Scheme 3**

[6] H. Greuter and D. Belluš, *J. Heterocyclic Chem.*, 1977, **14**, 203.
[7] H. Kurihara and H. Mishima, *J. Heterocyclic Chem.*, 1977, **14**, 1077.
[8] H. Beschke and H. Friedrich, *Chem.-Ztg.*, 1977, **101**, 377.
[9] E. Stark, K. Spohn, C. Skoetsch, E. Pech, E. Lorch, J. Haeufel, G. Bouchon, and E. Breitmaier, *Chem.-Ztg.*, 1977, **101**, 161.
[10] S. Kambe, K. Saito, A. Sakurai, and T. Hayashi, *Synthesis*, 1977, 841.
[11] L. E. Overman and S. Tsuboi, *J. Amer. Chem. Soc.*, 1977, **99**, 2813.

However, with $R^2$ = H or Ph, a competing cyclization of the pseudo-urea imino-nitrogen at the triple bond to give (ultimately) oxazoles becomes important.

Improved syntheses of 4-amino- and 4-hydroxy-2-pyridone have been reported[12] and are outlined in Scheme 4.

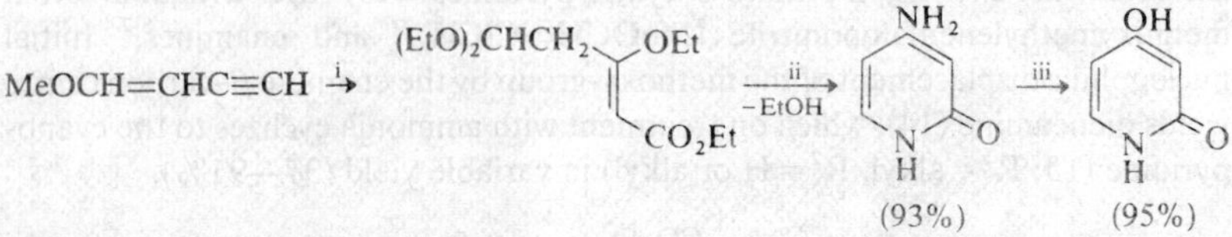

Reagents: i, NaOEt, $(EtO)_2CO$, EtOH; ii, $NH_3$; iii, $NaNO_2$, conc. $H_2SO_4$ or $CF_3CO_2H$.

**Scheme 4**

3,5-Dinitrophenyl azide and 2,5-diethyl-3,4-diphenylcyclopentadienone, in boiling toluene, yield 1-(3′,5′-dinitrophenyl)-3,6-diethyl-4,5-diphenyl-2-pyridone (33%),[13] a reaction analogous to that undergone by tetracylone with sulphonyl azides. In contrast, the dienone reacts with phenyl azide to give only the bridged ketone (18). The mechanisms of these unusual reactions are under investigation.

Pyridine, with copper sulphate (mono- or penta-hydrate) at 300 °C for 6—8 hours, undergoes nucleophilic hydroxylation to 2-pyridone in 95% yield.[14] 2-Quinolone and 1-isoquinolone are obtained in a similar manner but in much lower yield (25% and 17% respectively). Transamination of enamino-ketones $RCH_2COC(Ph)=CHNEt_2$ (prepared from the enamine $PhCH=CHNEt_2$ and acid chlorides $RCH_2COCl$) with methylamine, followed by treatment with dimethylformamide dimethyl acetal $[(MeO)_2CHNMe_2]$, provides a new synthesis of 4-pyridones (19).[15a] Replacement of the simple acid chloride $RCH_2COCl$ by ethyl malonyl chloride ($R = CO_2Et$) allows the synthesis of 4-pyridone-3-carboxylic acid and its derivatives (19; $R = CO_2Et$, $CO_2H$, $CONHNH_2$, or $CONHMe$).[15b]

Cobalt-catalysed cyclotrimerization of alkynes $R^1C{\equiv}CR^2$ with isocyanates $R^3NCO$ and the carbodi-imide $PhN=C=NPh$ constitutes a new synthesis of the 2-oxo- and 2-imino-1,2-dihydropyridines (20; X = O) and (20; X = NPh) respectively.[16] Attempts to extend the process to the preparation of pyridine-2-thiones by using an isothiocyanate–alkyne mixture failed in that preferential desulphurization of the isothiocyanate took place. Chloro-cyanines, *e.g.* (21), prepared by the action of phosgeniminium chloride, $Me_2\overset{+}{N}{=}CCl_2\ Cl^-$, on $NN$-dimethyl-crotonamides, on treatment with aniline, yield pyridine-imines, *e.g.* [20; $R^1$ (6-position) = $NMe_2$, $R^3$ = Ph, X = NPh].[17] In an extension of some work reported

[12] P. D. Cook, R. T. Day, and R. K. Robins, *J. Heterocyclic Chem.*, 1977, **14**, 1295.
[13] A. Hassner, D. J. Anderson, and R. H. Reuss, *Tetrahedron Letters*, 1977, 2463.
[14] P. Tomasik and A. Woszczyk, *Tetrahedron Letters*, 1977, 2193.
[15] (a) R. F. Abdulla, T. L. Emmick, and H. M. Taylor, *Synthetic Comm.*, 1977, **7**, 305; (b) R. F. Abdulla, K. H. Fuhr, and H. M. Taylor, *ibid.*, p. 313.
[16] P. Hong and H. Yamazaki, *Tetrahedron Letters*, 1977, 1333.
[17] M. Huys-Francotte, Z. Janousek, and H. G. Viehe, *J. Chem. Res. (S)*, 1977, 100.

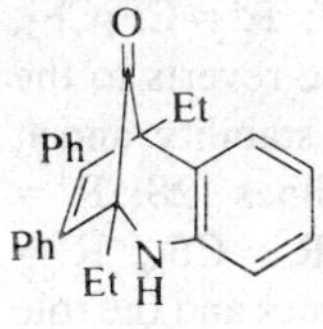

(18) 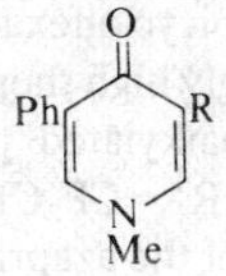          (19) 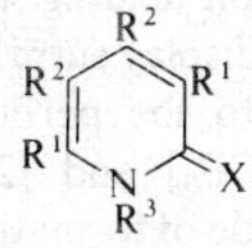          (20) 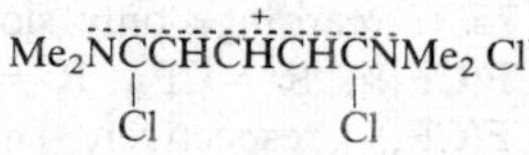          (21)

last year,[18] it has been shown that isothiocyanates and isoselenocyanates react with glutacondialdehyde anion to give 1-substituted-3-formyl-2(1*H*)-pyridine-thiones and -selenones (22; X = S) and (22; X = Se) respectively in practicable yields.[19]

1,2-Dihydropyridines, *e.g.* (23), isomerize to 1,4-dihydropyridines, *e.g.* (24), in benzene at 100 °C in the presence of the rhodium complex [RhCl(PPh)$_3$].[20] The use of dihydropyridines in synthesis and biosynthesis has been reviewed.[21]

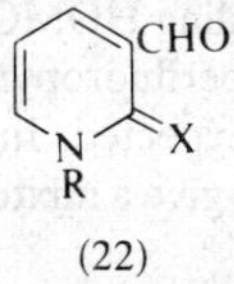

(22) 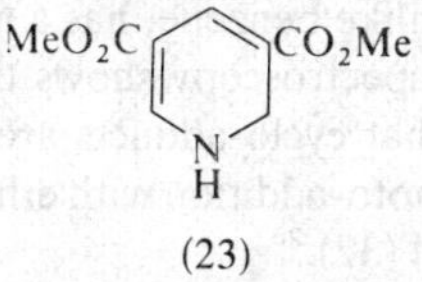                    (23) 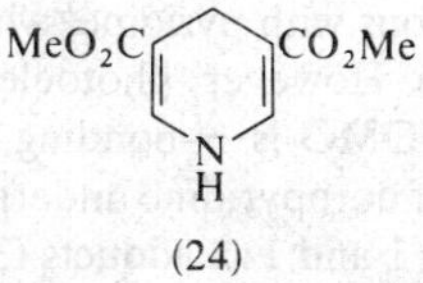                    (24)

Further adaptation of the Hantzsch dihydropyridine synthesis[22] has opened up synthetic routes to annelated 1,4-dihydropyridines, *e.g.* (25), in which the nitrogen is at a bridgehead position,[23a] and to 2-amino-1,4-dihydropyridine-3,5-dicarboxylates of type (26).[23b] The former are prepared by Michael addition of $\beta$-amino-acrylates (27) to alkylideneacetoacetic esters $R^1CH=C(CO_2Me)COCH_3$, whereas the latter involve a similar reaction of the acetoacetate with an amidinoacetic ester $(NH_2)_2C=CHCO_2R^2$.

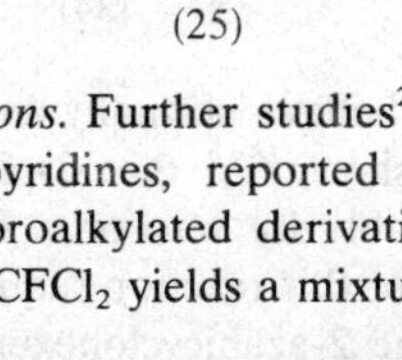

(25) 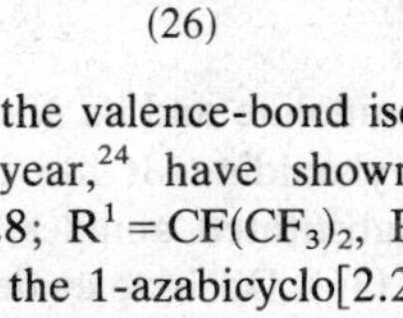                    (26) 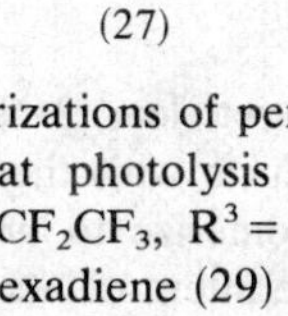                    (27)

*Reactions.* Further studies[25] on the valence-bond isomerizations of perfluoro-alkyl-pyridines, reported last year,[24] have shown that photolysis of the perfluoroalkylated derivative [28; $R^1 = CF(CF_3)_2$, $R^2 = CF_2CF_3$, $R^3 = CF_3$] in $CF_2ClCFCl_2$ yields a mixture of the 1-azabicyclo[2.2.0]hexadiene (29) and the

[18] R. K. Smalley in 'Aromatic and Heteroaromatic Chemistry', ed. O. Meth-Cohn and H. Suschitzky, (Specialist Periodical Reports), The Chemical Society, London, 1978, Vol. 6, p. 81.
[19] J. Becher, E. G. Frandsen, C. Dreier, and L. Henriksen, *Acta Chem. Scand.* (*B*) 1977, **31**, 843.
[20] U. Eisner and M. M. Sadeghi, *Tetrahedron Letters*, 1978, 299.
[21] J. P. Kutney, *Heterocycles*, 1977, **7**, 593.
[22] See ref. 18, p. 84.
[23] H. Meyer, F. Bossert, and H. Horstmann, *Annalen*, 1977, (*a*) p. 1888; (*b*) p. 1895.
[24] See ref. 18, p. 86.
[25] R. D. Chambers and R. Middleton, *J.C.S. Perkin I.* 1977, 1500.

azaprismanes [30; $R^1 = R^2 = CF(CF_3)_2$, $R^3 = CF_2CF_3$] and [30; $R^1 = CF_2CF_3$, $R^2 = R^3 = CF(CF_3)_2$]. Whereas, on heating, the bicyclohexadiene reverts to the pyridine (28), the azaprismanes display surprisingly high thermal stability, and at 175 °C rearrange only slowly to the perfluoroalkylated pyridines [28; $R^1 = CF(CF_3)_2$, $R^2 = CF_3$, $R^3 = CF_2CF_3$] and [28; $R^1 = CF_2CF_3$, $R^2 = CF_3$, $R^3 = CF(CF_3)_2$], respectively. The mode of formation of the azaprismanes and the role played by 2-azabicyclohexadiene intermediates are discussed.

(28)                          (29)                          (30)

Although photo-cycloadditions to benzene are well established, analogous reactions with pyridine, which, unlike benzene, has a non-bonding HOMO, are scarce. However, photoelectron spectroscopy shows that for perfluoropyridine the HOMO is $\pi$-bonding, and that cyclo-adducts are to be expected. In fact, pentafluoropyridine undergoes photo-addition with ethylene to give a mixture of the 1:1 and 1:2 adducts (31) and (32).[26]

The first examples of *meta*-bridging of pyridines and quinolines by ambident nucleophiles, *e.g.* amidines and dicarbanions, have been reported.[27] 3,5-Dinitropyridine and the amidine $PhCH_2C(NMe_2)=NH$ yield the 2,4-adduct (33), a reaction in accord with the 2,4-cycloadditions reported previously with 3,5-dinitrobenzenes. In contrast, with dibenzyl ketone in the presence of triethylamine, the dinitropyridine yields not the expected 2,4-adduct but the 2,6-*meta*-bridged product (34). The reasons for this change in reactivity pattern

(31)                  (32)                  (33)                  (34)

are not yet clear. 3-Nitroquinoline and its *N*-oxide behave normally and form 2,4-adducts with the benzylamidine. It has been shown[28a,b] that 2-picolines photoisomerize to *ortho*-substituted anilines, probably by way of a Dewar-pyridine intermediate, *e.g.* (36). Decisive evidence for this reaction sequence is provided by the isolation and characterization of the 2-azabicyclohexene (37) from the photo-rearrangement of the 2-pyridylacetic ester (35) (Scheme 5).[28c] Labelling studies suggest that the rearrangement of bicyclohexene (37) to the

[26] M. G. Barlow, D. E. Brown, and R. N. Haszeldine, *J.C.S. Chem. Comm.*, 1977, 669; *J.C.S. Perkin I*, 1978, 363.
[27] R. Bard, M. J. Strauss, and S. A. Topolosky, *J. Org. Chem.*, 1977, **42**, 2589.
[28] K. Takagi and Y. Ogata, *J.C.S. Perkin II*, 1977, (*a*) p. 1148; (*b*) p. 1980; (*c*) *J. Org. Chem.*, 1978, **43**, 944.

amino-ester (39) goes *via* intermediate (38) rather than directly from (36) by a 1,3-sigmatropic shift.

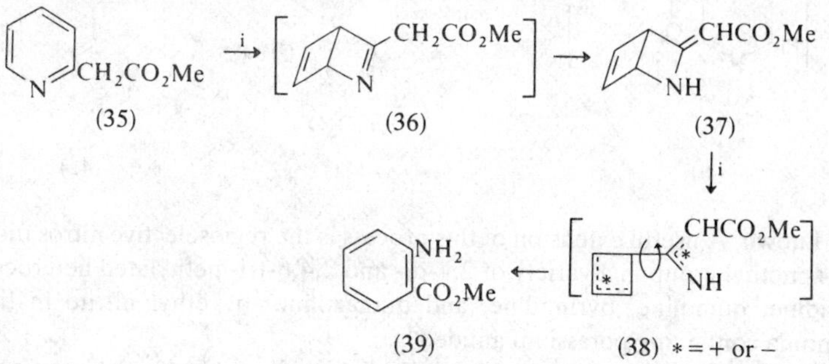

Reagents: i, $h\nu$ (253.7 nm), $H_2O$, NaOH.

**Scheme 5**

Some new examples of direct and indirect electrophilic substitution of pyridine have appeared. Of particular interest are the benzylation and cyclohexylation of 3-ethoxypyridine, which appear to be the first examples of the direct alkylation of the pyridine nucleus under Friedel–Crafts conditions.[29] Under optimum conditions, *i.e.* at 180 °C, for 5 hours, and using a 1:1:3 pyridine–alkyl chloride–aluminium chloride mixture, 2-benzyl-3-ethoxy- and 5-cyclohexyl-3-ethoxy-pyridine are obtained in 90% and 60% yield respectively. Direct acylation of pyridine at the 3-position has also been achieved for the first time, using iron pentacarbonyl as catalyst.[30] The procedure, which is particularly useful for the one-pot synthesis of 2-aryl-5-formyl-pyridines, involves initial phenylation of pyridine with phenyl-lithium in ether at room temperature, followed by treatment of the reaction mixture with iron pentacarbonyl in THF at −65 °C. Hydrolysis with acetic acid yields 2-phenyl-5-formylpyridine (73%). Only poor yields of pyridine-3-aldehyde have so far been obtained using pre-formed 3-pyridyl-lithium. The lithium acyltetracarbonylferrate (40) is suspected as the key intermediate, but as yet no supporting evidence has been obtained. A full report on the general synthesis of 3-acyl-2-alkyl-pyridines by [2,3] sigmatropic rearrangement of α-pyrrolidinyl-2-alkyl-pyridines, *e.g.* (41), noted last year,[31] has appeared.[32]

Regioselective metallation of the pyridine nucleus at the 3-position is now possible by treating the isoxazoline (42), prepared in the normal manner from pyridine-4-carboxylic acid and 2-amino-2-methylpropan-3-ol, with methyl-lithium at −78 °C.[33] Subsequent reaction with a variety of common electrophiles (*e.g.* $Et_2CO$, EtI, MeI, $CH_2$=$CHCH_2Br$, PhCHO, and $HCONMe_2$) furnishes the expected 3-substituted pyridines in practicable yields (55—83%). The preparation of aldoximes by the action of pentyl nitrite on 'active' methyl groups is

[29] F. M. Saidova and E. A. Filatova, *J. Org. Chem.* (*U.S.S.R.*), 1977, **13**, 1231.
[30] C.-S. Giam and K. Ueno, *J. Amer. Chem. Soc.*, 1977, **99**, 3166.
[31] See ref. 18, p. 87.
[32] E. B. Sanders, H. V. Secor, and J. I. Seeman, *J. Org. Chem.*, 1978, **43**, 324.
[33] A. I. Meyers and R. A. Gobel, *Tetrahedron Letters*, 1978, 227.

(40)                              (41)                              (42)

well known. A useful extension of this process is the regioselective nitrosation of the 4-methyl group in a variety of 2,4-di- and 2,4,6-tri-methylated heterocycles (pyridine, quinoline, pyrimidine, and quinazoline) by ethyl nitrite in liquid ammonia containing potassium amide.[34]

2,3,5,6-Pentabromo-4-pyridone, available in 90% yield from the bromination of 3,5-dibromo-4-pyridone in 80% sulphuric acid, is converted into pentabromopyridine in near quantitative yield by hot phosphorus oxybromide.[35] An improved procedure for the preparation of (2-pyridyl)diphenylphosphines by direct nucleophilic displacement of halide from 2-halogeno-pyridines, using lithium diphenylphosphide ($LiPPh_2$), has been reported.[36] The reaction fails with 3- and 4-chloropyridine. Pentachloro- and tetrachloro-4-iodopyridine react with sodium dimethylphosphite to yield not the anticipated tetrachloro-4-pyridyl phosphonate but 2,3,5,6-tetrachloropyridine.[37] The first examples of photo-stimulated $S_{RN}1$ reactions at the pyridine nucleus have been observed in the reaction of 2-halogeno-pyridines with potassium enolates in liquid ammonia.[38] Experiments show that the order of reactivity for 2-halogeno-pyridines is 2-Br > 2-Cl > 2-F, while with bromo-pyridines the order is 2-Br > 3-Br > 4-Br. The reaction of 2-bromopyridine with potassio-acetone is recommended as a useful large-scale preparation of 2-acetonylpyridine (43). Pyridine-borane ($C_5H_5N\cdot BH_3$), which is readily prepared from pyridine hydrochloride and sodium borohydride, is advocated as a good new reagent for the selective reduction of oximes to hydroxylamines.[39] Other reducible groups, *e.g.* $CO_2R$, CN, $NO_2$, $CONH_2$, and Cl, are not affected; yields are excellent (>85%).

The first characterizable pyridine nitrile oxide (44) has been isolated as a moderately stable crystalline solid.[40] It decomposes rapidly (15—20 min) at room temperature but is stable indefinitely at −200 °C. As expected, in solution, at room temperature, it undergoes slow dimerization to the furoxan (45).

The production of nicotinamide from β-picoline has been reviewed.[41]

[34] H. Yamanaka, H. Abe, T. Sakamoto, H. Hiranuma, and A. Kamata, *Chem. and Pharm. Bull. (Japan)*, 1977, **25**, 1821.
[35] S. D. Moshchitskii, A. Zeikans, A. A. Kisilenko, and V. P. Kukhar, *Khim. geterotsikl. Soedinenii*, 1978, 70.
[36] G. R. Newkome and D. C. Hager, *J. Org. Chem.*, 1978, **43**, 947.
[37] S. D. Moshchitskii, L. S. Sologub, A. F. Pavlenko, and V. P. Kukhar, *J. Gen. Chem. (U.S.S.R.)*, 1977, **47**, 1164.
[38] A. P. Komin and J. F. Wolfe, *J. Org. Chem.*, 1977, **42**, 2481.
[39] Y. Kikugawa and M. Kawase, *Chem. Letters*, 1977, 1279.
[40] M. Majewski, B. Serafin, and T. Oklesinska, *Roczniki Chem.*, 1977, **51**, 975.
[41] H. Beschke, H. Friedrich, H. Schaefer, and G. Schreyer, *Chem.-Ztg.*, 1977, **101**, 384.

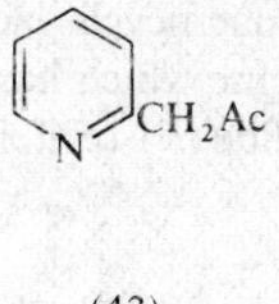

(43) 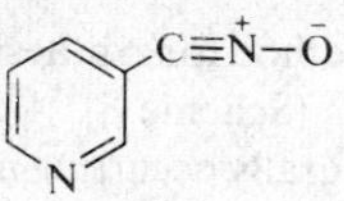                    (44) 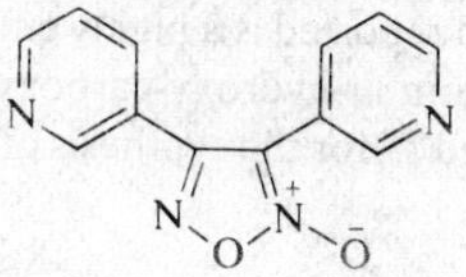                         (45)

Further examples[42] illustrating the use of quaternized 2-halogeno-pyridines for the preparation of simple aliphatic compounds have been reported, and include stereospecific two-step syntheses of primary alkylamines[43a] (Scheme 6; path *a*) and thiols[43b] (Scheme 6; path *b*) from alcohols, and high-yield syntheses of

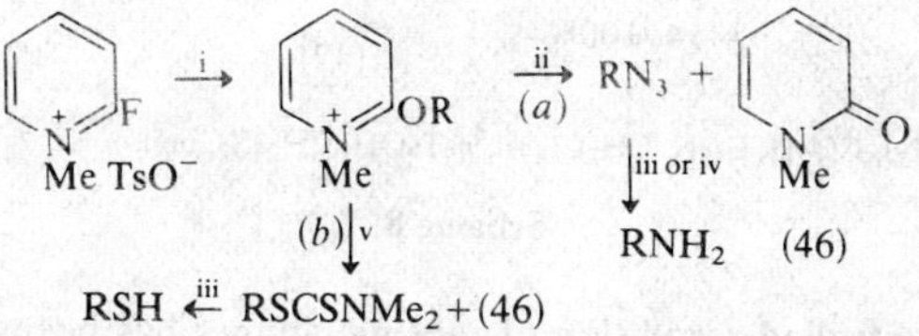

Reagents: i, ROH, Et$_3$N, CHCl$_3$; ii, LiN$_3$, HMPA; iii, LiAlH$_4$; iv, H$_2$, Pd; v, Me$_2$NCSS$^-$ Na$^+$, DMF.

**Scheme 6**

carbodi-imides[44a] (Scheme 7; path *a*) and isothiocyanates[44b] (Scheme 7; path *b*) from thioureas and triethylammonium dithiocarbamates, respectively.

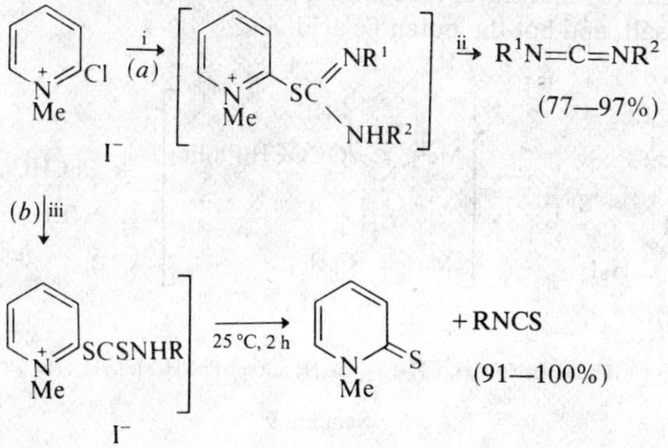

Reagents: i, R$^1$NHCSNHR$^2$, MeCN, Et$_3$N; ii, Et$_3$N; iii, RNHCSS$^-$ $^+$NHEt$_3$.

**Scheme 7**

[42] See ref. 18, p. 88.
[43] (*a*) K. Hojo, S. Kobayashi, K. Soai, S. Ikeda, and T. Mukaiyama, *Chem. Letters*, 1977, 635; (*b*) K. Hojo, H. Yoshino, and T. Mukaiyama, *ibid.*, p. 437.
[44] T. Shibanuma, M. Shiono, and T. Mukaiyama, *Chem. Letters*, 1977, (*a*) p. 575; (*b*) p. 573.

Also reported is a highly efficient method for synthesizing macrocyclic lactones (47) from $\omega$-hydroxy-carboxylic acids (Scheme 8)[45a], a scheme which has been adapted[45b] for the synthesis of the naturally occurring macrolide $(\pm)$-recifeiolide.

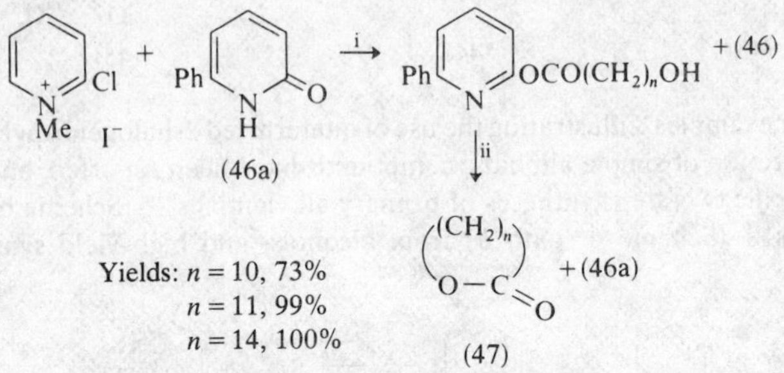

$$\text{Yields: } n = 10, 73\%$$
$$n = 11, 99\%$$
$$n = 14, 100\%$$

Reagents: i, $HO(CH_2)_nCO_2H$, $Et_3N$, $CH_2Cl_2$; ii, $p$-TsOH, $CH_2Cl_2$, boil.

**Scheme 8**

Kinetically controlled resolution of racemic amines has been achieved using mixtures of $(S)$- or $(R)$-2-chloro-1-(1'-cyclohexylethyl)-3-ethyl-6-methyl-pyridinium tetrafluoroborate (48) and $(S)$-(+)-2-phenylbutanoic acid.[46] For example, $(\pm)$-2-phenylethylamine with the $(S)$-enantiomer of (48) in dichloromethane–triethylamine solution (Scheme 9) gives amide (49) that is enriched with 23% of the $(S,S)$-diastereomer. The $(R)$-pyridinium salt is less efficient and gives only 14% of the $(S,R)$-amide. Experiments show $(a)$ that with the achiral pyridinium salt equal amounts of $(S,S)$- and $(S,R)$-amide are produced, and $(b)$ that chiral recognition is due mainly to the asymmetry of the pyridinium salt, and not the butanoic acid.

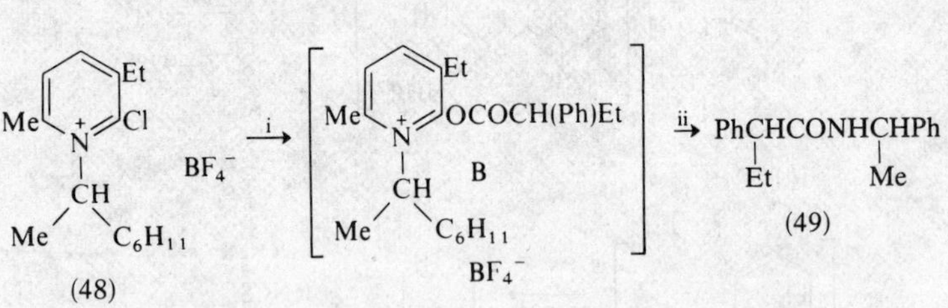

Reagents: i, $(S)$-(+)PhCH(Et)CO$_2$H, $CH_2Cl_2$, $Et_3N$; ii, $(\pm)$-PhCH(Me)NH$_2$, stir for 15 h.

**Scheme 9**

Surprisingly, 1-acyl-pyridinium salts, with Grignard reagents and with organocadmiums, undergo addition at the 2-(major product) and 6-positions rather than at the 1-acyl carbon centre.[47] Formation of acyl-pyridinium is very

[45] (a) T. Mukaiyama, K. Narasaka, and K. Kikuchi, *Chem. Letters*, 1977, 441; (b) K. Narasaka, M. Yamaguchi, and T. Mukaiyama, *ibid.*, p. 959.

[46] T. Mukaiyama, M. Onaka, and M. Shiono, *Chem. Letters*, 1977, 651.

[47] R. E. Lyle, J. L. Marshall, and D. L. Comins, *Tetrahedron Letters*, 1977, 1015.

rapid, and hence the acyl halide may be added to a mixture of pyridine and the organometallic compound. The process is useful for the synthesis of 1-acyl-dihydropyridines, particularly the 1-ethoxycarbonyl derivatives (50; $R^2$ = OEt), which appear to be more stable than the $N$-acetyl and $N$-benzoyl derivatives (50; $R^2$ = Me) and (50; $R^2$ = Ph) respectively.

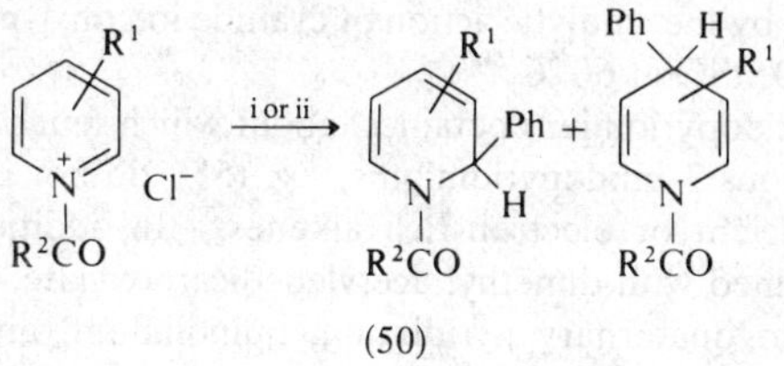

(50)

Reagents: i, PhMgBr or Ph$_2$Cd.

### Scheme 10

3-Aminothien-4-yl-pyridinium iodides (52), readily prepared from 1-cyano-methylpyridinium chloride (51) and carbon disulphide, on successive treatment with methyl iodide, methylamine, and 10% hydrochloric acid, afford ultimately 3,4-diamino-thiophens (53) in >80% yield, *via* the sequence of reactions outlined in Scheme 11.[48]

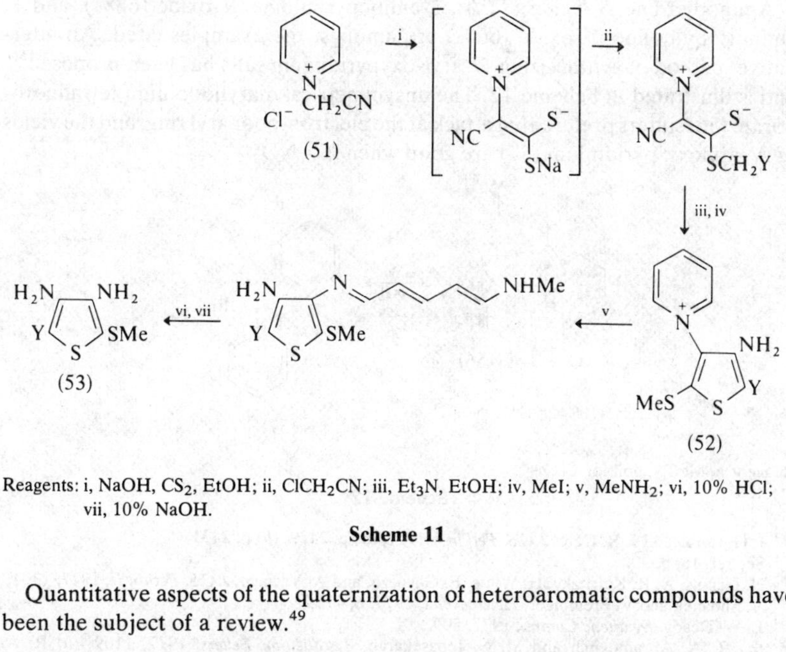

Reagents: i, NaOH, CS$_2$, EtOH; ii, ClCH$_2$CN; iii, Et$_3$N, EtOH; iv, MeI; v, MeNH$_2$; vi, 10% HCl; vii, 10% NaOH.

### Scheme 11

Quantitative aspects of the quaternization of heteroaromatic compounds have been the subject of a review.[49]

[48] Y. Tominaga, H. Fujito, Y. Matsuda, and G. Kobayashi, *Heterocycles*, 1977, **6**, 1871.
[49] J. A. Zoltewicz and L. W. Deady, *Adv. Heterocyclic Chem.*, 1978, **22**, 71.

Dimerization of 1-methylpyridinium salts to the paraquat intermediate (54) has been effected using diphenyl phosphite anion, $(PhO)_2PO^-$, in a variety of solvents (*e.g.* DMF, liquid $NH_3$, EtOH, and $H_2O$).[50a] Optimum yields are achieved with the phosphite in DMF at 70—75 °C for 4 hours. Diphenyl-thiophosphonite $(Ph_2PS^-)$ and diphenylphosphonite $(Ph_2PO^-)$ anions are also effective catalysts. The dihydrobipyridylidene (54) is also formed, in almost quantitative yield, by the catalytic action of cyanide ion on 1-methylpyridinium-4-carboxylate in DMSO at 60 °C.[50b]

Unlike the 3-oxidopyridinium betaines, about which much was reported last year,[51] the analogous 3-imidopyridiniums, *e.g.* (55), do not give cyclo-adducts with electron-deficient or electron-rich alkenes.[52] In addition, only resinous products are obtained with dimethyl acetylenedicarboxylate. The formation of pseudo-bases from quaternary pyridinium, quinolinium, and isoquinolinium cations has been reviewed.[53]

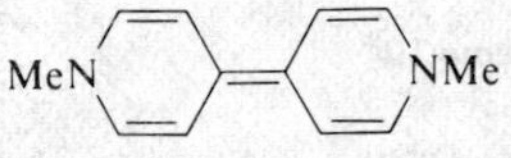

(54) 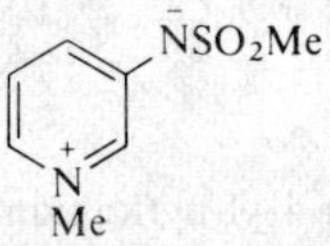　　　　　　　　　(55)

Direct formation of *N*-oxides of amino-aza-heterocycles can be troublesome. Welcome, therefore, is a report that *m*-chloroperbenzoic acid in acetone can be used as oxidizing agent without the need to protect the amino-function.[54] 2-Aminopyridine *N*-oxide (71%), 2-aminopyrimidine *N*-oxide (68%), and 1-aminoisoquinoline *N*-oxide (66%) are amongst the examples cited. An alternative method of synthesizing *O*-aryl-oxypyridinium salts has been proposed[55a] and is illustrated in Scheme 12. The unsymmetrical diaryliodonium tetrafluoroborate (56) suffers preferential attack at the electron-poor aryl ring, and the yields of *N*-aryloxy-pyridinium (57) are good when $X = NO_2$.

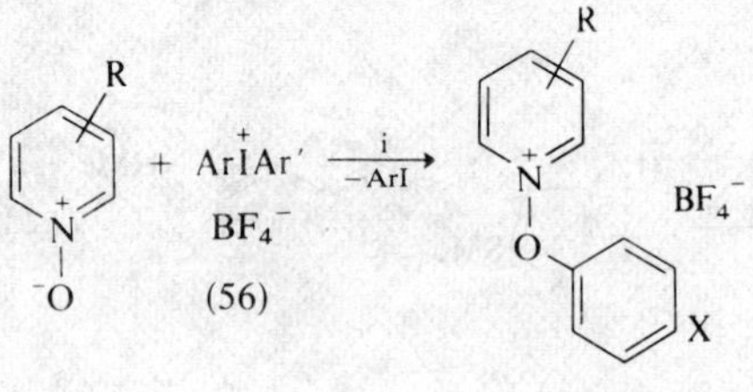

(57)

Reagent: i, MeCN, boil for 1 h.

**Scheme 12**

[50] J. G. Carey and J. R. Case, *J.C.S. Perkin I*, 1977, (a) p. 2429; (b) p. 2431.
[51] See ref. 18, p. 90.
[52] N. Dennis, A. R. Katritzky, H. Wilde, E. Gavuzzo, and A. Vaciago, *J.C.S. Perkin II*, 1977, 1304.
[53] V. Šimánek and V. Preininger, *Heterocycles*, 1977, **6**, 475.
[54] L. W. Deady, *Synthetic Comm.*, 1977, 509.
[55] (a) R. A. Abramovitch and M. N. Inbasekaran, *Tetrahedron Letters*, 1977, 1109; (b) R. A. Abramovitch, G. Alvernhe, and M. N. Inbasekaran, *ibid.*, p. 1113; (c) R. A. Abramovitch and M. N. Inbasekaran, *J.C.S. Chem. Comm.*, 1978, 149.

Attempts to prepare the 2-chloro-derivatives (57; R = 2-Cl; X = NO$_2$ or Cl) by heating pyridones (58; R = NO$_2$ or Cl) with phosphorus oxychloride or thionyl chloride failed, but gave instead the pyrido[2,3-*b*]benzofurans (59; R = NO$_2$ or Cl) in 70% and 42% yield respectively. Possible reaction pathways are elaborated in Scheme 13.

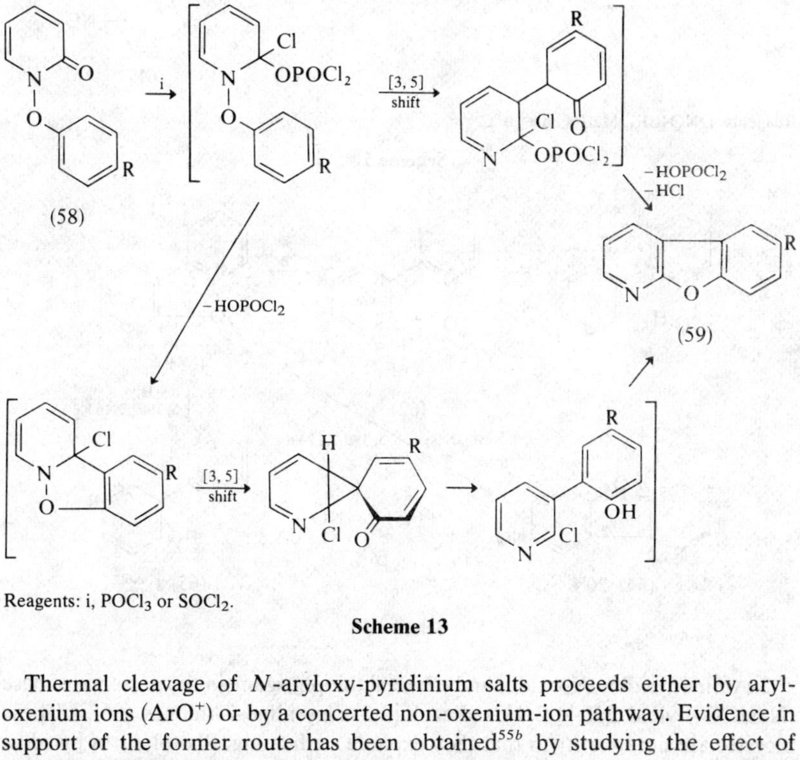

Reagents: i, POCl$_3$ or SOCl$_2$.

**Scheme 13**

Thermal cleavage of *N*-aryloxy-pyridinium salts proceeds either by aryl-oxenium ions (ArO$^+$) or by a concerted non-oxenium-ion pathway. Evidence in support of the former route has been obtained[55b] by studying the effect of substituents (R) upon the isomer ratios of substituted anisoles formed when *N*-aryloxy-pyridinium tetrafluoroborates (57; X = NO$_2$) decompose in anisole. The ratio of 4-nitro-2′-methoxydiphenyl ether to 4-nitro-4′-methoxydiphenyl ether produced remained constant (28:72) for a wide range of substituents X, a result in keeping with a stepwise process involving an oxenium ion. Confirmation of this result comes from a study of similar substitutions of anisole by aryloxenium ions (60) generated by a different route, as illustrated in Scheme 14.

Aryloxenium ions have also been invoked[55c] to explain the thermal iso-merization of benzisoxazolo[2,3-*a*]pyridinium tetrafluoroborate (63; R = NO$_2$) to 7-nitrobenzofuro[3,2-*b*]pyridine (64). In contrast, under similar conditions the unsubstituted benzisoxazolopyridinium salt (63; R = H), which is formed in high yield (80%) by intramolecular phenylation of *N*-oxide (62) (Scheme 15), yields only 2-(*o*-hydroxyphenyl)pyridine (65) (18%) and much tar. The *p*-nitro-substi-tuent is thought not only to enhance the electrophilicity of the oxenium moiety but also to destabilize the alternative resonance form (61) of the oxenium ion.

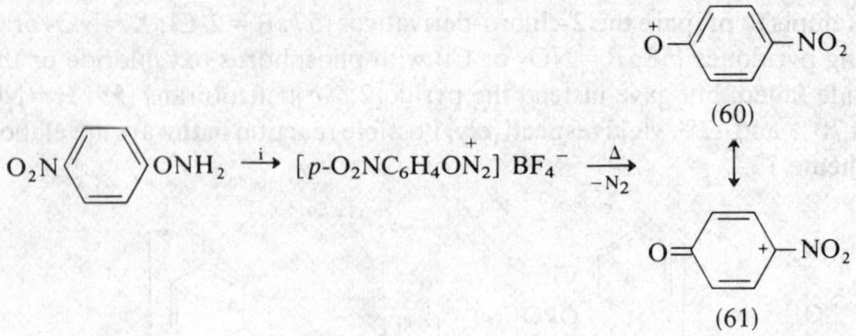

(60)

$$O_2N\langle\ \rangle ONH_2 \xrightarrow{i} [p\text{-}O_2NC_6H_4O\overset{+}{N}_2]\ BF_4^- \xrightarrow[-N_2]{\Delta}$$

(61)

Reagents: i, NO₂BF₄, MeCN, at −10 °C.

**Scheme 14**

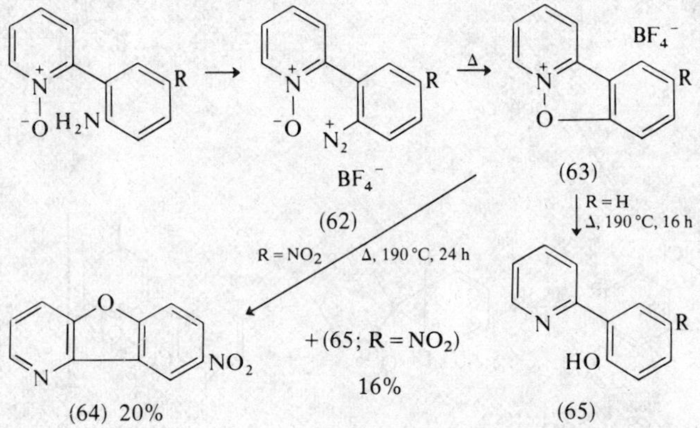

(62)

(63)

(64) 20%

+(65; R = NO₂)

16%

(65)

**Scheme 15**

2-Aminopyridine *N*-oxides and ethereal thiophosgene in the presence of base (NaHCO₃) yield 2*H*-[1,2,4]oxadiazolo[2,3-*a*]pyridine-2-thiones (66).[56] These bicycles are noteworthy in that they explode on heating. Photolysis of pyridine *N*-oxide at 10 K in an argon matrix produces a species which, by i.r. and u.v. spectra, has been identified tentatively as the isocyano-aldehyde (67) or (68).[57]

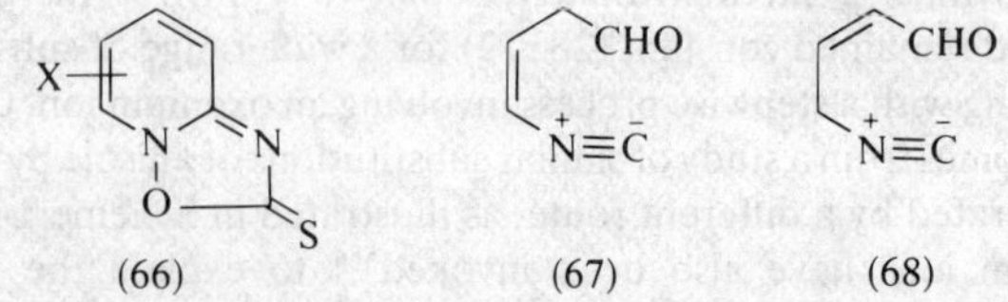

(66)                    (67)                    (68)

Such an intermediate could well explain the large amounts of polymer which attend the photolysis of pyridine *N*-oxide at ambient temperature. Rearrangements of tertiary amine oxides have been reviewed.[58]

[56] D. Rousseau and A. Taurins, *Canad. J. Chem.*, 1977, **55**, 3736.
[57] O. Buchardt, J. J. Christensen, C. Lohse, J. J. Turner, and I. R. Dunkin, *J.C.S. Chem. Comm.*, 1977, 837.
[58] S. Oae and K. Ogino, *Heterocycles*, 1977, **6**, 583.

The new synthesis of aldehydes RCHO from alkyl halides $RCH_2X$ outlined in Scheme 16[59] is reminiscent of the aldehyde→nitrile conversion reported last year.[60]

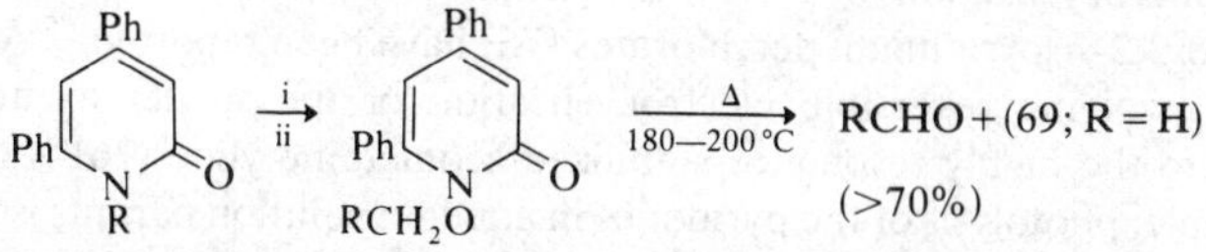


(69; R = OH)

Reagents: i, NaOMe, MeOH; ii, $RCH_2X$.

**Scheme 16**

Oxidation of *N*-aminopyridinium tosylates (70) with aqueous bromine yields, as expected, the tetrazenes (71).[61a] However, if R = Me or Bu$^t$, pyridiniopyridinium salts (72) are obtained, albeit in low yields. During the so-far unsuccessful search for a better oxidizing agent for this unusual coupling reaction, it has been found[61b] that *N*-aminopyridinium salts (70), with lead tetra-acetate in acetic acid, yield 1-(acetamido)-2-pyridones (73). A similar oxidation prevails for the quinolinium and isoquinolinium analogues.

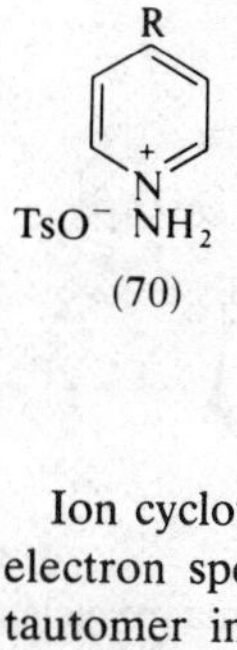


(70)

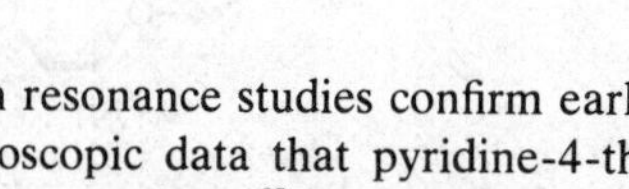


(71) R = OMe, SMe, $NMe_2$, or NMePh

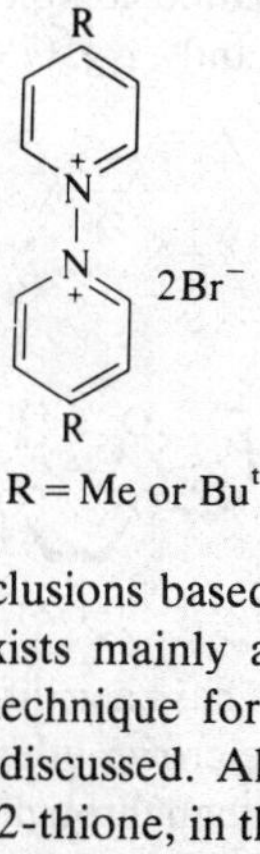


(72) R = Me or Bu$^t$

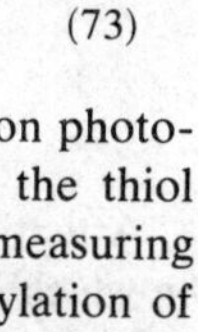


(73)

Ion cyclotron resonance studies confirm earlier conclusions based on photoelectron spectroscopic data that pyridine-4-thione exists mainly as the thiol tautomer in the gas phase.[62] The potential of this technique for measuring tautomeric equilibrium constants in the gas phase is discussed. Alkylation of ambident anions of the type $[N{=\!\!\!=}C{=\!\!\!=}S]^-$, *e.g.* pyridine-2-thione, in the presence of a phase-transfer catalyst ($Bu_4N^+ Br^-$) leads exclusively to the *S*-alkylated products.[63] Yields are generally superior to those obtained by more conventional methods, and easier experimental work-up procedures are claimed for a variety of systems, including pyridine-2-thione, pyrimidine-2-thione, and benzoxazoline-2-thione. 'Energies and Alkylations of Tautomeric Heterocyclic

[59] M. J. Cooke, A. R. Katritzky, and G. H. Millet, *Heterocycles*, 1977, **7**, 227.
[60] See ref. 18, p. 94.
[61] (*a*) D. G. Doughty and E. E. Glover, *J.C.S. Perkin I*, 1977, 1593; (*b*) J. T. Boyers and E. E. Glover, *ibid.*, p. 1960.
[62] C. B. Theissling and N. M. M. Nibbering, *Tetrahedron Letters*, 1977, 1777.
[63] H. J. M. Dou, P. Hassanaly, J. Kister, and J. Metzger, *Phosphorus, Sulfur, and Related Elements*, 1977, **3**, 355.

Compounds: Old Problems – New Answers' is the title of a review[64] which contains much data on the exhaustively investigated hydroxypyridine–pyridone equilibrium.

Novel photocyclizations of 1-vinyl-2-pyridones (74) in aqueous perchloric acid to oxazolo[3,2-a]pyridinium perchlorates (75) have been reported.[65] A reaction pathway involving reversible electrocyclization of the singlet excited vinyl-pyridone to the highly resonance-stabilized azomethine ylide (76) is favoured. For example, photolysis of the pyridones in aqueous solution permits isolation of the alcohols (77; R = Me) and (77; R = H) in 87.9% and 81% yield respectively.

(74)                    (76)                    (75) R = Me, 45%                    (77)
                                                     R = H, 50%

There is continuing interest in the synthesis and complexing properties of crown ethers[66a] and cryptands[67a] based on pyridine units. For example, cryptand (78) is an especially strong ligand for binding alkali-metal cations, and it forms stable crystalline complexes with lithium and sodium perchlorate.[67a] Pyridino-phane cryptands, *e.g.* (79), also form stable Na$^+$ and K$^+$ complexes.[67b]

(78)                                        (79)

Attention has been directed towards the synthesis of 2,6-pyridino-macrocycles in which the crown-ether oxygens are bonded directly to the pyridine ring, *e.g.* (80), rather than through a —CH$_2$— group,[66b] and also towards linear crown-type polyethers.[67b] Of particular interest is the enhanced ability to donate hydride ion shown towards sulphonium salts by the crown-ether NAD(P)H mimic (81).[68]

The synthesis of macrocycles composed of pyridine, furan, and thiophen subunits has been reviewed.[69]

[64] P. Beak, *Accounts Chem. Res.*, 1977, **10**, 186.
[65] P. S. Mariano, A. A. Leone, and E. Krochmal. jnr., *Tetrahedron Letters*, 1977, 2227.
[66] (a) M. Newcomb, J. M. Timko, D. M. Walba, and D. J. Cram, *J. Amer. Chem. Soc.*, 1977, **99**, 6392; (b) G. R. Newkome, A. Nayak, G. L. McClure, F. Danesh-Khoshboo, and J. Broussard-Simpson, *J. Org. Chem.*, 1977, 1500.
[67] (a) E. Buhleier, W. Wehner, and F. Vögtle, *Chem. Ber.*, 1978, **111**, 200; (b) B. Tummler, G. Maass, E. Weber, W. Wehner, and F. Vögtle, *J. Amer. Chem. Soc.*, 1977, **99**, 4683.
[68] T. J. van Bergen and R. M. Kellogg, *J. Amer. Chem. Soc.*, 1977, **99**, 3882.
[69] G. R. Newkome, J. D. Sauer, J. M. Roper, and D. C. Hager, *Chem. Rev.*, 1977, **77**, 513.

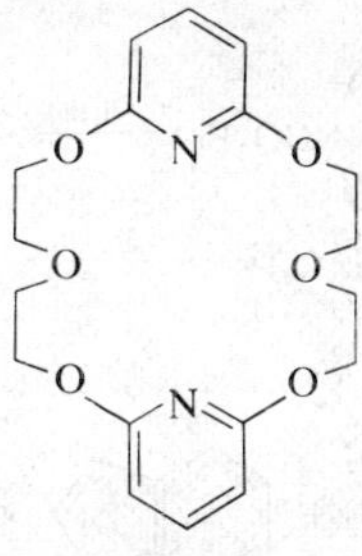

(80) 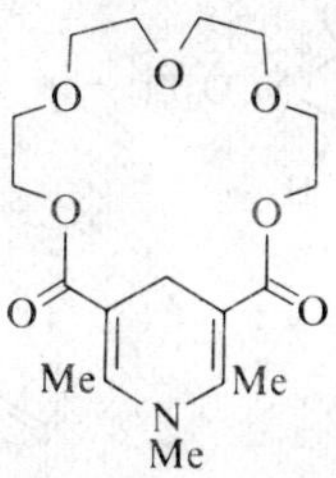                    (81)

Thermal rearrangement of propargyl 4-pyridyl ether at 550 °C affords a mixture of cyclobuta[*b*]- (83) and cyclobuta[*c*]-pyridine (84) in 35% and 17% yield respectively.[70] The two products are thought to arise by scrambling of the diradical intermediate (82), produced as indicated in Scheme 17.

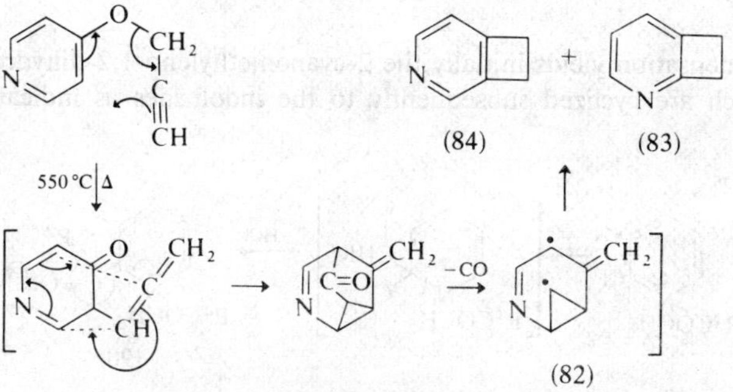

**Scheme 17**

*N*-Ethoxycarbonylpyrindine (86; R = CO₂Et), a deep-purple liquid, has been synthesized from the aziridino-azocine (85) by the reaction sequence outlined in Scheme 18.[71a] Reduction with lithium aluminium hydride furnishes a mixture of the isomeric hydropyrindines (87) and (88).

In view of the known decrease in aromaticity of other 4π heterocycles brought about by the attachment of an electron-withdrawing substituent on nitrogen, it is surprising that the pyrindine (86; R = CO₂Et) shows no loss of aromaticity as measured by ¹H n.m.r. spectroscopy.[71b] In fact, the *N*-ethoxycarbonyl and *N*-methyl derivatives (86; R = CO₂Et) and (86; R = Me) and the anion (89) all display similar strongly diatropic proton shifts. From these data it has been concluded that the magnitude of the contribution of the lone pair to the π-system is independent of the substituent on nitrogen.

Indolizines (91) are available by condensing 2-chloro-pyridinium salts with malononitrile or ethyl cyanoacetate, as outlined in Scheme 19.[72] In the absence of

[70] J. M. Riemann and W. S. Trahanovsky, *Tetrahedron Letters*, 1977, 1867.
[71] (*a*) A. G. Anastassiou, S. J. Girgenti, R. C. Griffith, and E. Reichmanis, *J. Org. Chem.*, 1977, **42**, 2651; (*b*) A. G. Anastassiou, E. Reichmanis, and S. J. Girgenti, *J. Amer. Chem. Soc.*, 1977, **99**, 7392.
[72] H. Pauls and F. Kröhnke, *Chem. Ber.*, 1977, **110**, 1294.

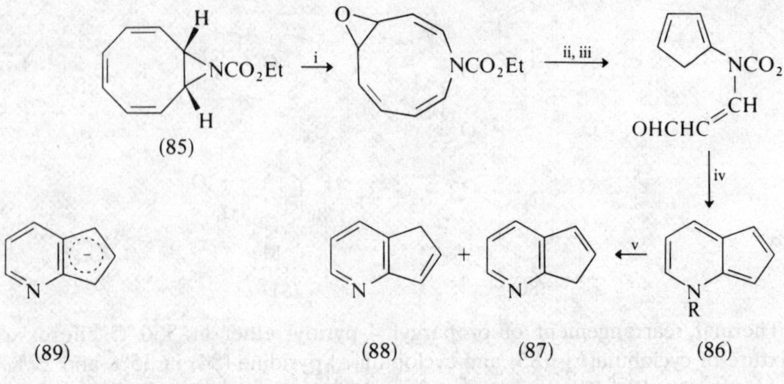

Reagents: i, *m*-ClC₆H₄CO₂OH, CH₂Cl₂ at 0 °C; ii, *hν*, Me₂CO at 10 °C; iii, Δ, 50 °C; iv, Al₂O₃
at −15 °C or Δ at 70 °C; v, LiAlH₄, Et₂O.

**Scheme 18**

base, condensation yields initially the 2-cyanomethyleno-1,2-dihydropyridines
(90), which are cyclized subsequently to the indolizines as indicated in the
Scheme.

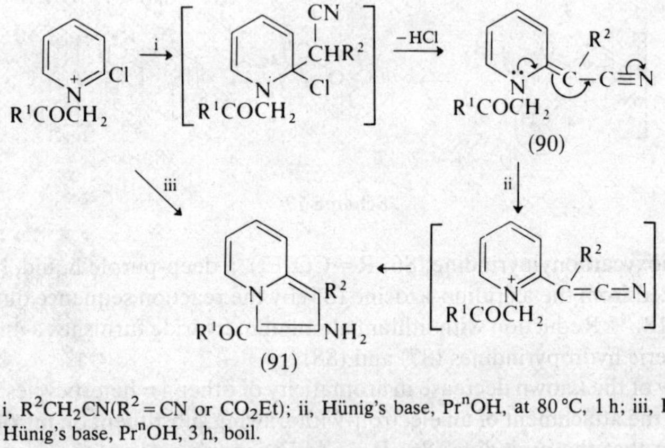

Reagents: i, R²CH₂CN(R² = CN or CO₂Et); ii, Hünig's base, PrⁿOH, at 80 °C, 1 h; iii, R²CH₂CN,
Hünig's base, PrⁿOH, 3 h, boil.

**Scheme 19**

2-Alkylidene-1-ethoxycarbonylmethyl-1,2-dihydropyridines, *e.g.* (92), cyclize
either to 3-ethoxycarbonyl-indolizines (93) or to 2-acetoxy-1-ethenyl-indolizines
(94), depending on the reaction conditions (Scheme 20).[73]

The chemistry of cyclazines[74a] and the synthesis of cyclazines and related
*N*-bridged annulenes[75] have been comprehensively reviewed.

[73] A. Kakehi, S. Ito, T. Maeda, R. Takeda, M. Nishimura, and T. Yamaguchi, *Chem. Letters*, 1978, 59.
[74] (*a*) A. Taurins, *Chem. Heterocyclic Compounds*, 1977, **30**, 245; (*b*) G. Maury, *ibid.*, p. 179; (*c*) H. L.
     Blewitt, *ibid.*, p. 117.
[75] W. Flitsch and U. Krämer, *Adv. Heterocyclic Chem.*, 1978, **22**, 321.

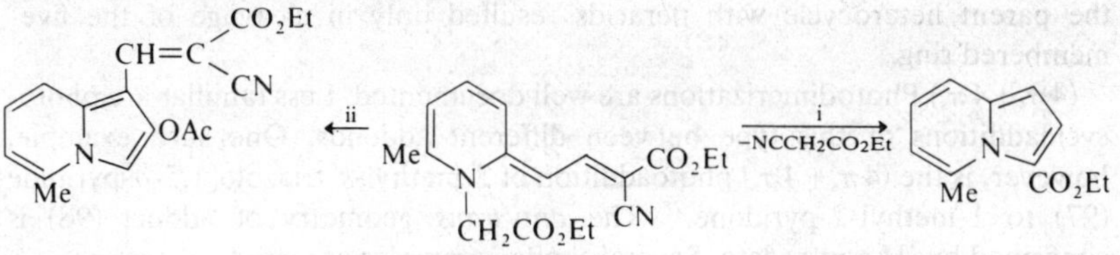

(94)    (92)    (93)

Reagents; i, Δ, xylene; ii, Ac₂O.

**Scheme 20**

Earlier work has demonstrated that pyridinium *N*-ylides and keten thioacetals and related compounds yield allyl ylides and indolizines.[76a] The extension of this reaction to pyridinium *N*-imines provides a useful synthesis of pyrazolo-[1,5-*a*]pyridines (95), as shown in Scheme 21.[76b]

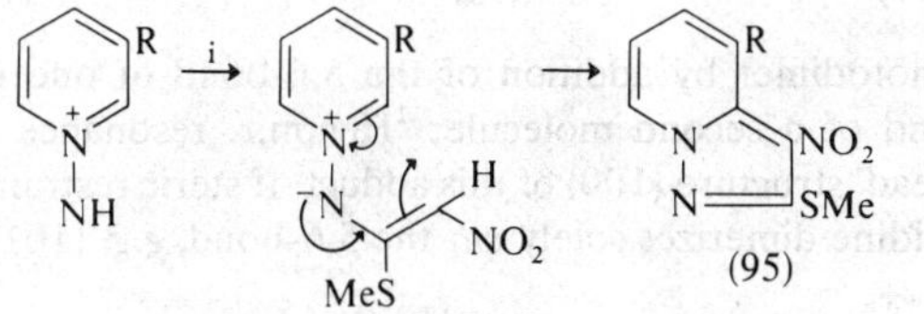

(95)

Reagents: i, (MeS)₂C=CHNO₂, EtOH.

**Scheme 21**

Imidazo[1,2-*a*]pyridine 1-oxide (96), the first example of an *N*-oxide of a π-excessive nitrogen-bridged polyaza-indene, has been synthesized by the route outlined in Scheme 22.[77] Attempts to prepare the *N*-oxide by direct oxidation of

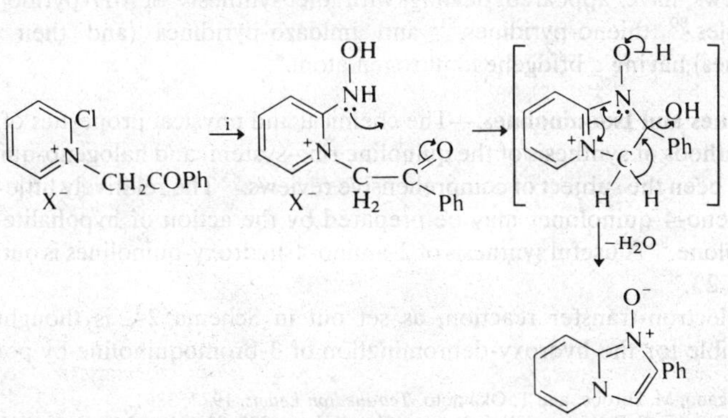

(96)

Reagents: i, anhydrous NH₂OH·HCl.

**Scheme 22**

[76] H. Fujito, Y. Tominaga, Y. Matsuda, and G. Kobayashi, (*a*) *Heterocycles*, 1976, **4**, 939; (*b*) *ibid.*, 1977, **6**, 379.
[77] E. S. Hand and W. W. Paudler, *J. Org. Chem.*, 1978, **43**, 658.

the parent heterocycle with peracids resulted only in cleavage of the five-membered ring.

$(4\pi_s + 4\pi_s)$ Photodimerizations are well documented. Less familiar are photocycloadditions of this type between different addends. One such example, however, is the $(4\pi_s + 4\pi_s)$ photoaddition of 2-methyl-*s*-triazolo[1,5-*a*]pyridine (97) to 1-methyl-2-pyridone.[78] The *anti-trans* geometry of adduct (98) is confirmed by [1]H n.m.r. data. Several similar examples are cited.

Photodimerizations of the isomeric 1,2,4-*s*-triazolo[4,3-*a*]pyridines, *e.g.* (99), have also been reported.[79a] The 3-methyl derivative yields a thermally labile

(97)                            (98)                            (99)

cyclobutane photodimer by addition of the 5,6-bond of one molecule to the 7,8-double-bond of a second molecule; [1]H n.m.r. resonance studies confirm the 'head-to-head' structure (100) of this adduct. If steric restraints are imposed, the triazolopyridine dimerizes solely *via* the 5,6-bond, *e.g.* (101) → (102).[79b]

(100)                            (101)                            (102)

Reviews have appeared dealing with the synthesis of 6*H*-pyrido[4,3-*b*]-carbazoles,[80] thieno-pyridines,[81] and imidazo-pyridines (and their benzo-analogues) having a bridgehead nitrogen atom.[82]

**Quinolines and Isoquinolines.**—The chemical and physical properties of quinoline, methods of synthesis of the quinoline ring-system, and halogeno-quinolines have all been the subject of comprehensive reviews.[83] The relatively little-known 3-halogeno-4-quinolones may be prepared by the action of hypohalite on the 4-quinolone.[84] A useful synthesis of 2-amino-4-hydroxy-quinolines is outlined in Scheme 23.[85]

An electron-transfer reaction, as set out in Scheme 24, is thought to be responsible for the hydroxy-debromination of 3-bromoquinoline by potassium

[78] T. Nagano, M. Hirobe, and T. Okamoto, *Tetrahedron Letters*, 1977, 3891.
[79] K. T. Potts, E. G. Brugel, and W. C. Dunlap, *Tetrahedron*, 1977, **33**, (*a*) p. 1247; (*b*) p. 1253; (*c*) K. T. Potts, W. C. Dunlap, and F. S. Apple, *ibid.*, p. 1263.
[80] M. Sainsbury, *Synthesis*, 1977, 437.
[81] J. M. Barker, *Adv. Heterocyclic Chem.*, 1977, **21**, 65.
[82] I. A. Mazur, B. E. Mandrichenko, and R. I. Katkevich, *Russ. Chem. Rev.*, 1977, **46**, 634.
[83] *Chem. Heterocyclic Compounds*, 1977, Vol. 32.
[84] J. Renault, P. Mailliet, S. Renault, and J. Berlot, *Synthesis*, 1977, 865.
[85] S. B. Kadin and C. H. Lamphere, *Synthesis*, 1977, 500.

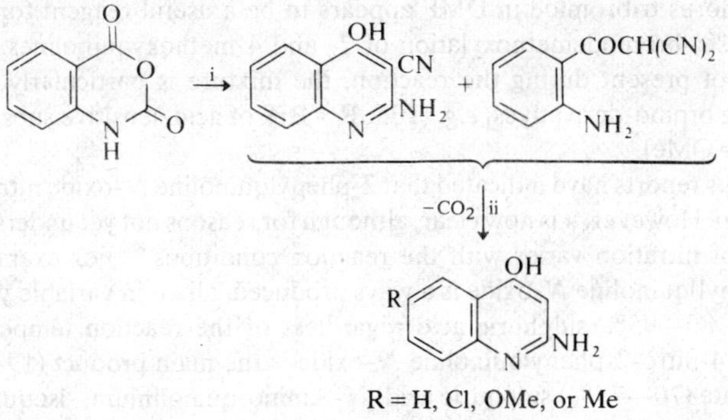

Reagents: i, CH$_2$(CN)$_2$, EtOH; ii, 48% HBr or 6M-KOH, boil.

**Scheme 23**

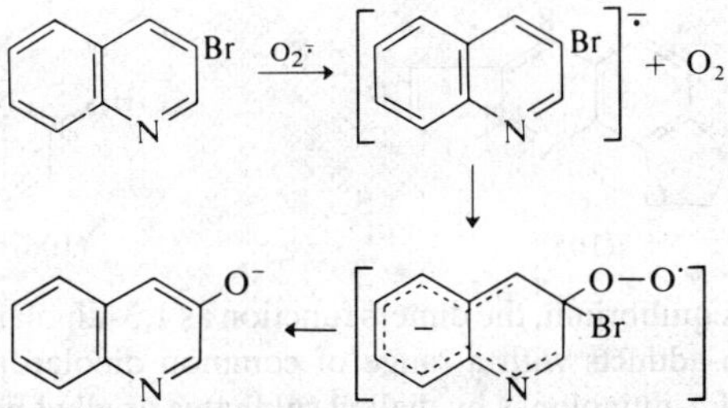

**Scheme 24**

superoxide in DMSO containing 18-crown-6.[86] Debromination does not take place in benzene, or in DMSO in the absence of the crown ether. Similar reactions with 2-chloroquinoline and 2-bromoisoquinoline are reported, but in these cases reaction could be by an ordinary $S_N$Ar-type substitution.

Methylsulphinylmethyl carbanion, CH$_3$SOCH$_2^-$, has been used for the methylation of electron-deficient heterocycles. However, with 2-methylquinoline (103) (Scheme 25) it gives the 2,4-bridged quinolinophane (104) (60%) rather than, as expected, 2,4-dimethylquinoline.[87]

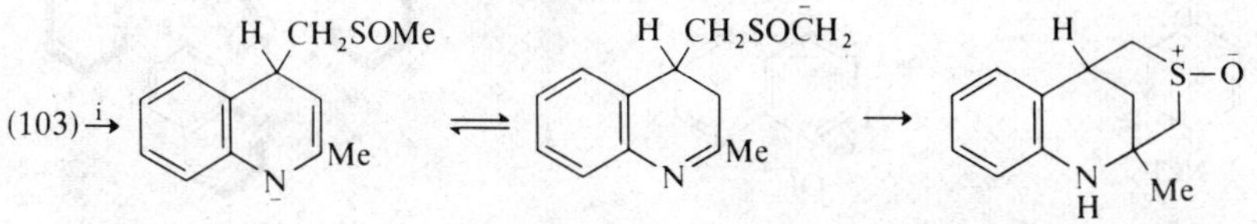

Reagent: NaH, DMSO at 70 °C, 3.5 h.

**Scheme 25**

[86] T. Yamaguchi and H. C. van der Plas, *Rec. Trav. chim.*, 1977, **96**, 89.
[87] H. Kato, I. Takeuchi, Y. Hamada, M. Ono, and M. Hirota, *Tetrahedron Letters*, 1978, 135.

Phosphorus tribromide in DMF appears to be a useful reagent for the high-yield (90%) bromodemethoxylation of 2- and 4-methoxyquinolines.[88] As free HBr is not present during the reaction, the mixture is particularly useful for preparing bromo-derivatives, *e.g.* (105; R = Br), of acid-sensitive substrates, *e.g.* (105; R = OMe).

Previous reports have indicated that 2-phenylquinoline *N*-oxide nitrates at the 4-position. However, it is now clear, although for reasons not yet understood, that the site of nitration varies with the reaction conditions.[89] For example, 2-(3-nitrophenyl)quinoline *N*-oxide is always produced, albeit in variable yield (27—78%), in 80—95% sulphuric acid regardless of the reaction temperature. In contrast, 4-nitro-2-phenylquinoline *N*-oxide is the main product (19—35%) in more dilute (70—75%) sulphuric acid. *N*-Amino-quinolinium, -isoquinolinium, and -phenanthridinium salts, with base, give not the anticipated *N*-imides but their dimers, *e.g.* (106).[90] Despite the fact that the *N*-imide monomers cannot be

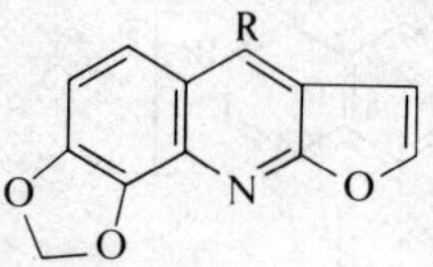

(105) 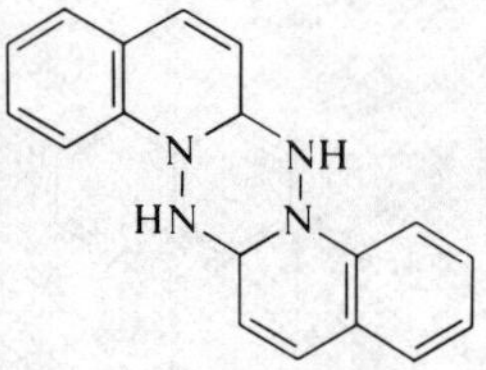                                                              (106)

detected in the equilibrium, the dimers function as 1,3-dipolar azomethine imines and yield cyclo-adducts with a range of common dipolarophiles. The *N*- and *O*-alkylation of 4-quinolones by dialkyl sulphates or alkyl halides is much more efficient in the presence of the phase-transfer catalyst ($Bu_4^nN^+OH^-$).[91] Yields, even with long alkyl chains, are excellent. The species responsible for the intense blue colour generated in the colour test for chinoform (5-chloro-7-iodo-8-hydroxyquinoline) with quinone dichloro di-imide (107) has been shown by i.r., u.v., $^1H$ n.m.r., and mass spectroscopy to have the quinone-di-imine structure (108).[92]

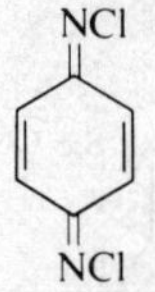

(107) 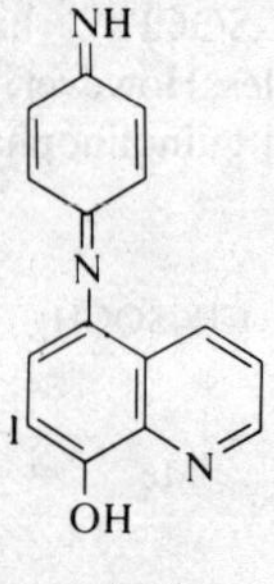                        (108)

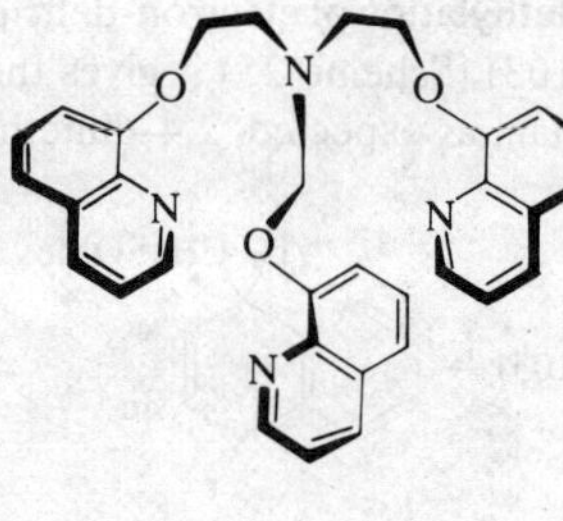

[88] T. Yajima and K. Munakata, *Chem. Letters*, 1977, 891.
[89] M. Tamane, S. Takeo, and H. Noda, *Chem. and Pharm. Bull. (Japan)*, 1977, **25**, 1256.
[90] R. Huisgen, R. Grashey, and R. Krischke, *Annalen*, 1977, 506.
[91] J. Renault, P. Mailliet, J. Berlot, and S. Renault, *Compt. rend.*, 1977, **285**, C, 199.
[92] M. Yamamoto, J. Sugimura, and T. Uno, *Chem. and Pharm. Bull. (Japan)*, 1977, **25**, 47.

A new non-cyclic cryptate (109) for alkali-metal ions has been prepared[93] whose phase-transfer activity exceeds that of dibenzo-18-crown-6. Its high selectivity for bivalent cations, particularly $Ba^{2+}$ and $Sr^{2+}$, and also its marked preference for $Na^+$ over $K^+$ and $Li^+$ have led the authors to speculate that cryptates of this type may have the 'greatest complexation capacity of all neutral ligands so far investigated'.

Evidence has been produced[94] which suggests that cycloprop[*b*]indoles and other products from the photolysis of ethyl 2-cyano-1,2-dihydroquinoline-1-carboxylates, *e.g.* (110), arise not, as was assumed previously, *via* an aza-hexatriene intermediate but *via* the excited singlet state of (110) or a dipolar species of form (111). In contrast, the long-assumed (but recently doubted) intermediacy of 1,2-dihydroquinolines in the Doebner–von Miller quinoline synthesis has been confirmed.[95] The isolable dihydroquinolines are produced in high yield on allowing the arylamine and acetaldehyde to stand at room temperature for 24 hours.

The yields of polynuclear heterocycles, *e.g.* (112), available by condensing quinone-methides, *e.g.* (113), with arylamines are much improved ($36 \rightarrow 77\%$) if the hemi-acetal (114) is used as the quinone-methide precursor in place of the Mannich base (115).[96]

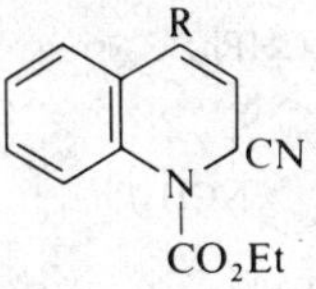

(110) 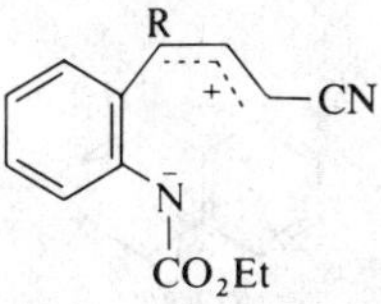      (111) 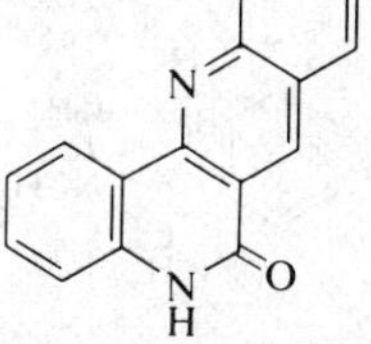      (112)

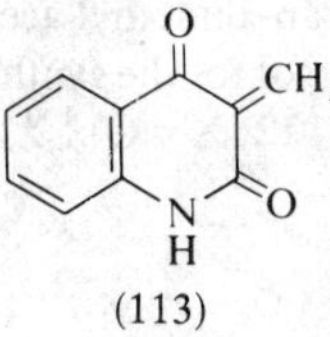

(113) 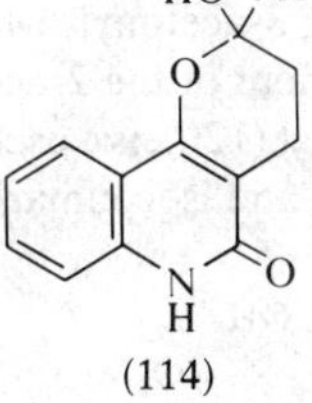      (114) 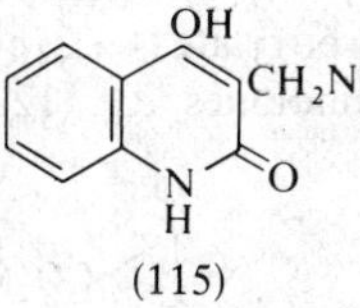      (115)

1,2,4-Triazolo[4,3-*a*]quinolines[79c] undergo similar photodimerizations to the 1,2,4-triazolo[4,3-*a*]pyridines reported on page 164. However, whereas the triazolo-quinolines, *e.g.* (116), yield *cis*-fused head-to-tail cyclobutane dimers, *e.g.* (117), the 1,2,4-triazolo[3,4-*a*]isoquinolines (118) form cisoid head-to-head cyclo-dimers of type (119). Co-dimerization of the two systems has also been noted.

[93] F. Vögtle, W. M. Müller, W. Wehner, and E. Buhleier, *Angew. Chem. Internat. Edn.*, 1977, **16**, 548.
[94] M. Ikeda, S. Matsugashita, and Y. Tamura, *Heterocycles*, 1978, **9**, 281.
[95] G. A. Dauphinee and T. P. Forrest, *Canad. J. Chem.*, 1978, **56**, 632.
[96] J. L. Asherson and D. W. Young, *J.C.S. Chem. Comm.*, 1977, 916.

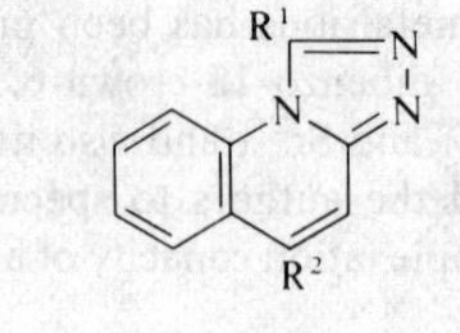

(116)                                    (117)

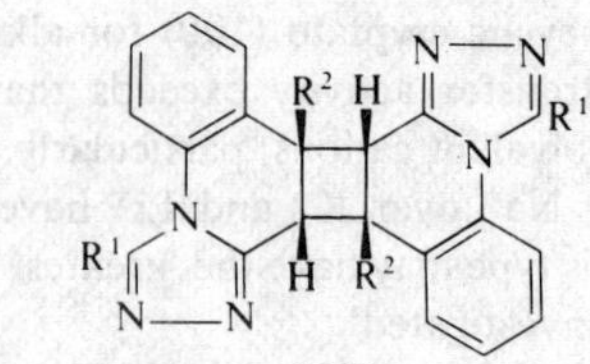

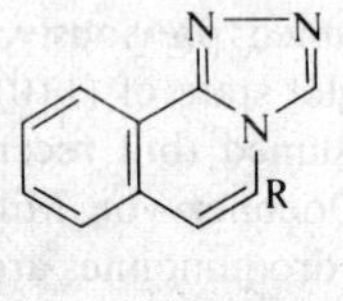

(118)                                    (119)

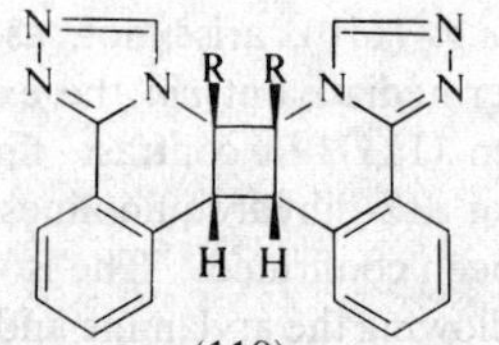

The chemistry of 1*H*-pyrrolo-[2,3-*b*]-[97a] and -[3,2-*b*]-quinolines[97b] and of benzo[*a*]quinolizines[98] has been reviewed.

A new synthesis of isoquinolines by intramolecular, palladium-catalysed reaction of aryl halides with alkenes is typified by the example given in Scheme 26.[99]

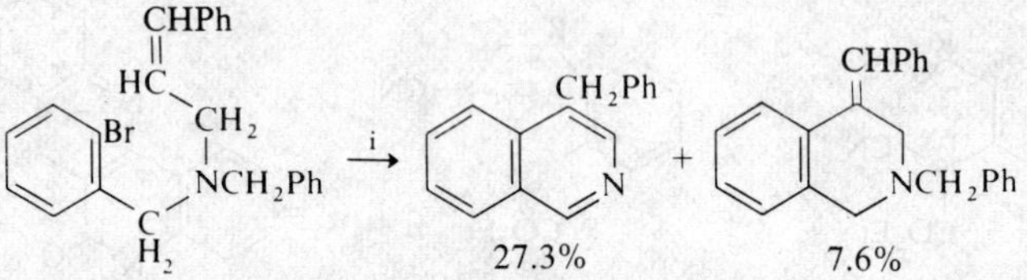

Reagents: i, Me$_2$NCH$_2$CH$_2$NMe$_2$, Pd(OAc)$_2$, PPh$_3$, N$_2$, 69 h, at 125 °C.

**Scheme 26**

Scheme 27 shows how dimethyl 3-acetonyl-6-hydroxyphthalate (121), prepared by acid-catalysed rearrangement of the 2-acetonylfuran–dimethyl acetylenedicarboxylate (4 + 2) cyclo-adduct (120), is a useful precursor for the synthesis of isoquinolones, *e.g.* (122; X = N), and isocoumarins, *e.g.* (122; X = O).[100]

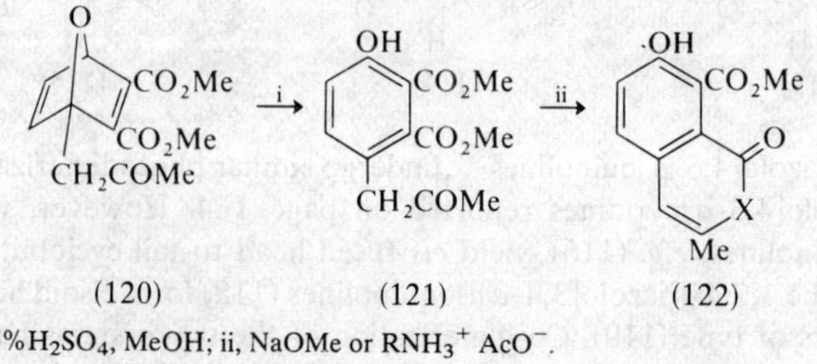

Reagents: i, 95% H$_2$SO$_4$, MeOH; ii, NaOMe or RNH$_3$$^+$ AcO$^-$.

**Scheme 27**

[97] M. A. Khan and J. F. de Rocha, *Heterocycles*, 1977, **6**, (*a*) p. 1229; (*b*) p. 1927.
[98] F. D. Popp and R. F. Watts, *Heterocycles*, 1977, **6**, 1189.
[99] M. Mori, K. Chiba, and Y. Ban, *Tetrahedron Letters*, 1977, 1037.
[100] L. M. Gomes and M. Aicart, *Compt. rend.*, 1977, **285**, *C*, 571.

Alkylthio-vinyl isocyanates, *e.g.* (123; $R = Pr^i$), prepared as shown in Scheme 28, behave as equivalents of acyl isocyanates.[101] On prolonged (8 h) heating at 150 °C they yield methylthio-uracils, *e.g.* (124) (100%), and with enamines ($EtCH=CHNR_2$) they give methylthio-dihydropyridones, *e.g.* (125). In contrast to the former reaction, the arylthio-vinyl isocyanate (123; $R = Ph$) undergoes quantitative intramolecular cyclization at 150 °C to 3-methylthio-isoquinolone (126). Interestingly, attempts to remove the SMe group with either $CuCl_2$–CuO or $Br_2$–CuO resulted only in chlorination or bromination at the 4-position, the latter in quantitative yield.

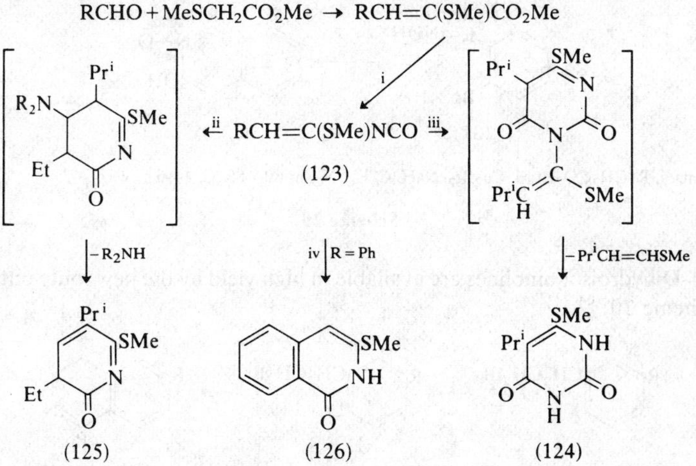

$$RCHO + MeSCH_2CO_2Me \rightarrow RCH=C(SMe)CO_2Me$$

Reagents: i, *via* $CO_2H$, COCl, and $CON_3$; ii, $EtCH=CHNC_5H_{10}$, $C_6H_6$; iii, Δ, 150 °C, 8 h; iv, Δ, 150 °C, 2 h.

**Scheme 28**

Further investigations on the classical Pictet–Gams isoquinoline synthesis have revealed that whereas the hydroxy-amides (127; $R^1 = $ Me or Et), with $P_2O_5$ in boiling decalin, undergo conventional cyclization to the 3-substituted iso-quinolines (128; $R^1 = $ Me or Et, $R^2 = $ H), the amides (127; $R^1 = Bu^n$, $CH_2Ph$, Ph, or $p$-MeOC$_6$H$_4$) yield only the rearranged 4-alkylated isoquinolines (128; $R^1 = $ H, $R^2 = Bu^n$, $CH_2Ph$, Ph, or $p$-MeOC$_6$H$_4$).[102] A convincing explanation of this rearrangement has been offered, based on an oxazoline intermediate (isolable under more mild conditions) which, as the oxazolium salt (129), suffers ring-opening with migration of group $R^1$ to produce the isomeric amide (130; $R^1$ as in 127), cyclization of which furnishes the 4-substituted isoquinoline. Significantly, rearrangement is observed only with groups ($R^1$) of good migrating aptitude.

$$PhCH(OH)CH(R^1)NHCOPh$$

(127)

[101] K. Takaki, A. Okamura, Y. Ohshiro, and T. Agawa, *J. Org. Chem.*, 1978, **43**, 402.
[102] N. Ardabilchi, A. O. Fitton, J. R. Frost, and F. Oppong-Boachie, *Tetrahedron Letters*, 1977, 4107.

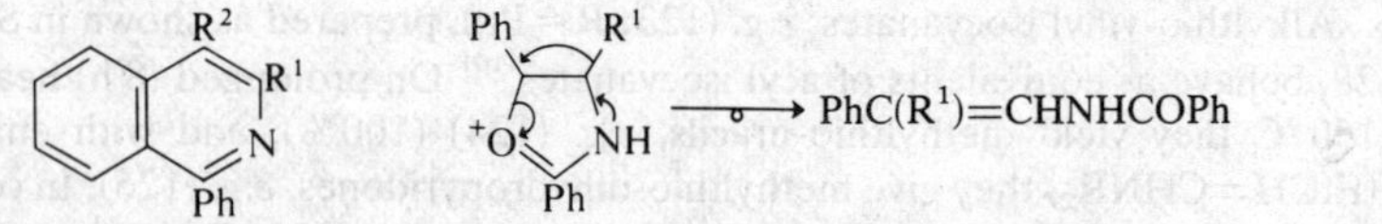

(128)                    (129)                    (130)

A modified Hurtley reaction (Scheme 29) on *o*-bromo-aldo- or -keto-oximes (131; $R^1$ = H or Me) provides ready access to 3-hydroxy-isoquinoline *N*-oxides (132; $R^1$ = H or Me; $R^2$ = Ac, Bz, $CO_2Et$, or CN).[103]

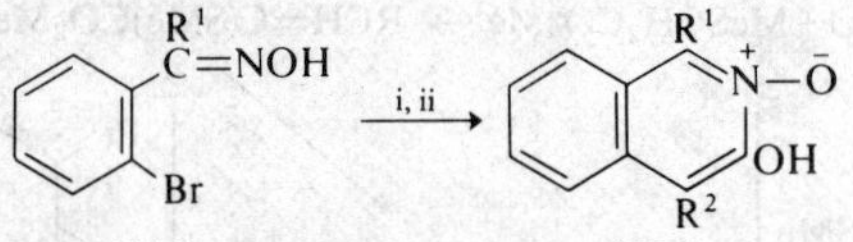

(131)                    (132)

Reagents: i, $R^2CH_2CO_2Et$; ii, $Cu_2Br_2$, NaH, $C_6H_6$, $N_2$, at 60—80 °C, 4—6 h.

**Scheme 29**

3,4-Dihydroisoquinolines are available in high yield by the new route outlined in Scheme 30.[104]

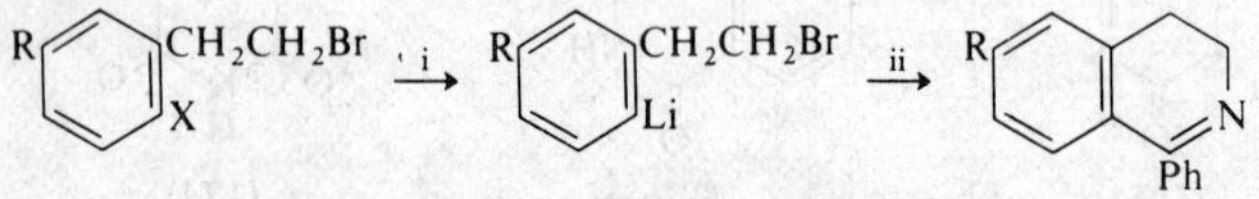

R = H or MeO; X = Br or I

Reagents: i, $Bu^nLi$, $C_6H_{14}$ or $Bu^tLi$, $C_6H_{14}$, $Et_2O$, at −100 °C; ii, PhCN, THF, at −100 °C.

**Scheme 30**

Phase-transfer catalysts improve substantially the yields of the Reissert reaction.[105] For example, isoquinoline, benzoyl chloride, and aqueous potassium cyanide, in the presence of benzyltriethylammonium chloride, give *N*-benzoyl-1-cyano-1,2-dihydroisoquinoline in 82% yield (previous yield 38%). $^{13}$C n.m.r. studies on the adducts of quinolines and isoquinolines with acetylenedicarboxylates have necessitated revision of the structures of many of these adducts.[106] For example, the so-called 'azepine adduct' from 1-methylisoquinoline has been re-formulated as the cyclobutapyrrole (133).

A review covering the use of acetylenedicarboxylic esters in synthesis during the period 1972—1976, and which mentions many heterocyclic systems, has been published.[107]

[103] A. McKillop and D. P. Rao, *Synthesis*, 1977, 760.
[104] C. A. Hergrueter, P. D. Brewer, J. Tagat, and P. Helquist, *Tetrahedron Letters*, 1977, 4145.
[105] T. Koizumi, K. Takeda, K. Yoshida, and E. Yoshii, *Synthesis*, 1977, 497.
[106] R. M. Acheson and G. Proctor, *J.C.S. Perkin I*, 1977, 1924.
[107] M. Baumgarth, *Chem.-Ztg.*, 1977, **101**, 118.

$N$-Amination of 3-acylamido-isoquinolines, using $O$-mesitylhydroxylamine, followed by polyphosphoric-acid-induced cyclization of the resulting $N$-amino-isoquinolinium salts, provides a route to the unusually stable $o$-quinonoid $s$-triazolo[1,5-$b$]isoquinolines (134).[108] The thermal stability (stable in boiling xylene) of these systems has been attributed to contributing meso-ionic structures.

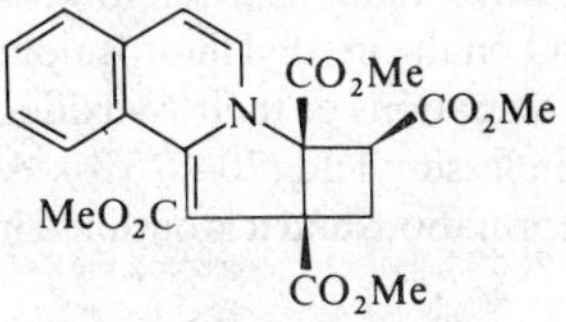

(133)

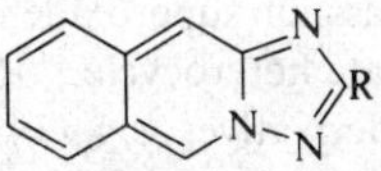

(134)

**Acridines, Phenanthridines, and Other Systems with One Nitrogen Atom.**—The synthesis of 9-substituted acridines during the period 1970—1976 has been reviewed.[109] The efficiency of the established procedure for the cyclization of $o$-aryl-formanilides, *e.g.* (135), to benzophenanthridines, *e.g.* (136), has been

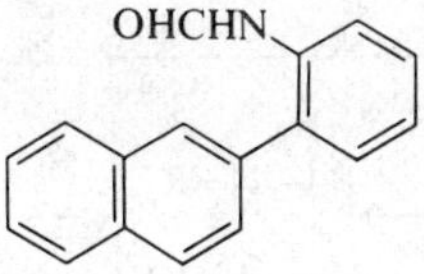

(135)

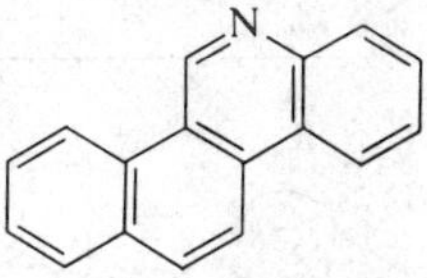

(136)

greatly improved (97%) by using equivalent amounts of phosphorus oxychloride and stannic chloride as cyclizing agent.[110a] Previously, a 10-times excess of $POCl_3$ was used. The capability of isocyanides to function as ground-state electrophilic carbenoid species has been demonstrated in the thermolysis of isocyano-phenyl-naphthalenes, *e.g.* (137), in n-tetradecane.[110b] Insertion of the isocyano-carbon into the C—C and C—H bonds of the adjacent phenyl ring produces benzo-cycloheptindole (138) (3%) and benzophenanthridine (139) (31%).

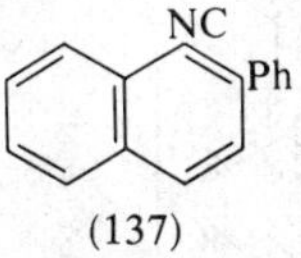

(137)

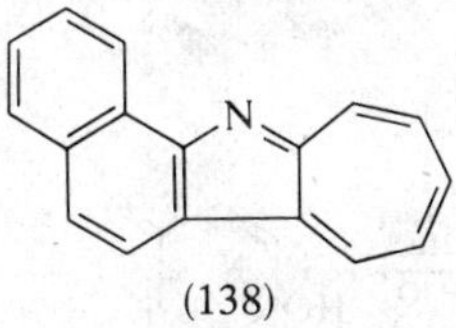

(138)

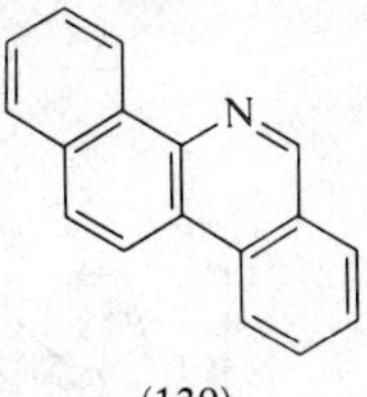

(139)

In addition to their other notable successes mentioned earlier in this Report, phase-transfer catalysts also appear to be beneficial in the preparation of arene and aza-arene oxides under mild conditions.[111] The epoxidations are, however,

[108] Y. Tamura, M. Iwaisaki, Y. Miki, and M. Ikeda, *Heterocycles*, 1977, **6**, 949.
[109] S. Skonieczny, *Heterocycles*, 1977, **6**, 987.
[110] J. H. Boyer and J. R. Patel, (*a*) *Synthesis*, 1978, 205; (*b*) *J.C.S. Chem. Comm.*, 1977, 855.
[111] S. Krishnan, D. G. Kuhn, and G. A. Hamilton, *J. Amer. Chem. Soc.*, 1977, **99**, 8121.

very sensitive to pH; for example, oxidation of bénzo[*f*]quinoline to epoxide
(140) with sodium hypochlorite occurs in high yield (80%) at pH 8—9 in the
presence of tetra-n-butylammonium bisulphate. At higher pH no epoxidation
takes place, whereas at pH < 8 alternative (but unspecified) oxidations arise.
Curiously, *N*-oxides are not formed, and phenanthridine yields neither the
*N*-oxide nor the epoxide but phenanthridone, possibly *via* an oxaziridine intermediate. Phenanthridones and acridones are available in quantitative yield by the
action of potassium superoxide in DMSO on the methyl fluorosulphonate salts of
the appropriate heterocycle.[112] Similar oxidations of their *N*-oxides produce the
*N*-hydroxy-derivatives, *e.g.* (141), but in lesser yield (70—75%). A mechanistic
rationale has been presented for amide formation, and it is outlined in Scheme 31.

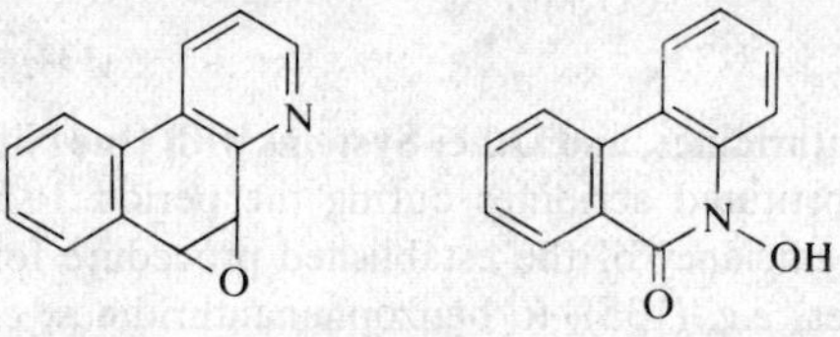

(140)                                      (141)

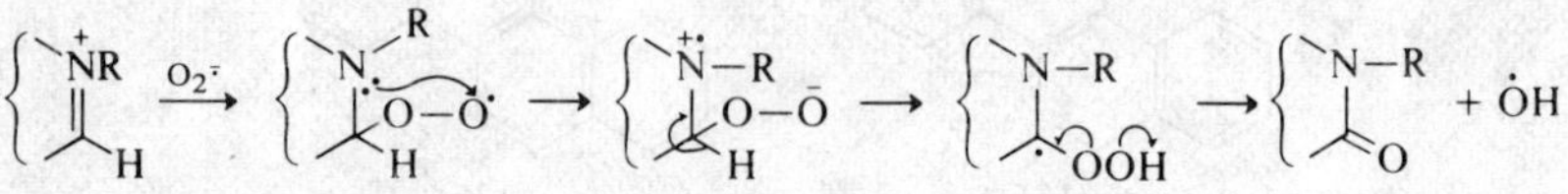

**Scheme 31**

**Pyrans and Pyrones.**—A novel one-pot synthesis of maltol (142; R = Me) and
related γ-pyrones from furfuryl alcohols has been formulated (Scheme 32).[113]

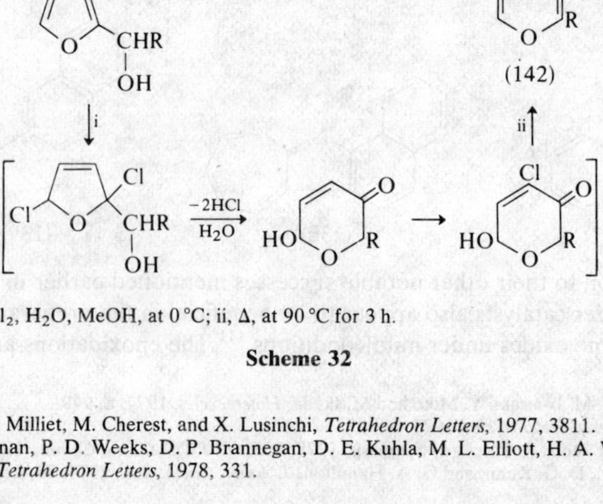

(142)

Reagents: i, 2Cl₂, H₂O, MeOH, at 0 °C; ii, Δ, at 90 °C for 3 h.

**Scheme 32**

[112] A. Picot, P. Milliet, M. Cherest, and X. Lusinchi, *Tetrahedron Letters*, 1977, 3811.
[113] T. M. Brennan, P. D. Weeks, D. P. Brannegan, D. E. Kuhla, M. L. Elliott, H. A. Watson, and B. Wlodecki, *Tetrahedron Letters*, 1978, 331.

Oxidative cyclization of 1,3,4,6-tetraketones $RCOCH_2COCOCH_2COR$ by lead tetra-acetate to a 4-hydroxy-$\alpha$-pyrones (144) has long been thought to involve oxidative fission of the tetraketone to 2 moles of acyl-keten ($RCOCH=C=O$), which then dimerize. However, incontrovertible evidence has now been presented[114] for an intramolecular reaction (Scheme 33) involving a 1,2-acyl migration in the initially formed carbene (143).

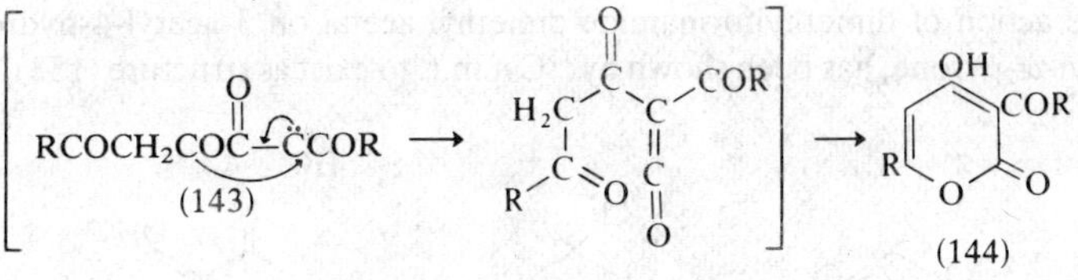

**Scheme 33**

Careful re-investigation of the photolysis products of 3,5-dimethyl-$\gamma$-pyrone in trifluoroethanol (reported last year[115]) by preparative t.l.c. ($SiO_2$) has revealed the presence of two minor, but mechanistically significant, products.[116] One, identified as 1,3-dimethyl-6-oxabicyclo[3.1.0]pent-3-en-2-one (145), is the first isolable example of the hitherto elusive cyclopentadienone epoxides, which have long been postulated as intermediates in these photolyses. This dienone epoxide is the photo-precursor of the final product, *i.e.* the isomeric 3,6-dimethyl-$\alpha$-pyrone, but not of the trifluoroethanol adduct (146). This latter product, as suggested previously, derives directly from the zwitterionic precursor of the bicyclopent-3-en-2-one (145). The other product proved to be cyclopent-1-ene-3,5-dione (147), a photo-rearrangement product of (145).

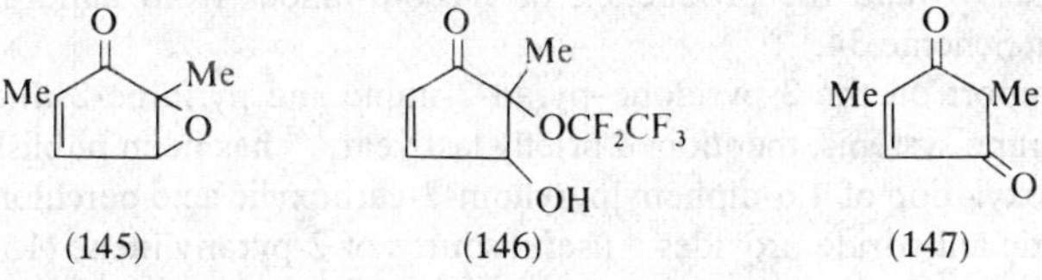

The photo-rearrangements undergone by 2,3-dimethyl-4-hydroxy- and 3-ethyl-4-hydroxy-2-methyl-pyrylium salts in 96% sulphuric acid[117] are similar to rearrangements of (145). The major products are the furyl cations (148; R = Me) and (148; R = Et) respectively, along with the isomeric 2-hydroxy-pyrylium salts (149) and (150).

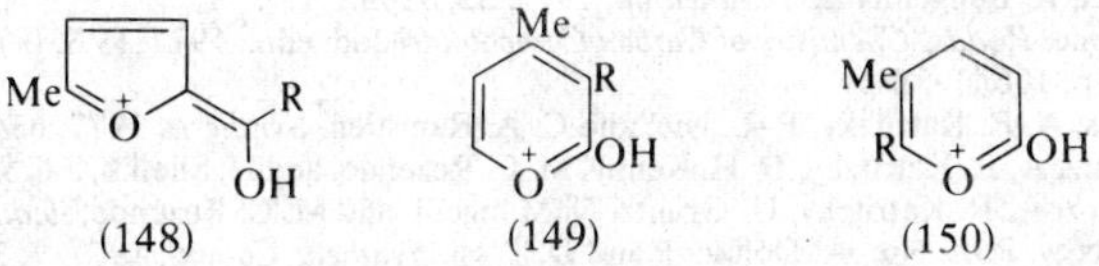

[114] C. W. Bird and P. Thorley, *Chem. and Ind.*, 1977, 872.
[115] See ref. 18, p. 106.
[116] J. A. Barltrop, A. C. Day, and C. J. Samuel, *J.C.S. Chem. Comm.*, 1977, 598.
[117] J. W. Pavlik, D. R. Bolin, K. C. Bradford, and W. G. Anderson, *J. Amer. Chem. Soc.*, 1977, **99**, 2816.

Perchloro-2-pyrone and Grignard reagents RMgX, at low temperatures (−60 to −10 °C), yield 6-alkyl-trichloro-2-pyrones (151).[118] Further treatment of (151; R = Me) with methylmagnesium bromide brings about ring-opening to the allene-diol Me$_2$C(OH)C(Cl)=C=C(Cl)C(OH)Me$_2$, which also results from the action of methyl-lithium on the perchloro-2-pyrone. Curiously, Grignard reagents other than MeMgBr yield not allene-diols but dihydropyrones (152).

4-Hydroxy-5-oximino-7-methyl-5$H$-pyrano[2,3-$b$]pyridine 8-oxide, formed by the action of dimethylformamide dimethyl acetal on 3-acetyl-4-hydroxy-6-methyl-$\alpha$-pyrone, has been shown by $^{13}$C n.m.r. to exist as structure (153), one of

(151)                                (152)                                (153)

five possible tautomeric forms.[119] The reactions of $\gamma$-pyrones with hydroxylamine, hydrazine, and their derivatives, some of which were reported last year,[120] have been reviewed.[121] A survey of six-membered rings containing oxygen has also appeared.[122a]

Continuation of the work reported last year[123] on the conversion of 2,4,6-triphenylpyrylium salts into the corresponding 2,4,6-triphenylpyridiniums has yielded additional synthetically useful procedures. Examples are the conversion of alkyl, aryl, and heteroaryl primary amines into iodides,[124] esters,[125a] and thiocyanates[125b] and the production of carbodi-imides from amidrazones, as outlined in Scheme 34.[126]

A full report on the 2-pyridone–pyran-2-imine and pyridine-2-thione–thiopyran-2-imine systems, mentioned briefly last year,[127] has been published.[128]

Decarboxylation of 4,6-diphenylpyrylium-2-carboxylic acid perchlorate (154) in hot acetic anhydride provides a useful source of 2-pyranylidene (155), which has been trapped using phenylacetylene[129a] and ferrocene.[129b] The ferrocene adduct (156; X = O$^+$), when heated with ammonia in a sealed tube at 100 °C for 7 hours, yields 2-ferrocenyl-4,6-diphenylpyridine (156; X = N).

[118] A. Roedig, M. Försch, H. Abel, and S. Bauer, *Chem. Ber.*, 1977, **110**, 1000.
[119] K. Roth and W. Lowe, *Z. Naturforsch.*, 1977, **32b**, 1175.
[120] See ref. 18, p. 112.
[121] C. Morin and R. Beugelmans, *Tetrahedron*, 1977, **33**, 3183.
[122] R. Livingstone, *Rodds' Chemistry of Carbon Compounds*, 2nd. edn., 1977, **IVE**, (a) p. 1; (b) p. 347.
[123] See ref. 18, p. 106.
[124] N. F. Eweiss, A. R. Katritzky, P.-L. Nie, and C. A. Ramsden, *Synthesis*, 1977, 634.
[125] (a) U. Gruntz, A. R. Katritzky, D. H. Kenny, M. C. Rezende, and H. Sheikh, *J.C.S. Chem. Comm.*, 1977, 701; (b) A. R. Katritzky, U. Gruntz, N. Mongelli, and M. C. Rezende, *ibid.*, 1978, 133.
[126] A. R. Katritzky, P.-L. Nie, A. Dondoni, and D. Tassi, *Synthetic Comm.*, 1977, **7**, 387.
[127] See ref. 18, p. 107.
[128] A. S. Afridi, A. R. Katritzky, and C. A. Ramsden, *J.C.S. Perkin I*, 1977, 1436.
[129] (a) Yu. P. Andreichikov, N. V. Kholodova, and G. N. Dorofeenko, *J. Org. Chem. (U.S.S.R.)*, 1977, **13**, 1443; (b) V. V. Krasnikov, Yu. P. Andreichikov, N. V. Kholodova, and G. N. Dorofeenko, *ibid.*, p. 1444.

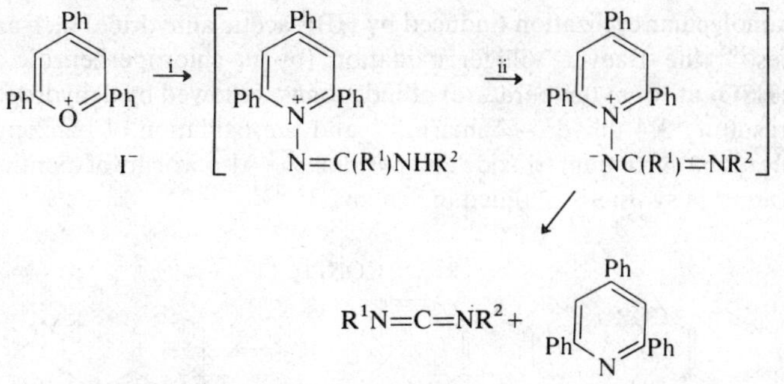

Reagents: i, $R^1C(NHR^2)=NNH_2$, EtOH, at 20 °C, for 12 h; ii, KOH, MeOH, pH 9; iii, Δ, 150—200 °C.

**Scheme 34**

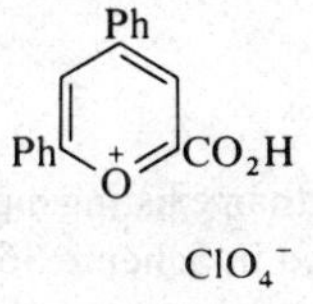

The action of mixtures of perchloric acid and an acid anhydride, or alternatively perchloric acid, a carboxylic acid, and polyphosphoric acid, on ketones, lactones, and carboxylic acids, mentioned briefly last year,[130] has been utilized for the preparation of various fused pyrylium systems, including the bis-indeno[3,4:5,6]pyrylium salt (157),[131a] acyloxy-benzofuro[3,2-*c*]pyrylium systems (158),[132] and other benzopyrylium salts.[131b,133]

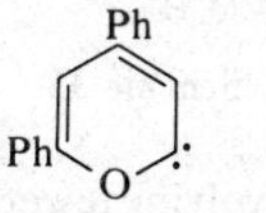

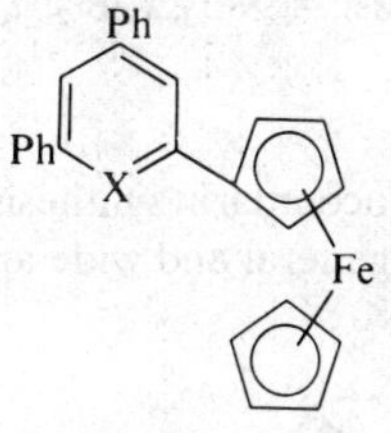

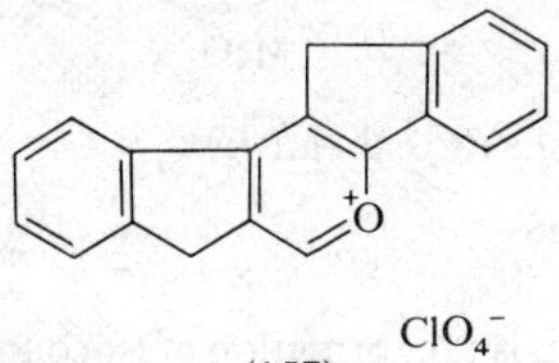

**Coumarins, Chromones, Xanthones, and their Thio-analogues.**—Several new methods of preparing coumarins have been reported, and include the action of the Wittig reagent $Ph_3P=CHCO_2Et$ on *o*-hydroxy-aldehydes or -ketones,[134]

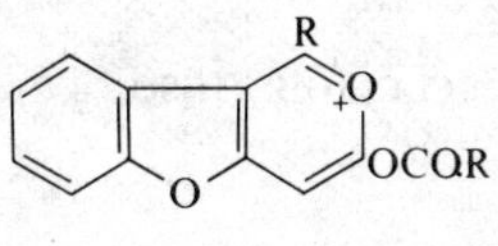

[130] See ref. 18, p. 106.
[131] D. V. Pruchkin, E. V. Kuznetsov, and G. N. Dorofeenko, *Khim. geterotsikl. Soedinenii*, 1978, (*a*) p. 275; (*b*) p. 1479.
[132] G. N. Dorofeenko, V. G. Korobkova, and V. I. Volbushko, *Khim. geterotsikl. Soedinenii*, 1977, 553.
[133] E. V. Kuznetsov, I. V. Shcherbakova, and G. N. Dorofeenko, *Khim. geterotsikl. Soedinenii*, 1977, 1481.
[134] R. S. Mali and V. J. Yadav, *Synthesis*, 1977, 464.

the intramolecular cyclization (induced by HBr–acetic anhydride) of 1-aryl-1,3-diketones,[135] the Baeyer–Villiger oxidation (by *m*-chloroperbenzoic acid in $Ac_2O–H_2SO_4$ at room temperature) of indanones, followed by dehydrogenation of the resulting 3,4-dihydro-coumarin,[136] and the oxidation of benzopyrylium perchlorates by chromium trioxide and pyridine.[137] Also worthy of mention is the novel coumarin synthesis outlined in Scheme 35.[138]

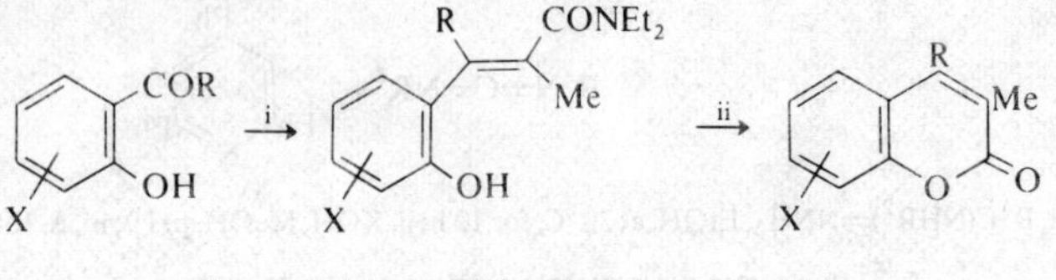

X = OH, Me, or OMe; R = Me or $CH_2OMe$

Reagents: i, $MeC\equiv CNEt_2$, $MgBr_2$; ii, $Me_2CO$, $H^+$.

**Scheme 35**

A new isocoumarin synthesis, involving fewer stages than existing methods and promising general and wide applicability, is exemplified in Scheme 36.[139]

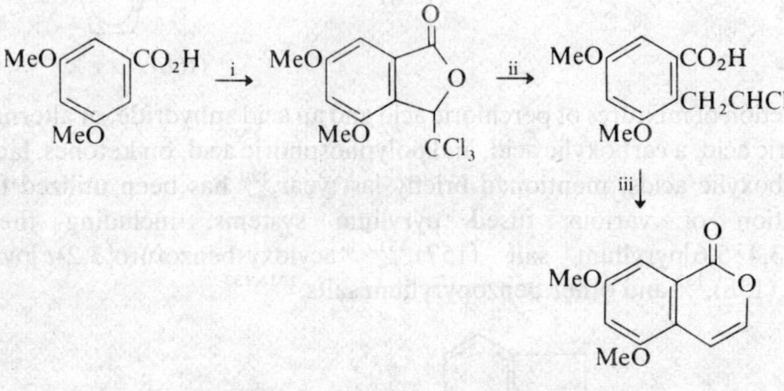

Reagents: i, $CCl_3CH(OH)_2$, $H_2SO_4$; ii, Zn, AcOH; iii, EtOH, conc. HCl.

**Scheme 36**

More simple, but so far of less general use, is the formation of isocoumarins by cyclization of *o*-alkenyl-benzoic acids [(159) → (160)], using lithium tetra-chloropalladate in aqueous dioxan at room temperature.[140]

Ring-opening of 4-methylcoumarin under alkaline conditions is a useful new approach to the synthesis of the acid-sensitive *o*-isopropenyl-phenol (161).[141] A

[135] V. K. Ahluwalia and D. Kumar, *Indian J. Chem.*, 1977, **15B**, 514.
[136] A. Chatterjee, S. Bhattacharya, J. Banerji, and P. C. Ghosh, *Indian. J. Chem.*, 1977, **15B**, 214.
[137] P. Bouvier, J. Andrieux, H. Cunha, and D. Molho, *Bull. Soc. chim. France*, 1977, 1187.
[138] S. I. Pennanen, *Heterocycles*, 1977, **6**, 1181.
[139] B. K. Sarkhel and J. N. Srivastava, *Indian J. Chem.*, 1977, **15B**, 103.
[140] A. Kasahara, T. Izumi, K. Sato, M. Maemura, and T. Hayasaka, *Bull. Chem. Soc. Japan*, 1977, **50**, 1899.
[141] V. V. Dhekne, B. D. Kulkarni, and A. S. Rao, *Indian J. Chem.*, 1977, **15B**, 755.

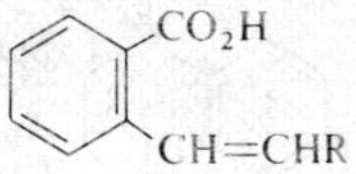

(159) 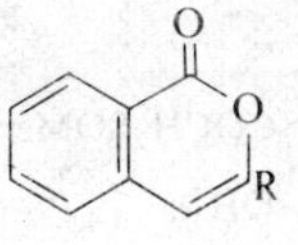　　　　　(160) R = H　46%　　　　(161)
　　　　　　　　　　　　 R = Ph　42%

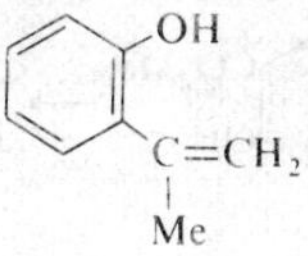

new superior method for *O*-methylation of 4-hydroxy-2-pyrones and 4-hydroxy-coumarins uses dimethyl sulphate and sodium hydride in hexamethylphosphortriamide (HMPT) as the methylating mixture.[142] The general chemistry of chromones and its derivatives has been comprehensively reviewed.[143]

Condensation of *o*-acetoxybenzoyl chloride with lithium enolates, prepared from *O*-trimethylsilyl ethers as shown in Scheme 37, provides a useful new route to chromones.[144]

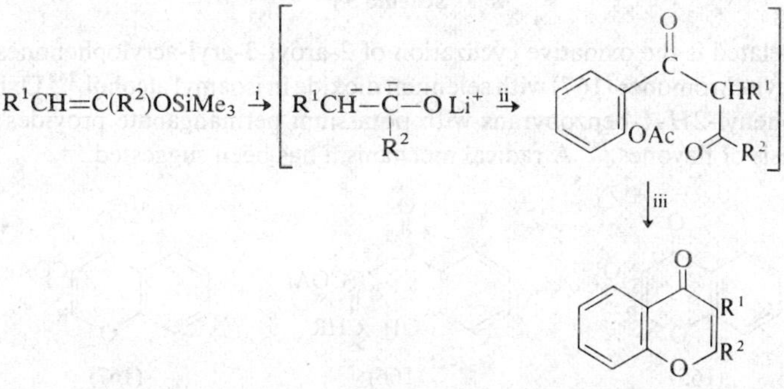

$$R^1CH=C(R^2)OSiMe_3 \xrightarrow{i} \left[ R^1CH-\overset{=}{\underset{R^2}{C}}-O^-Li^+ \xrightarrow{ii} \right]$$

(80—90%)

Reagents: i, PhLi; ii, *o*-AcOC$_6$H$_4$COCl, MeOCH$_2$CH$_2$OMe at $-70\,°C$; iii, H$^+$.

**Scheme 37**

Equally useful is the synthesis of chromone-2-carboxylic acids illustrated in Scheme 38.[145] The Knoevenagel condensation products (162) react with boron trifluoride etherate to yield chromone-2-carboxylates (163), whereas with *N*-ethylpiperidine in toluene the coumarones (164; R$^2$ = CHCO$_2$Bu) are obtained.

*o*-Hydroxy-ω-nitroacetophenone cyclises to 2-oximinocoumarone (164; R$^1$ = H, R$^2$ = NOH) in DMSO–pyridine or in POCl$_3$–DMF, whereas in a mixture of acetic anhydride and sulphuric acid the 3-nitro-chromone (165; R = Me) is obtained in 80% yield.[146] 3-Nitrochromone (165; R = H) results from cyclization of the hydroxynitroacetophenone with trimethyl orthoformate and a catalytic amount of pyridine. Similar results were reported last year by other authors.[147]

[142] E. Suzuki, B. Katsuragawa, and S. Inoue, *Synthesis*, 1978, 144.
[143] *Chem. Heterocyclic Compounds*, 1977, Vol. 31.
[144] T. Watanabe, Y. Nakashita, S. Katayama, and M. Yamauchi, *J.C.S. Chem. Comm.*, 1977, 493.
[145] G. Pifferi, G. Gaviraghi, M. Pinza, and P. Ventura, *J. Heterocyclic Chem.*, 1977, **14**, 1257.
[146] K. V. Rao and V. Sundaramurthy, *Indian J. Chem.*, 1977, **15B**, 236.
[147] See ref. 18, p. 108.

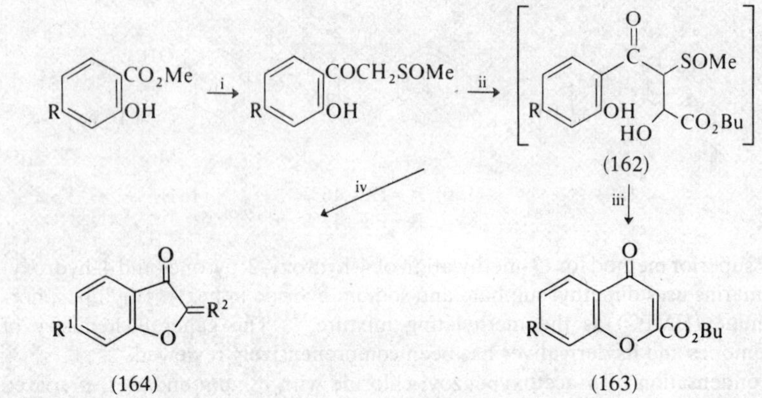

(162)                                    (163)

(164)                                    (163)

Reagents: i, $\bar{C}H_2SOMe$; ii, OHCCO$_2$Bu, NaOAc; iii, Et$_2$O·BF$_3$; iv, N-ethylpiperidine, PhMe.

**Scheme 38**

Also related is the oxidative cyclization of 2-aroyl-3-aryl-acrylophenones (166) to 3-acyl-chromones (167) with selenium dioxide in isoamyl alcohol.[148] Oxidation of 2-phenyl-2H-1-benzopyrans with potassium permanganate provides a new synthesis of flavones.[149] A radical mechanism has been suggested.

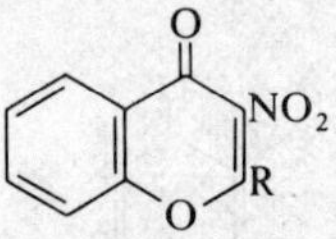

(165) 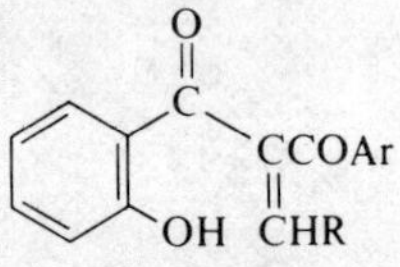                        (166) 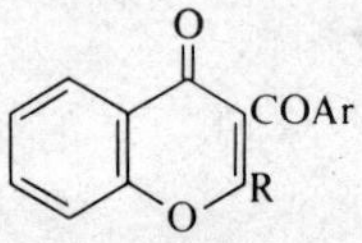                        (167)

Cycloalkeno-chromones (170; n = 1, 2, or 3) are available in high yield (90%) by base-catalysed condensation of aldehyde (168) with the appropriate cyclo-alkanone, followed successively by oxidation (CrO$_3$) and cyclization by acid (HCl) of the resulting aldol (169).[150]

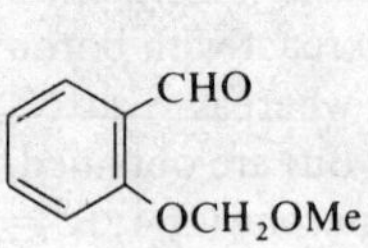

(168) 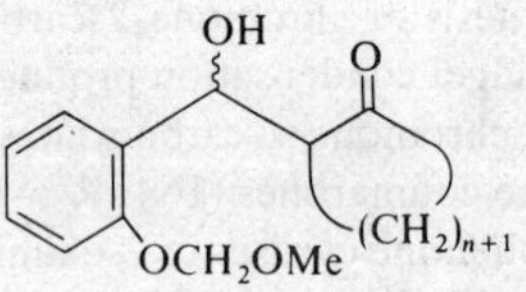                        (169) 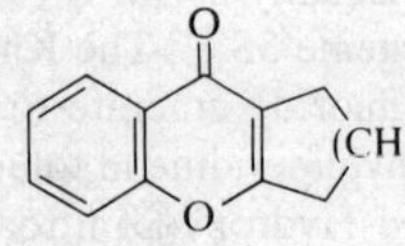                        (170)

An alternative synthesis of xanthone-2-carboxylic acids (173) has been proposed[151] which involves the photo-Fries rearrangement of o-methoxy-benzoates (171) followed by base-catalysed (2M-KOH) cyclization of the benzophenones (172) so formed. Overall yields of 20% are claimed.

[148] K. A. Thakar and V. N. Ingle, Indian J. Chem., 1977, **15B**, 571.
[149] Y. Ashihara, Y. Nagata, and K. Kurosawa, Bull. Chem. Soc. Japan, 1977, **50**, 3298.
[150] T. Watanabe, S. Katayama, Y. Nakashita, and M. Yamauchi, Chem. and Pharm. Bull. (Japan), 1977, **25**, 2778.
[151] R. Graham and J. R. Lewis, Chem. and Ind., 1977, 798.

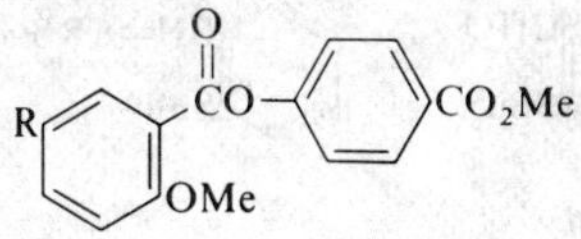

(171)

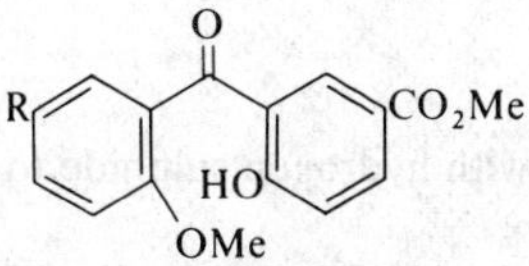

(172) 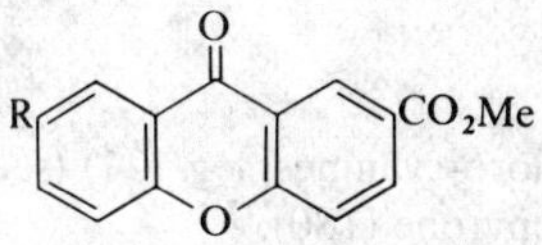                                        (173)

Tetrahydrothiopyran-4-ones (174), readily prepared as shown in Scheme 39 (path *a*) are useful precursors of 4*H*-thiopyran-4-ones, *e.g.* (175).[152] Dehydrogenation, however, has to be carried out in a stepwise manner, as shown, since direct oxidation of (174) to (175), using $SeO_2$ in boiling toluene, fails. An alternative preparation of the 2,6-diphenyl derivative (175; R = Ph) which avoids using the toxic $SeO_2$ involves Pummerer rearrangement of the dihydrothiopyran *S*-oxide (176), as also shown in Scheme 39 (path *b*).[153]

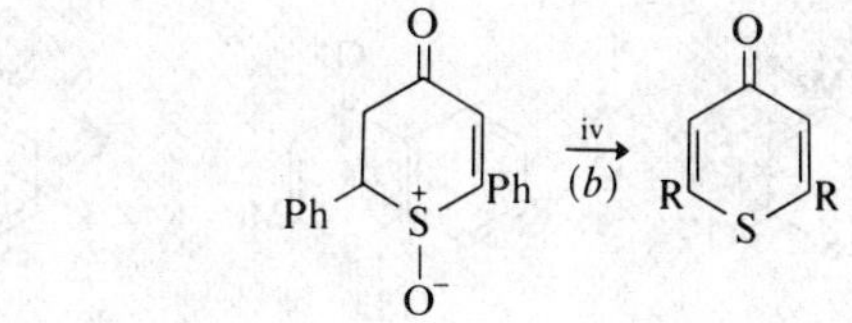

(174)

(176)                                        (175)

Reagents: i, $H_2S$, NaOAc; ii, *N*-chlorosuccinimide, $C_5H_5N$; iii, $SeO_2$, PhMe; iv, $(CF_3CO)_2O$, $CH_2Cl_2$.

**Scheme 39**

4-Thiochromanones, on photolysis (350 nm) in methanol, undergo not photoisomerization, such as is observed with 4-isothiochromanones, but photodehydrogenation to 4-thiochromones in moderate (50—55%) yield.[154] 2-(t-Butylthio)-benzaldehyde (177), prepared in near quantitative yield from *o*-nitrobenzaldehyde as indicated in Scheme 40, is a useful new synthon for thiocoumarins (178; $R^1 = CO_2H$, $CONH_2$, $CO_2Et$, CN, Bz, Ph, or 2-thienyl)[155a] and benzo[1,2-*b*: 5,4-*b'*]bisthiopyrans (179).[155b]

[152] C. H. Chen, G. A. Reynolds, and J. A. van Allan, *J. Org. Chem.*, 1977, **42**, 2777.
[153] C. H. Chen, *Heterocycles*, 1977, **7**, 231.
[154] A. Couture, A. Lablache-Combier, and T. Q. Minh, *Tetrahedron Letters*, 1977, 2873.
[155] O. Meth-Cohn and B. Tarnowski, *Synthesis*, 1978, (*a*) p. 56; (*b*) p. 58.

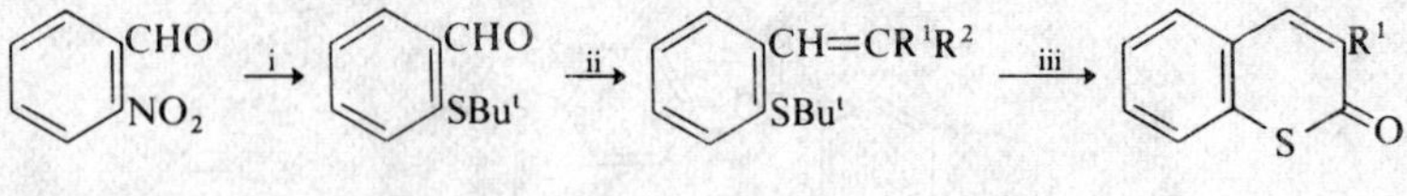

(177)                                                              (178)

Reagents: i, $Bu^tSH$, DMF, at 100 °C; ii, $CH_2R^1R^2$ ($R^2 = CO_2H$ or CN); iii, PPA.

**Scheme 40**

Chloro-cyanines, *e.g.* (21) (see p. 153), react with hydrogen sulphide to yield dithiopyrone (180).[17]

Apparently, the cyclization of $S$-phenyl-3-oxobutanethioates (181) to $2H$-1-benzothiopyran-2-ones (182) in polyphosphoric acid is successful only with the $m$-methoxy-derivative (181; $R = m$-MeO).[156] In all other cases the isomeric thiochromones, *e.g.* (183), are formed. Spectroscopic methods of distinguishing between these hitherto difficultly distinguishable isomers are discussed. In the $^1H$ n.m.r. spectrum the thiochromones show characteristic deshielding of the 5-proton by the *peri* carbonyl group. Equally diagnostic are their mass spectra, which show fragmentation of the parent ions to the ketens (184) by a retro-Diels–Alder reaction.

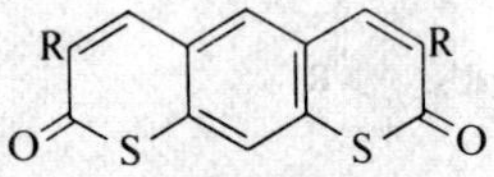

(179) 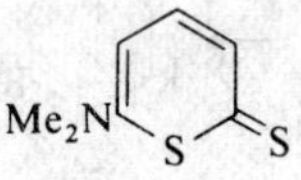                        (180) 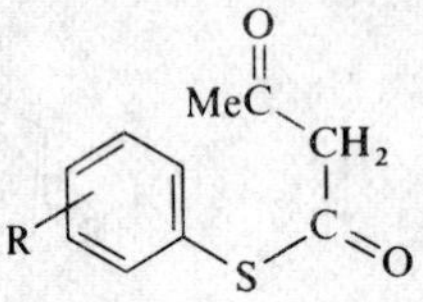                        (181)

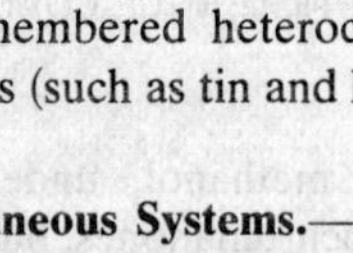

(182) 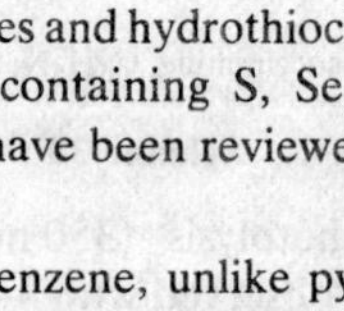                        (183) 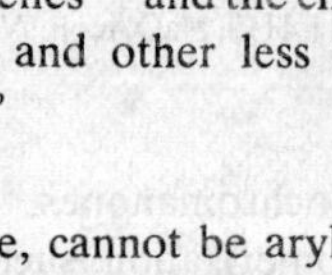                        (184)

The chemistry of thiochromenes and hydrothiochromenes[157] and the chemistry of six-membered heterocycles containing S, Se, Te, and other less familiar elements (such as tin and lead) have been reviewed.[122b]

**Miscellaneous Systems.**—Arsabenzene, unlike pyridine, cannot be arylated or alkylated directly by the usual methods. Useful, therefore, is the acid-catalysed rearrangement of 1-aryl-4-methoxy-arsacyclohexadienes, which provides access to 2-aryl-4-substituted arsabenzenes as shown in Scheme 41.[158]

Simple entry into 2- and 3-substituted arsabenzenes, the latter being inaccessible by previously reported methods, is now possible *via* the arsabenzene–

[156] H. Nakazumi and T. Kitao, *Bull. Chem. Soc. Japan*, 1977, **50**, 939.
[157] S. K. Klimenko, V. G. Kharchenko, and T. V. Stolbova, *Khim. geterotsikl. Soedinenii.*, 1978, 3.
[158] G. Märkl and R. Liebl, *Angew. Chem. Internat. Edn.*, 1977, **16**, 637; G. Märkl and J. B. Rampal, *Tetrahedron Letters*, 1977, 3449.

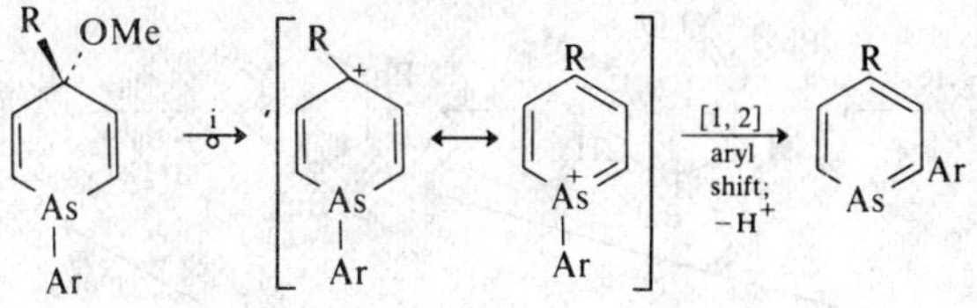

Reagents: i, *p*-TSA or Et$_2$O·BF$_3$, C$_6$H$_6$.

**Scheme 41**

acetylenedicarboxylate cyclo-adducts, *e.g.* (184; R$^1$ = R$^2$ = CO$_2$Me).[159] On thermolysis at 400 °C (g.l.p.c.), the adducts extrude acetylene to give as major product (80%) the dicarboxylate (185). Surprisingly, a small quantity (5%) of dimethyl phthalate is also produced, by loss of HC≡As from the cyclo-adduct (184). As expected, arsabenzene and methyl propiolate yield two adducts (184; R$^1$ = CO$_2$Me, R$^2$ = H) and (184; R$^1$ = H, R$^2$ = CO$_2$Me), in the ratio of 3:2. Pyrolysis, as before, furnishes a 3:2 mixture of 3-methoxycarbonyl- and 2-methoxycarbonyl-arsabenzene.

Investigations continue into the chemistry of arsabenzenes bearing common functional groups. For example, Wittig reactions on arsabenzaldehyde (186; X = O), although slow (30 h), give the 4-vinyl derivatives (186; X = CHCOR; R = Me, Ph, or OMe) in practicable yields.[160] Attack by a carbanion at arsenic is not observed.

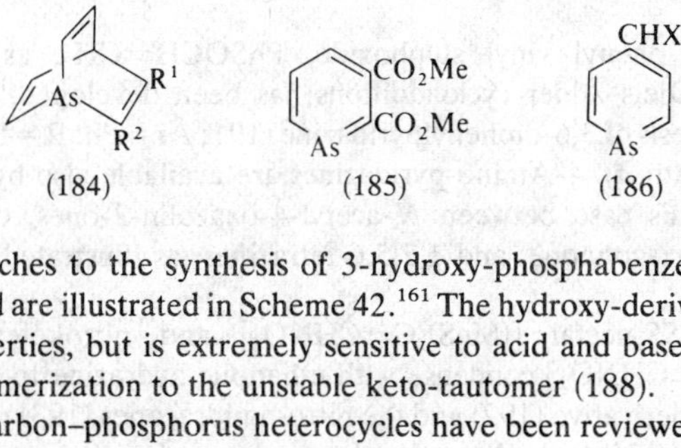

(184)          (185)          (186)

Two approaches to the synthesis of 3-hydroxy-phosphabenzenes have been developed, and are illustrated in Scheme 42.[161] The hydroxy-derivative (187) has phenolic properties, but is extremely sensitive to acid and base, which readily promote tautomerization to the unstable keto-tautomer (188).

Polycyclic carbon–phosphorus heterocycles have been reviewed.[162]

## 2 Systems containing Two Heteroatoms, and their Benzo-derivatives

**Diazines.**—*Pyridazines, Cinnolines, and Phthalazines.* Azines, *e.g.* (189), derived from α-diketones and either β-keto-esters or β-diketones, exist predominantly as the enamine tautomers, and as such are useful precursors for a new general pyridazine synthesis (Scheme 43).[163]

[159] A. J. Ashe, III, and H. S. Friedman, *Tetrahedron Letters*, 1977, 1283.
[160] G. Märkl, J. B. Rampal, and V. Schöberl, *Tetrahedron Letters*, 1977, 2701.
[161] G. Märkl, G. Adolin, F. Kees, and G. Zander, *Tetrahedron Letters*, 1977, 3445.
[162] S. D. Venkataramu, G. D. Macdonell, W. R. Purdum, M. El-Deek, and K. D. Berlin, *Chem. Rev.*, 1977, **77**, 121.
[163] S. Evans and E. E. Schweizer, *J. Org. Chem.*, 1977, **42**, 2321.

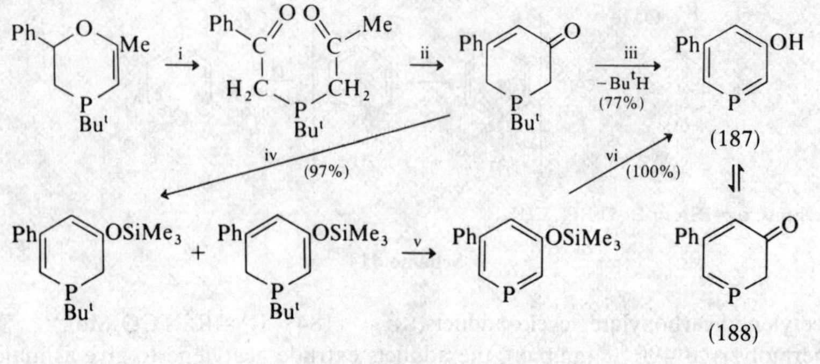

Reagents: i, HCl; ii, NaOEt, at 25 °C, 3 h; iii, Δ, 350 °C; iv, Me₃SiCl; v, Δ, 250 °C; vi, acid-free MeOH.

**Scheme 42**

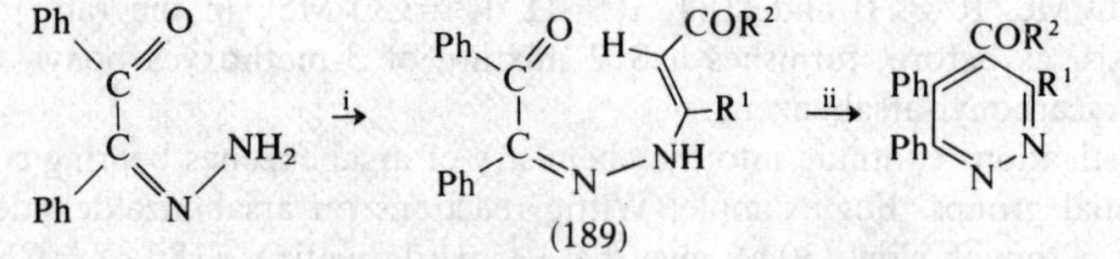

Reagents: i, $R^1COCH_2COR^2$ ($R^1$ = Me or Ph, $R^2$ = OEt or Me); ii, KOH, EtOH, boil.

**Scheme 43**

The use of phenyl vinyl sulphoxide, $PhSOCH=CH_2$, as an acetylene equivalent in Diels–Alder cycloadditions has been developed,[164] an example being the synthesis of 3,6-diphenylpyridazine (191; Ar = Ph; R = H) in 97% yield (Scheme 44; path *a*). 4-Amino-pyridazines are available also by cycloaddition reactions, in this case between *N*-acetyl-4-oxazolin-2-ones, or (better) the unsubstituted oxazolinone, and 1,2,5,6-tetrazines, as illustrated in Scheme 44 (path *b*).[165]

Nitroketen *SS*-acetal [$(MeS)_2C=CHNO_2$] and nitroketen thioaminals [$RHN(MeS)C=CHNO_2$] condense with ethanolic hydrazine to give the nitro-thioacetic acid derivative (192) and the nitro-amidrazones (193) respectively, all of which are useful intermediates for the synthesis of 4-nitro-pyridazines (194) (Scheme 45).[166]

Oxidative cyclization of 1,4-diketone dioximes was cited in last year's Report[167] as an improved route to 3,6-diaryl-pyridazine 1,2-dioxides. Related is the oxidative cyclization of 1,3,4,6-tetraketones (195) to the explosive 3,6-diacyl-4,5-dioxo-4,5-dihydropyridazine 1,2-dioxides (196), using dinitrogen tetroxide ($N_2O_4$).[168] A full report on the unexpected nuclear alkylation of

[164] L. A. Paquette, R. E. Moerck, B. Harirchian, and P. D. Magnus, *J. Amer. Chem. Soc.*, 1978, **100**, 1597.
[165] J. A. Deyrup and H. L. Gingrich, *Tetrahedron Letters*, 1977, 3115.
[166] H. Hamberger, H. Reinshagen, G. Schulz, and G. Sigmund, *Tetrahedron Letters*, 1977, 3619.
[167] See ref. 18, p. 116.
[168] B. Unterhalt and U. Pindur, *Arch. Pharmazie*, 1977, **310**, 264.

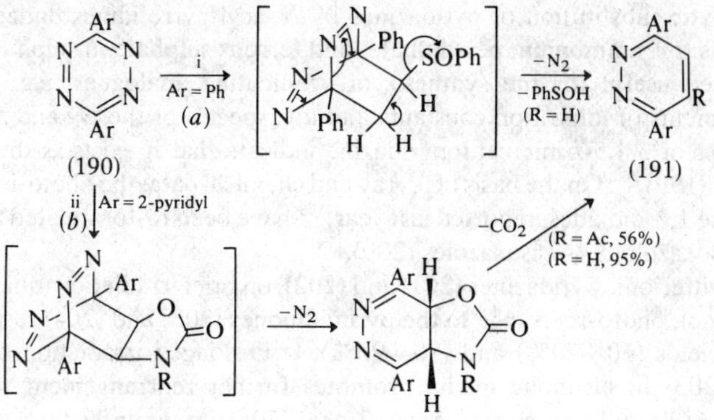

(190) (191)

Reagents: i, $PhSOCH{=}CH_2$, PhMe, boil; ii; 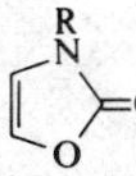 (R = Ac or H); *for* R = Ac, PhMe at 110 °C for 46 h; *for* R = H, EtOAc at 25 °C for 24 h.

**Scheme 44**

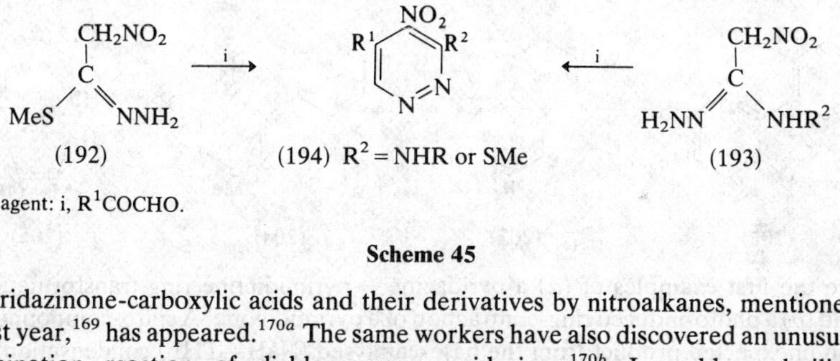

(192) (194) $R^2$ = NHR or SMe (193)

Reagent: i, $R^1COCHO$.

**Scheme 45**

pyridazinone-carboxylic acids and their derivatives by nitroalkanes, mentioned last year,[169] has appeared.[170a] The same workers have also discovered an unusual amination reaction of dichloro-cyano-pyridazines.[170b] In cold or boiling methanol, with a secondary amine, 3,6-dichloro-4-cyanopyridazine (197) suffers nucleophilic displacement of the 3- (major reaction) and 6-chloro groups. However, with primary amines, in cold (0—5 °C) methanol, nuclear amination (20—35%) takes place at the 5-position. Curiously, nuclear amination is not observed with 3-chloro-4-cyanopyridazine.

$ArCOCH_2COCOCH_2COAr$

(195)

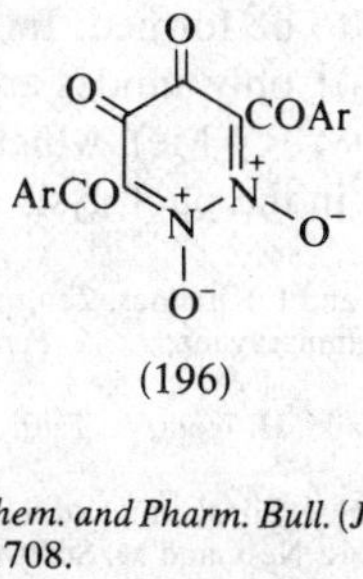

(196)

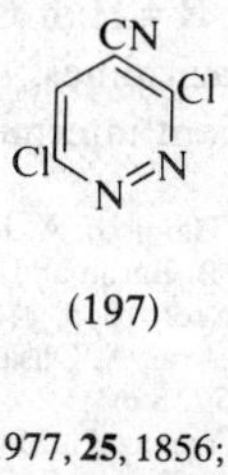

(197)

[169] See ref. 18, p. 117.

[170] (*a*) M. Yanai, S. Takeda, and M. Nishikawa, *Chem. and Pharm. Bull. (Japan)*, 1977, **25**, 1856; (*b*) M. Yanai, S. Takeda, and T. Mitsuoka, *ibid.*, p. 1708.

Homolytic substitution of pyridazines by *N*-acyl-pyrrolidines under Minisci conditions (*i.e.* ammonium persulphate, and ferrous sulphate in sulphuric acid), has proved useful for the synthesis of azanicotine analogues, *e.g.* (198).[171] Measurements of ionization constants and u.v. spectra of the *N*- and *S*-methyl derivatives of 3,4,5-trimercaptopyridazine indicate that it exists as the N(2)-H tautomer (199).[172] On the basis of *X*-ray and chemical data, the photo-isomers of pyridazine 1,2-dioxides, reported last year,[173] have been re-formulated as 3a,6a-dihydroisoxazolo[5,4-*d*]isoxazoles (200).[174]

The zwitterionic pyridazines (201) and (202), on brief irradiation (quartz filter) in methanol, photo-rearrange to the pyrimidinones (203) and (204) respectively in high yields (60—70%) and (80—90%).[175] Prolonged irradiation of pyrimidones (203) in alcoholic media promotes further rearrangement to the 3-alkoxy-1,6-diaza-4-oxo-spiro[4,5]dec-2-enes (205). Apparently, these reactions

(198)          (199)          (200)          (201)

(202)          (203)          (204)          (205)

are the first examples of (*a*) a pyridazine → pyrimidinone ring-transformation and (*b*) a photo-induced ring-contraction of a pyrimidinone. A spiro-compound is the unexpected product from the base-catalysed (NaH—THF) condensation of diethyl pyridazine-4,5-dicarboxylate (206; R = CO$_2$Et) with 1,3-diphenyl-guanidine.[176] Instead of intramolecular nucleophilic attack at the adjacent ester carbonyl, which would lead to formation of a pyridazino-azepinedione, substitution occurs at the electron-deficient 4-position (Scheme 46) to produce the 1,3,7,8-tetra-aza-spiro[4,5]dec-6,9-ene (207). Investigations reveal that a strongly electron-withdrawing group (CO$_2$Et or CN) is necessary at the 5-position if a spiro-compound is to be formed. In the absence of such a substituent [*e.g.* (206; R = H or CONH$_2$)], only amides are formed.

Diaziridines, *e.g.* (208; R = Me), which so far have been considered only as transient intermediates in the photolysis of 3-oxidopyridazinium betaines, have

[171] G. Heinisch, A. Jentzsch, and I. Kirchner, *Tetrahedron Letters*, 1978, 619.
[172] G. B. Barlin and P. Lakshminarayana, *J.C.S. Perkin I*, 1977, 1038.
[173] See ref. 18, p. 118.
[174] H. Arai, A. Ohsawa, K. Saiki, H. Igeta, A. Tsuji, T. Akimoto, and Y. Iitaka, *J.C.S. Chem. Comm.*, 1977, 856.
[175] T. Yamazaki, M. Nagata, S.-I. Hirokami, and S. Miyakoshi, *Heterocycles*, 1977, **8**, 377.
[176] G. Adembri, S. Chimichi, R. Nesi, and M. Scotton, *J.C.S. Perkin I*, 1977, 1020.

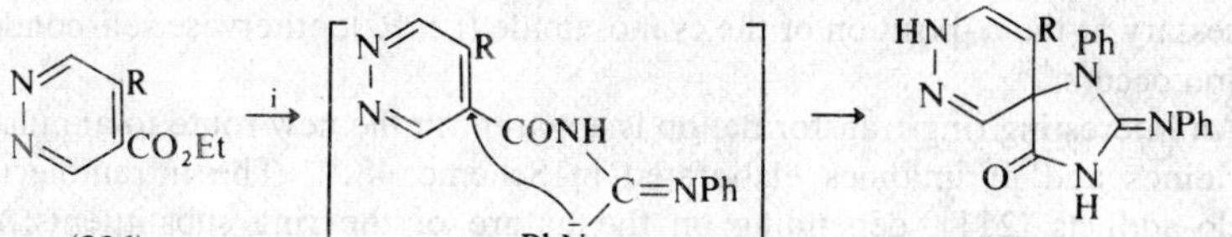

Reagents: i, NaH, THF, PhNHC(NHPh)=NH.

**Scheme 46**

now been isolated and characterized.[177] In acetonitrile solution the corresponding phthalazinium betaines behave similarly, whereas in nucleophilic solvents, *e.g.* methanol or ethylamine, the diaziridine intermediate (208; R = H) ring-opens to the phenyl-diaziridines (209; R' = OMe or NHEt).

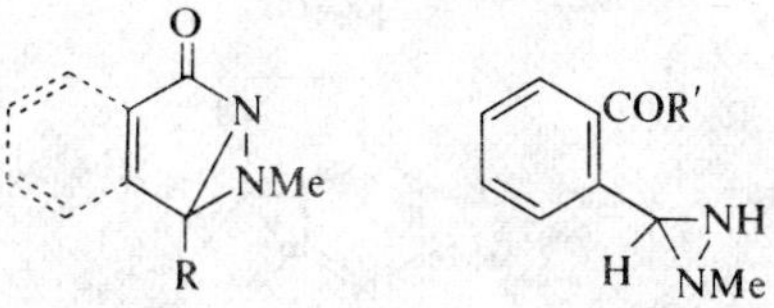

*Pyrimidines and Quinazolines.* Acid-catalysed condensation of urea and propiolic acid in boiling benzene has been proposed as a new, viable, commercial process for the production of uracil.[178] Also new is the synthesis of pyrimidinones and pyrimidines from a mixture of amides as outlined in Scheme 47.[179] Isolation of the pyrimidinone (210) is dependent on the nature of $R^2$. Generally, if this substituent

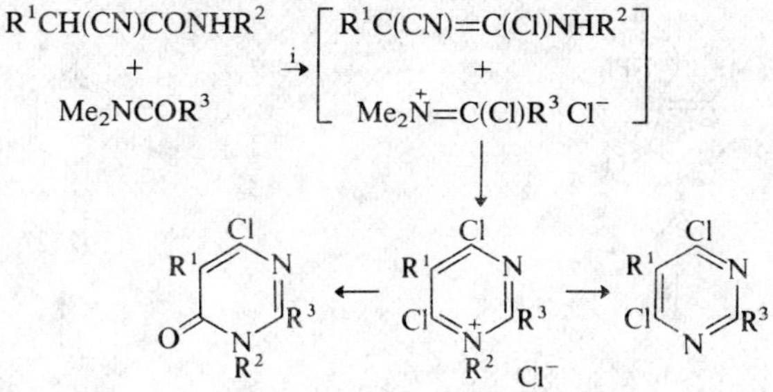

Reagent: i, POCl₃.

**Scheme 47**

is 5-alkyl (*e.g.* cyclohexyl) then only the dichloro-pyrimidine is isolated. The process is limited in its general applicability in that an alkyl or aryl group is

[177] Y. Maki, M. Kawamura, H. Okamoto, M. Suzuki, and K. Kaji, *Chem. Letters*, 1977, 1005.
[178] R. J. De Pasquale, *J. Org. Chem.*, 1977, **42**, 2185.
[179] R. L. N. Harris and J. L. Huppatz, *Angew. Chem. Internat. Edn.*, 1977, **16**, 779.

necessary at the $\alpha$-position of the cyano-amide (*i.e.* $R^1$), otherwise self-condensation occurs.[180]

An interesting ring-transformation is apparent in the new route to annulated pyridines and pyrimidines elaborated in Scheme 48.[181] The intramolecular cyclo-adducts (211), depending on the nature of the ring substituents, can eliminate either of the bridging groups, as exemplified in the reaction scheme.

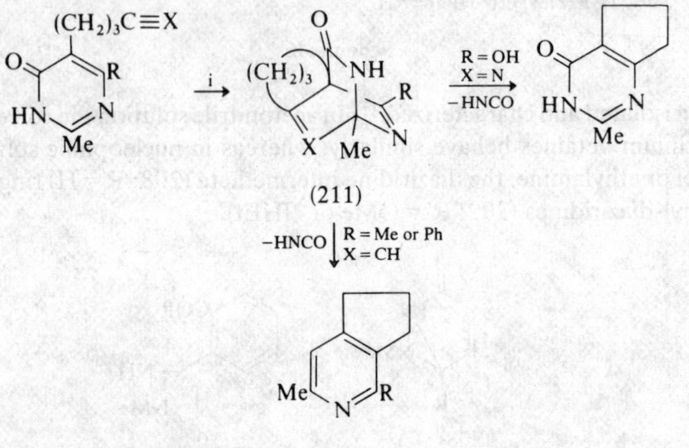

Reagents: i, MeCN or DMF, 180—200 °C.

**Scheme 48**

A novel base-catalysed ring-expansion of 4-amino-antipyrines to 5-amino-4(3$H$)-pyrimidinones, *e.g.* (212), has been announced,[182] and is outlined in Scheme 49.

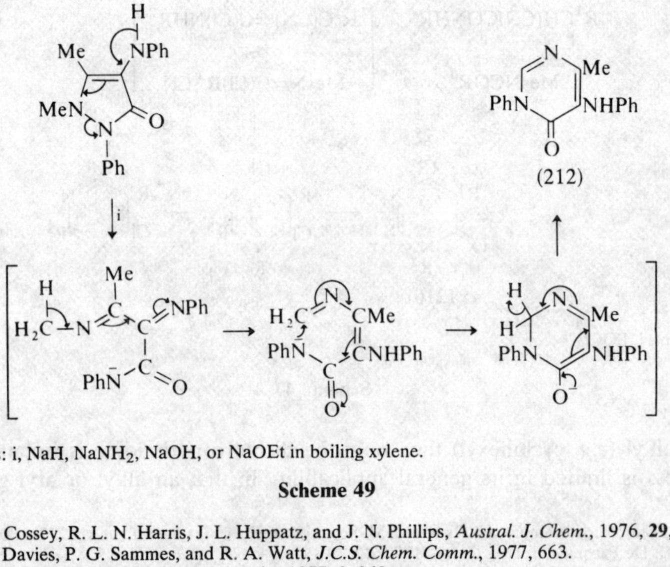

Reagents: i, NaH, NaNH$_2$, NaOH, or NaOEt in boiling xylene.

**Scheme 49**

[180] A. L. Cossey, R. L. N. Harris, J. L. Huppatz, and J. N. Phillips, *Austral. J. Chem.*, 1976, **29**, 1039.
[181] L. B. Davies, P. G. Sammes, and R. A. Watt, *J.C.S. Chem. Comm.*, 1977, 663.
[182] T. Ueda, N. Oda, and I. Ito, *Heterocycles*, 1977, **8**, 263.

5,6-Dihydro-4-pyrimidinones result from the cycloaddition of diphenylcyclopropenones with guanidines and with amidines.[183] A general route to the hitherto inaccessible 2-(pyrimidin-2'-yl)acetic acid (214) and its 4-substituted derivatives has been described,[184] the first stage of which involves self-condensation of malonamide to the dihydroxypyrimidine derivative (213). Heteroarylation[185] and 2-hydroxymethylation[186] of simple pyrimidines have been achieved in high yield.

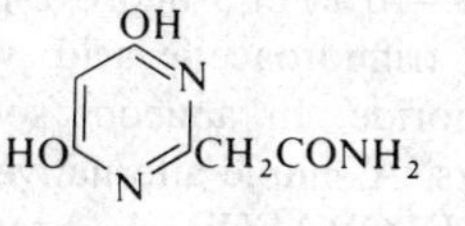

$$(213) \qquad\qquad (214)$$

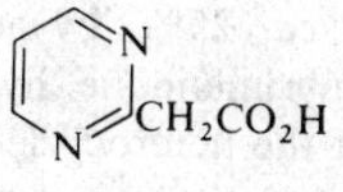

The former process involves photo-coupling of the heteroarene (furan, thiophen, or *N*-methylpyrrole) with a 5-iodopyrimidine in acetonitrile; the latter a radical substitution with aqueous methanol under Minisci conditions.

Although predicted theoretically some years ago, unequivocal experimental evidence for the so-called 'oxa-benzidine' rearrangement is still awaited. However, to the list of reactions suspected of involving this type of rearrangement must now be added the process outlined in Scheme 50, which is useful for the direct arylation of pyrimidine.[187] Similar reactions occur with 2-chloro-3-nitro-pyridine and with 2-chloropyrazine, but not with 2-chloro- or 2-fluoro-pyridine.

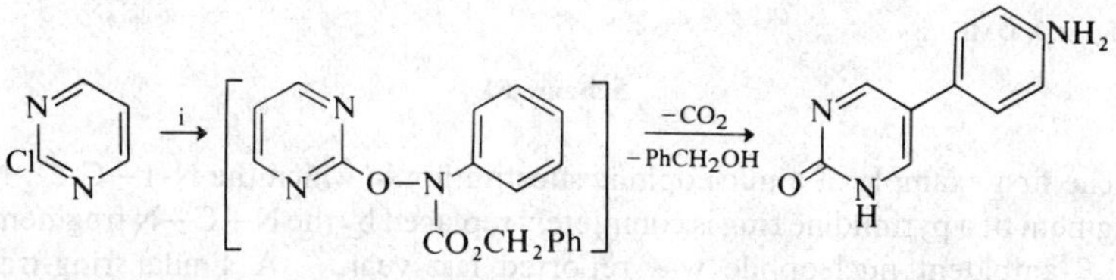

Reagents: i, HON(Ph)CO$_2$CH$_2$Ph, KOH, H$_2$O.

**Scheme 50**

Coupling of diazotized *p*-chloroaniline with 4-methyl- and with 4,6-dimethyl-pyrimidine occurs at the 4-methyl group, and not, as previously reported, at the 5-position.[188] Pyrimidines bearing unsaturated carbon side-chains are uncommon species. Timely, therefore, are reports on the synthesis of the hitherto undescribed 2-isopropenyl-pyrimidines by dehydrobromination of 2-(2'-bromoprop-2'-yl)-pyrimidines[189] and of 2-alkynyl-pyrimidines by coupling of 2-iodo-pyrimidines with monosubstituted alkynes in the presence of bis(triphenylphosphine)palladium(II) chloride[Pd(PPh$_3$)Cl$_2$], cuprous iodide, and triethylamine.[190] Nitration of some pyrimidine bases and nucleotides has been

[183] T. Eicher, G. Franke, and F. Abdesaken, *Tetrahedron Letters*, 1977, 4067.
[184] D. J. Brown and P. Waring, *Austral. J. Chem.*, 1977, **30**, 621.
[185] D. W. Allen, D. J. Buckland, B. G. Hutley, A. C. Oades, and J. B. Turner, *J.C.S. Perkin I*, 1977, 621.
[186] T. Sakamoto, K. Kanno, T. Ono, and H. Yamanaka, *Heterocycles*, 1977, **6**, 525.
[187] T. Sheradsky and E. Nov, *J.C.S. Perkin I*, 1977, 1296.
[188] D. T. Hurst and M. L. Wong, *J.C.S. Perkin I*, 1977, 1985.
[189] D. J. Brown and P. Waring, *Austral. J. Chem.*, 1977, **30**, 1785.
[190] K. Edo, H. Yamanaka, and T. Sakamoto, *Heterocycles*, 1978, **9**, 271.

effected in high yield at the 5-position, using nitronium tetrafluoroborate in sulpholane at ambient temperature.[191] Nitration of uridine and 2'-deoxyuridine is accompanied by extensive rupturing of the glycoside link and the formation of 5-nitrouracil. However, 2-deoxyuridine-5'-monophosphate yields a mixture of the 5-nitro-compound and its 3'-O-nitrate derivative. Direct fluorination of 2-pyrimidinone in acetic acid solution has been claimed, using a mixture of fluorine and nitrogen.[192] The yield (5—10%) of 5-fluoro-2-pyrimidinone can be optimized to *ca.* 25%, by adding trifluoroacetic acid, which suppresses formation of 2-pyrimidinone hydrofluoride. In aqueous solution, complete decomposition of the heterocycle occurs. A simple alternative synthesis of the clinically effective anti-tumour agent FTORAFUR, 1-(2-tetrahydrofuryl)-5-fluorouracil, (215) has been developed[193] (Scheme 51). Of the various alkoxy-tetrahydrofurans tried, the t-butyl derivative (216; R = Bu$^t$) gave the best yields.

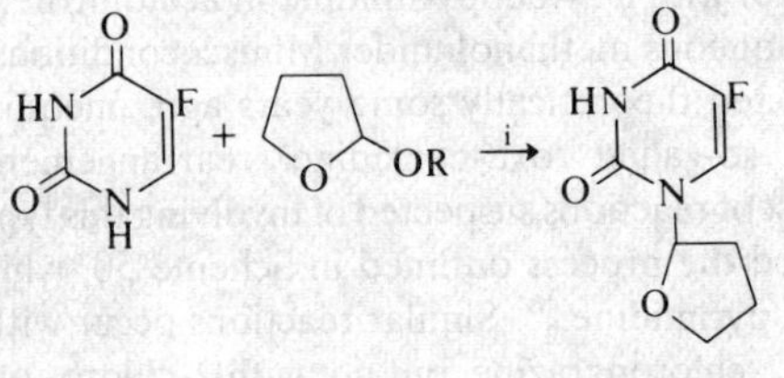

(216)                              (215)

Reagent: i, DMF, Δ.

### Scheme 51

The first example of a nucleophilic substitution in which the N-1—C-2—N-3 fragment of a pyrimidine ring is completely replaced by the N—C—N fragment of a 1,3-ambident nucleophile was reported last year.[194] A similar ring-transformation has now been observed[195] in the conversion of 1,3-dimethyluracil into isocytosine (217), using guanidine as the nucleophile (Scheme 52). Analogous

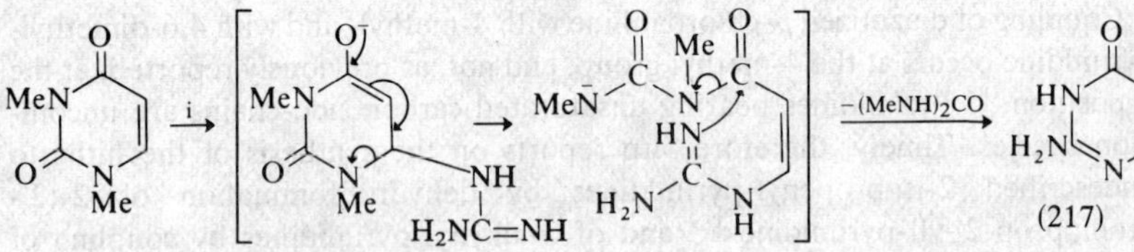

(217)

Reagents: i, (NH$_2$)$_2$C=NH, EtOH, 18 h.

### Scheme 52

[191] G.-F. Huang and P. F. Torrence, *J. Org. Chem.*, 1977, **42**, 3821.
[192] D. Cech, H. Beerbaum, and A. Holý, *Coll. Czech. Chem. Comm.*, 1977, **42**, 2694.
[193] T. Kametani, K. Kigasawa, M. Hiiragi, K. Wakisaka, O. Kusama, H. Sugi, and K. Kawasaki, *Heterocycles*, 1977, **6**, 529; *J. Heterocyclic Chem.*, 1977, **14**, 473.
[194] G. M. Brooke in 'Aromatic and Heteroaromatic Chemistry', ed. O. Meth-Cohn and H. Suschitzky, (Specialist Periodical Reports), The Chemical Society, London, 1978, Vol. 6, p. 214.
[195] K. Hirota, K. A. Watanabe, and J. J. Fox, *J. Heterocyclic Chem.*, 1977, **14**, 537; *J. Org. Chem.*, 1978, **43**, 1193.

reactions with the less basic urea and thiourea also occur, but only in the presence of sodium methoxide. In contrast, the reaction fails completely with formamidine, acetamidine, and 1,1-dimethylurea, and also with 1- and 3-methyluracil.

Carbene adducts, *e.g.* (218), of uracils and uridines have been described.[196] The dichlorocarbene adduct is obtained only with (bromodichloromethyl)phenyl-mercury as the carbene source; the reaction fails using $CHCl_3$–KOH and with $CCl_3CO_2Na$. The reactions of carbenes with pyrimidines and isoquinoline derivatives have been reviewed.[197] Also reviewed are transformations of oxygen heterocycles into pyrimidines,[198] and ring-modifying reactions of pyrimidines containing quaternary nitrogen.[199]

(218)

Imidoyl isocyanides $PhN=C(R)NC$, formed by the action of silver cyanide on imidoyl bromides in chloroform, differ from the structurally related and better known $N$-acyl isocyanides in that, on heating, they undergo isomerization to the thermodynamically more stable isomeric imidoyl cyanides $ArN=C(Ph)CN$ rather than trimerization.[200] Also formed, in minor amount (10—30%), is 2,2′-diphenyl-4,4′-bisquinazoline (220), probably *via* head-to-tail dimerization of the isocyanide followed by a double electrocyclic ring-closure of the extended $\pi$-system (219). Attempts to transform $\beta$-styryl isocyanides into 2,2′-

(219)　　　　(220)

bisisoquinolines. have so far failed. Related is the quantitative conversion of 2-azidoalkyl isocyanates of type (221) into 2-trifluoromethylquinazolin-4-ones.[201] The azido-isocyanates, prepared as shown in Scheme 53, at 170—180 °C undergo rearrangement with loss of nitrogen to the imidoyl isocyanates (222), which by electrocyclic ring-closure produce the quinazolinones (223).

[196] H. P. M. Thiellier, G. J. Koomen, and U. K. Pandit, *Tetrahedron*, 1977, **33**, 1493; 2603; 2609.
[197] U. K. Pandit, *Heterocycles*, 1977, **8**, 609.
[198] K. Takagi and M. Hubert-Habart, *Bull. Soc. chim. France*, 1977, 369.
[199] H. C. van der Plas, *Heterocycles*, 1978, **9**, 33.
[200] G. Höfle and B. Lange, *Angew. Chem. Internat. Edn.*, 1977, **16**, 727.
[201] V. I. Gorbatenko, V. N. Fetyukhin, N. V. Mel'nichenko, and L. I. Samarai, *J. Org. Chem. (U.S.S.R.)*, 1977, **13**, 2161.

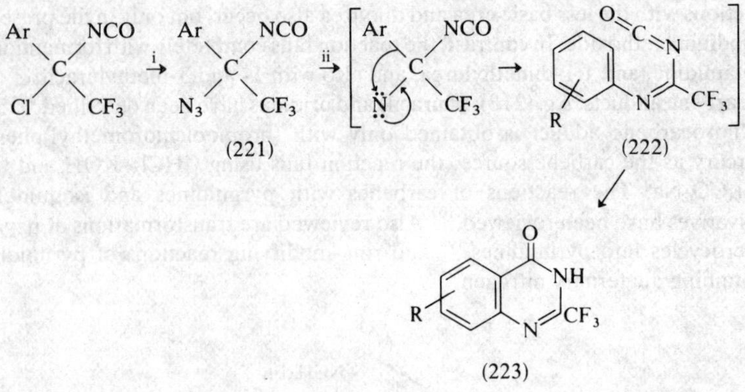

(221)                                                                (222)

(223)

Reagents: i, $Me_3SiN_3$; ii, $\Delta$, 170—180 °C.

**Scheme 53**

In contrast, diphenylimidoyl isocyanate, $PhC=N(Ph)NCO$, generated either from the imidoyl chloride and silver cyanate in acetonitrile solution, or, more unusually, by thermolysis of the thiatriazole (224), cyclizes to 2-phenyl-quinazolin-4-one in only poor yield (36%).[202] 2-Amino-quinazolines are accessible by the new route outlined in Scheme 54.[203]

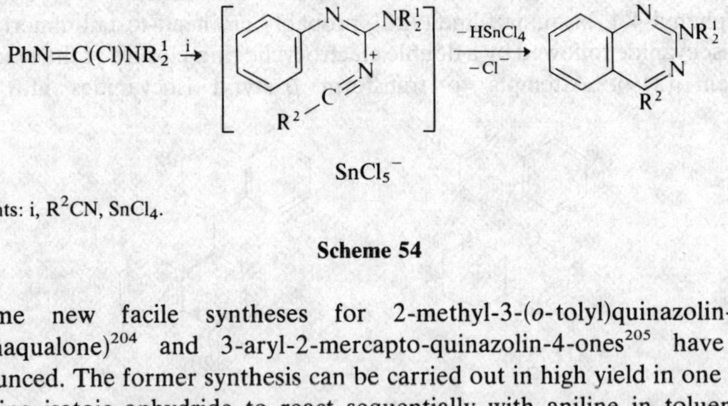

Reagents: i, $R^2CN$, $SnCl_4$.

**Scheme 54**

Some new facile syntheses for 2-methyl-3-(o-tolyl)quinazolin-4-one (methaqualone)[204] and 3-aryl-2-mercapto-quinazolin-4-ones[205] have been announced. The former synthesis can be carried out in high yield in one pot by allowing isatoic anhydride to react sequentially with aniline in toluene and pentane-2,5-dione in ethanolic hydrogen chloride. The latter preparation consists of condensing aryl isothiocyanates with anthranilic acid in hot acetic acid–sodium acetate mixture.

*Pyrazines, Quinoxalines, and Phenazines.* 2,3,5,6-Tetraphenylpyrazine is a by-product (10%) from the Lewis-acid-catalysed ($Et_2O\cdot BF_3$) reaction of 2H-

[202] A. Holm, C. Christophersen, T. Ottersen, H. Hope, and A. Christensen, *Acta Chem. Scand.* (*B*), 1977, **31**, 687.
[203] V. G. Pavra, R. Madroñero, and S. Vega, *Synthesis*, 1977, 345.
[204] M. S. Manhas, S. G. Amin, and V. V. Rao, *Synthesis*, 1977, 309.
[205] K. F. Kottke, K. F. Friedrich, D. Knoke, and H. Kuhmstedt, *Pharmazie*, 1977, **32**, 540.

azirines with ethyl cyanoacetate.[206] More interesting is the formation of the 2,11-diazatricyclo-octadecatriene (226) (60%) by the action of $Et_2O \cdot BF_3$ on the bicyclic azirine (225).

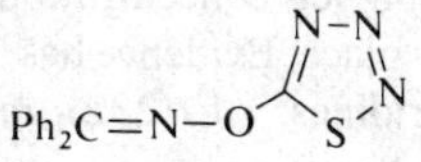

$$Ph_2C=N-O \quad \text{(224)} \qquad \text{(225)} \qquad \text{(226)}$$

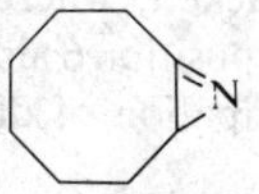

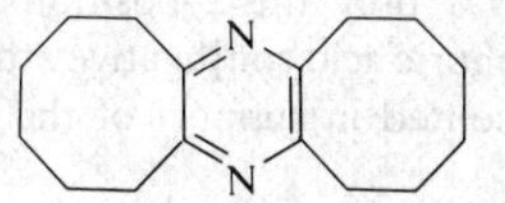

An old report concerning the formation of 2,3-dihalogeno-6-phenyl-pyrazines when $\alpha$-amino-$\alpha$-phenylacetonitrile reacts with chloral or bromal has been shown to be in error.[207a] The products are in fact $N$-(2,2-dihalogenoethenyl)-1-imino-1-phenyl-acetonitriles $PhC(CN)=NCH=CX_2$ (X = Cl or Br). Catalytic reduction (Pd/C) of oxadiazolopyrazines (227; X = O), prepared by condensing 3,4-diaminofurazan with $\alpha$-diketones, provides a useful source of 2,3-diamino-pyrazines (228).[207b] In marked contrast, analogous reductions of the tri-azolopyrazines (227; X = NPh) furnish only the tetrahydrotriazolopyrazines (229).

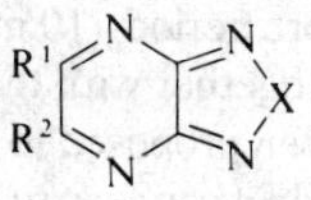

$$\text{(227)} \qquad \text{(228)} \qquad \text{(229)}$$

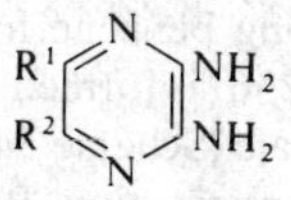

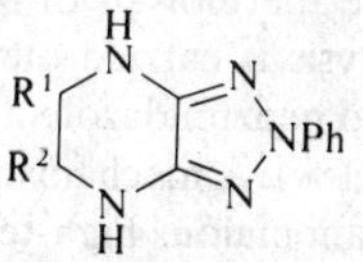

In concentrated sulphuric acid solution, chlorinated pyrazines and quinoxalines are regiospecifically oxidized by peroxysulphuric acid (Caro's acid) at the nitrogen adjacent to the halogen-bearing carbon.[208] This procedure does in fact provide the first direct, high-yield synthesis of the hitherto difficultly accessible 2-chloropyrazine 1-oxides. Pyrazine-based cyclophanes, *e.g.* (230), have been prepared[209] and, under standard conditions, methylate only at N-4. This property has been used to construct the 1,3,5-bridged cyclophane (231) from (230) and 1,4-di-iodobutane.

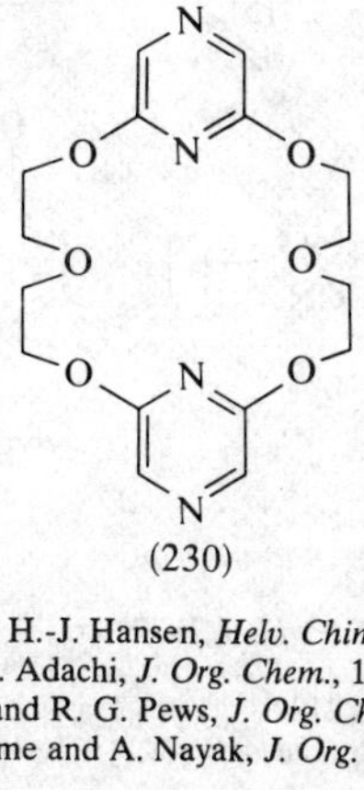

$$\text{(230)} \qquad \text{(231)}$$

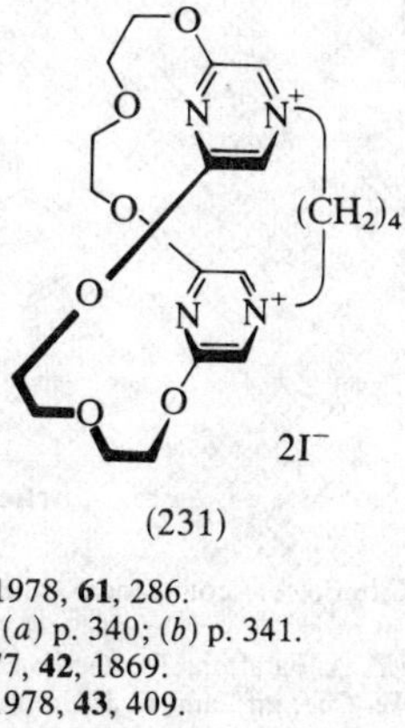

[206] H. Bader and H.-J. Hansen, *Helv. Chim. Acta*, 1978, **61**, 286.
[207] N. Sato and J. Adachi, *J. Org. Chem.*, 1978, **43**, (*a*) p. 340; (*b*) p. 341.
[208] C. E. Mixan and R. G. Pews, *J. Org. Chem.*, 1977, **42**, 1869.
[209] G. R. Newkome and A. Nayak, *J. Org. Chem.*, 1978, **43**, 409.

Further studies[210] on the homolytic alkylation of quinoxaline by *N*-chloro-amines reported last year[211] have revealed that the position of alkylation is strongly affected by the acidity of the medium. For example, at acid concentration <50% only the 2-position is attacked, whereas at higher concentrations of sulphuric acid conjugative attack at position 6 also takes place. Evidence has been presented in support of the participation of di-oxaziridines, *e.g.* (232), in the

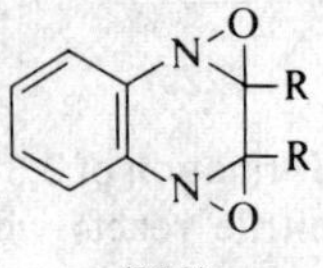

(232)

photochemical rearrangement of 2,3-disubstituted quinoxaline 1,4-dioxides to 1,3-disubstituted benzimidazolones.[212] The thermally induced ring-contraction of 2-azidoquinoxaline mono- and di-oxides to 2-cyano-1-hydroxy-benz-imidazoles and their 3-oxides, respectively, is well known. Surprising, therefore, is the isolation in high yield (94%) of the 3*H*-2,1,4-benzoxadiazine 4-oxide (233) from the thermolysis of quinoxaline dioxide (234) in boiling toluene.[213] If the thermolysis is carried out in boiling benzene for a short period (10 min), the expected benzimidazole dioxide (236) is formed (45%) together with the oxadi-azine (41%). A mechanistic rationale (Scheme 54) has been proposed to account for the anomalous high-temperature reaction. Both products appear to arise by competing cyclization of the nitroso-intermediate (235), the benzimidazole diox-ide (236) rearranging to the oxadiazine at the higher reaction temperature.

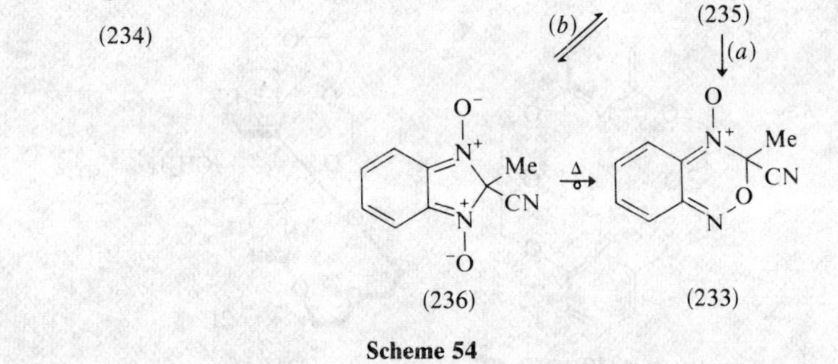

(234)                              (235)

(236)                              (233)

**Scheme 54**

[210] T. Caronna, A. Citterio, T. Crolla, and F. Minisci, *J.C.S. Perkin I*, 1977, 865.
[211] See ref. 18, p. 130.
[212] A. A. Jarrar and Z. A. Fataftah, *Tetrahedron*, 1977, **33**, 2127.
[213] J. P. Dirlam, B. W. Cue, jnr., and K. J. Gombatz, *J. Org. Chem.*, 1978, **43**, 76.

Titanous chloride in methanol or THF is useful for the deoxygenation of quinoxaline 1,4-dioxides.[214]

Tetra-azabinaphthylene (biquinoxalylene) (237) has been prepared (60%) by flash vacuum pyrolysis (800 °C; 0.02 Torr) of the pentacycle (238),[215] and also by reduction of the squaric acid bis-amidine derivative (239),[216] prepared as indicated in Scheme 55.

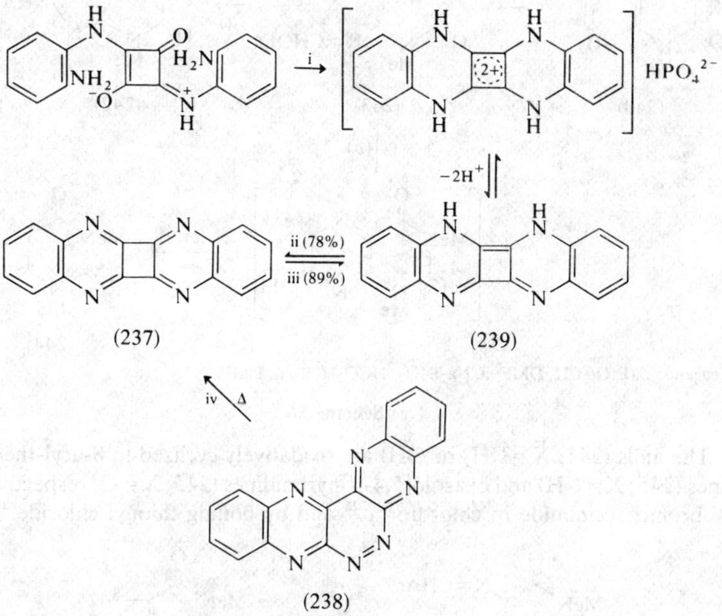

(237)          (239)

(238)

Reagents: i, PPA; ii, I₂, DMF; iii, Na₂S₂O₄, 2M-NH₃; iv, 860 °C, 0.015 Torr.

**Scheme 55**

Developments in quinoxaline chemistry during the period 1963—1975 have been reviewed,[217] as have recent advances in the chemistry of phenazine N-oxides.[218]

**Purines, Pteridines, and Related Systems.**—This year has seen the development, particularly by Japanese workers, of several new and modified procedures for the synthesis of purines and other fused pyrimidine ring systems. Especially useful as precursors for these syntheses are the arylidene derivatives (241; X = NH₂ or OH) and the 6-amino-azo-compounds (240; X = N). The amino-azo-compounds are cyclized directly to purine derivatives (242) on fusion with arylamines,[219a] or,

[214] B. W. Cue, jnr., J. P. Dirlam, and E. A. Glazer, *Org. Prep. Proced. Internat.*, 1977, **9**, 263.
[215] S. Kanoktanaporn and J. A. H. MacBride, *Tetrahedron Letters*, 1977, 1817.
[216] S. Hünig and H. Pütter, *Chem. Ber.*, 1977, **110**, 2532.
[217] G. W. H. Cheeseman and E. S. G. Werstiuk, *Adv. Heterocyclic Chem.*, 1978, **22**, 367.
[218] S. Pietra, G. F. Bettinetti, A. Albini, and G. Minoli, *Khim. geterotsikl. Soedinenii*, 1977, 1587.
[219] M. Higuchi and F. Yoneda (a) *Heterocycles*, 1977, **6**, 1901; (b) *J.C.S. Perkin I*, 1977, 1336; (c) *Chem. and Pharm. Bull.* (*Japan*), 1977, **25**, 2797.

better, in a two-stage process as outlined in Scheme 56 (path *a*).[220] On prolonged heating in DMF the 5-arylideneamino-6-ethoxymethyleneamino-uracils (243; X = CH) undergo electrocyclic ring-closure (Scheme 56; path *b*) and so provide an unequivocal synthetic route to 6-aryl-pteridines (244).[219b]

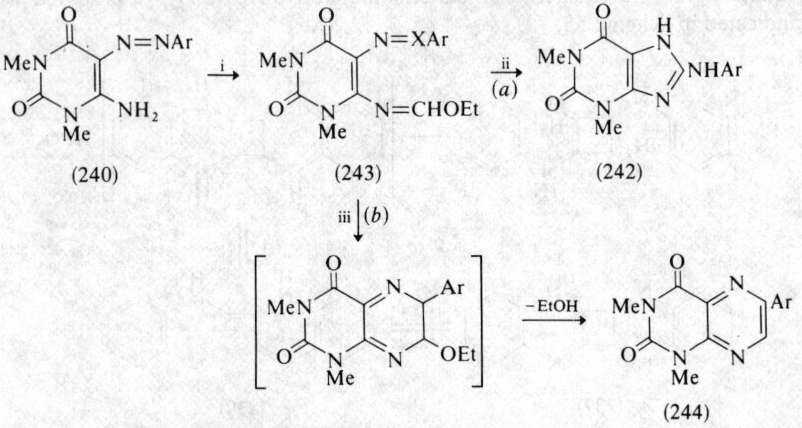

(240)                              (243)                              (242)

(244)

Reagents: i, (EtO)$_3$CH, DMF; ii, Na$_2$S$_2$O$_4$, HCO$_2$H; iii, Δ, DMF.

**Scheme 56**

The anils (241; X = NH$_2$ or OH) are oxidatively cyclized to 8-aryl-theophyllines (245; X = NH) and oxazolo[5,4-*d*]pyrimidines (245; X = O) respectively by *N*-bromosuccinimide in chloroform[221] and by boiling thionyl chloride.[222a,223a]

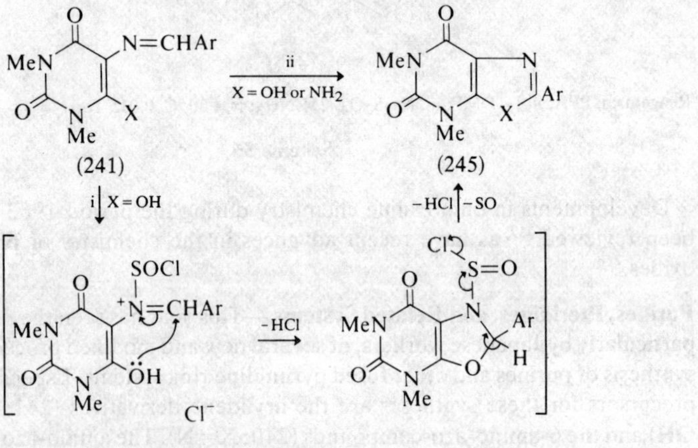

(241)                              (245)

Reagents: i, SOCl$_2$ at 55 °C for 30 min; ii, NBS, CHCl$_3$.

**Scheme 57**

[220] K. Senga, M. Ichiba, H. Kanazawa, and S. Nishigaki, *Synthesis*, 1977, 264.
[221] K. Senga, J. Sato, K. Shimizu, and S. Nishigaki, *Heterocycles*, 1977, **6**, 1919.
[222] K. Senga, J. Sato, and S. Nishigaki, *Heterocycles*, 1977, **6**, (a) p. 689; (b) p. 945.
[223] K. Senga, K. Shimizu, and S. Nishigaki, (a) *Chem. and Pharm. Bull. (Japan)*, 1977, **25**, 495; (b) *Heterocycles*, 1977, **6**, 1907.

The latter cyclizations probably involve sulphinyl chloride intermediates, as exemplified in Scheme 57.

Similar intermediates are probably involved in the thionyl-chloride-promoted cyclization of 6-(benzylidene-1-methylhydrazino)-1,3-dimethyl-uracils (246; $R^1 = Me$; $R^2 = H$) to pyrazolo[3,4-$d$]pyrimidinediones (247).[222b] Less easily explained is the cyclization of hydrazones (246; $R^1 = Me$; $R^2 = H$) to the oxazolo-[5,4-$d$]pyrimidines (245; $X = O$) by sodium nitrite in boiling acetic acid.[224a] Undoubtedly, nitrosation at the 5-position is the initial step in these reactions, but thereafter events remain unclear. Interestingly, the ketone hydrazones (246; $R^1 = H$, $R^2 = Me$), on similar treatment, followed by reduction with sodium dithionite, furnish 6-aryl-1,3-dimethyl-lumazines (248) in low yields (10—15%).[224b]

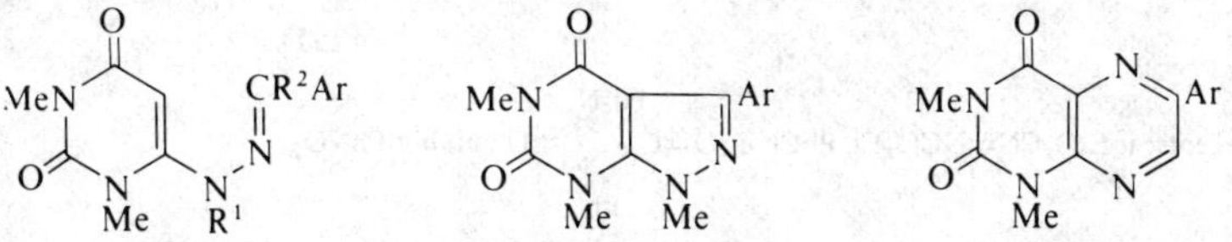

(246)　　　　　(247)　　　　　(248)

A full account of the ring-closure of pyrimidin-6-yl hydrazones (246; $R^1 = R^2 = $ H) to pyrazolo[3,4-$d$]pyrimidines (250), with aldehydes in boiling DMF (Scheme 58), noted last year,[225] has been published.[226] Apparently, cyclization involves the diazahexatrienes (249), which are isolable from the direct action of aldehydes $R^2$CHO on the hydrazone in hot ethanol.

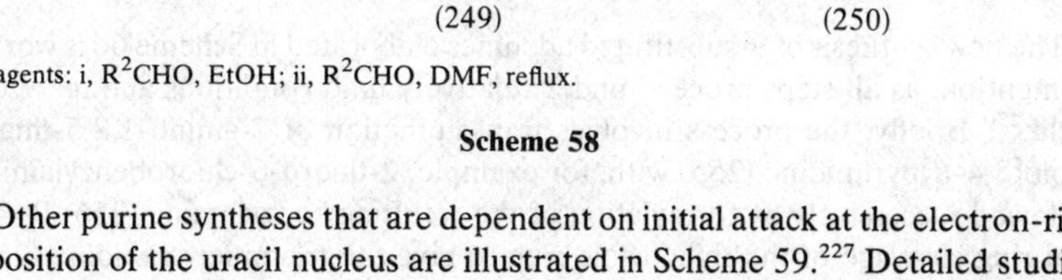

(246; $R^1 = R^2 = $ H)　$\xrightarrow{i}$　(249)　$\rightarrow$　(250)

Reagents: i, $R^2$CHO, EtOH; ii, $R^2$CHO, DMF, reflux.

**Scheme 58**

Other purine syntheses that are dependent on initial attack at the electron-rich 5-position of the uracil nucleus are illustrated in Scheme 59.[227] Detailed studies show that anils (251) are intermediates in these oxidative cyclizations and that boiling nitrobenzene is the most efficient oxidant.

The hitherto unknown 3,9-dialkyl-hypoxanthines (253) have been synthesized by heating 1-alkyl-5-alkylamino-imidazole-4-carboxyamides (252) with triethyl orthoformate in acetic anhydride.[228] Attempts to benzylate 9-benzyltheophylline

[224] K. Senga, Y. Kanamori, and S. Nishigaki, *Heterocycles*, 1977, **6**, (*a*) p. 1925; (*b*) p. 693.
[225] See ref. 18, p. 134.
[226] F. Yoneda, T. Nagamatsu, T. Nagamura, and K. Senga, *J.C.S. Perkin I*, 1977, 765.
[227] F. Yoneda, M. Higuchi, and S. Matsumoto, *J.C.S. Perkin I*, 1977, 1754; F. Yoneda, M. Kawamura, S. Matsumoto, and M. Higuchi, *ibid.*, p. 2285.
[228] T. Itaya and K. Ogawa, *Heterocycles*, 1977, **6**, 965.

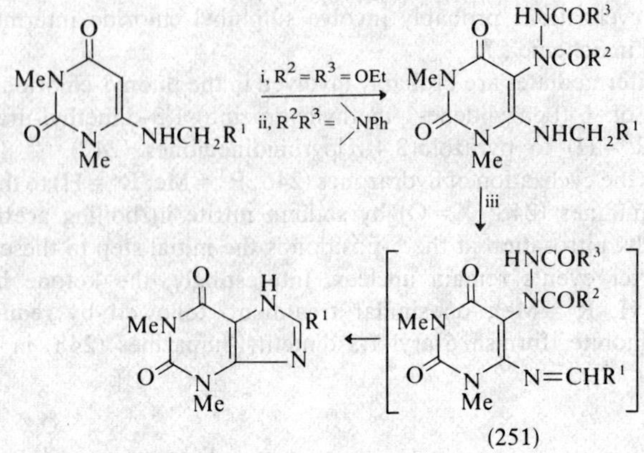

(251)

Reagents: i, $EtO_2CN=NCO_2Et$, PhCl; ii, $O=C\overset{N=N}{\underset{\underset{Ph}{N}}{\diagdown}}C=O$ ; iii, hot $PhNO_2$.

**Scheme 59**

with benzyl bromide in DMF gave not the expected 7,9-dibenzyl derivative but the rearranged 7-benzyl isomer (254).[229] Further studies demonstrated that this rearrangement is not thermal (no reaction in hot DMF) but is acid-catalysed by the HBr present in the benzyl bromide–DMF mixture.

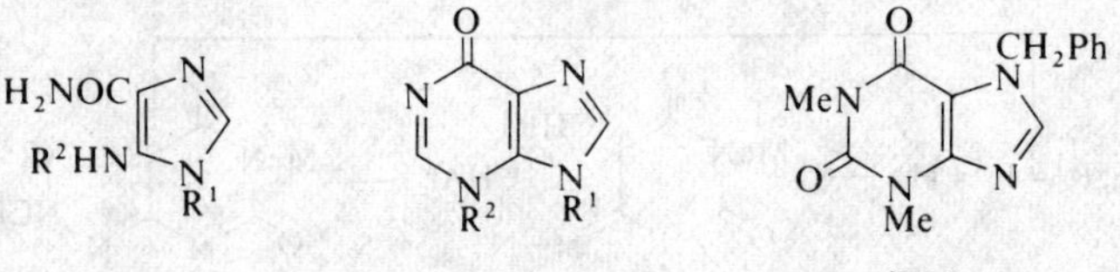

(252)                       (253)                       (254)

The new synthesis of 9-substituted adenines elaborated in Scheme 60 is worthy of mention, as all steps proceed under relatively mild conditions and in >90% yield.[230] Briefly, the process involves transamination of 7-amino-1,2,5-thiadi-azolo[3,4-d]pyrimidine (255) with, for example, 2-fluoro-6-chlorobenzylamine, followed successively by formylation of the resulting benzylamine (256; R = H) and ring cleavage of the thiadiazole ring to give the triaminopyrimidine (257), which cyclizes *in situ* to the 9-substituted adenine (258), which is a highly active coccidiostat.

The hitherto difficultly accessible pyrazolo[4,3-d]pyrimidine 1-oxides (260) are now available by a new, one-step synthesis from 6-(bromomethyl)-1,3-dimethyl-5-nitro-uracils (259) (Scheme 61).[231] Condensation in ethyl acetate at 0 °C allows isolation of the alkylamino-intermediates (261), which in boiling ethanol cyclize to the N-oxides (260).

[229] J. H. Lister, *Heterocycles*, 1977, **6**, 383.
[230] G. D. Hartman, S. E. Biffar, L. M. Weinstock, and R. Tull, *J. Org. Chem.*, 1978, **43**, 960.
[231] S. Senda, K. Hirota, T. Asao, and Y. Yamada, *J.C.S. Chem. Comm.*, 1977, 556.

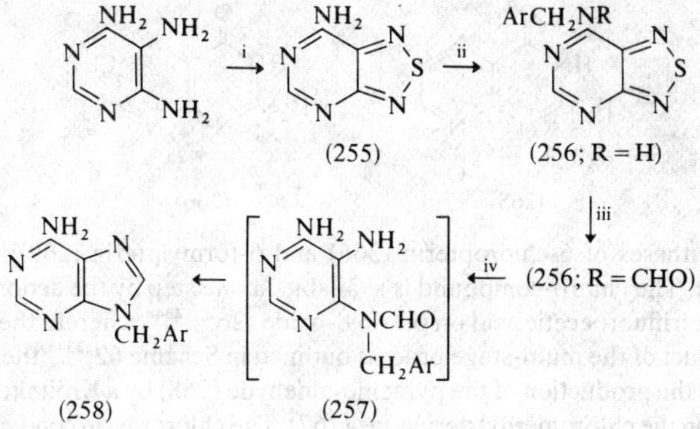

(255)  (256; R = H)

(256; R = CHO)

(258)  (257)

Reagents: i, $SOCl_2$; ii, 2-Cl-6-F-$C_6H_3CH_2NH_2$ (=$ArCH_2NH_2$ throughout); iii, $HCO_2H$, $Ac_2O$; iv, Raney Ni, EtOH, $H_2O$.

**Scheme 60**

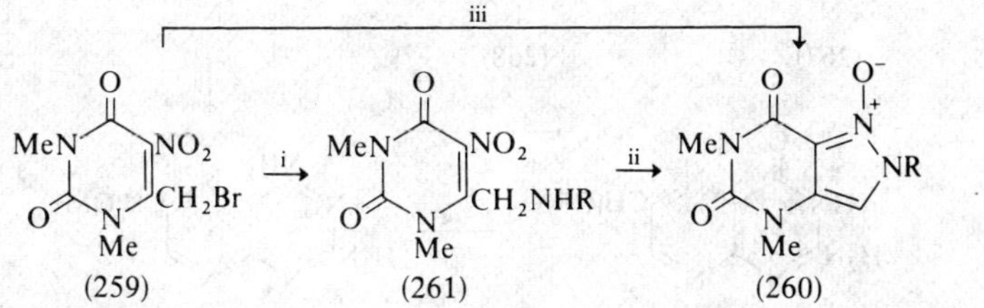

(259)  (261)  (260)

Reagents: i, $RNH_2$, EtOAc, at 0 °C, 0.5 h; ii, EtOH, reflux 12 h; iii, $RNH_2$, EtOH, boil for 4 h.

**Scheme 61**

Most unusual is the synthesis of *vic*-triazolo[4,5-*d*]pyrimidine (262) by heating 6-azido-1,3-dimethyluracil with potassium carbonate in DMF.[232] Whereas intramolecular cyclization of an azide function onto C=N to give tetrazoles is well known, the corresponding formation of triazoles by addition to C=C is extremely rare. Also noteworthy are the 5-diazo-6-amino-uracils (263; R = Me, Et, or Ph), which, although stable in boiling acetic acid, cyclize (with loss of nitrogen) in hot DMF to indolo[2,3-*d*]pyrimidines (264).[233] Far superior are the photodecompositions of (263) which furnish the indolo-pyrimidines in quantitative yield.

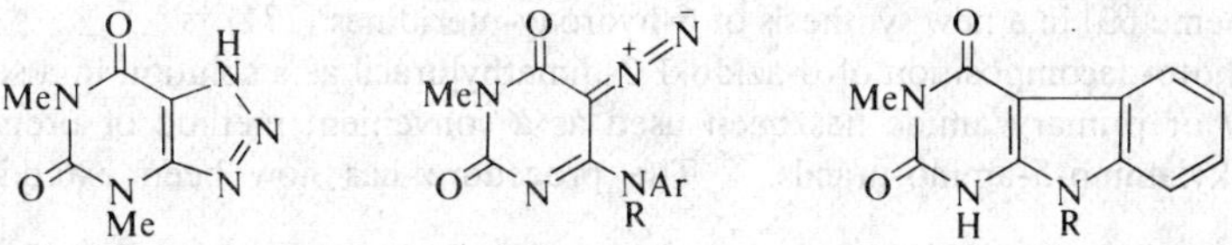

(262)  (263)  (264)

[232] K. Senga, M. Ichiba, and M. Nishigaki, *Heterocycles*, 1977, **6**, 1915.
[233] Y. Sakuma and F. Yoneda, *Heterocycles*, 1977, **6**, 1911.

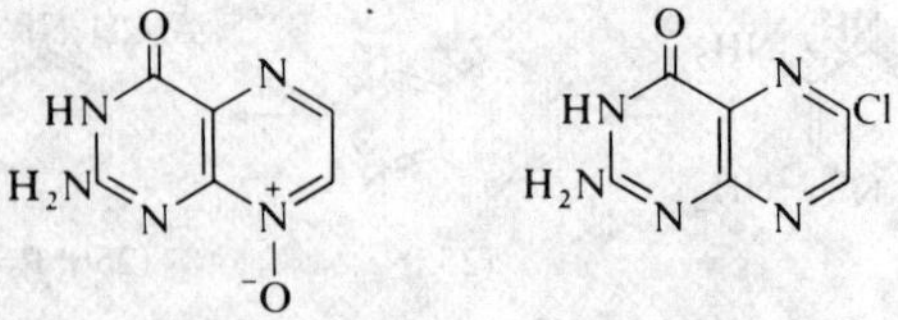

(265)                                    (266)

New syntheses of 6-chloropterin (266) and 6-formylpterin (269) have been elaborated. The chloro-compound is available in one step by the action of acetyl chloride in trifluoroacetic acid on pterin 8-oxide (265),[234a] whereas the aldehyde is the product of the multi-stage process outlined in Scheme 62,[234b] the first stage of which is the production of the pyrazine-aldehyde (268) by a Kröhnke aldehyde synthesis on the chloromethyl derivative (267). The chloropterin condenses easily with arylthiols and alkyl mercaptans, but not with alkylamines.

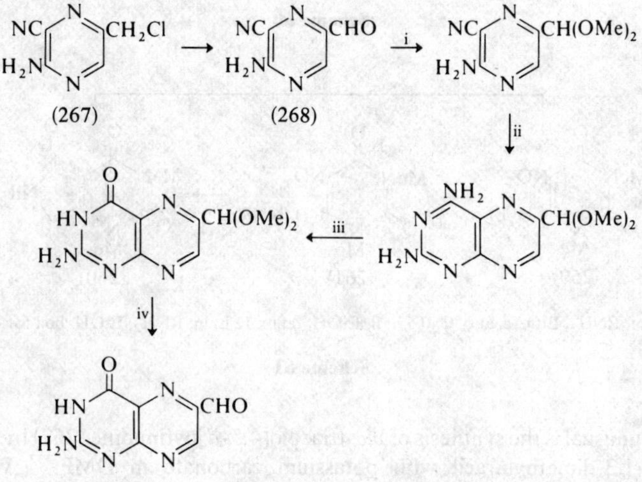

(267)                  (268)

(269)

Reagents: i, MeOH, *p*-TSA; ii, $(NH_2)_2C=NH$, boil; iii, 5% NaOH; iv, $HCO_2H$ or $CF_3CO_2H$.

**Scheme 62**

6-Amino-1,3-dimethyl-5-nitrosouracil (270), reported last year[235] as a highly versatile precursor for the purine and pteridine ring systems, has again found use (Scheme 63) in a new synthesis of 6-hydroxy-pteridines (271).[223b]

Photo-decomposition of 6-azido-1,3-dimethyluracil as a solution in a secondary or primary amine has been used as a convenient method of preparing 6-alkylamino-5-amino-uracils.[236] The procedure has now been extended[237]

[234] (a) E. C. Taylor, and R. Kobylecki, *J. Org. Chem.*, 1978, **43**, 680; (b) E. C. Taylor, R. N. Henrie, II, and R. C. Portney, *ibid.*, p. 736.
[235] See ref. 18, p. 131.
[236] S. Senda, K. Hirota, M. Suzuki, and K. Maruhashi, *J.C.S. Chem. Comm.*, 1976, 731.
[237] S. Senda, K. Hirota, T. Asao, and K. Maruhashi, *J. Amer. Chem. Soc.*, 1977, **99**, 7358.

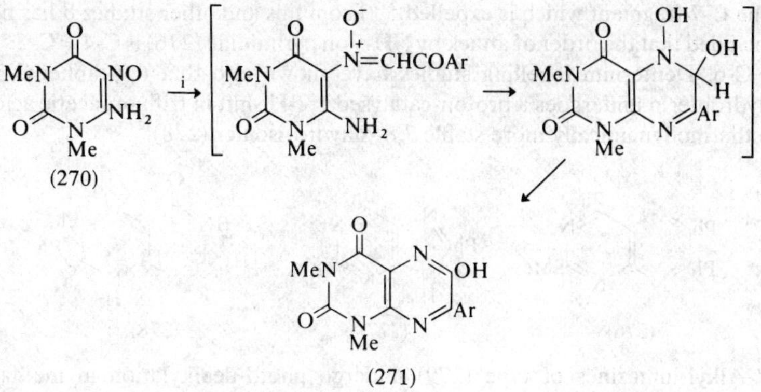

(270) → (271)

Reagents: i, $C_5H_5\overset{+}{N}CH_2COAr\ Br^-$, $H_2O$, NaOH, $C_5H_5N$.

**Scheme 63**

(Scheme 64) to the synthesis of dihydrolumazines, *e.g.* (272), and fervenulins, *e.g.* (273), by using amino-esters (or -ketones) and acylhydrazones in place of the alkylamine.

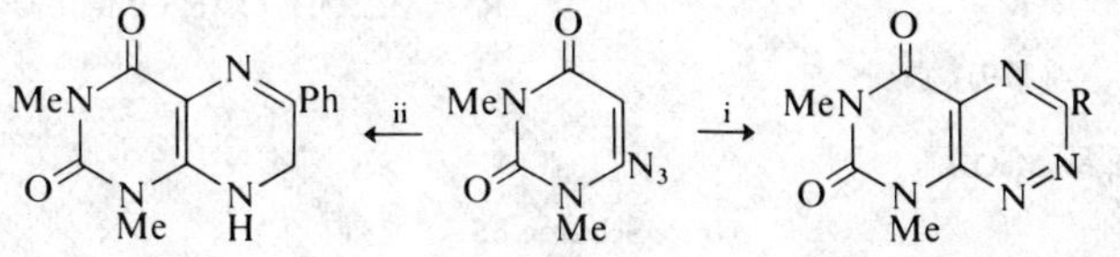

(272) ← ii ← ... → i → (273)

Reagents: i, $RCONHNH_2$, THF, $O_2$, *hv*, 3 h; ii, $PhCOCH_2NH_2$, THF, *hv*, 3 h.

**Scheme 64**

Ultraviolet and infrared spectroscopic data for a series of 1-alkyl-adenines reveal that the imino-form (274) predominates in a non-polar environment, whereas the percentage of amino-form (275) increases with increasing dielectric

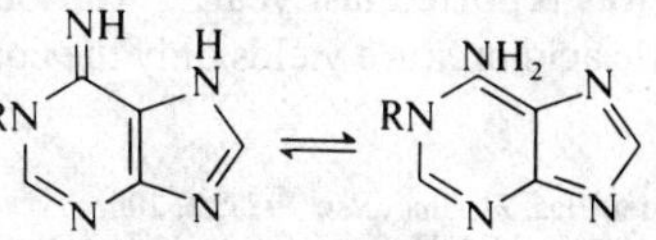

(274) ⇌ (275)

constant of the solvent.[238] In aqueous solution the amino-form is dominant. 3-Alkyl-adenines exist solely as the amino-tautomers. Deuterium labelling studies on the ring-contraction of diphenyl-(methylthio)pteridine (276) to the (methylthio)-purine (277) with $KNH_2$ in liquid $NH_3$ at $-33\ °C$ demonstrate that it

[238] M. Dreyfus, G. Dodin, O. Bensaude, and J. E. Dubois, *J. Amer. Chem. Soc.*, 1977, **99**, 7027.

is the C-7 fragment which is expelled.[239] From this and other studies it has been concluded that the order of attack by $NH_2^-$ on pyrimidine (276) is C-4 $\geqslant$ C-2 $>$ C-7 $>$ C-6. Deuterium labelling studies have shown also that 6,7-diphenyl-5,6-dihydropterin undergoes a proton-catalysed 1,2-H shift in trifluoroacetic acid to the thermodynamically more stable 7,8-dihydro-isomer (278).[240]

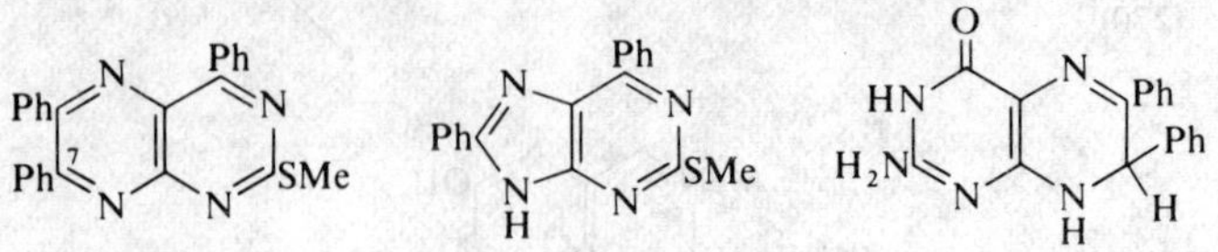

(276)                              (277)                              (278)

8-Alkyl-lumazines of type (279) undergo photo-dealkylation in methanol solution as illustrated in Scheme 65.[241]

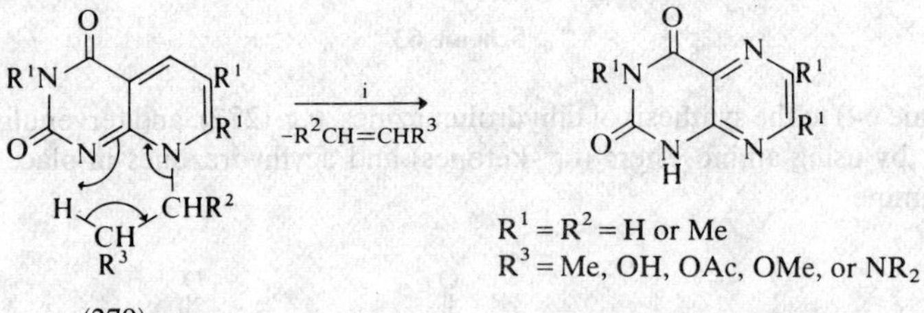

$R^1 = R^2 = H$ or Me
$R^3 =$ Me, OH, OAc, OMe, or $NR_2$

(279)

Reagents: i, *hν*, MeOH.

**Scheme 65**

6-(*N*-Alkylanilino)-uracils (280) have again been used[242] extensively as starting materials for the synthesis of isoalloxazines (281),[243,244a] 8-chloro-flavins (282),[245a] isoalloxazine 5-oxides (283),[244a] 10-aryl-isoalloxazines (284),[245b] and 5-deaza-alloxazine (285).[246] These reactions are summarized in Scheme 66. The chlorination of *N*-oxides by acyl halides is well documented, and good evidence has been presented for the participation of *N*-oxides of type (283) in the reaction to produce (282). However, the process is unusual in that chlorination does not occur with POCl$_3$ alone. The use of *N*-nitrosodimethylamine as a novel nitrosating agent (reagent ii) was reported last year.[247] Curiously, nitrosation using the standard NaNO$_2$–acetic acid mixture yields only the corresponding isoalloxazine 5-oxides (283).

[239] J. Nagel and H. C. van der Plas, *Heterocycles*, 1977, **7**, 205.
[240] P. K. Sengupta, H. A. Breitschmid, J. H. Bieri, and M. Viscontini, *Helv. Chim. Acta*, 1977, **60**, 922.
[241] V. J. Ram, W. R. Knappe, and W. Pfeiderer, *Tetrahedron Letters*, 1977, 3795.
[242] See ref. 18, p. 132.
[243] F. Yoneda and Y. Sakuma, *Heterocycles*, 1977, **6**, 431; F. Yoneda, K. Shinozuka, Y. Sakuma, and K. Senga, *ibid.*, p. 1179.
[244] F. Yoneda, Y. Sakuma, and K. Shinomura, (*a*) *J.C.S. Perkin I*, 1978, 348; (*b*) *J.C.S. Chem. Comm.*, 1977, 175.
[245] F. Yoneda, Y. Sakuma, and K. Shinozuka, (*a*) *J.C.S. Chem. Comm.*, 1977, 681; (*b*) *Heterocycles*, 1978, **9**, 7.
[246] K. Senga, K. Shimuzu, S. Nishigaki, and F. Yoneda, *Heterocycles*, 1977, **6**, 1361.
[247] See ref. 18, p. 134.

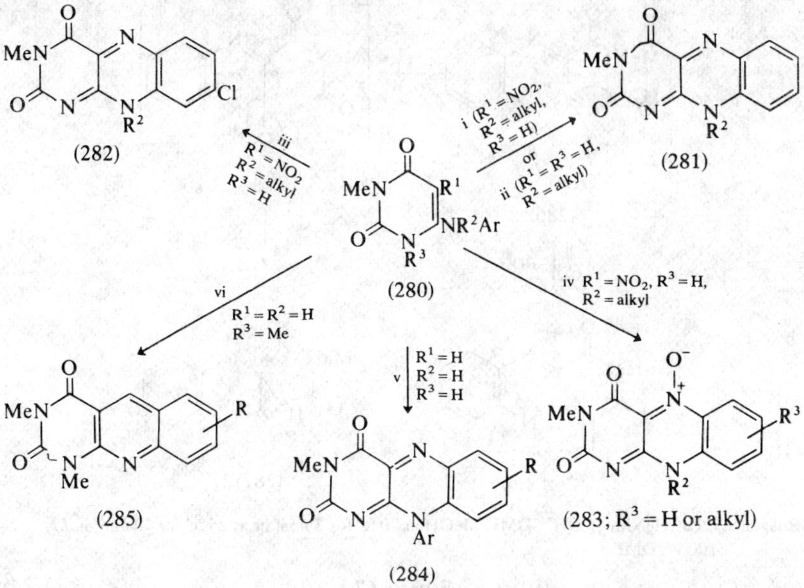

(282)  (280)  (281)

(285)  (284)  (283; $R^3$ = H or alkyl)

Reagents: i, EtO$_2$CN=NCO$_2$Et, Ph$_3$P or C$_5$H$_5$N; ii, Me$_2$NNO, POCl$_3$; iii, POCl$_3$, DMF, at 90 °C, 1 h; iv, excess of conc. H$_2$SO$_4$; v, ArNO, Ac$_2$O, under reflux, 20 min; vi, Me$_2$NCH(OMe)$_2$, at 95 °C for 1.5 h.

**Scheme 66**

3,10-Dimethylisoalloxazine (286; R = Me) undergoes demethylation (10-position) (30%) and ring contraction (35%), to produce the spiro-hydantoin (287; R = Me), with trimethylbenzylammonium hydroxide in MeOH–DMF at 70 °C (for 3 minutes) or at room temperature (for 40 minutes).[244b] Under the same conditions, other 10-alkyl-isoalloxazines yield almost exclusively the spiro-hydantoins (287; R = Et, Pr$^n$, or Bu$^n$). In contrast, irradiation of the dialkyl derivatives in DMF solution containing the phase-transfer catalyst results only in dealkylation. Hydroxide ion is necessary for the success of both reactions. These intriguing transformations have been rationalized on the basis of calculations of the π-electron density which show that in the excited state the π-electron density at position 10a increases while that at position 10 decreases relative to the electron densities prevailing at these positions in the ground state. Hence it has been concluded that, in the dark (*i.e.* ground-state) reaction, attack by OH$^-$ takes place at the 10a-position (Scheme 67; path *a*) to yield spiro-compound (287), whereas under photolytic (excited-state) conditions dealkylation results from nucleophilic attack of OH$^-$ at the 10-position (path *b*).

Prolonged treatment (several days) of alloxazines with peroxytrifluoroacetic acid at room temperature yields the 5,10-dioxides, the u.v. spectra of which support the tautomeric forms (288; R = H or Me).[248] 5-Deazariboflavine (289) is of proven value in flavine biochemistry. Of interest, therefore, is a report on the

[248] K. L. Perlman, W. E. Pfeiderer, and M. Gottlieb, *J. Org. Chem.*, 1977, **42**, 2203.

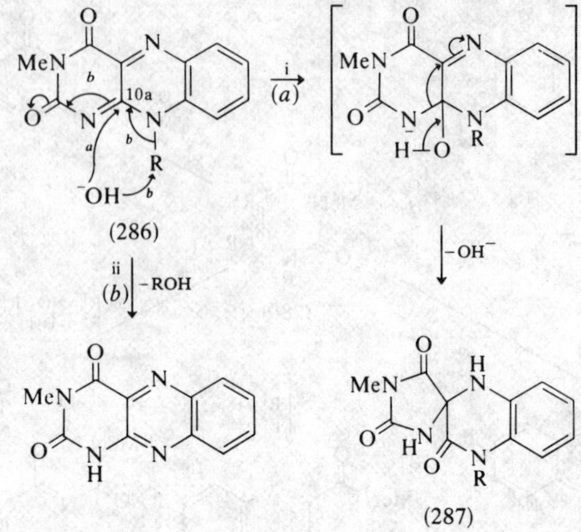

(286)

(287)

Reagents: i, $PhCH_2NMe_3^+$ $OH^-$, DMF, MeOH at 70 °C for 3 min, or at 25 °C for 40 min; ii, $h\nu$, 100 W, DMF.

**Scheme 67**

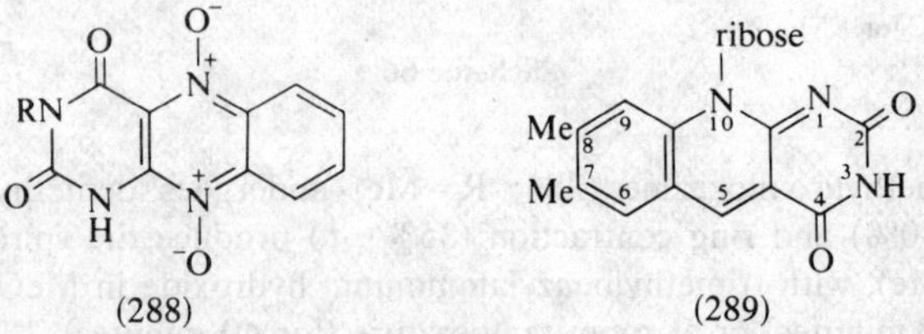

(288)                                         (289)

preparation of several new deaza-isosteres of riboflavine and lumichrome.[249] Included are the 1- and 3-deaza-, the 1,5-di-, and the 1,3,5-tri-deaza-derivatives.

Two new methods for the synthesis of pyrimido[4,5-e]pyridazines (290), related to the antibiotic fervenulin, have been developed (Scheme 68).[250]

Also new is the synthesis of benzo[g]pteridines (293) outlined in Scheme 69.[251] Formation of the key intermediate quinoxaline (292) involves a novel aromatization of the chlorotetrahydro-derivative (291), a tentative mechanism for which is illustrated.

The bis(dimethyluracil) sulphide (294; R = Me) undergoes photocyclization in THF to a mixture of 1,4-dithiino[2,3-d:5,6-d']dipyrimidine (295; R = Me) (18%), dipyrimidino[2,3-d;4,5-d']thiophen (297) (8.2%), and its dihydro-derivative (296) (9%).[252] The tricycle (295; R = Me) is an obvious dimerization product of two uracil-6-thiolyl radicals, whereas the pyrimidino-thiophens are thought to arise *via* the unusual rearrangement–electrocyclic cyclization

[249] W. T. Ashton, D. W. Graham, R. D. Brown, and E. F. Rogers, *Tetrahedron Letters*, 1977, 2551.
[250] S. Nishigaki, M. Ichiba, J. Sato, K. Senga, M. Noguchi, and F. Yoneda, *Heterocycles*, 1978, 9, 11.
[251] E. C. Taylor and J. V. Berrier, *Heterocycles*, 1977, 6, 449.
[252] T. Itoh and H. Ogura, *Tetrahedron Letters*, 1977, 2595.

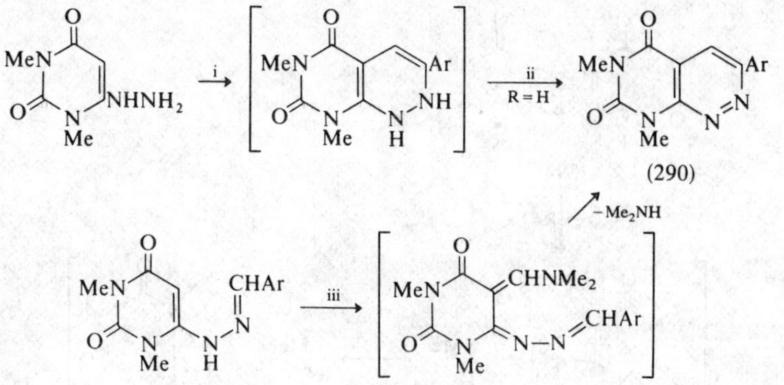

Reagents: i, $ArCOCH_2Br$; ii, $EtO_2CN=NCO_2Et$; iii, $Me_2NCH(OEt)_2$.

**Scheme 68**

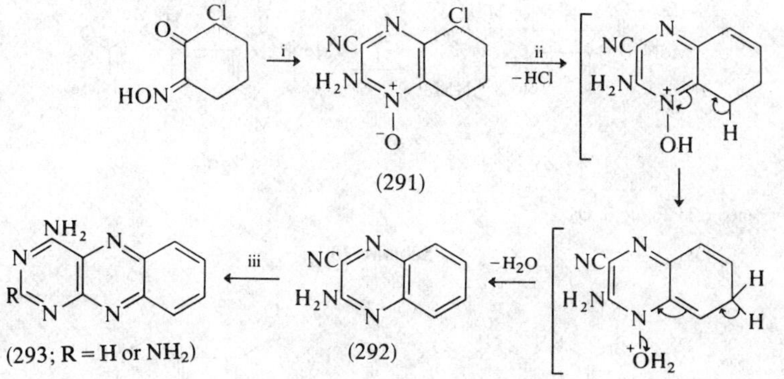

Reagents: i, $(NC)_2C=CHNH_2$; ii, AcOH, $>100\,°C$; iii, $HC(OEt)_3$ or $(NH_2)_2C=NH$.

**Scheme 69**

sequence outlined in Scheme 70. Curiously, the $N$-ethyl analogue (294; R = Et), under similar photolytic conditions, yields only dithiino-dipyrimidine (295; R = Et) (11%) and the 'normal' cyclization product, *i.e.* (298) (13%).

Thermal decomposition of (*o*-azidophenylthio)uracil (299) furnishes a 2:1 mixture of the dihydro-5-thia- (300) and dihydro-10-thia-isoalloxazines (301).[253] The formation of the former product is analogous to the preparation of phenothiazines from *o*-azidodiphenyl sulphides, and presumably involves rearrangement *via* a spiro-intermediate. However, the direct insertion of the nitrene intermediate derived from azide (299) into the uracil ring to give the 1,4-benzothiazine system (301) is unique. More surprising is the production of the unrearranged 10-thiaisoalloxazine (301) (70%) as the only product from the photolysis of (299) in methanol solution.

[253] T. Hiramitsu and Y. Maki, *J.C.S. Chem. Comm.*, 1977, 557.

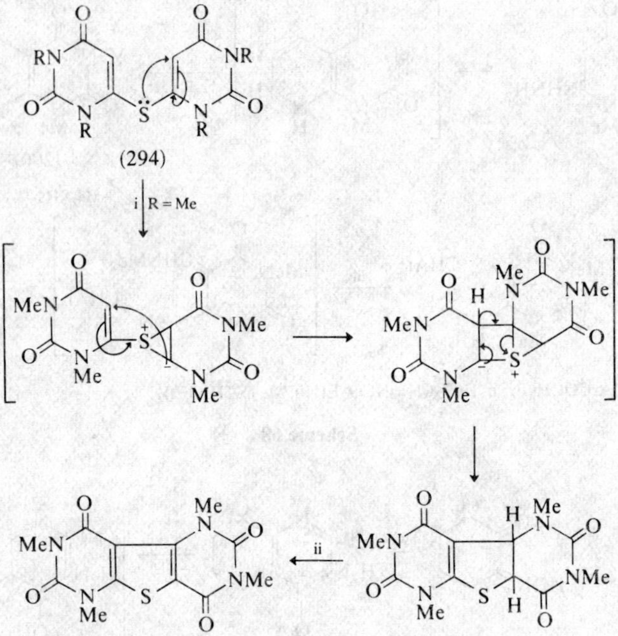

(294)

i | R = Me

(297)                                        (296)

Reagents: i, THF, *hν*; ii, O₂.

**Scheme 70**

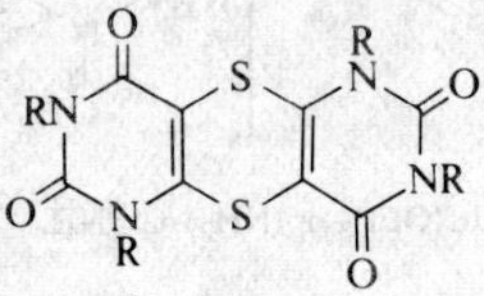

(295) 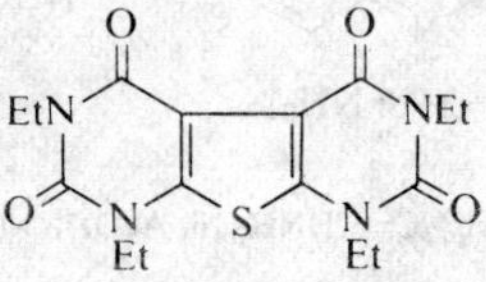                                        (298)

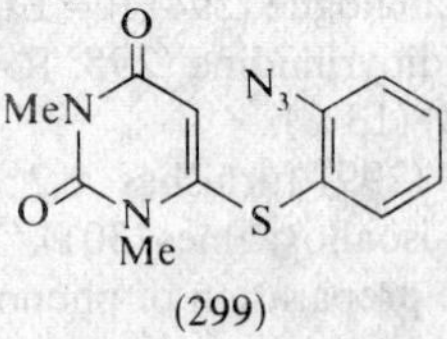

(299) 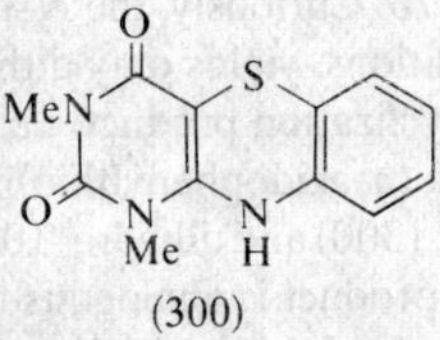                    (300) 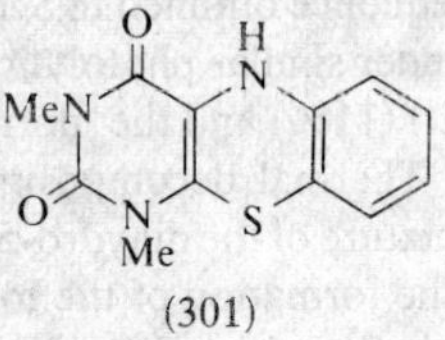                    (301)

**Other Systems with Two Nitrogen Atoms.**—New cyclic neutral ligands of varying ring size based on the 1,10-phenanthroline, *e.g.* (302), and 2,2′-bipyridyl systems have been obtained in remarkably good yields and without recourse to high-dilution techniques.[254] However, complexes are isolated only with Ag⁺ and Hg²⁺ ions. In contrast, the acyclic phenanthroline ligand (303; X = O), containing

²⁵⁴ E. Buhleier and F. Vögtle, *Annalen*, 1977, 1080.

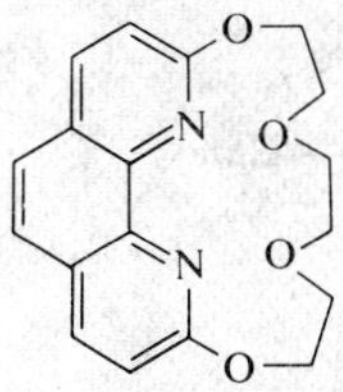

(302) 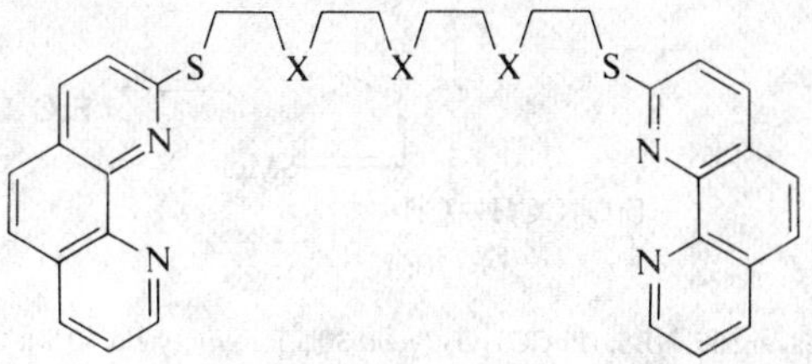                              (303)

mixed oxygen and sulphur donor sites, forms stable complexes with several heavy metals, *e.g.* $Co^{2+}$. Replacement of the oxygen sites by a methylene bridge, *i.e.* (303; $X = CH_2$), destroys the complexing ability. The chemistry of phenanthrolines has been reviewed.[255]

The nitro-enamine (304), formed in the one-pot condensation of 2-aminopyridine with triethyl orthoformate and ethyl nitroacetate, cyclizes in polyphosphoric acid to the $4H$-pyrido[1,2-*a*]pyrimidinone (305; $R^1 = NO_2$, $R^2 = H$).[256] This bicyclic system is also obtainable by the action of trichloroacetic anhydride in boiling toluene on 2-benzylideneaminopyridine, as illustrated in Scheme 71.[257] Analogous reactions also occur with 2-arylideneamino-quinolines and -isoquinolines.

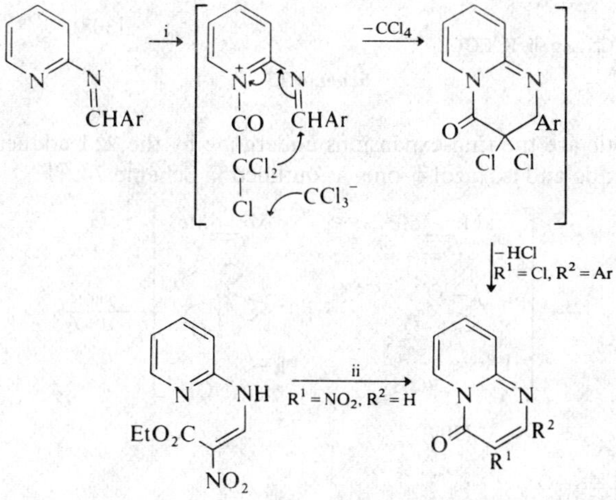

(304)                              (305)

Reagents: i, $(CCl_3CO)_2O$, PhMe, at 100—105 °C; ii, PPA.

**Scheme 71**

[255] L. A. Summers, *Adv. Heterocyclic Chem.*, 1978, **22**, 1.
[256] O. S. Wolfbeis, *Chem. Ber.*, 1977, **110**, 2480.
[257] T. Morimoto and M. Sekiya, *Chem. and Pharm. Bull. (Japan)*, 1977, **25**, 1607.

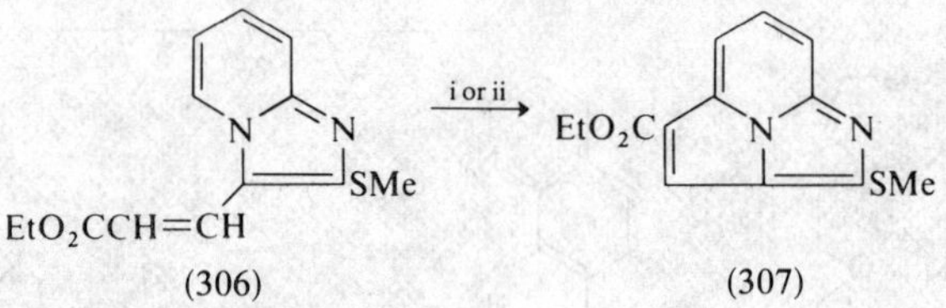

(306)                                          (307)

Reagents: i, NBS, (PhCO)$_2$O$_2$ (yield 30%); ii, $h\nu$ (yield 40%).

**Scheme 72**

Azaindolizines, *e.g.* (306), the starting materials for two new synthetic routes to azacycl[3.2.2]azines, *e.g.* (307), (see Scheme 72),[258] are the subject of two comprehensive reviews.[74b,c] Also reviewed is the chemistry of pyrrolo-diazines possessing a bridgehead nitrogen.[259]

**Oxazines, Thiazines, and their Benzo-derivatives.**—Organo-zinc compounds feature in the conversion of $\alpha$-bromo-esters into 1,3-oxazin-6-ones (308) (Scheme 73).[260]

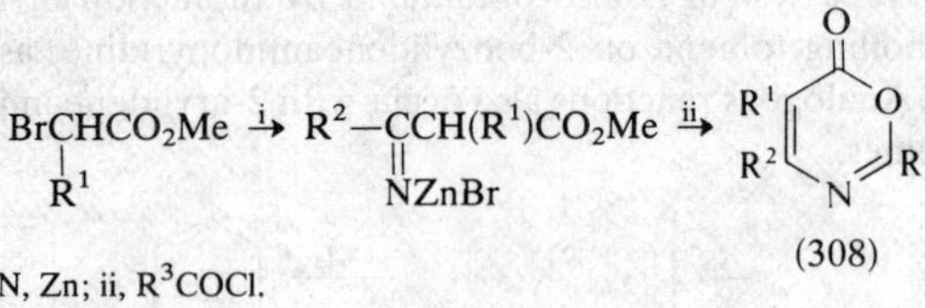

Reagents: i, R$^2$CN, Zn; ii, R$^3$COCl.

**Scheme 73**

More exotic are the ring-expansions undergone by the 2:1 adducts (309) of arylnitrile oxide and isoxazol-5-one, as outlined in Scheme 74.[261]

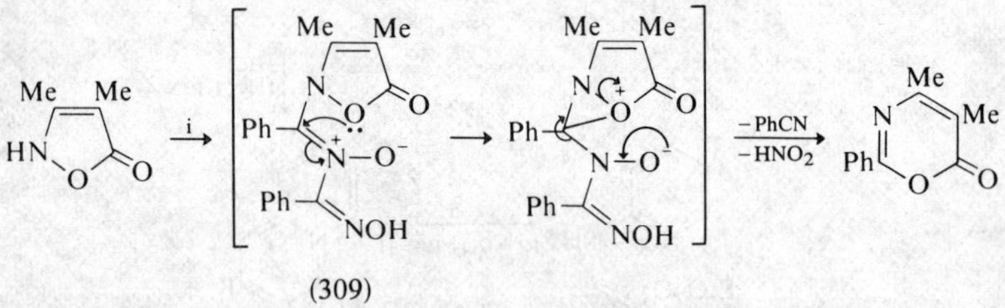

(309)

Reagent: i, PhC≡$\overset{+}{N}$—$\overset{-}{O}$.

**Scheme 74**

Thermolysis of furandione (310) affords the dibenzoylketen (311), which enters into (4 + 2) cycloadditions with nitriles to give the 1,3-oxazin-4-ones (312).[262]

258 K. Kurata, H. Awaya, Y. Tominaga, Y. Matsuda, and G. Kobayashi, *Heterocycles*, 1977, **8**, 293.
259 D. E. Kuhla and J. G. Lombardino, *Adv. Heterocyclic Chem.*, 1977, **21**, 1.
260 I. I. Lapkin, V. I. Semenov, and M. I. Belonovich, *Zhur. org. Khim.*, 1977, **13**, 1328.
261 G. L. Vecchio, F. Foti, G. Grassi, and F. Risitano, *Tetrahedron Letters*, 1977, 2119.
262 E. Ziegler, G. Kollenz, G. Kriwetz, and W. Ott, *Annalen*, 1977, 1751.

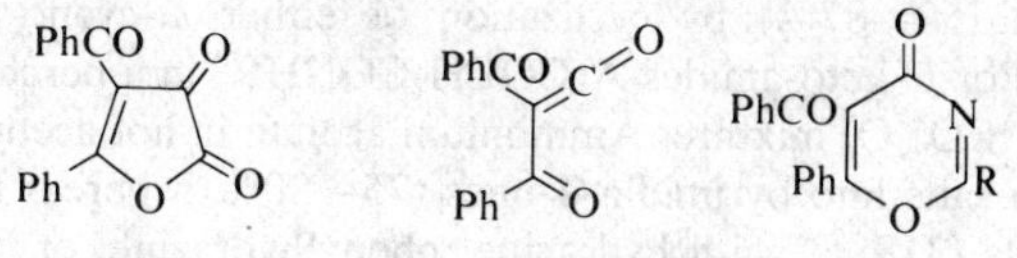

(310)　　　　　(311)　　　　　(312)

Continuation of the study of ring transformations of 1,3-oxazin-4-ones has led to the discovery of new routes to 5-acetyl-pyrimidinones (313)[263a] (Scheme 75; path *a*), 1,2,4-oxadiazoles (314)[263b] (path *b*), and isoxazoles (315) (path *c*).[263b] In buffered hydroxylamine hydrochloride solution, nucleophilic ring-opening of the oxazinone (path *a*) leads ultimately to oxadiazole (314). In contrast, hydroxylamine hydrochloride effects protonation of the oxazinone ring (316) and subsequent ring-opening and ring-closure to the isoxazoles (315), as illustrated in path (*c*).

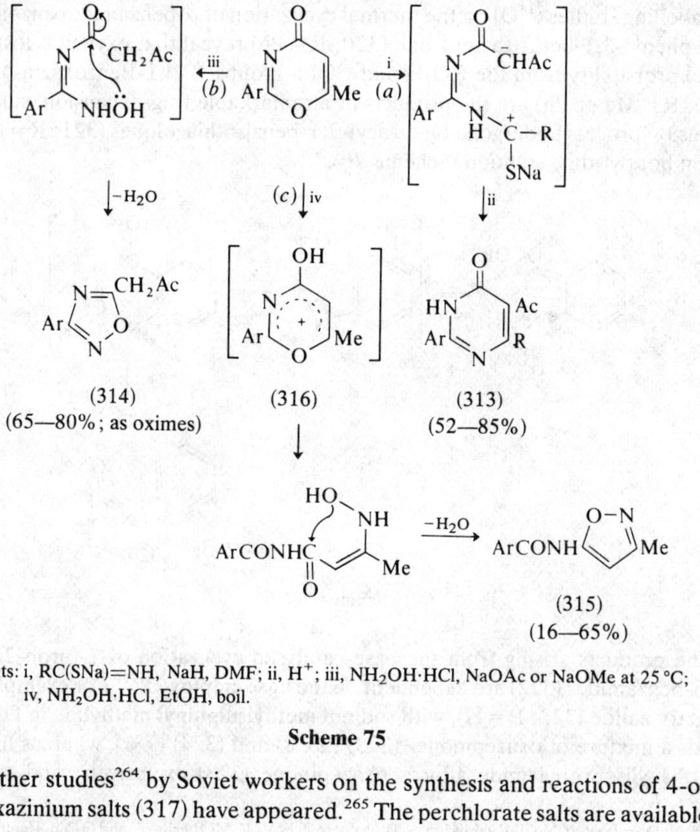

(314)　　　　　(316)　　　　　(313)
(65—80%; as oximes)　　　　　(52—85%)

(315)
(16—65%)

Reagents: i, RC(SNa)=NH, NaH, DMF; ii, H⁺; iii, NH₂OH·HCl, NaOAc or NaOMe at 25 °C; iv, NH₂OH·HCl, EtOH, boil.

**Scheme 75**

Further studies[264] by Soviet workers on the synthesis and reactions of 4-oxo-1,3-oxazinium salts (317) have appeared.[265] The perchlorate salts are available in

[263] (a) Y. Yamamoto, Y. Azuma, and T. Kato, *Heterocycles*, 1977, **6**, 1610; (b) Y. Yamamoto and Y. Azuma, *ibid.*, 1978, **9**, 185.
[264] See ref. 18, p. 137.
[265] G. N. Dorofeenko, Yu, I. Ryabukhin, and V. D. Karpenko, *Khim. geterotsikl. Soedinenii*, 1977, 704; *J. Org. Chem. (U.S.S.R.)*, 1977, **13**, 2288.

variable yield (50—87%) by cyclization of either $\omega$-cyano-acetophenones $ArCOCH_2CN$ or $\beta$-keto-amides $ArCOCH_2CONHR^3$ in a perchloric acid–acid anhydride $[(R^1CO)_2O]$ mixture. Ammonium acetate in hot acetic acid converts the oxazinium salts into pyrimidin-4-ones (75—80%) whereas the benzo-1,3-oxazinium salts (318),[266] with hydrazine, phenylhydrazine, or hydroxylamine, ring-contract to the *o*-hydroxyphenyl derivatives (319; X = NH), (319; X = NPh), and (319; X = O) respectively.

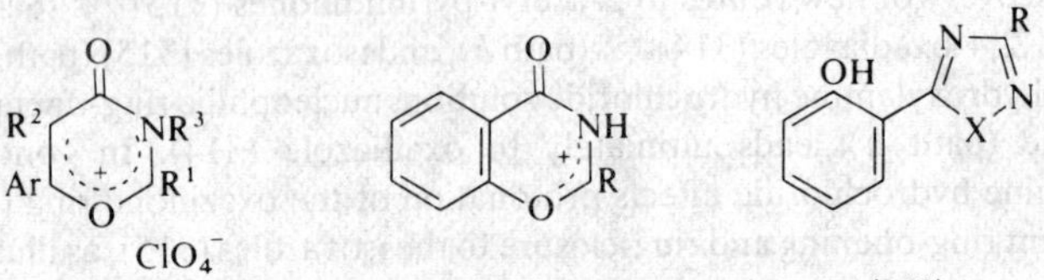

(317)        (318)        (319)

Labelling studies ($^{18}O$) on the thermal cyclization of *o*-benzamidobenzoic acid to 2-phenyl-3,1-benzoxazin-4-one (320; R = Ph) reveal that oxygen is lost with equal probability from the $CO_2H$ and COPh groups.[267] 3,1-Benzoxazin-4-ones (320; R = Me or Ph) are the products of a remarkable ring-expansion, sulphur-extrusion process undergone by *O*-acyl-2,1-benzisothiazolones (321; R = Me or Ph) in hot pyridine solution (Scheme 76).[268]

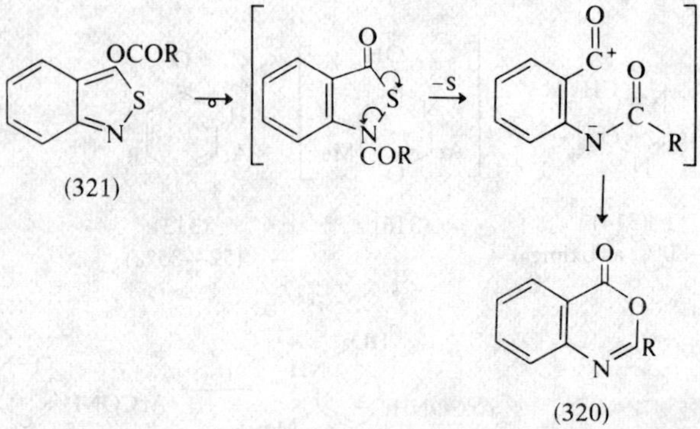

(321)

(320)

**Scheme 76**

The products arising from the base-catalysed cyclization of 2-(prop-2-ynyl-oxy)-benzamides (322) are dependent on the base employed.[269] For example, the primary amide (322; R = H), with sodium methylsulphinyl methylide in DMSO, yields a mixture of oxazepinones (323) (34%) and (324) (7%), whereas lithium cyclohexylisopropylamide affords oxazepinone (323) exclusively, and sodium

[266] G. N. Dorofeenko, Yu. I. Ryabukhin, S. B. Bulgarevich, V. V. Mezheritskii, and O. Yu. Ryabukhina, *J. Org. Chem. (U.S.S.R.)*, 1977, **13**, 2289.

[267] R. V. Poponova, B. M. Bolotin, L. S. Zeryukina, and R. U. Safina, *Khim. geterotsikl. Soedinenii*, 1977, 614.

[268] M. Davis and S. P. Pogany, *J. Heterocyclic Chem.*, 1977, **14**, 267.

[269] V. Scherrer, M. Jackson-Mülly, J. Zsindely, and H. Schmid, *Helv. Chim. Acta*, 1978, **61**, 716.

isopropoxide the 1,3-benzoxazinone (325) (34%). Allenyloxybenzamide (326) has been isolated as a stable intermediate from similar cyclizations of the *N*-methylamide (322; R = Me).

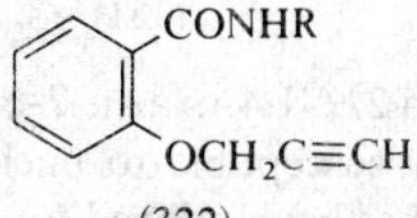

(322) 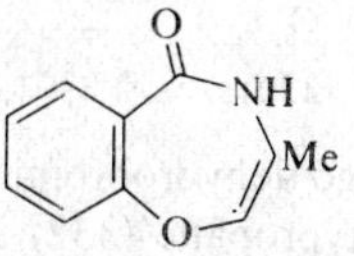          (323) 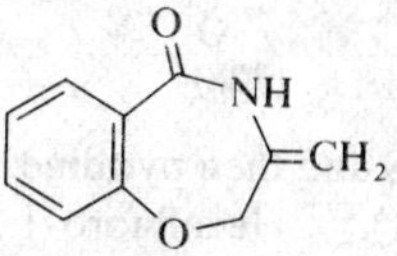          (324)

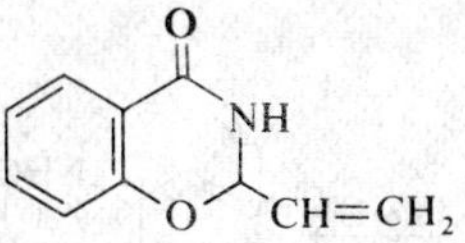

(325) 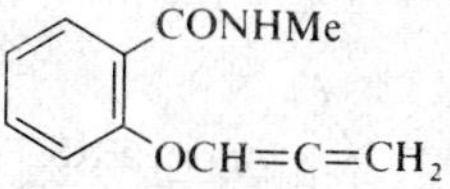          (326)

In contrast, with the *N*-phenyl- and *N*-(α-naphthyl)-amides (322; R = Ph) and (322; R = α-C₁₀H₇) the reaction takes a different course (Scheme 77) to produce chromones (327).

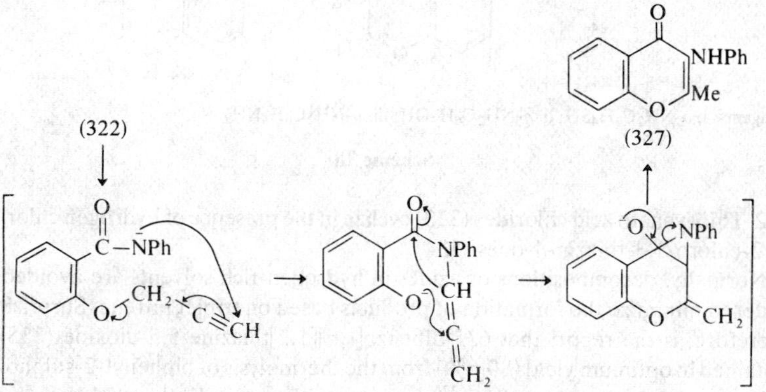

**Scheme 77**

Flash vacuum pyrolysis (625 °C; 0.1—1 Torr) of 2,3-benzoxazin-1-one (328) produces biphenylene (7%) and 2-phenyl-3,1-benzoxazin-4-one (320; R = Ph) (11%) rather than the hoped-for[270] benzazete-derived products.[271] The azirino-α-lactone (329) has been suggested as a possible reaction intermediate. The gas-phase thermal and photochemical decompositions of nitrogen, oxygen, and sulphur heterocycles have been reviewed.[272]

3,4-Dihydro-2*H*-1,4-oxazin-2-ones (330), which are the cyclization products of α-amino-acid acetonyl or phenacyl ester hydrobromides, *e.g.* BrH·NH₂CHR¹CO₂CH₂COR² (R¹ = alkyl, R² = Me or Ph), when treated with

[270] See ref. 18, p. 141.
[271] K. L. Davies, R. C. Storr, and P. J. Whittle, *J.C.S. Chem. Comm.*, 1978, 9.
[272] S. Braslavsky and J. Heicklen, *Chem. Rev.*, 1977, **77**, 473.

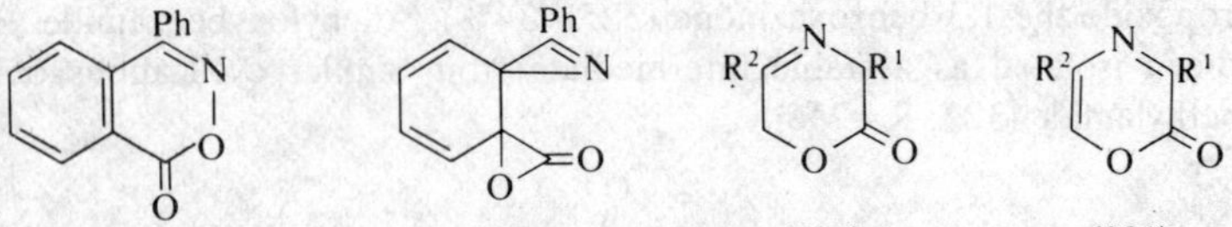

(328)            (329)            (330)            (331)

NBS and then pyridine, undergo dehydrobromination to $2H$-1,4-oxazin-2-ones (331).[273] Hexafluoro-1,2-epoxypropane (332) is highly susceptible to nucleophilic attack at the $\beta$-carbon centre, and this property has been utilized for the synthesis of a variety of perfluoroalkylated heterocycles.[274] Representative examples are given in Scheme 78.

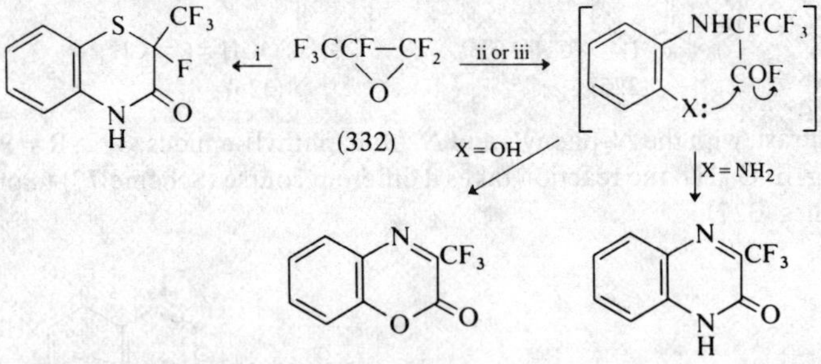

Reagents: i, $o$-NH$_2$C$_6$H$_4$SH; ii, $o$-NH$_2$C$_6$H$_4$OH; iii, $o$-NH$_2$C$_6$H$_4$NH$_2$.

**Scheme 78**

2-Thiocyanato acid chlorides (333) cyclize in the presence of hydrogen chloride to 2-chloro-1,3-thiazin-4-ones (334).[275]

Normally, decompositions of azides in hydrogen-rich solvents are avoided in order to minimize the formation of products based on triplet nitrene. Surprising, therefore, is the report that $6H$-dibenzo[$c,e$][1,2]thiazine 5,5-dioxide (335) is obtained in optimum yield (80.6%) from the thermolysis of biphenyl-2-sulphonyl azide in cyclohexane at 120 °C.[276] Decomposition of $o$-(2-thienyl)benzenesulphonyl azide under similar conditions furnishes the thieno[3,2-$c$]benzo-1,2-thiazine dioxide (336), in practicable yield (61%). Whereas, in POCl$_3$-pyridine

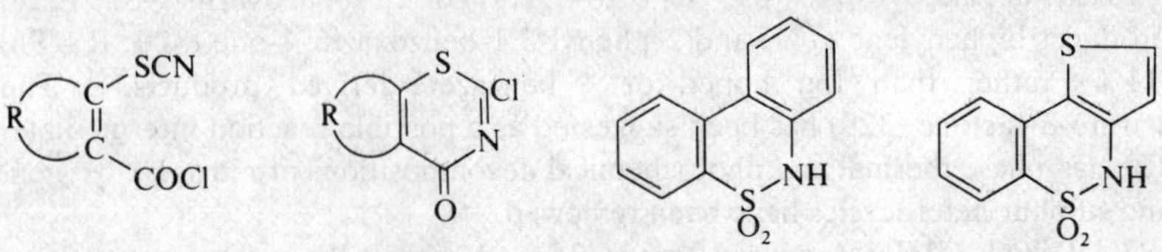

(333)            (334)            (335)            (336)

[273] G. Schulz and W. Steglich, *Chem. Ber.*, 1977, **110**, 3615.
[274] N. Ishikawa and S. Sasaki, *Bull. Chem. Soc. Japan*, 1977, **50**, 2164.
[275] G. Simchen and G. Entenmann, *Annalen*, 1977, 1249.
[276] R. A. Abramovitch, T. Chellathurai, I. T. McMaster, T. Takaya, C. I. Azogu, and D. P. Vanderpool, *J. Org. Chem.*, 1977, **42**, 2914.

mixture at 100 °C, amides (337) cyclize exclusively to 2$H$-1,3-benzothiazines (338),[277a] in phosphorus oxychloride a mixture of the 2$H$-isomers and 4$H$-isomers (339) is obtained.[277b]

(337)          (338)          (339)

The chemistry and applications of aza- and thia-analogues of phenoxazines and related compounds have been reviewed.[278]

## 3 Systems containing Three and Four Heteroatoms

**Triazines.**—Oxidation of the 2-amino-4,7-methano-indazole (340) with lead tetra-acetate produces the 5,8-bridged tetrahydrobenzotriazine (341) in 75.5% yield.[279] $\alpha$-Diketone oxime-hydrazones, *i.e.* HON=C(R$^2$)C(R$^1$)=NNH$_2$, condense with orthoformates R$^3$C(OEt)$_3$ to yield $\alpha$-(1-ethoxyalkylidenehydrazono)-oximes HON=C(R$^2$)C(R$^1$)=NN=C(R$^3$)OEt, which on thermolysis at 200 °C ring-close to 1,2,4-triazine 4-oxides (342).[280] The $N$-oxides are also accessible directly by heating the oximino-hydrazones with an imidate hydrochloride R$^3$(RO)C=NH·HCl.

(340)          (341)          (342)

There is continuing interest in the synthetic potential of the (4 + 2) cyclo-adducts of polyazabenzenes. For example, 1,2,4-triazines bearing dialkylamino- or alkoxy-groups at the 3- and/or 5-positions react with ynamines to yield cyclo-adducts of type (343; X = NMe$_2$ or OMe), which by loss of nitrogen give substituted pyridines (344).[281a] However, 1,2,4-triazines containing two or three dialkylamino-groups react only with electron-deficient dienophiles, *e.g.* dimethyl acetylenedicarboxylate. In addition to the anticipated dimethyl 2,6-bis-dimethylaminopyridine-3,4-dicarboxylate, pyrrolo[2,1-*f*][1,2,4]triazine (345) is formed, albeit in minor amounts. Anomalous behaviour is shown by 3,5-bis-dimethylamino-6-methoxy-1,2,4-triazine, which with the acetylene-dicarboxylate forms the 2,5-cyclo-adduct (346) rather than the 3,6-adducts,

[277] J. Szabó, L. Fodor, I. Varga, E. Vinkler, and P. Sohár, *Acta. Chim. Acad. Sci. Hung.*, 1977, **92**, (*a*) p. 317; (*b*) p. 403.
[278] C. O. Okafor, *Heterocycles*, 1977, **7**, 391.
[279] I. Ito, N. Oda, S.-I. Nagai, and Y. Kudo, *Heterocycles*, 1977, **8**, 319.
[280] V. Böhnisch, G. Burzer, and H. Neunhoeffer, *Annalen*, 1977, 1713.
[281] (a) H. Neunhoeffer and B. Lehmann, *Annalen*, 1977, 1413; (*b*) H. Ewald, B. Lehmann, and H. Neunhoeffer, *ibid.*, p. 1718.

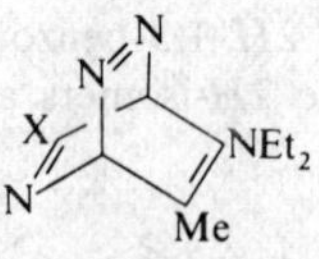

(343) 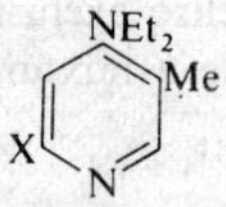                    (344) 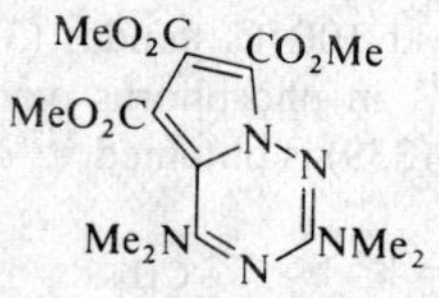                    (345)

*e.g.* (343), obtained with other triazines. Adduct (346), on heating, extrudes MeOCN to form pyrimidine (347).

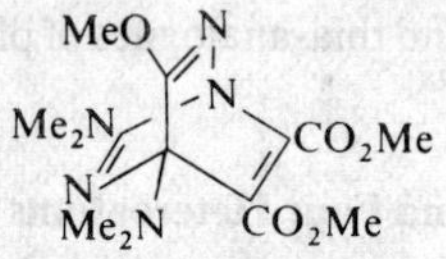

(346) 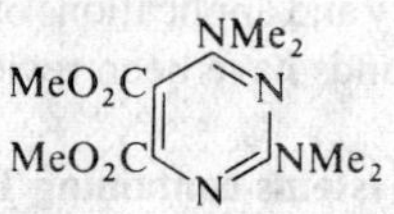                                    (347)

Generally, cycloadditions of this type are carried out in acetic acid solution. However, it has now been found[281b] that, in the absence of solvent, other products, *e.g.* (349) and (350), are formed in minor amounts (13% and 8% respectively), probably by cycloaddition of the 1,4-dipolar intermediate (348) to the alkyne and carbonyl bonds respectively of the acetylenedicarboxylate.

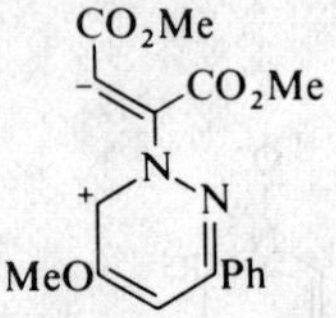

(348) 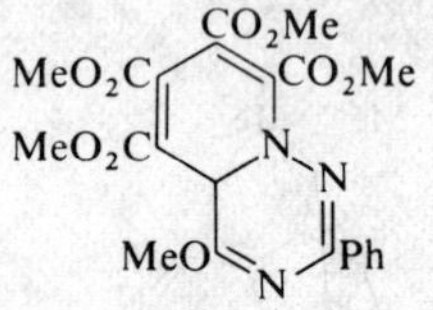                    (349) 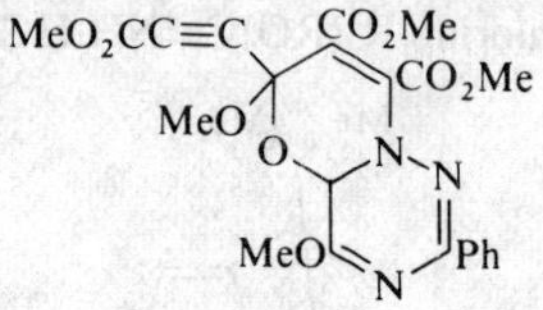                    (350)

Isonitrosoacetanilide is the starting material for a new, simple synthesis of 6-aza-2-thiouracil (351) (Scheme 79).[282]

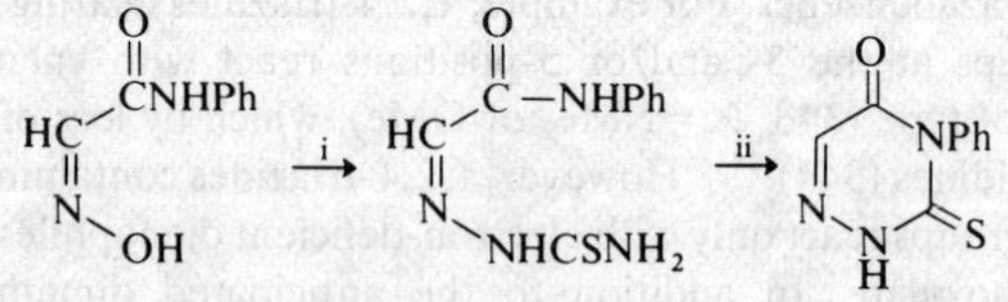

(351)

Reagents: i, NH₂CSNHNH₂, HCl; ii, OH⁻.

**Scheme 79**

Equally facile is the preparation of benzo-1,2,4-triazines (352) outlined in Scheme 80, the *o*-amino-phenylhydrazine being available in high yield (80—85%) from the electrolytic reduction of benzotriazole in 4M-HCl at 0 °C.[283]

[282] G. Dolescall and G. Vankai, *Acta Chim. Acad. Sci. Hung.*, 1977, **92**, 323.
[283] M. Falsig and P. E. Iversen, *Acta Chem. Scand. (B)*, 1977, **31**, 15.

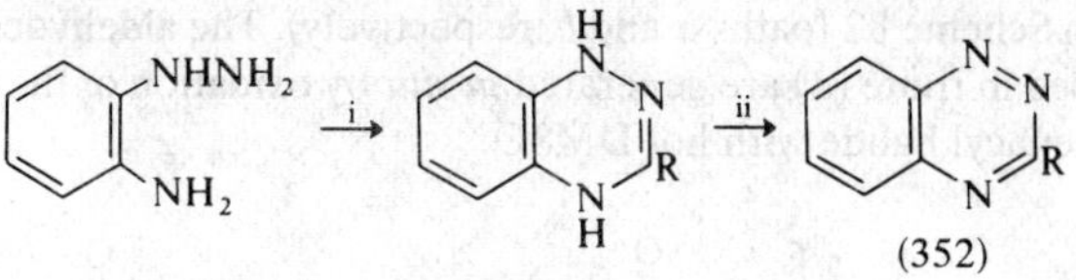

(352)

Reagents: i, RC(OR')₃, C₅H₅N, at 25 °C; ii, K₃Fe(CN)₆.

**Scheme 80**

In a previous section (p. 194), the thermal electrocyclization of 5-arylazo-6-ethoxymethyleneamino-uracils to 8-arylamino-theophyllines was noted. Related are the 6-dimethylaminomethyleneamino-derivatives (353), which on irradiation under anaerobic conditions also cyclize to 8-arylamino-theophyllines (242).[219c] However, in the presence of oxygen, photocyclization of (353) affords 6-aryl-6,7-dihydro-6-azalumazin-7-ones (355). A mechanistic rationale has been proposed (Scheme 81) involving the initial formation of the 5,6-dihydroazalumazine (354) followed either by formation of hydroperoxide (path *a*) or ring contraction (path *b*), depending on the reaction conditions. Thermolysis of (353) produces a mixture of the 5,6-dihydro-azalumazine (354) and theophylline (242).

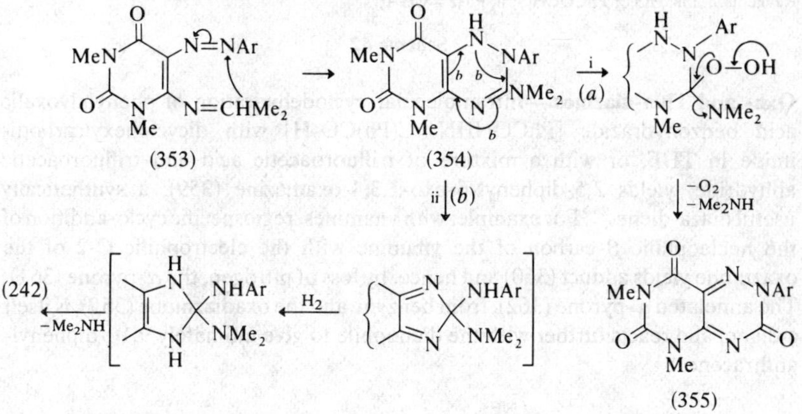

Reagents: i, *hv*, O₂; ii, *hv*, N₂.

**Scheme 81**

A full account of the methods of direct synthesis of the previously inaccessible fervenulin 4-oxides, reported last year,[284] has appeared.[285] Also noted last year[286] was the value of 6-hydrazino-1,3-dimethyl-5-nitrosouracil (356) as a precursor in purine and pteridine syntheses. Extension of this work has now provided access to 3-aryl-fervenulins (357)[287a] and pyrimido[5,4-*e*]-*as*-triazines (358),[287b] as

[284] See ref. 18, p. 143.
[285] M. Ichiba, S. Nishigaki, and K. Senga, *J. Org. Chem.*, 1978, **43**, 469.
[286] See ref. 18, p. 144.
[287] (*a*) M. Ichiba, S. Nishigaki, and K. Senga, *Heterocycles*, 1977, **6**, 1921; (*b*) K. Senga, M. Ichiba, Y. Kanamori, and S. Nishigaki, *ibid.*, 1978, **9**, 29.

elaborated in Scheme 82 (paths *a* and *b*, respectively). The aldehydes or α-keto-aldehydes used in route (*a*) are generated *in situ* by oxidation of the appropriate benzyl or phenacyl halide with hot DMSO.

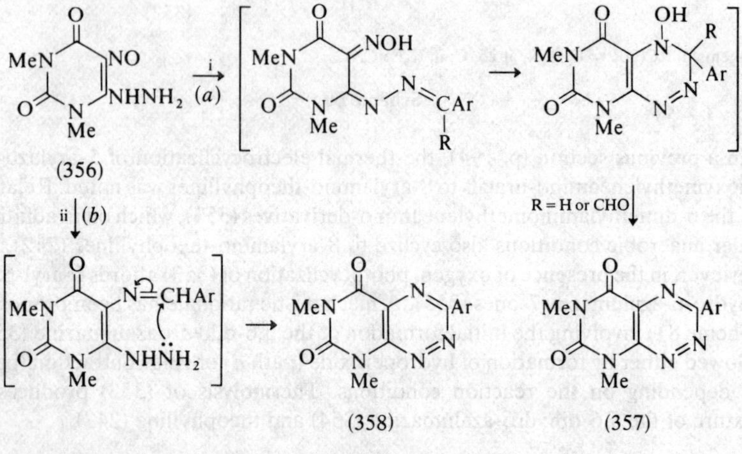

(356)

Reagents: i, PhCHO or PhCOCHO; ii, Ph₃P=CHAr.

**Scheme 82**

**Oxa- and Thia-diazines.**—Intramolecular cyclodehydration of phenylglyoxalic acid benzohydrazide [PhCONHN=C(Ph)CO₂H] with dicyclohexylcarbodi-imide in THF, or with a mixture of trifluoroacetic acid and trifluoroacetic anhydride, yields 2,5-diphenyl-6-oxo-1,3,4-oxadiazine (359), a synthetically useful diaza-diene.[288] For example, with ynamines, regiospecific cyclo-addition of the nucleophilic β-carbon of the ynamine with the electrophilic C-2 of the oxazinone yields adduct (360) and hence, by loss of nitrogen, the α-pyrone (361). The annelated α-pyrone (362), from benzyne and the oxadiazinone (359), is itself a diene, and reacts further with the dienophile to give ultimately 9,10-diphenyl-anthracene.

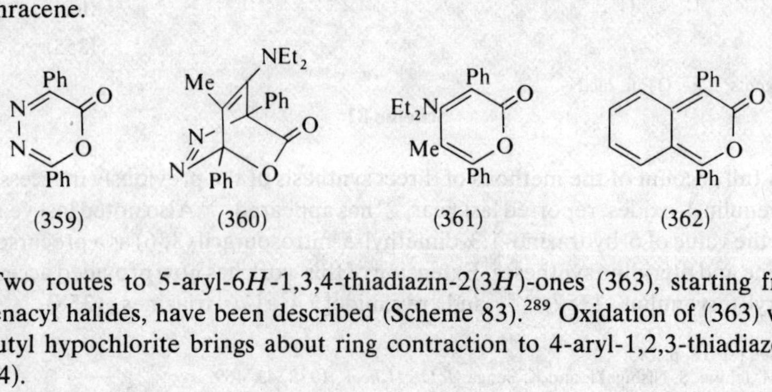

(359)            (360)            (361)            (362)

Two routes to 5-aryl-6*H*-1,3,4-thiadiazin-2(3*H*)-ones (363), starting from phenacyl halides, have been described (Scheme 83).[289] Oxidation of (363) with t-butyl hypochlorite brings about ring contraction to 4-aryl-1,2,3-thiadiazoles (364).

[288] W. Steglich, E. Buschmann, G. Gansen, and L. Wilschowitz, *Synthesis*, 1977, 252.
[289] G. Ege, P. Arnold, G. Jooss, and R. Noronha, *Annalen*, 1977, 791.

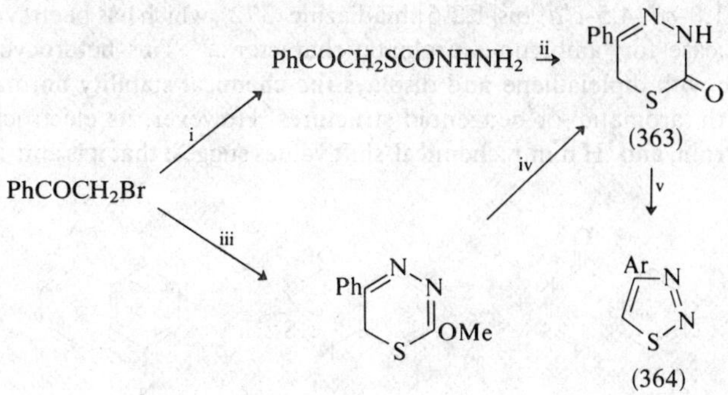

PhCOCH₂SCONHNH₂ $\xrightarrow{ii}$

Ph$\underset{}{\overset{N-NH}{\diagdown}}$

(363)

PhCOCH₂Br

PhCOCH₂Br $\xrightarrow{i}$

$\xrightarrow{iv}$

$\xrightarrow{iii}$

Ph$\underset{S}{\overset{N-N}{\diagdown}}$OMe

(364)

Ar$\underset{S}{\overset{N}{\diagdown}}$N

Reagents: i, K⁺ S̄CONHNH₂, DMF; ii, Me₂CO, H⁺; iii, MeOCSNHNH₂; iv H₂O; v, BuᵗOCl.

**Scheme 83**

Intermolecular cyclodehydration of a thioamide (RCSNH₂) and perfluoro-acetone in trifluoroacetic anhydride–pyridine constitutes a convenient synthesis of 6*H*-1,3,5-oxathiazines (365).[290] On vacuum thermolysis (140 °C, 20 Torr) the oxathiazines lose trifluoroacetone to give thiocarboxy-amides (367), probably *via* ring-opening of their valence tautomers, the 2*H*-1,3-thiazetes (366). These thioamides are reactive dienes, and as such are useful for the preparation of 4*H*-1,3-thiazines (368) and 4*H*-1,3,5-thiadiazines (369), as exemplified in Scheme 84.

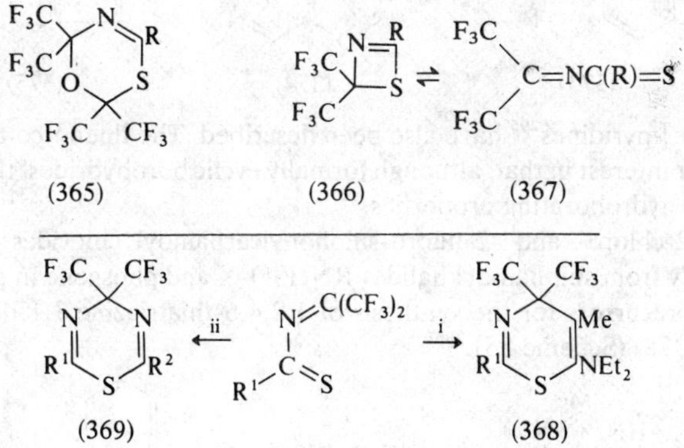

(365)

(366)

(367)

(369)

(368)

Reagents: i, MeC≡CNEt₂; ii, R²CN.

**Scheme 84**

2,1,3-Benzothiadiazinylium salts (371), prepared by alkylation of the benzo-2,1,3-thiadiazinone (370) with Meerwein's reagent (R₃O⁺ BF₄⁻; R = Me or Et), show marked dienophilic character at the N—S bond, although M.O. calculations suggest that they are heteroaromatic.[291] Less clear is the character of the

[290] K. Burger, R. Ottlinger, and J. Albanbauer, *Chem. Ber.*, 1977, **110**, 2114.
[291] W. Kosbahn and H. Schäfer, *Angew. Chem. Internat. Edn.*, 1977, **16**, 780.

naphtho[1,8-*cd* : 4,5-*c'd'*]bis[1,2,6]thiadiazine (372), which has been synthesized as a test case for 'ambiguous aromatic character'.[292] This heterocycle is iso-electronic with dipleiadiene and displays the chemical stability normally asso-ciated with 'aromatic' or benzenoid structures. However, its electrochemistry, u.v. spectrum, and $^1$H n.m.r. chemical-shift values suggest that it is anti-aromatic.

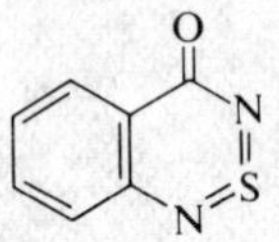

(370) 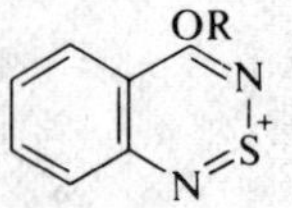          (371) 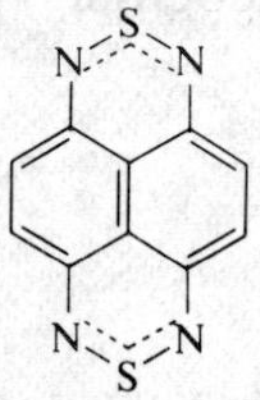          (372)

**Other Systems.**—Interest continues in the chemistry of boron–nitrogen hetero-cycles, and several new syntheses have been noted this year. Direct cyclization of arylsulphonylhydrazones ArCH=NNHSO$_2$Ar with hot (60 °C) boron tribromide affords a novel route to the aromatic 2,3,1-diazaborines (373).[293] Similar treat-ment of the $\beta$-naphthyl hydrazone yields only the linear isomer (373a). Syntheses of borazarofuro- (374)[294] and borazarothieno-[3,2-*c*]- (375) and borazaro-

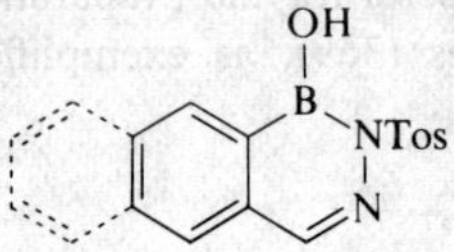

(a)   (373) 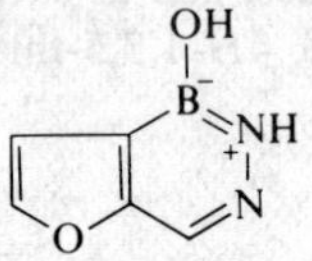          (374) 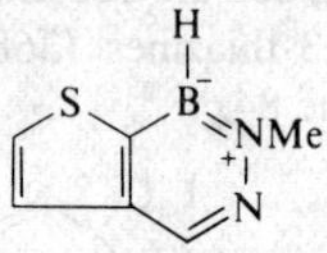          (375)

thieno[2,3-*c*]-pyridines[295] have also been described. The thieno-compounds are of particular interest in that, although formally cyclic borohydrides, they show no reducing or hydroborating properties.

2-Alkyl-2-chloro- and -2-fluoro-sulphonylcarbamoyl chlorides (376), pre-pared simply from sulphamoyl halides RNHSO$_2$X and phosgene in pyridine,[296a] are useful precursors for the synthesis of 1,2,4,6-thiatriazine 1,1-dioxides, *e.g.* (377) and (378) (Scheme 85).[296b]

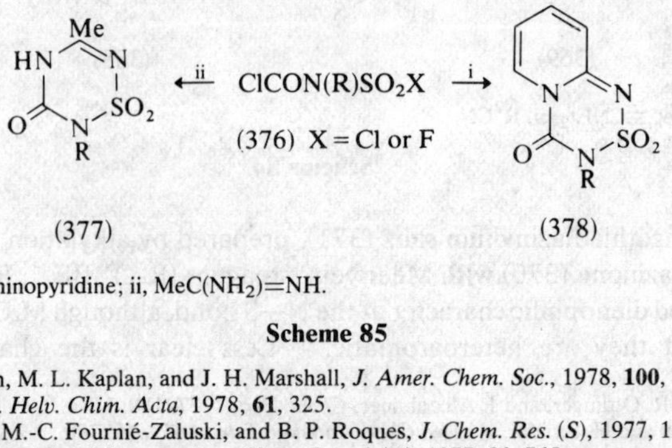

ClCON(R)SO$_2$X

(376)  X = Cl or F

(377)          (378)

Reagents: i, 2-aminopyridine; ii, MeC(NH$_2$)=NH.

**Scheme 85**

[292] R. C. Haddon, M. L. Kaplan, and J. H. Marshall, *J. Amer. Chem. Soc.*, 1978, **100**, 1235.
[293] B. W. Müller. *Helv. Chim. Acta*, 1978, **61**, 325.
[294] D. Florentin, M.-C. Fournié-Zaluski, and B. P. Roques, *J. Chem. Res. (S)*, 1977, 158.
[295] S. Gronowitz and A. Maltesson, *Acta Chem. Scand. (B)*, 1977, **31**, 765.
[296] D. Bartholomew and I. T. Kay, *J. Chem. Res. (S)*, 1977, (*a*) p. 237; (*b*) p. 238; (*c*) p. 239.

Attempts to cyanomethylate the thiatriazoline dioxide (379; X = SO$_2$), using chloroacetonitrile, resulted in ring-expansion of the triazoline to 1,2,4,6-thiatriazin-5-one 1,1-dioxide (380; X = SO$_2$), as outlined in Scheme 86.[296c] By analogy, the triazoline-dione (379; X = CO) was correctly predicted to undergo ring-expansion to the 1,3,5-triazine-dione (380; X = CO).

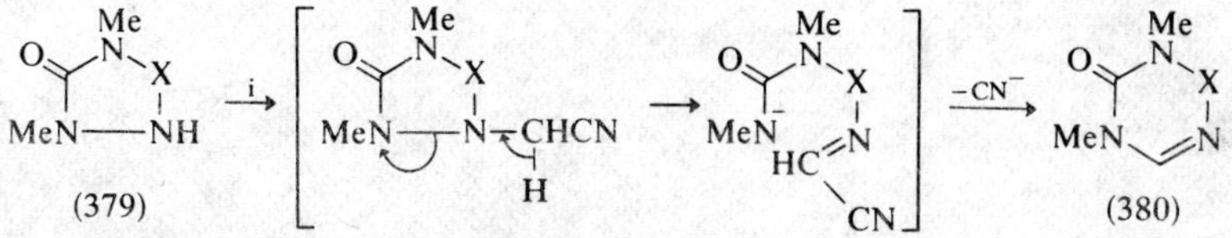

Reagents: i, ClCH$_2$CN, base.

**Scheme 86**

Dichloro-1,2,4,6-thiatriazines (381) are formed in moderate yields (40—50%) by the prolonged action of sulphur dichloride on *N*-cyano-amidines H$_2$NC(R)=NCN.[297]

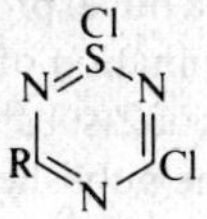

(381)

[297] P. P. Kornuta, L. I. Derii, and E. A. Romanenko, *Khim. geterotsikl. Soedinenii*, 1978, 273.

# 5

# Seven-membered Ring Systems

BY G. R. PROCTOR

The organisation of this chapter is the same as that adopted in the previous Report.

## 1 Tropones

A new synthesis of tropones has been announced which involves cyclocoupling of polybromo-ketones and 1,3-dienes, promoted by iron carbonyl.[1] Stipitatic acid and hinokitiol have been made by a novel procedure[2] involving intermediates (1) and (2) followed by Hofmann elimination of the bridge in (2).

The [13]C n.m.r. spectra of a series of 2-substituted tropones have been analysed, using a combination of methods; the published assignments for tropolone acetate were corrected and substituent parameters were defined and compared with corresponding parameters for monosubstituted benzenes.[3] Studies of nucleophilic displacement reactions of chlorine or of tosyloxy-groups in troponoids[4] reveal that with either $Me_2NH$ in DMSO or NaSMe in ethanol, substituents in the 3-position were replaced faster than those in 2- or 4-positions: neither M.O. calculations nor e.s.r. studies predicted this. 1,3-Diaza-azulenes are obtained in acceptable yields from the reaction of benzamidine and 2-(ethylthio)tropone:[5] in related work,[6] the interesting structure (3) has been recognized.

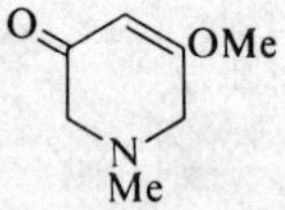

(1) 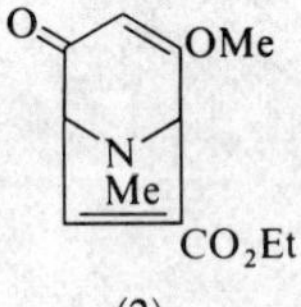          (2) 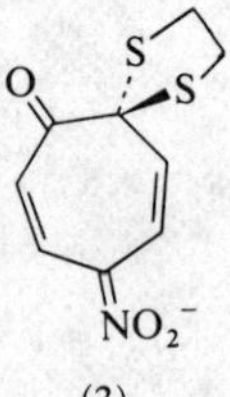          (3)

Cycloadditions and insertions on tropone continue to be studied. A series of methylene insertions leads to intermediates from which the carbene (4) was

[1] H. Takaya, Y. Hayakawa, S. Makino, and R. Noyori, *J. Amer. Chem. Soc.*, 1978, **100**, 1778.
[2] Y. Tamura, T. Saito, H. Kiyokawa, L. C. Chen, and H. Ishibashi, *Tetrahedron Letters*, 1977, 4075.
[3] J. F. Bagli and M. St. Jacques, *Canad. J. Chem.* 1978, **56**, 578.
[4] M. Cavazza, M. P. Columbini, M. Martinelli, L. Nucci, L. Pardi, F. Pietra, and S. Santucci, *J. Amer. Chem. Soc.*, 1977, **99**, 5997.
[5] M. Cavazza, R. Cabrino, F. Delcima, and F. Pietra, *J.C.S. Perkin I*, 1978, 609.
[6] M. Cavazza, C. A. Veracini, G. Morganti, and F. Pietra, *J.C.S. Chem. Comm.*, 1978, 167.

trapped.[7] Tropone reacts with diphenylnitrilimine,[8] giving a variety of products, the major one being the fused pyrazole (5). Phthalimidonitrene cycloaddition takes place at the tropone 4,5-bond *via* the more stable transition state in which there are two favourable secondary orbital interactions.[9] $\pi$-Electrons in the side-chain of 2-methoxy-5-styryltropone have been shown to participate in cycloaddition with maleic anhydride.[10]

The 4,5-dehydrotropone (6), from tropone, has been trapped with dienes in (2 + 4) cycloadditions, and the products have been further elaborated in interesting ways.[11]

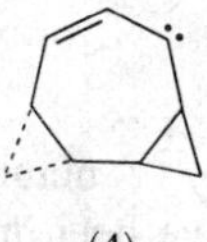

(4) 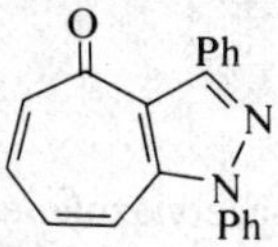          (5) 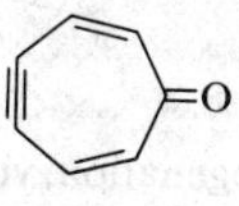          (6)

## 2 Tropolones

The biphenylene analogue (7),[12] its boron trifluoride complex, and the cyclobutatropolone (8) have been obtained by a ring-expansion reaction. Both n.m.r. and methylation studies support the tautomeric structures shown, both systems being less acidic than tropolone.[13] The 7,10-dithiasesquifulvalene-3,4-quinone (9)[14] has been made, and shows some dipolar character [cf. (9a)].

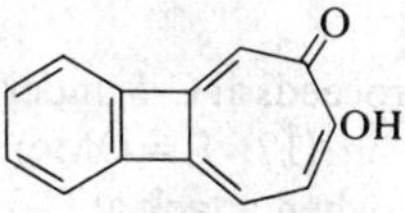

(7) 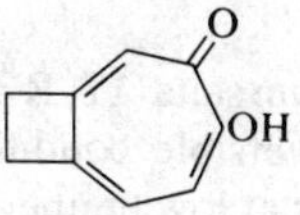          (8) 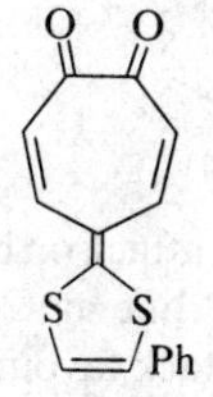          (9) 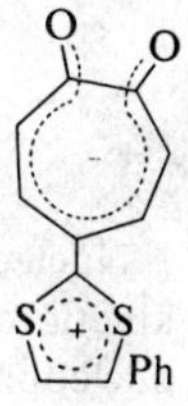          (9a)

## 3 Tropylium Salts and Tropilidenes

Diacetylmethyl-tropylium salts (10) have been made[15] and studied, treatment with triethylamine causing cyclization [to (11)] reminiscent of the behaviour of phenolic tropylium salts.[16] Acetonyl-tropylium perchlorate has been deprotonated to a substance (12), which underwent cycloaddition with acetylenic esters

[7] M. Oda, Y. Ito, and Y. Kitahara, *Tetrahedron Letters*, 1978, 977.
[8] D. Mukherjee, C. R. Watts, and K. N. Houk, *J. Org. Chem.*, 1978, **43**, 817.
[9] D. W. Jones, *J.C.S. Chem. Comm.*, 1978, 404.
[10] I. Saito, Y. Watanabe, and S. Ito, *Tetrahedron Letters*, 1977, 3049.
[11] T. Nakazawa, Y. Niimoto, and I. Murata, *Tetrahedron Letters*, 1978, 569.
[12] M. Sato, H. Fujino, S. Ebine, and J. Tsunetsugu, *Tetrahedron Letters*, 1978, 143.
[13] M. Sato, S. Ebine, and J Tsunetsugu, *J.C.S. Chem. Comm.*, 1978, 215.
[14] K. Takahashi, K. Morita, and K. Takase, *Chem. Letters*, 1977, 1505.
[15] K. Komatsu, S. Tanaka, S. Saito, and K. Okamoto, *Bull. Chem. Soc. Japan*, 1977, **50**, 3425.
[16] P. Bladon, P. L. Pauson, G. R. Proctor, and W. J. Rodger, *J. Chem. Soc. (C)*, 1966, 926.

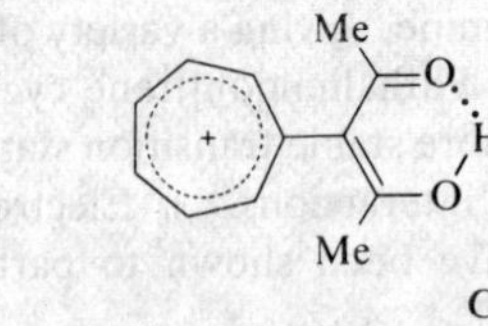

(10) 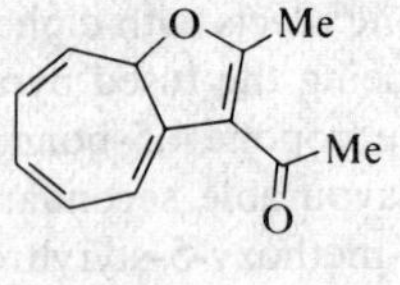                                                    (11)

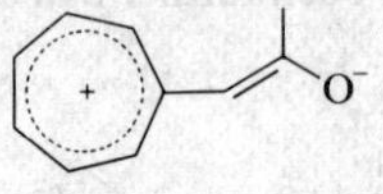

(12) 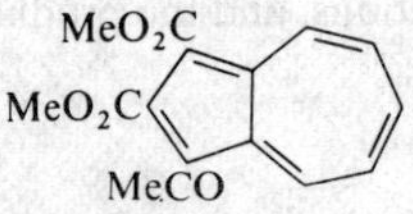                                                    (13)

and dehydrogenation, yielding 1-acetylazulenes, *e.g.* (13).[17] A number of bridged tropylium salts, 'tropyliocyclophanes', such as (14), have been made by ring-expansion of cyclophanes.[18] The fusion of thiophens to tropylium rings has been achieved, the *b*-fused systems, *e.g.* (15),[19] being more stable than those involving *c*-fusion, *e.g.* (16),[20] presumably since greater delocalization of electrons is possible without disruption of the thiophen rings.

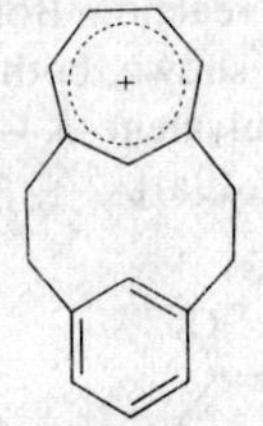

(14) 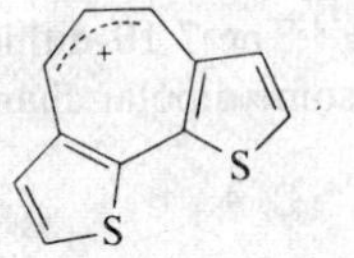                        (15) 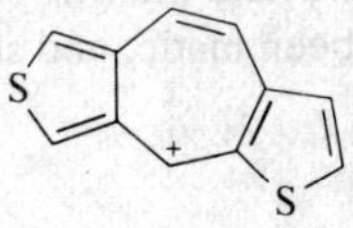                        (16)

Nucleophilic attack on benzotropylium salts (17; R = Cl) proceeds at C-5 under kinetic control but at C-7 under reversible conditions;[21] in (17; R = OMe), nucleophilic attack favours C-5 except at low temperatures, when attack at C-7 yields acetals; these equilibrate at higher temperatures.[22]

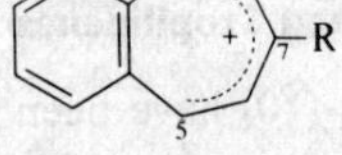

(17)

7-Phenylsulphonylcycloheptatriene[23] and 7-phenyl-7-acetoxycyclohepta-triene[24] have been synthesised, the latter giving phenylcarbene on ther-

17 K. Hafner, M. Romer, W. Ausderfunten, K. Komatsu, S. Tanaka, and K. Okamoto, *Annalen*, 1978, 376.
18 H. Horita, T. Otsubo, and S. Misumi, *Chem. Letters*, 1977, 1309.
19 S. Gronowitz and P. Pedaja, *Tetrahedron*, 1978, **34**, 587.
20 B. Tomtov and S. Gronowitz, *J. Heterocyclic Chem.*, 1978, **15**, 285.
21 B. Fohlisch, C. Fischer, E. Widmann, and E. Wolf, *Tetrahedron*, 1978, **34**, 533.
22 B. Fohlisch, C. Fischer, and W. Rogler, *Chem. Ber.*, 1978, **111**, 213.
23 R. A. Abramovitch, V. Alexanian, and J. Roy, *J.C.S. Perkin I*, 1977, 1928.
24 R. W. Hoffman, R. Schuttler, and I. H. Loof, *Chem. Ber.* 1977, **110**, 3410.

molysis. A specific synthesis for 3-substituted tropilidenes has been developed[25] which involves the reaction of cyclohepta-2,6-dienone with organolithium compounds and controlled dehydration of the products. A number of tropilidenes, *e.g.* (18), were obtained, along with other products from the reaction of

(18)

dichloromethyl methyl ether with 2,5-dimethylhexa-1,5-diene.[26] The cycloheptatriene ring has been directly formylated;[27] it reacted with phosphorus-containing carbenes,[28] and it reacted intramolecularly with diazo-substituted side-chains.[29] Cycloheptatrienes also undergo photochemically induced sigmatropic 1,5-hydrogen shifts, and the reversible interconversion of the various pairs of isomers has been studied.[30]

The *syn*- and *anti*-bishomocycloheptatrienes (19) and (20) undergo allylic chlorination, but in the latter case only one of the possible positions (∗) is substituted.[31]

(19)          (20)

Cycloheptatrienylidene (21) has been generated from trimethylsilyltropylium tetrafluoroborate and trapped with activated alkenes.[32] The unusual spiro-system (22) has now been shown to be surprisingly stable,[33] and the existence of 4,5-benzo-1,2,4,6-cycloheptatrienes, *e.g.* (23; R = Me), has been inferred by two groups of workers[34,35] from the nature of products obtained on dimerization.

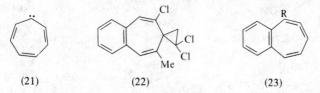

(21)                    (22)                    (23)

[25] K. Takeuchi, T. Maeda, and K. Okamoto, *Bull. Chem. Soc. Japan*, 1977, **50**, 2817.
[26] C. F. Garbers, H. C. S. Spies, H. E. Visagie, J. C. A. Boeyens, and A. A. Chalmers, *Tetrahedron Letters*, 1978, 81.
[27] T. Asao, S. Kuroda, and K. Kato, *Chem. Letters*, 1978, 41.
[28] G. Haas and M. Regitz, *Chem. Ber.*, 1978, **111**, 1733.
[29] T. Miyashi, Y. Nishizawa, T. Sugiyama, and T. Mukai, *J. Amer. Chem. Soc.*, 1977, **99**, 6109.
[30] W. Abraham, E. Henke, and D. Kreysig, *Tetrahedron Letters*, 1978, 345.
[31] M. R. Detty and L. A. Paquette, *J. Org. Chem.*, 1978, **43**, 1118.
[32] M. Reiffen and R. W. Hoffman, *Tetrahedron Letters*, 1978, 1107.
[33] B. Muller and P. Weyerstahl, *Annalen*, 1977, 982.
[34] E. E. Waali, J. M. Lewis, D. E. Lee, E. W. Allen, and A. K. Chappell, *J. Org. Chem.*, 1977, **42**, 3460.
[35] C. Mayor and W. M. Jones, *Tetrahedron Letters*, 1977, 3855.

### 4 Azulenes

A close study of deuteriated azulenes[36] showed that, in the mass spectrometer, 72% of ions suffer complete hydrogen scrambling prior to expulsion of $-CH_2-$. The rearranged ions are the same as those in the spectrum of naphthalene. The reaction of sodium cyclopentadienide with pyrylium salts has been used to make several more azulenes,[37] *e.g.* (24), and chlorination of 4,6,8-trimethylazulene in $H_2SO_4$ has been studied.[38]

The azulenophanes, *e.g.* (25),[39] and (26)[40] have been made. Their syntheses required elaboration and ring-expansion of pre-formed [2,2]cyclophanes on the one hand and, on the other, the coupling of an azulene derivative with the appropriate dithiol followed by photolytic desulphurisation.

### 5 Azepines and Benzazepines

Further work[41] on the photolysis of azidobenzene derivatives has been reported; in alcohols, structures such as (27; X = H, Cl, Br, or Me) were obtained. 2-Alkoxy-azepines are produced by deoxygenation with tributylphosphine of nitrobenzene in alcohols.[42] There is an interesting claim to have isolated 'azatropolones' (28) and (29) by thermolysis of compounds (30) and (31) respectively, themselves the photo-products of the interaction of (32) with

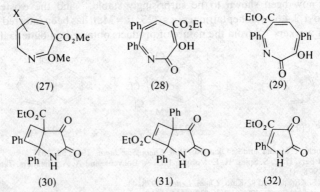

[36] R. Stoize and H. Budzikiewicz, *Monatsh.*, 1978, **109**, 325.
[37] Yu. N. Porshnev, E. M. Tereshchenko, and M. I. Cherkashin, *Zhur. org. Khim.*, 1978, **14**, 263.
[38] Yu. N. Porshnev, V. I. Erikhov, and N. A. Andronova, *Zhur. org. Khim.*, 1977, **13**, 2405.
[39] T. Kawashima, T. Otsubo, Y. Sakata, and S. Misumi, *Tetrahedron Letters*, 1978, 1063.
[40] K. Fukazawa, M. Aoyagi, and S. Ito, *Tetrahedron Letters*, 1978, 1067.
[41] R. Purvis, R. K. Smalley, W. A. Strachan, and H. Suschitzky, *J.C.S. Perkin I*, 1978, 191.
[42] M. Hasaki, K. Fukui, and J. Kita, *Bull. Chem. Soc. Japan*, 1977, **50**, 2013.

phenylacetylene,[43] the substances (28) and (29) are hydrolytically unstable, giving pyridine carboxylates,for which process mechanisms have been propounded.[43]

The substance (33), once thought to have been an azepinone, has been shown to arise from compound (34) by acidic catalysis which leads initially to the cyano-octanedione diester (35), now isolated.[44]

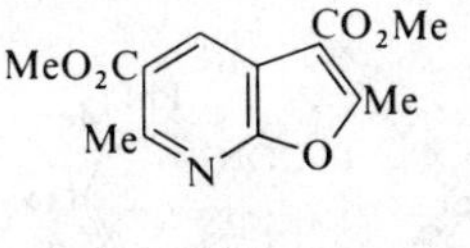

(33) 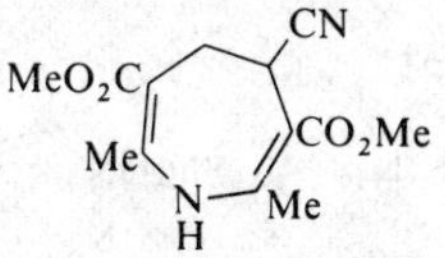          (34) 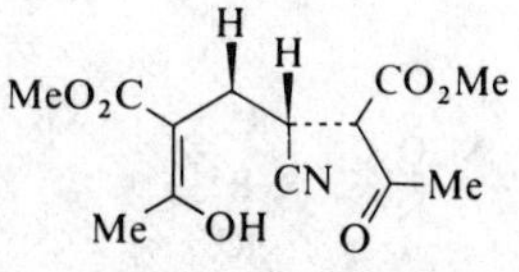          (35)

The 2-benzazepinedione derivatives (36; *cis* and *trans*) were isolated from a photochemically induced reaction of butadiene with *N*-methylphthalimide,[45] while various 1-benzazepine derivatives have been obtained[46] by sensitised photolysis of compound (37) at −78 °C. In particular, the tetraene (38) was isolated and at 180 °C it was converted into compound (39), while *o*-chloranil dehydrogenated it to the benzazepine (40).

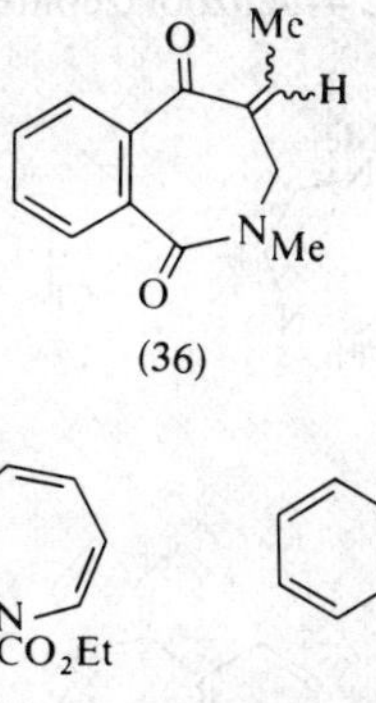

(36) 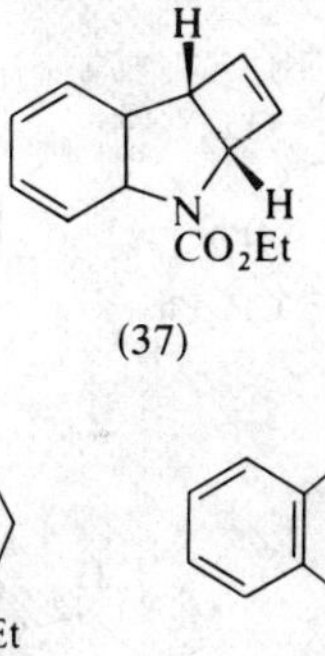          (37)

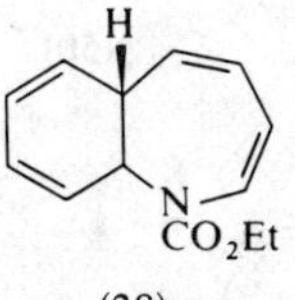

(38) 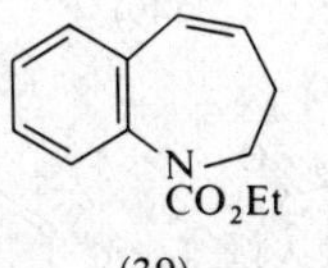          (39) 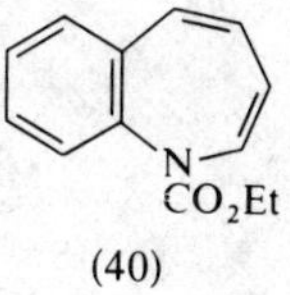          (40)

## 6 Diazepines and Benzodiazepines

The 1,2-diazepines (41; $R^1$ and $R^2$ = Me, Ph, *etc.*), obtained from the reaction of hydrazine with pyrylium salts, have been shown[47] to give 1,4-diazepines (42) on photolysis, while pyrolysis of the diazanorcaradienes (43) gives the 1,2-diazepines (44), presumably *via* the intermediate (45).[48] In a related case, thermolysis of (46) caused conversion into a 1,3-diazepine (47), for which a mechanism has been presented.[49]

[43] T. Sano, Y. Horiguchi, and Y. Tsuda, *Heterocycles*, 1978, **9**, 731.
[44] B. Gregory, E. Bullock, and T. S. Chen, *Canad. J. Chem.*, 1977, **55**, 4061.
[45] P. H. Mazzocchi, M. J. Bowen, and N. K. Narain, *J. Amer. Chem. Soc.*, 1977, **99**, 7063.
[46] A. G. Anastassiou, E. Reichmanis, S. J. Girgenti, and M. Schaeferridder, *J. Org. Chem.*, 1978, **43**, 315.
[47] G. Reissenwerber and J. Sauer, *Tetrahedron Letters*, 1977, 4389.
[48] H. D. Fuhlhuber and J. Sauer, *Tetrahedron Letters*, 1977, 4393.
[49] J. A. Moore, W. J. Freeman, R. C. Gearhart, and H. B. Yokelson, *J. Org. Chem.*, 1978, **43**, 787.

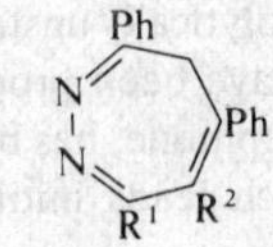

(41) 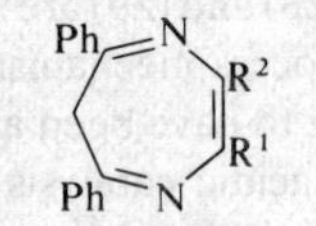          (42) 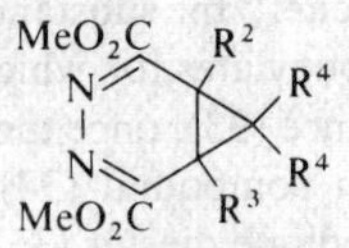          (43)

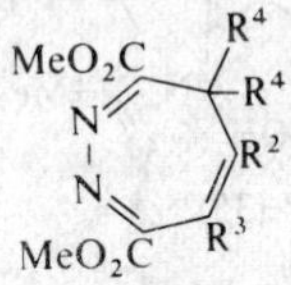

(44) 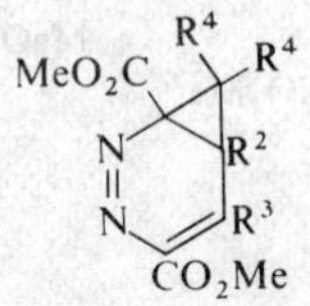          (45) 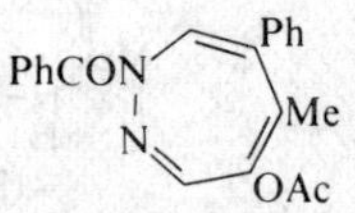          (46) 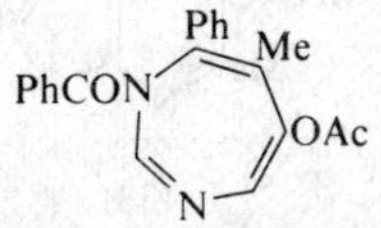          (47)

A variety of 1,5-benzodiazepin-2-ones (48) were obtained from the reaction of phenylketen dimer with certain *o*-phenylenediamines.[50]

Interest in 1,4-diazepines continues. For example, new syntheses have been devised for 'Medazepol' (49),[51] triazolobenzodiazepines, *e.g.* (50),[52] thienodiazepinones, *e.g.* (51),[53] and also imidazo-2,4-benzodiazepines (52).[54]

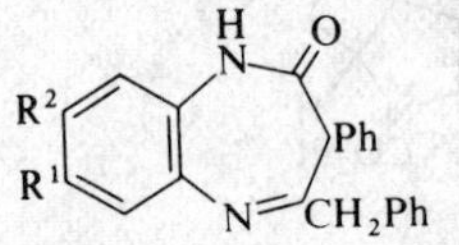

(48) 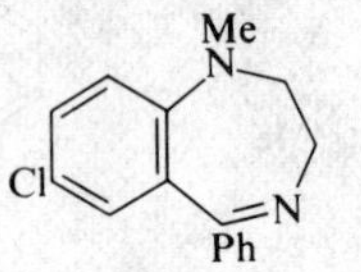          (49) 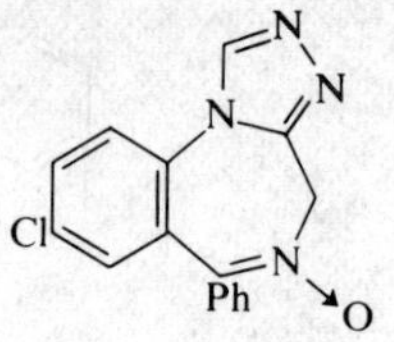          (50)

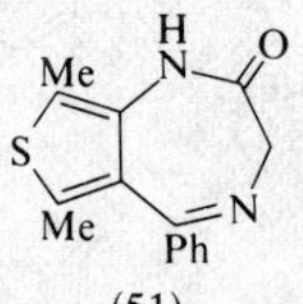

(51) 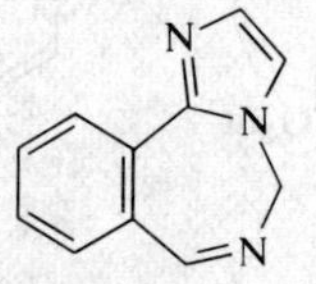          (52)

## 7 Miscellaneous

The keten (53) has been converted into the heptafulvalene analogues (54; R = H, Me, Ph, or CO$_2$Me), shown to be 'polyolefinic' by comparison of their physical and chemical properties with those of heptafulvalene.[55] On the other hand, cyclohepta[*c,d*]phenalen-6-one (55) is observed to exhibit some aromatic properties in the ground state, since a significant downfield shift in the positions of all

[50] Z. Subovics, G. Feher, and L. Toldy, *Acta Chim. Acad. Sci. Hung.*, 1977, **92**, 293.
[51] M. Mihalik, V. Sunijic, F. Kajfez, and M. Zink, *J. Heterocyclic Chem.*, 1977, **14**, 941.
[52] A. Walser and G. Zenchoff, *J. Heterocyclic Chem.*, 1978, **15**, 161.
[53] K. Grohe and H. Heitler, *Annalen*, 1977, 1947.
[54] R. I. Fryer, and J. V. Earley, *J. Heterocyclic Chem.*, 1977, **14**, 1435.
[55] H. Kato, Y. Kitahara, N. Morita, and T. Asao, *Chem. Letters*, 1977, 873.

vinyl protons in the $^1H$ n.m.r. spectrum occurs in (55) compared with the dihydro-derivative (56).[56]

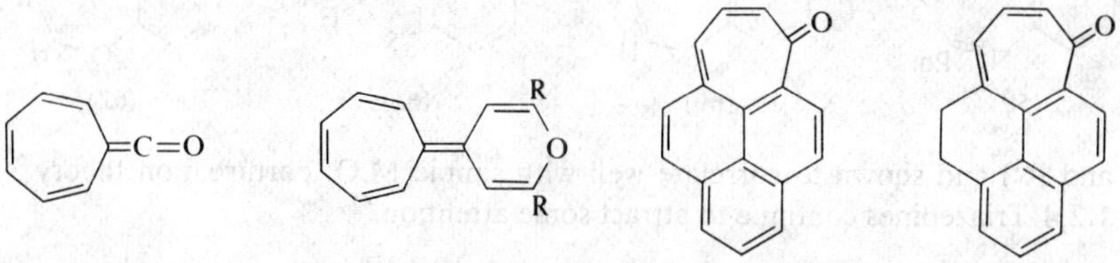

(53)         (54)         (55)         (56)

Pleiadiene-7,8-dione (57) has been synthesised[57] and found to exhibit dipolar character. From the $^{13}C$ and $^1H$ n.m.r. spectra it was deduced that 8-phenyl- and 8,8-diphenyl-heptafulvene[58] are not dipolar in character, and the spin couplings found in other non-benzenoid aromatics were interpreted.[59] Steric crowding in the 'quinarene' (58) is responsible for a greater than usual separation of charge and increase in single-bond character of the inter-ring bonds.[60]

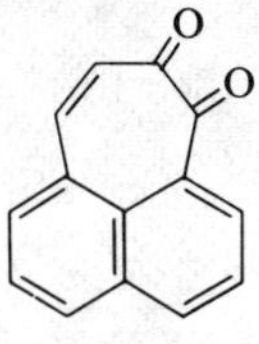

(57) 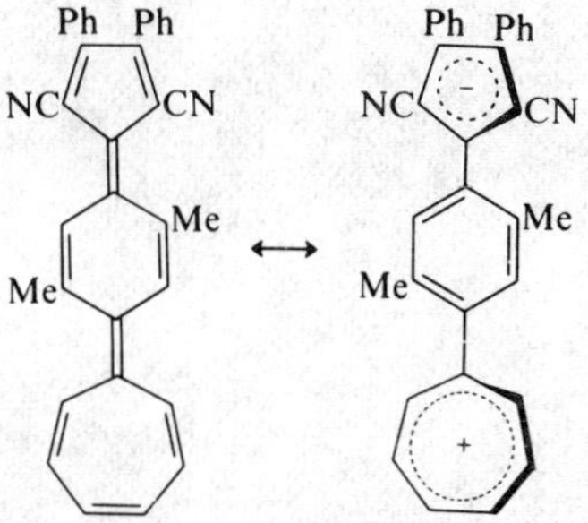                        (58)

Tropone *N*-arylimines (8-azaheptafulvenes) have been further studied and shown to undergo cycloaddition with dimethyl acetylenedicarboxylate and phenylsulphene ($PhCH=SO_2$).[61,62]

Photolysis of certain quinoline *N*-oxides gives rise to benzoxazepines, *e.g.* (59), whose further photolytic and other reactions have been studied;[63] in aprotic solvents, indoles may be obtained.[64] Photolysis of benzothiepins (60; H = CN, H, OAc, or OMe) caused ring-contraction as the primary reaction [to (61)[65]], while in similar reactions on 1-benzoxepins (62) the tendency for ring-contraction has been related to the $\pi$-donor and -acceptor properties of various substituents ($R^2$

[56] K. Yamamoto, Y. Kayane, and I. Murata, *Bull. Chem. Soc. Japan*, 1977, **50**, 1964.
[57] J. Tsunetsugu, M. Sato, M. Kanda, M. Takahashi, and S. Ebine, *Chem. Letters*, 1977, 885.
[58] K. Komatsu, M. Fujimori, and K. Okamoto, *Tetrahedron*, 1977, **33**, 2791.
[59] S. Braun and J. Kinkeldei, *Tetrahedron*, 1977, **33**, 3127.
[60] K. Takahashi, T. Sakae, and K. Takase, *Chem. Letters*, 1978, 237.
[61] K. Sanechíka, S. Kajigaeshi, and S. Kanemasa, *Chem. Letters*, 1977, 861.
[62] T. Iwasaki, S. Kajigaeshi, and S. Kanemasa, *Bull. Chem. Soc. Japan*, 1978, **51**, 229.
[63] M. Somei, R. Kitamura, H. Fujii, K. Hashiba, S. Kawai, and C. Kaneko, *J.C.S. Chem. Comm.*, 1977, 899.
[64] R. Kitamura, H. Fujii, K. Hashiba, N. Somei, and C. Kaneko, *Tetrahedron Letters*, 1977, 2911.
[65] H. Hofmann and H. Gaube, *Annalen*, 1977, 1874.

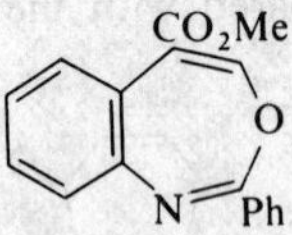

and $R^1$) and shown to correlate well with simple M.O. perturbation theory.[66] 1,2,4-Triazepines continue to attract some attention.[67]

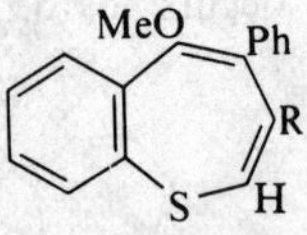

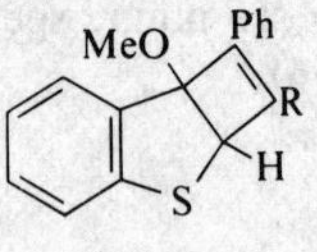

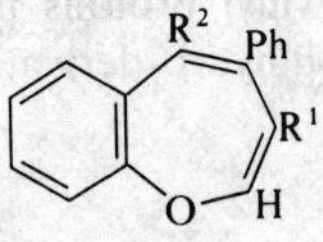

[66] H. Hofmann and P. Hofmann, *Annalen*, 1977, 1597.
[67] A. Hasnaoui, J. P. Lavergne, and P. Viallefont, *J. Heterocyclic Chem.*, 1978, **15**, 71.

# 6
# Medium-sized Rings and Macrocycles

BY O. METH-COHN

## 1 Reviews

The chemistry of annulenes in general has been covered in the second series of International Reviews of Science,[1] while the special features of homoaromatic systems of all kinds are the subject of a fascinating review by Paquette.[2] A recent book deals at length with the chemistry of cyclo-octatetraene.[2a]

## 2 General Topics

There seems no shortage of contributions which reaffirm that planar annulenes of the $(4n + 2)\pi$-electron type are aromatic while the $4n$ series are anti-aromatic.[3] The major area of disagreement seems to be over the magnitude of this property. Thus Hess and Schaad's continued studies[4] of resonance energy per $\pi$-electron of the annulenes containing 12, 14, 16, and 18 $\pi$-electrons correlate excellently with the rate of formation of these annulenes by Diels–Alder methods (as reported by Sondheimer last year[5]), the $(4n + 2)$ series being formed rapidly relative to the $4n$ group. However, Herndon[6] concludes that the resonance energies of the same series of annulenes confer only slight aromaticity or anti-aromaticity, certainly far less than that arrived at by studies of heats of combustion. He utilized the relationship $\Delta G^* = 27.8 - 0.55(\Delta RE)$ (where $\Delta RE$ is the difference in resonance energy between the $\pi$-systems of Sondheimer's reactant and adduct) to arrive at the appropriate resonance energies.

Another approach to this problem employs graph theory, whereby topological resonance energies (considered to be conceptually advantageous over Dewar resonance-energy data) are derived for annulenes and their ions.[7] Even more easily grasped is the importance of symmetrical Kekulé structures in determining the degree of aromaticity of annulenes, as measured by n.m.r. spectroscopy. Thus

[1] P. Skrabal, *Internat. Rev. Sci., Org. Chem., Series Two*, 1976, **3**, 329.
[2] L. A. Paquette, *Angew. Chem. Internat. Edn.*, 1978, **17**, 106.
[2a] G. I. Fray and R. G. Saxton, 'The Chemistry of Cyclo-octatetraene and its Derivatives', Cambridge University Press, 1978.
[3] M. Cocordano, *Compt. rend.*, 1977, **284**, *C*, 441.
[4] B. A. Hess and L. J. Schaad, *J.C.S. Chem. Comm.*, 1977, 243.
[5] See Volume 6 of these Reports, p. 154.
[6] W. C. Herndon, *J.C.S. Chem. Comm.*, 1977, 817.
[7] I. Gutman, M. Milun, and N. Trinajstic, *Croat. Chem. Acta*, 1977, **49**, 441 (*Chem. Abs.*, 1978, **87**, 151 588).

the dibenzopyrene (1) (on which a second benzo-ring is fused at '*a*') is highly aromatic compared to the monobenzo-analogue (1), or the even less aromatic '*b*'-fused dibenzopyrene.[8]

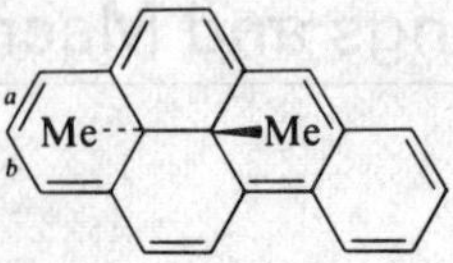

(1)

### 3 Eight-membered Systems

**Homocyclic Systems.**—The photo-initiated cycloaddition of benzenes and alkynes continues to be a route of choice for cyclo-octatetraenes (cot's). Thus while hexafluorobenzene and various phenylacetylenes photo-add, giving the cot's (3) and (4) by thermal reorganisation of the isolable bicyclic intermediate (2),[9]

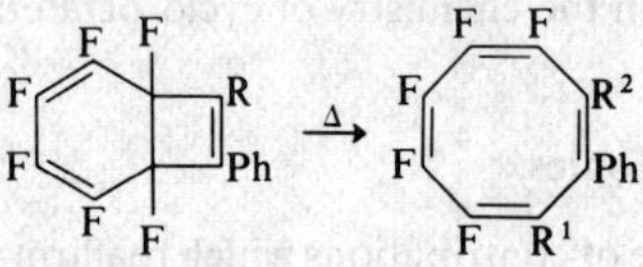

(2)

(3) $R^1 = F$
(4) $R^2 = F$

propyne adds (without isolation of an intermediate) to give an analogous methyl hexafluorocyclo-octatetraene of uncertain assignment.[10] But-2-yne, however, gives products (5)—(7) in mobile thermal and photoequilibrium (Scheme 1).[10] The

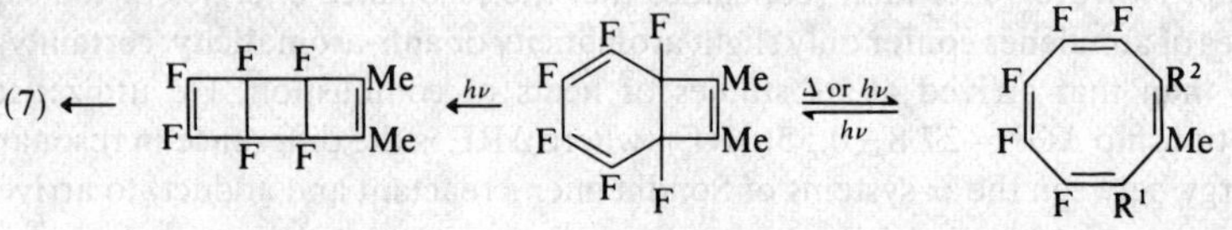

(6)                                          (5)                          (7) $R^1 = F$, $R^2 = Me$
                                                                              *or* $R^1 = Me$, $R^2 = F$

**Scheme 1**

reaction of benzene and methyl phenylpropiolate is again believed to give firstly a bicyclic addition product [*cf.* (2) or (5)] which undergoes further (2 + 2) cycloaddition, giving the prismane (8), as shown in Scheme 2. In the presence of a triplet sensitiser the cycloaddition proceeds directly to the cot (9), also derived by similar irradiation of the prismane (8).[11] Specific syntheses of cot derivatives are still under active scrutiny, and a very mild and efficient bromination–dehy-

[8] R. H. Mitchell, R. J. Caruthers, and L. Mazuch, *J. Amer. Chem. Soc.*, 1978, **100**, 1007.
[9] B. Sket and M. Zupan, *J. Amer. Chem. Soc.*, 1977, **99**, 3504.
[10] D. Bryce-Smith, A. Gilbert, B. H. Orger, and P. J. Twitchett, *J.C.S. Perkin I*, 1978, 232.
[11] A. H. A. Tinnemans and D. C. Neckers, *Tetrahedron Letters*, 1978, 1713.

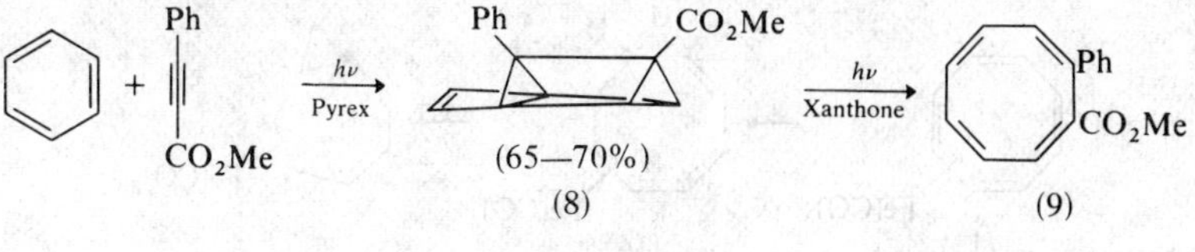

**Scheme 2**

drobromination (using LiF, $Li_2CO_3$, HMPA, and powdered glass!) gives access to the tri- and tetra-methylcyclo-octatetraenes (11).[12] The potentially diatropic cyclo-octatriene-1,4-dione is still an unstudied curiosity, and yet another likely precursor, the acetal (12) (in which the 2,3-bond could be *trans*) proved an

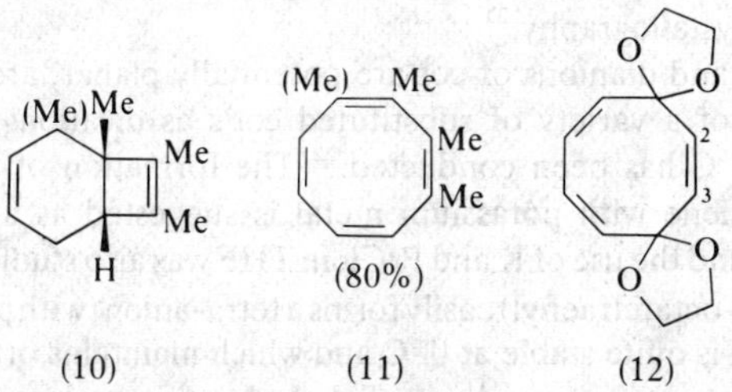

(10)          (11)          (12)

unsuitable precursor owing to its sensitivity to acetal-cleaving reagents.[13] The elusive cyclo-octatriene-1,2-dione, as well as its 3-bromo derivative (which evaded isolation last year[14]), has now been pinned down by careful use of the same type of method (dehydrobromination) as had previously proved ineffective. Both compounds exist in equilibrium with a bicyclic dione, and while the bromo derivative rapidly gives benzocyclobutenedione by loss of HBr at $-20\,°C$, the parent gives the same product together with cyclo-octa-4,6-diene-1,2-dione by disproportionation at ambient temperature.[15]

The fluxional behaviour of $[(\eta^6\text{-cot})Cr(CO)_3]$ has been shown[16] to proceed primarily by way of 1,3-shifts and to a lesser extent by 1,2-shifts, rather than by random shifts as had been suggested earlier.[17] An ingenious modified Forsen–Hoffman spin-saturation method was employed to analyse the exchange kinetics of the 4-signal $^{13}C$ n.m.r. spectrum.[16] The 'triple-decker' titanium complex $[Ti_2(cot)_3]$ has been shown to be paramagnetic, with two unpaired electrons. Two-electron reduction occurs readily to give a stable green dianion.[18] The organic chemistry of uranocene has been studied, including the preparation of alkoxy, amino, and aminoalkyl derivatives, and their interaction with alkyl-lithium, whereby alkyl-uranocenes were formed by way of an yne intermediate.[19] The course of protonation of cot–iron tricarbonyl complex (Scheme 3) has been

[12] P. F. King and L. A. Paquette, *Synthesis*, 1977, 279.
[13] P. A. Chaloner, A. B. Holmes, M. A. McKervey, and R. A. Raphael, *J.C.S. Perkin I*, 1977, 2524.
[14] See Volume 6 of these Reports, p. 157.
[15] M. Oda, S. Miyakoshi, and Y. Kitahara, *Chem. Letters*, 1977, 293.
[16] B. E. Mann, *J.C.S. Chem. Comm.*, 1977, 626.
[17] F. A. Cotton, D. L. Hunter, and P. Lahuerta, *J. Amer. Chem. Soc.*, 1974, **96**, 4723, 7926.
[18] S. P. Kolesnikov, J. G. Dobson, and P. S. Skell, *J. Amer. Chem. Soc.*, 1978, **100**, 999.
[19] C. A. Harman, D. P. Bauer, and S. R. Berryhill, *Inorg. Chem.*, 1977, **16**, 2143.

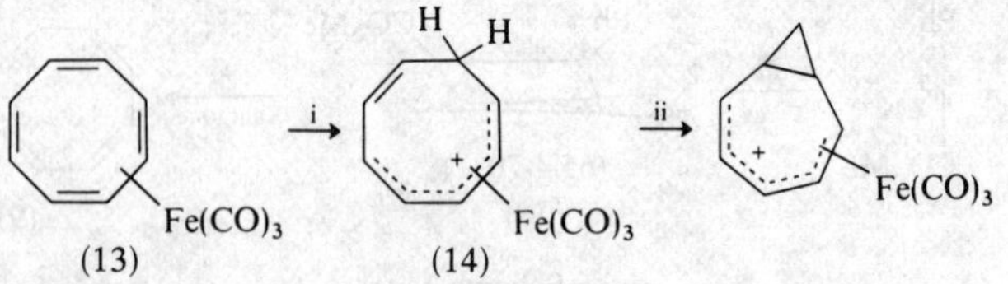

(13) → i → (14) → ii → Fe(CO)₃ structure

Reagents: i, FSO₃H, FSO₂Cl, at −120 °C; ii, warm to −60 °C.

**Scheme 3**

confirmed by low-temperature $^{13}$C n.m.r. studies in 'magic acid'.[20] The product from the action of acetyl chloride on the complex (13) has now been shown to give (14) by X-ray crystallography.[21]

The dications and dianions of cot are potentially planar, aromatic systems; a thorough study of a variety of substituted cot's as dications in $SbF_5$–$SO_2ClF$ solution at −78 °C has been conducted.[22] The formation of cot dianion from cyclo-octa-1,5-diene with potassium metal is suggested as a good laboratory method (30%), and the use of K and $PrCl_3$ in THF was also studied, being a milder route.[23] Bi(cyclo-octatetraenyl) easily forms a tetra-anion (with potassium in liquid ammonia) which is quite stable at 0 °C and which maintains orthogonal rings, as indicated by an n.m.r. study. A parallel study of its polarographic reduction reveals firstly a two-electron reduction wave followed by two one-electron reductions.[24]

Cot is generally a poor diene for cycloaddition purposes, and doubt has even been expressed as to its behaviour as a normal diene in cycloadditions. By making comparisons with cyclo-octa-1,3-diene (cod), recent work confirms the $_\pi 4_s + _\pi 2_s$ reaction of cot with various 1,3,4-triazole-2,5-diones (tad), since negligible solvent effects were noted and the cycloadditions were 100—1100 times slower than with cod. Apparently the cycloadditions require both cot and cod to become almost planar, as mirrored in their free energies of activation (9 kcal mol⁻¹ for cod and 13.7 kcal mol⁻¹ for cot at −40 and −10 °C respectively).[25] The same dienophile reacts readily with (cot)irontricarbonyl (13) to give the cyclo-adducts (15)

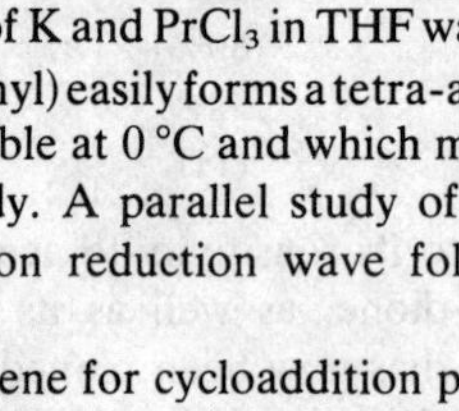

(15)

²⁰ G. Olah, G. Liang, and S. Yu, *J. Org. Chem.*, 1977, **42**, 4262.
²¹ A. D. Charles, P. Diversi, B. F. G. Johnson, K. D. Karlin, J. Lewis, A. V. Rivera, and G. M. Sheldrick, *J. Organometallic Chem.*, 1977, **128**, C31.
²² G. A. Olah, J. S. Staral, G. Liang, L. A. Paquette, W. P. Melega, and M. J. Carmody, *J. Amer. Chem. Soc.*, 1977, **99**, 3349.
²³ W. J. Evans, A. L. Wayda, C.-W. Wang, and W. M. Cwirla, *J. Amer. Chem. Soc.*, 1978, **100**, 333.
²⁴ L. A. Paquette, C. D. Ewing, and S. G. Traynor, *J. Amer. Chem. Soc.*, 1976, **98**, 279.
²⁵ H. Isaksen and J. P. Snyder, *Tetrahedron Letters*, 1977, 889.

and (16) in about a 2 : 1 ratio.[26] The first adduct has been converted into the interesting heterocycle (17), which, not surprisingly, happily evolves nitrogen at

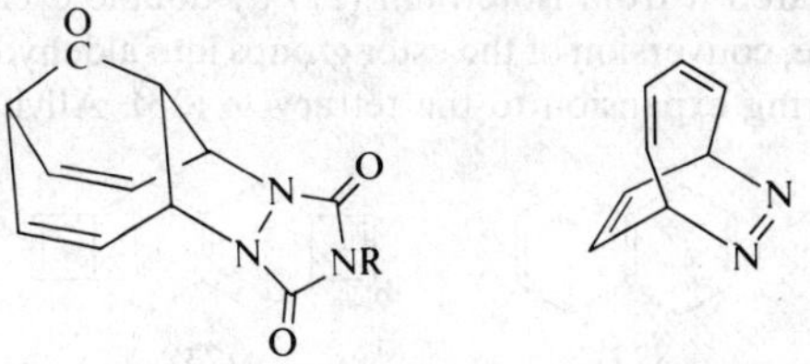

(16)                                            (17)

50 °C to give cot quantitatively.[27] Other interesting $(2+2+2)$ cycloadditions are noted (Scheme 4) for the 1,4-bis-methylene derivatives (18) and (19),[28] the latter being the newly reported parent system of derivatives described last year.[29]

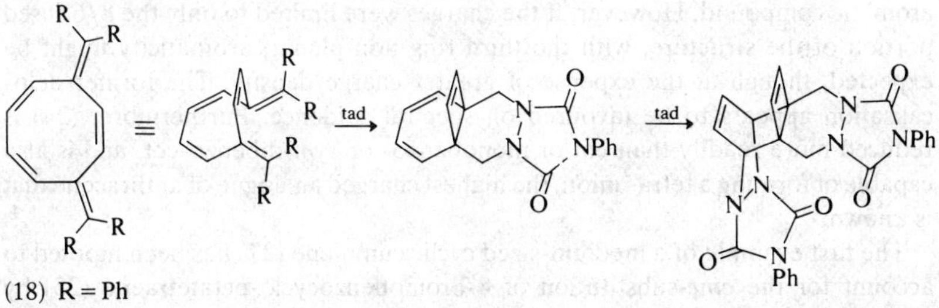

(18) R = Ph
(19) R = H

**Scheme 4**

When two potentially anti-aromatic systems are fused together, an aromatic periphery to the new product is possible. In general, a weak stabilization is indeed noted. New examples this year include the 8/4 fused system (20), which does little to clarify the situation since M.O. calculations and its n.m.r. spectrum are inconclusive, while its u.v. spectrum suggests delocalisation. The compound gives red, thermally stable but air-sensitive derivatives, and reacts with TCNE by a $(4+2)$ cycloaddition at the starred positions.[30] It would appear that the delocalisation energy is virtually balanced by the strain energy required for planarity. Octalene (21) has been another target molecule this year. Although cot-1,2-

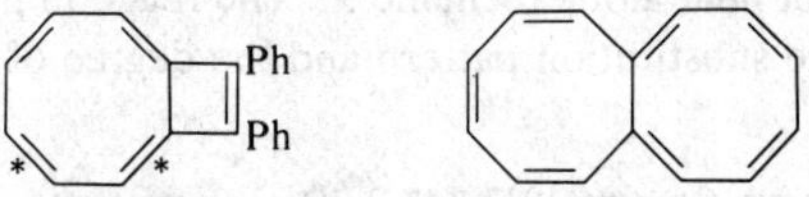

(20)                                            (21)

[26] H. Olsen, *Acta. Chem. Scand (B)*, 1977, **31**, 635.
[27] H. Olsen and J. P. Snyder, *J. Amer. Chem. Soc.*, 1978, **100**, 285.
[28] M. Oda, N. Fukazawa, and Y. Kitahara, *Tetrehedron Letters*, 1977, 3277.
[29] See Volume 6 of these Reports, p. 155.
[30] M. Oda, H. Oikawa, N. Fukazawa, and Y. Kitahara, *Tetrahedron Letters*, 1977, 4409.

dialdehyde (prepared by an improved synthesis from the corresponding diester) proved abortive in an attempted double Wittig condensation,[31] Vogel's group[32] successfully prepared it from isotetralin (22) by double cyclopropanation with ethyl diazoacetate, conversion of the ester groups into aldehyde tosylhydrazones, and subsequent ring expansion to the tetracycle (23). Allylic bromination and

(22)                                    (23)

dehybrobomination sequences led smoothly to octalene (21). Dynamic [13]C n.m.r. spectroscopy was used to unravel the complex conformational inversions and isomerisations.[33] Another interesting candidate for scrutiny was the potentially $20\pi$ dianion of the bis-cyclo-octatetraenobenzene (24). The charge could be totally delocalised were the dianion planar, but would thereby create an anti-aromatic compound. However, if the charges were limited to only the 8/6 fused portion of the structure, with the third ring non-planar, aromaticity might be expected, though at the expense of greater charge density. The former delocalisation appears to be favoured on spectral evidence. Furthermore, (24) is reduced more readily than cot or monobenzo- or *sym*-dibenzo-cot, and is also capable of forming a tetra-anion, the highest charged analogue of anthracene that is known.[34]

The first example of a medium-sized cyclic cumulene (27) has been mooted to account for the *cine*-substitution of 4-bromobenzocyclo-octatetraene (25) by potassium t-butoxide, giving the 5-t-butoxy derivative (26) in good yield. The intermediate was neither isolable nor trappable with other reagents.[35]

(24)                        (25) X = 4-Br                        (27)
                           (26) X = 5-OBu$^t$

Molybdenum hexacarbonyl has been added to the synthetic arsenal of reagents for the synthesis of specifically alkylated cots. Thus, for example, the propellane (28) isomerises at lower temperature and to a different product with this catalyst than by the action of heat alone (Scheme 5). The reaction pathway is, however, sensitive both to the substitution pattern and the degree of unsaturation of the starting material.[36]

[31] H. Alper and R. A. Partis, *Gazzetta*, 1977, **107**, 201.
[32] E. Vogel, H. Runzheimer, H. Volker, F. Hogrefe, B. Baasner, and J. Lex, *Angew. Chem. Internat. Edn.*, 1977, **16**, 871.
[33] J. F. M. Oth, K. Muellen, H. Runzheimer, H. Volker, P. Mues, and E. Vogel, *Angew. Chem. Internat. Edn.*, 1977, **16**, 872.
[34] L. A. Paquette, G. D. Ewing, S. Traynor, and J. M. Gardlik, *J. Amer. Chem. Soc.*, 1977, **99**, 6115.
[35] H. N. C. Wong, T.-L. Chan, and F. Sondheimer, *Tetrahedron Letters*, 1978, 667.
[36] L. A. Paquette, J. M. Photis, and R. P. Michael, *J. Amer. Chem. Soc.*, 1977, **99**, 7899.

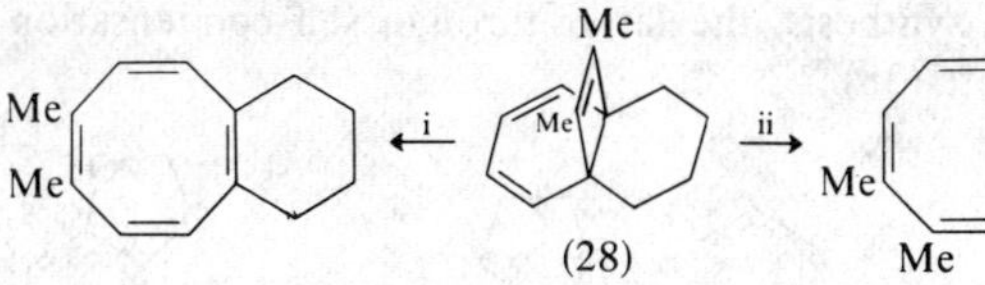

(28)

Reagents: i, Δ, ii, Mo(CO)₆, C₆H₆.

**Scheme 5**

The complex contortions of the benzo-homocyclo-octatetraene (29), which undergoes ready valence isomerism on irradiation, have been unravelled and compared with those of several analogues.[37] Another molecular acrobat is the readily made compound (30). It rearranges thermally to the isomer (31) (in which

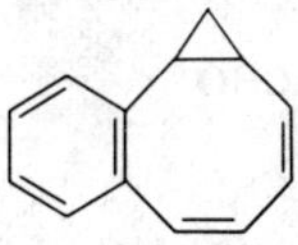

(29) 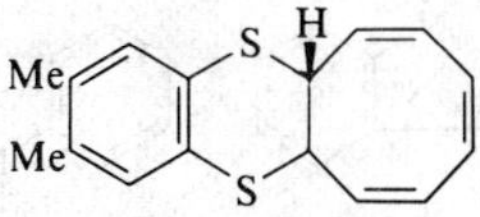                    (30) 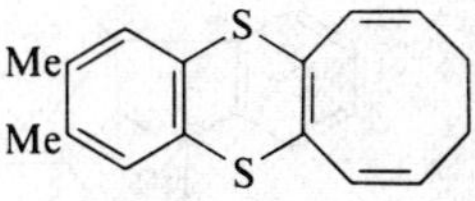                         (31)

the eight-membered ring exists in equilibrium with its 6/4-ring tautomer – a useful diene), easily dehydrogenates with DCDQ, and yields the interesting ring-expanded isomer (32) with strong base.[38] The dication of cot is known as a transient intermediate of difficult access, despite its potential aromaticity. The fused analogue (33) is reported as being easily prepared and stable at ambient temperature, having a $14\pi$-diatropic periphery.[39]

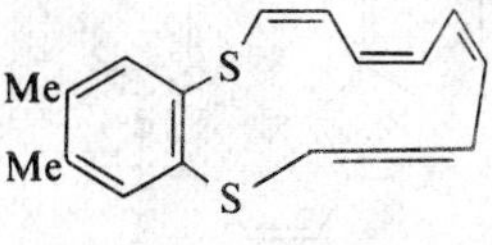

(32) 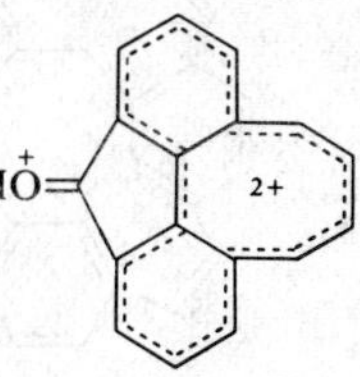                              (33)

Ring-expansions by use of acetylene dicarboxylates (ADC) in cycloadditions are well known and nicely examplified in the conversion of the enamine (34) into the dibenzo-cot (35). A fascinating application of this reaction to the synthesis of the anti-leukaemic lignan steganacin (36) has been published.[40] The marine natural product caulerpin, previously reported as a dinaphthopyrazine derivative, has now been shown to be the bis-indolo-cyclo-octatetraene (37), both by

[37] M. Kato, T. Chikamoto, and T. Miwa, *Bull. Chem. Soc. Japan*, 1977, **50**, 1082.
[38] L. Mögel, W. Schroth, and B. Werner, *J.C.S. Chem. Comm.*, 1978, 57.
[39] I. Willner, A. L. Gutman, and M. Rabinowitz, *J. Amer. Chem. Soc.*, 1977, **99**, 4167.
[40] D. Becker, L. R. Hughes, and R. A. Raphael, *J.C.S. Perkin I*, 1977, 1674; see also D. J. Haywood and S. T. Reid, *ibid.*, p. 2457.

degradation and by synthesis, the latter through self-condensation of the 3-formylindole-2-acetate (38).[41]

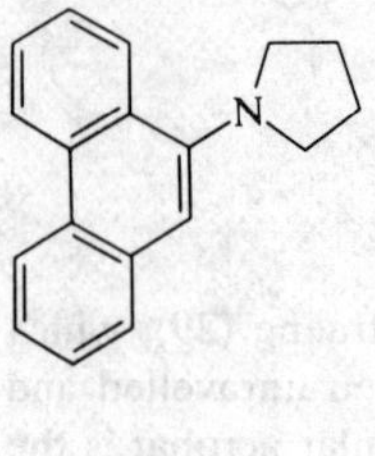

(34) 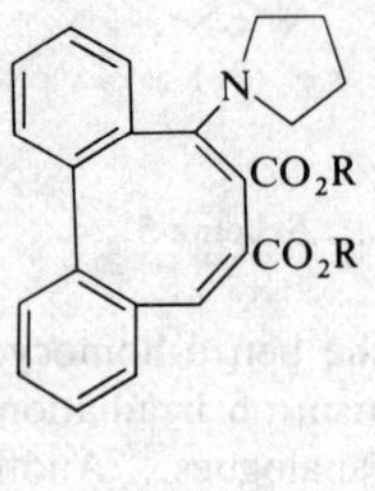                          (35) 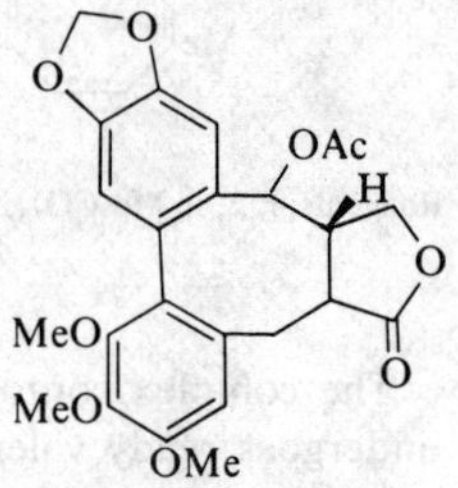                          (36)

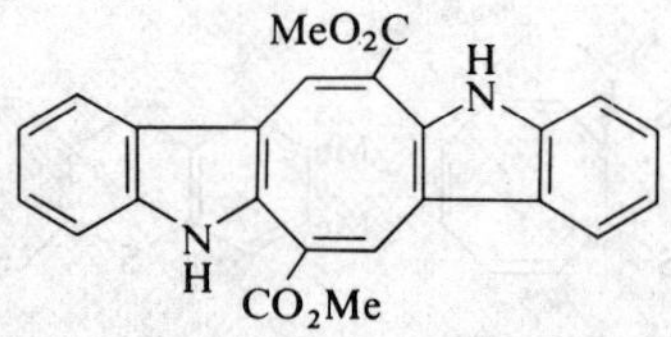

(37) 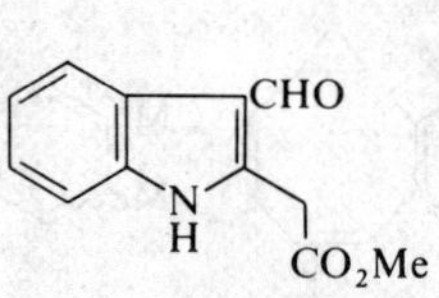                                      (38)

Highly fused derivatives of cot studied this year include the as yet unknown [8]circulene (39), which is predicted to exist in a conformationally mobile 'saddle' structure,[42] and a series of tetraphenylenes (40) containing one, two, three, or four further bridging groups (A—D are, e.g., $CH_2$ or CO). These were prepared from 2,2'-dilithiobiphenyls, and their chemistry has been reported.[43]

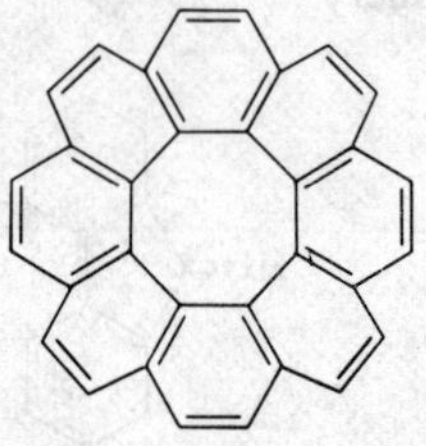

(39) 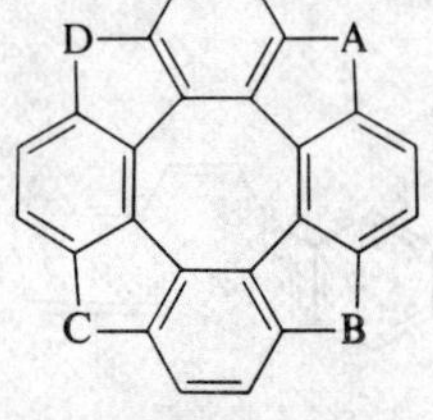                                      (40)

**Heterocyclic Systems.**—New systems this year include the pharmaceutically useful higher analogues (41) of the diazepine super-drugs,[44] the unusual dioxa-diaza-heterocycle (42), derived by irradiation of pyridazine dioxide,[45] and the benzodithiocine (43), which exists as its 6/4 valence isomer on photolysis, revertible to (43) on heating.[46]

[41] B. C. Maiti, R. H. Thomson, and M. Mahendran, J. Chem. Research (S), 1978, 126; J. Chem. Research (M), 1978, 1683; and B. C. Maiti, and R. H. Thomson, NATO Conf. Ser. (Ser.) 4, 1977, 1 (Marine Nat. Prod. Chem.), p. 159.
[42] T. Liljefors and O. Wennerström, Tetrahedron, 1977, 33, 2999.
[43] D. Hellwinkel, G. Reiff, and V. Nykodym, Annalen, 1977, 1013.
[44] DDSA Pharmaceuticals Limited, Netherlands P. 75 07 092 (Chem. Abs., 1977, 87, 85 067).
[45] H. Arai, A. Ohsawa, K. Saiki, and H. Igeta, J.C.S. Chem. Comm., 1977, 133.
[46] W. Schroth and L. Mögel, Z. Chem., 1977, 17, 441.

(41)    (42)    (43)

## 4 Nine-membered Systems

Cyclononatetraene anions are potentially $10\pi$ aromatic systems. Whereas potassium acetylcyclononatetraenide shows some aromatic character, the corresponding lithium salt does not, but it appears that the charge is localised at the acetyl group. With trimethylsilyl chloride, the anion is trapped as its valence isomer (44).[47] Significantly, in this respect, lithium cyclononatetraenide has been shown to give 1-azidoindene with tosyl azide,[48] and not, as long believed, diazocyclononatetraene.[49] The same lithium salt reacts with bromo- or acetoxy-tropylium borofluoride at $-60\,°C$ to give (45), which rapidly undergoes electrocyclization at $20\,°C$ to yield the useful monoheptafulvalene precursor (46).[50] Dibenzononalene (47) has been prepared from 2,2',6,6'-tetraformylbiphenyl, and it readily yields a diatropic dianion with butyl-lithium.[51]

(44)    (45) X = Br or Ac    (46)    (47)

Epoxidation of the valence isomer of N-ethoxycarbonyl-azonine (48) leads to the epoxide (49) in 55% yield; this gives a labile isomer, probably (50), that is readily transformed into difficultly accessible 1-pyrindenes (Scheme 6).[52] Cyclo-addition of $\alpha$-pyrone to the mono-*trans*-benzazonine (51) proceeds in two steps, giving the regiospecific adduct (52) at $80\,°C$ and the product (53) of further internal cycloaddition after loss of $CO_2$ at $140\,°C$.[53]

## 5 Ten-membered Systems

1,6-Bridged [10]annulenes (54) are known to undergo chemical attack from the opposite face to the bridging group. This is further reinforced by the formation of mono- and di-cyclo-adducts from $4+2$ interaction with 1,3,4-triazole-2,5-

[47] G. Boche and F. Heidenhain, *Angew. Chem. Internat. Edn.*, 1978, **17**, 283.
[48] E. E. Waali, J. L. Taylor, and N. T. Allison, *Tetrahedron Letters*, 1977, 3873.
[49] D. Lloyd and N. W. Preston, *Chem. and Ind.*, 1976, 1039.
[50] M. Neuenschwander and A. Frey, *Chimia (Switz.)*, 1977, **31**, 333.
[51] I. Willner and M. Rabinowitz, *J. Amer. Chem. Soc.*, 1977, **99**, 4507.
[52] A. G. Anastassiou, S. J. Girgenti, R. C. Griffiths, and E. Reichmanis, *J. Org. Chem.*, 1977, **42**, 2651.
[53] A. G. Anastassiou and R. Badri, *Tetrahedron Letters*, 1977, 4465.

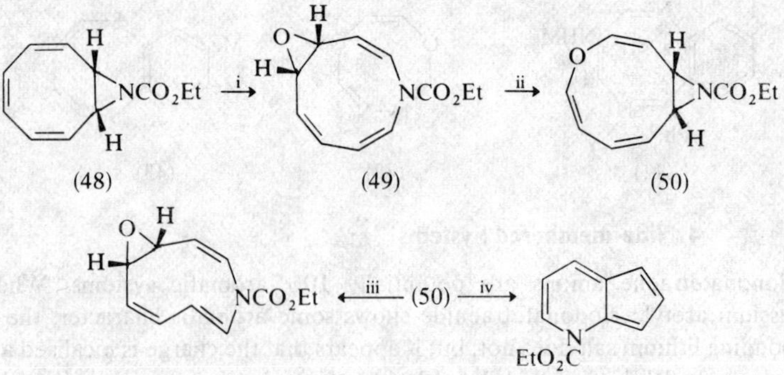

Reagents: i, MCPBA, 0 °C; ii, *hν*, Me₂CO, at −10 °C; iii, 20 °C; iv, Al₂O₃, at −15 °C.

**Scheme 6**

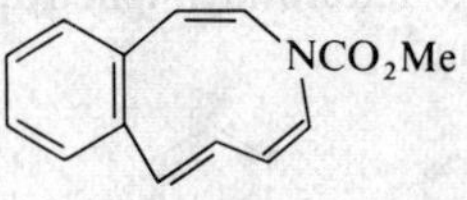

diones.[54] Furthermore, while *p*-dinitrobenzene forms both 1:1 and 2:1 complexes with *p*-bis-dimethylamino-benzene, the 2,7-dinitro-derivative of the [10]annulene (54; X = CH₂) only forms 1:1 complexes, only one face being available.[55] In this light, the chemistry of the first reported annulenophanes (55; Ar = *p*-phenylene or 1,5-naphthylene) should be interesting. They were prepared by established methods from the bis-mercaptomethyl-annulene and the bis-bromomethyl-arene, and showed clear evidence of π,π-interactions both in their u.v. and n.m.r. spectra.[56] The unstable orange 'azulenologue' (56) of the bridged [10]annulene (54) has also been reported, and, in accord with its proposed structure, shows only seven resonances in its ¹³C n.m.r. spectrum, a high-field methylene resonance and 'aromatic protons' in its ¹H n.m.r. spectrum, and a u.v. spectrum related to that of azulene.[57]

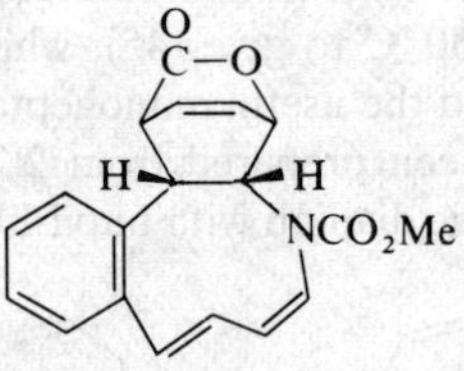

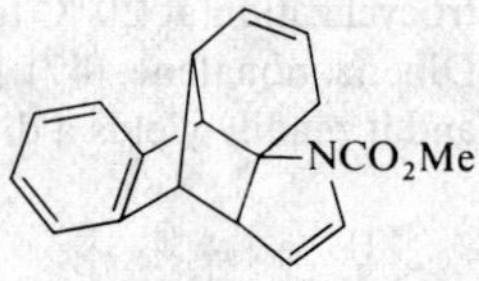

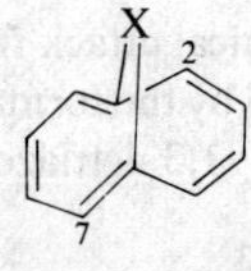

[54] P. Ashkenazi, D. Ginsburg, and E. Vogel, *Tetrahedron*, 1977, **33**, 1169.
[55] J. A. Chudek, R. Foster, and E. Vogel, *J.C.S. Perkin II*, 1977, 994.
[56] M. Matsumoto, T. Otsubo, Y. Sakata, and S. Misumi, *Tetrahedron Letters*, 1977, 4425.
[57] S. Masamune and D. W. Brooks, *Tetrahedron Letters*, 1977, 3239.

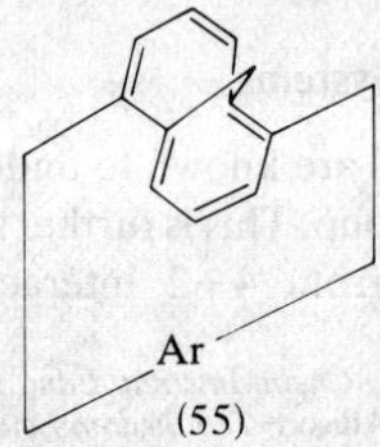

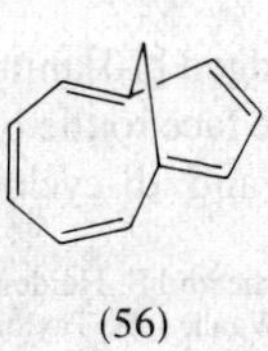

## 6 Twelve-, Thirteen-, and Fourteen-membered Systems

Tribenzo[12]annulene (57) undergoes photo-valence-isomerization to (58) by way of its *cis-trans-trans*-isomer. Further action of heat or light on (58) gives naphthalene and phenanthrene.[58] Introduction of an extra 13-Me group into the weakly paratropic 5,10-dimethyl derivative of [13]annulenone (59) causes the

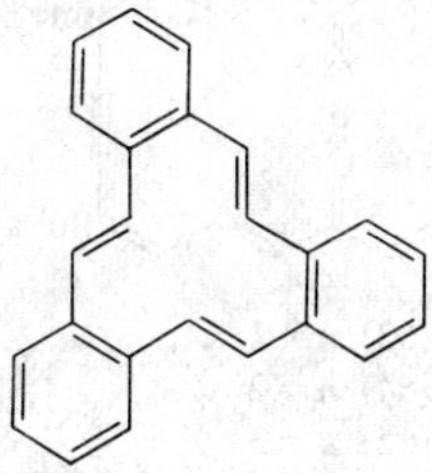

(57) 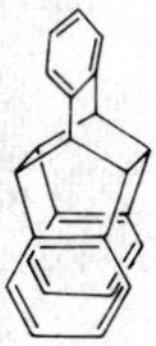     (58) 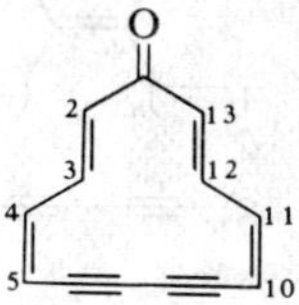     (59)

2,3-*trans* double bond to become conformationally unstable, preferring the inverted configuration with H-2 inwards and H-3 outwards.[59] Similar paratropicity (increased in acid media) is also reported for the 4,5-monobenzo-derivative of (59), which is decreased on further 10,11-benzo-fusion,[60] in line with the well-documented effect of annelation on tropicity.[61]

A tetrabenzo[13]annulenyl anion (60) has also been shown to reveal reduced diatropicity due to benzo-fusion, steric congestion being limited because three of the annular hydrogens are 'inside' the ring.[62,63]

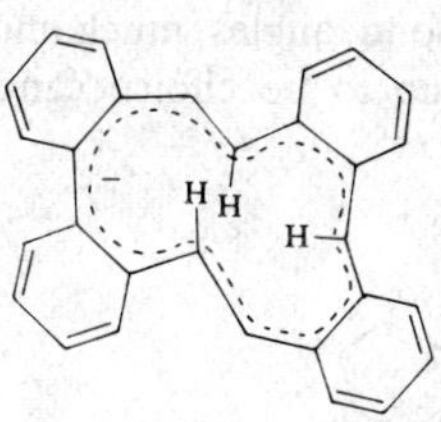

(60) 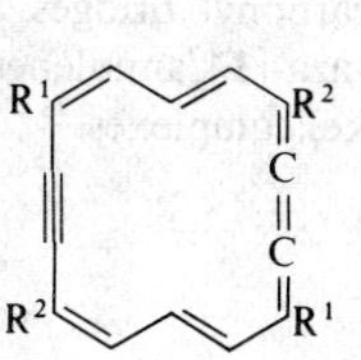     (61) $R^1$ and $R^2$ = Bu$^t$ or Ph

Further examples of the strongly diatropic 1,8-bis-dehydro[14]annulenes (61) have been synthesized,[64] as have a series of benzo-bis-dehydro-annulenes (62) with $14\pi$-, $16\pi$-, $18\pi$-, and $20\pi$-systems, starting from phthalaldehyde.[65] The higher vinylogues of naphthalene continue to be of interest, the [14][14] system (63), reported last year,[56] having been scrutinised by $^{13}$C n.m.r. spectroscopy

[58] M. W. Tausch, M. Elian, A. Bucur, and E. Cioranescu, *Chem. Ber.*, 1977, **110**, 1744.
[59] T. M. Cresp, J. Ojima, and F. Sondheimer, *J. Org. Chem.*, 1977, **42**, 2130.
[60] J. Ojima, Y. Yokoyama, and M. Enkaku, *Bull. Chem. Soc. Japan*, 1977, **50**, 1522.
[61] See Volume 6 of these Reports, pp. 161 and 164.
[62] I. Willner, A. Gamliel, and M. Rabinowitz, *Chem. Letters*, 1977, 1273.
[63] A. Gamliel, I. Willner, and M. Rabinowitz, *Synthesis*, 1977, 410.
[64] K. Fukui, T. Nomoto, S. Nakatsuji, S. Akiyama, and M. Nakagawa, *Bull. Chem. Soc. Japan*, 1977, **50**, 2758; *cf.* Vol. 6 of these Reports, p. 161.
[65] N. Darby, T. M. Cresp, and F. Sondheimer, *J. Org. Chem.*, 1977, **42**, 1960.
[66] See Volume 6 of these Reports, p. 162.

(using relaxation data to help assign resonances)[67] and by X-ray crystallography, which seems to confirm the suspicion that both rings are separately delocalised. The 'north' and 'south' sides of the system show bond-length alternation, while the central cumulene bonds reveal alkyne-type bond lengths, the outer bonds being longer than the inner.[68] Two other stable analogues [(64)[69] and (65)[70]] have

|     | n | m |
| --- | - | - |
| (62) a | 1 | 1 |
| b | 2 | 1 |
| c | 2 | 2 |
| d | 3 | 2 |

(63)  n = 1
(64)  n = 2
(65)  n = 3

been similarly prepared, being deep violet and deep green-blue respectively, all the available data supporting the view that independent ring currents are induced in each ring, unlike the case of naphthalene. The whole series, including an analogous stable deep green [18][18] analogue,[71] show predictably related u.v. spectra and the usual high-field inner-proton resonances and low-field outer-proton resonances in their n.m.r. spectra. The effect increases towards the central common bond, as ring-current effects of the adjacent ring become significant. The X-ray data on the doubly bridged [14]annulene (66) indicate that the strain caused by the carbonyl bridges affects bond angles much more than bond lengths.[72] Tetra-aza-[14]annulenes continue to be choice candidates for the formation of nickel complexes.[73]

(66)

## 7 Fifteen-, Sixteen-, and Seventeen-membered Systems

While the [15]annulenone (67; R = H or Me) is non-planar, fusion of a ring, *e.g.* as in (67; R—R = CO–O–CO or CO–O–CH₂), causes the carbonyl group to turn inwards and creates a planar, strongly diatropic system. Protonation similarly

[67] H. Nakanishi, S. Akiyama, and M. Nakagawa, *Chem. Letters*, 1977, 1515.
[68] Y. Kai, N. Yasuoka, N. Kasai, S. Akiyama, and M. Nakagawa, *Tetrehedron Letters*, 1978, 1703.
[69] S. Nakatsuji, S. Akiyama, and M. Nakagawa, *Tetrahedron Letters*, 1977, 3723.
[70] S. Nakatsuji, S. Akiyama, and M. Nakagawa, *Tetrahedron Letters*, 1978, 483.
[71] M. Osuka, Y. Yoshikawa, S. Akiyama, and M. Nakagawa, *Tetrahedron Letters*, 1977, 3719.
[72] R. Destro and M. Simonetta, *Acta Cryst.*, 1977, **B33**, 3219.
[73] G. P. Ferrara and J. C. Dabrowiak, *Inorg. Nuclear Chem. Letters*, 1978, **14**, 31; E. Lorch and E. Breitmaier, *Chem.-Ztg.*, 1977, **101**, 262.

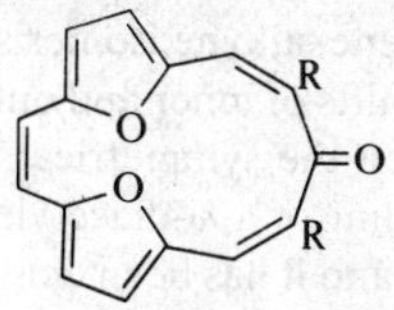

(67)

yields a stable $14\pi$ cationic system.[74] The paratropic [16]annulene (68) shows very similar properties to a similar tris-dehydro-analogue reported earlier.[75]

A series of [17]annulenes (70)—(73) are available from cyclo-octatetraene dimer by the action of a carbene,[76] an epoxidant,[77] or a nitrene[78] followed by

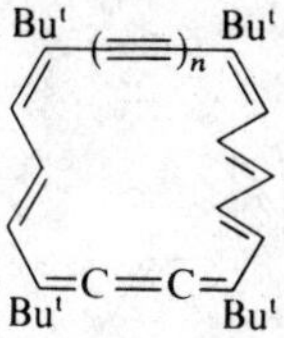

(68) $n = 1$
(69) $n = 2$

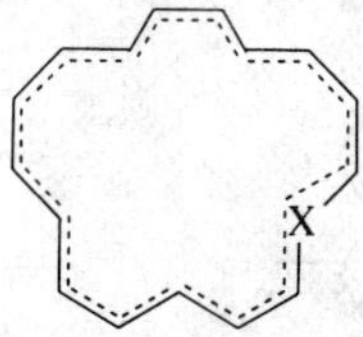

(70) $X = CH^- M^+$
(71) $X = O$
(72) $X = NCO_2Et$
(73) $X = NH$

low-temperature photo-induced ring expansion, using the now well-known approach of Schröder and Oth. Other new [17]annulenones include the mono- and di-benzo-derivatives (74), which show the usual decrease in paratropicity with increased annelation,[79] and the related paratropic furan-based derivative (75).[80]

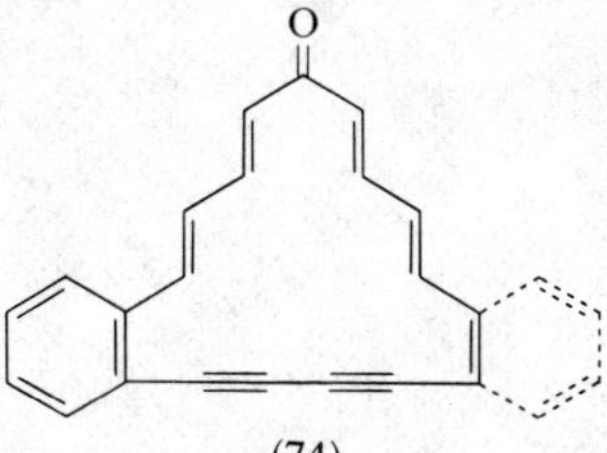

(74)

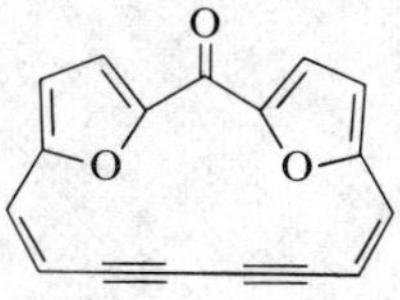

(75)

## 8 Eighteen-membered and Larger Systems

An unsymmetrical isomer (69) of the tris-dehydro-[18]annulene reported last year[81] shows a temperature-dependent n.m.r. spectrum (+125 to −80 °C),

[74] H. Ogawa, H. Kato, and Y. Taniguchi, *Chem. and Pharm. Bull. (Japan)*, 1977, **25**, 511.
[75] Y. Yoshikawa, S. Nakatsuji, F. Iwatani, S. Akiyama, and M. Nakagawa, *Tetrahedron Letters*, 1977, 1737.
[76] P. Hildenbrand, G. Plinke, J. F. M. Oth, and G. Schröder, *Chem. Ber.*, 1978, **111**, 107.
[77] G. Schröder, G. Plinke, and J. F. M. Oth, *Chem. Ber.*, 1978, **111**, 99.
[78] H. Rötele, G. Heil, and G. Schröder, *Chem. Ber.*, 1978, **111**, 84.
[79] J. Ojima, M. Ishiyama, and A. Kimura, *Bull. Chem. Soc. Japan*, 1977, **50**, 1584.
[80] H. Ogawa, H. Kato, and Y. Taniguchi, *Chem. and Pharm. Bull. (Japan)*, 1977, **25**, 517.
[81] See Volume 6 of these Reports, p. 163.

suggesting that there is cumulene–alkyne isomerism in $1:2$ ratio. Using $\Delta\tau$ (the difference between chemical shifts of inner and outer protons) as an index of ring current, the difference between the symmetrical and unsymmetrical isomers is insignificant.[82] Oxidative coupling of $\alpha,\omega$-diacetylenes has long been a method of choice for annulene synthesis, and it has been employed to synthesise the atropic systems (76)[83] and (77)[84], coupling occurring at the arrowed bond. However, the higher analogues (78) and (79) were not available by this method.[84] By treating aromatic dialdehydes with similar bis-Wittig reagents, a series of cyclophanes (80) has been prepared.[85]

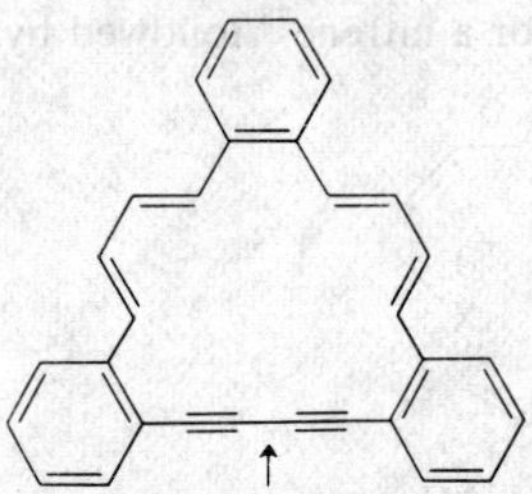

|  |  |  |
|---|---|---|
| (76) | (77) $m = n = 3$ | (80) Ar and Ar′ are |
|  | (78) $m = 3, n = 4$ | $p$-$C_6H_4$, |
|  | (79) $m = n = 4$ | 2,5-furyl, |
|  |  | or 2,5-thienyl |

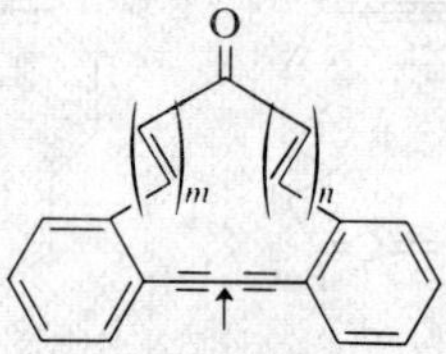

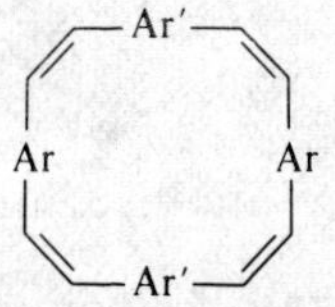

[82] M. Osuka, S. Akiyama, and M. Nakagawa, *Tetrahedron Letters*, 1977, 1649.
[83] J. Ojima, M. Enkaku, and C. Uwai, *Bull. Chem. Soc. Japan*, 1977, **50**, 933.
[84] J. Ojima, M. Enkaku, and M. Ishiyama, *J.C.S. Perkin I*, 1977, 1548.
[85] B. Thulin, O. Wennerström, I. Somfai, and B. Chmielarz, *Acta Chem. Scand.* (*B*), 1977, **31**, 135; H.-E. Högberg, B. Thulin, and O. Wennerström, *Tetrahedron Letters*, 1977, 931; A. Strand, B. Thulin, and O. Wennerström, *Acta Chem. Scand.* (*B*), 1977, **31**, 521.

# 7
# Electrophilic Substitution

BY D. J. CHADWICK

## 1 Introduction

The organization of this Chapter is similar to that adopted in preceding years, though the emphasis will necessarily have changed somewhat. Electrophilic substitutions in homocyclic and heterocyclic systems are discussed together under the appropriate sub-headings.

Several reviews have been published on electrophilic homocyclic and hetero-cyclic aromatic substitution.[1] Other reviews and books of relevance include polychloro-aromatic compounds,[2] annulenes and related compounds,[3] cyclo-butadiene–metal complexes,[4] substitution *via* heteroaromatic *N*-oxide re-arrangements,[5] $\pi$-excessiveness in heteroaromatic compounds,[6] and special topics in heterocyclic chemistry.[7] The book by Jones and Bean is a mine of information on pyrrole chemistry.[8]

**Theoretical Considerations.**—Several papers have appeared on the application of 1-arylethyl ester pyrolyses to the study of electrophilic aromatic reactivities. By this means, accurate $\sigma^+$ values have been determined for *para*-cyclohexyl and t-butyl substituents. The effect of *meta*-t-butyl substitution in the same reaction and in the protiodetritiation in TFA at 70 °C has been measured. The results indicate that the Baker–Nathan order in solvolysis reactions arises from steric hindrance to solvation of the transition state and that C–C is more important than C–H hyperconjugation.[9] This approach has also been used to study non-additivity of methyl[10] and chlorine[11] substituent effects.

The application of an extended Hammett equation based on the separation of enthalpy and entropy effects appears to provide a very good empirical description of substituent effects for systems in which a large deviation from the simple

[1] B. V. Smith, *Org. Reaction Mechanisms 1975*, (publ. 1977) p. 275; A. N. Kost and V. A. Budylin, *Zhur. Vses. Khim. O-va*, 1977, **22**, 315; T. Shimura, *Tokoshi Nyusu, Kagaku Kogyo Shiryo*, 1975, **10**, 67 (*Chem. Abs.*, 1978, **88**, 73 738).
[2] 'Polychloroaromatic Compounds', ed. H. Suschitzky, Plenum, London, 1974.
[3] P. Skrabal, *Internat. Rev. Sci.: Org. Chem.*, Ser. Two, 1976, **3**, 229.
[4] A. Efraty, *Chem. Rev.*, 1977, **77**, 691.
[5] R. A. Abramovitch and I. Shinkai, *Accounts Chem. Res.*, 1976, **9**, 192.
[6] A. F. Pozharskii, *Khim. geterotsikl. Soedinenii*, 1977, 723 (*Chem. Abs.*, 1977, **87**, 167 913).
[7] 'The Chemistry of Heterocyclic Compounds, Vol. 3: Special Topics in Heterocyclic Chemistry', ed. A. Weissberger and E. C. Taylor, Wiley, New York, 1977.
[8] R. A. Jones and G. P. Bean, 'The Chemistry of Pyrroles', Academic Press, London, 1977.
[9] E. Glyde and R. Taylor, *J.C.S. Perkin II*, 1977, 678.
[10] E. Glyde and R. Taylor, *J.C.S. Perkin II*, 1977, 1537.
[11] E. Glyde and R. Taylor, *J.C.S. Perkin II*, 1977, 1541.

Hammett equation is expected.[12] A new model system (1) has been proposed in order to define intrinsic polar substituent effects by $^{19}F$ and $^{13}C$ n.m.r. The chemical shift parameters from the alkyl-substituted compounds (1; $R^1$ = alkyl,

(1)

$R^2$ = H or 3- or 4-F) are claimed to indicate unambiguously that the polar inductive effect of alkyl groups attached to an $sp^3$-hybridized carbon is zero. Substituent-induced structural changes were identified as a definite problem in the use of 4-substituted quinuclidines as model systems for defining a scale of polar effects.[13] It has been argued that the rates of quaternization of a range of *meta*- and *para*-substituted *NN*-dimethylanilines provide evidence for involvement of *d*-orbitals in resonance in the *para* Cl, Br, and I compounds.[14] Because the effects of polysubstitution in benzene are not additive, and definitions for additivity for H-exchange and nitration are different, any theoretical treatment of the susceptibility of heteroaromatic compounds to electrophilic attack must only be made with explicit consideration of the electrophile.[15] Previous work has given the order of reactivity to electrophilic substitution: pyrrole > furan > tellurophen > selenophen > thiophen > benzene; solvolysis rates for the 1-aryl-ethyl acetates of benzene, thiophen, furan, and *N*-methylpyrrole provided the same reactivity order, and it was assumed that the two classes of reactions might be used interchangeably for evaluation of $\sigma^+$ values. New data for solvolyses including the selenophen and tellurophen derivatives show, however, that the two reactions do not give identical indications as to the relative reactivities of furan and tellurophen. The two reactions also show differing sensitivities to 2-methyl substitution*: for electrophilic reactions the order of sensitivity is furan > benzene > tellurophen > selenophen ≃ thiophen ≫ pyrrole    whereas    in    the solvolysis the order is furan > pyrrole > thiophen > benzene > selenophen > tellurophen: the reasons for these orders are obscure.[16]

The energies of various benzene–$H^+$ and benzene–$Li^+$ complexes relative to face structures have been calculated by the CNDO/2-FK method. The structure of lowest energy for the Li system is the hexagonal $\pi$-complex whereas this structure is a high-energy maximum for the H case.[17] A linear relation between energies of $\sigma$- and $\pi$-complex structures, calculated by CNDO/2 and MINDO/2 methods, has been demonstrated for protonated benzene, *o*- and *p*-xylene, toluene, fluorobenzene, and pyrrole.[18] The effect of $BF_3$ in facilitating hydrogen exchange between benzene and HF has been modelled by a STO-3G SCF

---

* This part of the paper is garbled in *Chemical Abstracts*.

[12] T. M. Krygowski and W. R. Fawcett, *J.C.S. Perkin II*, 1977, 2033.
[13] W. Adcock and T.-C. Khor, *J. Org. Chem.*, 1978, **43**, 1272.
[14] V. Baliah and V. M. Kanagasabapathy, *Indian J. Chem., Sect. B*, 1978, **16**, 64.
[15] A. R. Katritzky, *Cron. Chim.*, 1977, **53**, 32 (*Chem. Abs.*, 1978, **88**, 189 535).
[16] S. Clementi, F. Fringuelli, P. Linda, G. Marino, G. Savelli, A. Taticchi, and J. L. Piette, *Gazzetta*, 1977, **107**, 339.
[17] D. Heidrich and D. Deininger, *Tetrahedron Letters*, 1977, 3751.
[18] D. Heidrich and M. Grimmer, *Tetrahedron Letters*, 1977, 3565.

description of the reaction co-ordinate in the presence and absence of BF$_3$. The catalyst reduces the height of the reaction barrier by more than half.[19] The effect of solvation by TFA in the electrophilic substitution of toluene has been studied by calculation (INDO) of the relative energies of solvated *ortho-*, *meta*, and *para*-protonated toluenes.[20] STO-3G *ab initio* and electrostatic calculations of the energy changes associated with *meta-* and *para*-protonation of Ph(CH$_2$)$_n$NH$_3^+$ ($n = 0$—2) indicate that the deactivating and *meta*-directing properties of positively charged substituents are primarily due to field effects of the positive poles. The calculations also support $\pi$-polarization by positive poles. Calculated and experimental partial rate factors give parallel trends, though the calculations over-estimate substituent effects of positive poles and under-estimate the conjugative effect of NH$_3^+$ in PhNH$_3^+$.[21]

*Ab initio* calculations of protonation energies for the various sites in guanine and adenine show that protonation at N-7 is favoured in the former.[22] A CNDO/2 and MINDO/2 theoretical study of tautomerism in 1,2,4-triazole and its amino- and diamino-derivatives indicates that unsymmetrical triazole rings are favoured over symmetrical and exocyclic amino tautomers over imino forms.[23]

## 2 Reactions with Electrophilic Hydrogen

**Exchange Reactions.**—The protonation of toluene and ethylbenzene in super-acidic media [*e.g.* HF–SbF$_5$ (1:1)/SO$_2$FCl and HF–TaF$_5$ (30:1)] has been re-investigated by $^{13}$C n.m.r. spectroscopy. In both substrates, and both acids, *para*-protonation predominates, in agreement with earlier work.[24] The pro-tiodetritiation of variously tritiated 1,2-diphenylethanes and 9,10-dihydro-phenanthrenes has been investigated in a study of the effect of strain on aromatic reactivity. The overall reactivity of the phenanthrene relative to fluorene is compatible with differences in planarity. The ratio of the reactivities of the positions $\alpha$ and $\beta$ to the central ring of the former is higher than in the latter, and confirms that the low reactivity of the $\alpha$-positions of fluorene arises from an increase in strain on going to the transition state for $\alpha$-substitution.[25] Partial rate factors have been determined for the H/D exchange of *o-*, *m-*, and *p*-fluoro-toluenes with DBr.[26] In an analogous study of benzyl chloride and the 2-, 3-, 4-, and 3,4-di-methyl derivatives, the CH$_2$Cl group was found to activate the *para*-position slightly and to deactivate *ortho-* and *meta*-positions.[27]

The problem of kinetic solvent isotope effects in H$_2$O–D$_2$O mixtures on systems for which proton transfer from H$_3$O$^+$ and a subsequent reaction step are each partly rate-limiting has been considered, and an expression relating the

[19] G. Alagona, E. Scrocco, E. Silla, and J. Tomasi, *Theor. Chim. Acta*, 1977, **45**, 127.

[20] J. C. Rayez and J. J. Dannenberg, *Tetrahedron Letters*, 1977, 671.

[21] W. F. Reynolds, T. A. Modro, and P. G. Mezey, *J.C.S. Perkin II*, 1977, 1066.

[22] A. Pullman and A. M. Armbruster, *Compt. rend.* 1977, **284**, *D*, 231.

[23] V. V. Makarskii, V. A. Zubkov, V. A. Lopyrev, and M. G. Voronkov, *Khim. geterotsikl. Soedinenii*, 1977, 540 (*Chem. Abs.*, 1977, **87**, 67 696).

[24] D. Fărcaşiu, M. T. Melchior, and L. Craine, *Angew. Chem. Internat. Edn.*, 1977, **16**, 315.

[25] H. V. Ansell and R. Taylor, *J.C.S. Perkin II*, 1977, 866.

[26] P. P. Alikhanov, T. S. Amamchan, T. G. Bogatskaya, O. N. Guve, V. R. Kalinachenko, G. V. Motsarev, and L. M. Yakimenko, *Zhur. org. Khim.*, 1977, **13**, 565 (*Chem. Abs.*, 1977, **87**, 21 869).

[27] P. P. Alikhanov, V. R. Kalinachenko, T. S. Amamchan, V. R. Rozenberg, G. V. Motsarev, and L. M. Yakimenko, *Zhur. org. Khim.*, 1977, **13**, 758 (*Chem. Abs.*, 1977, **87**, 38 420).

isotope effect to the atom fraction of deuterium in the solvent has been derived for a symmetrical isotope-exchange process. This has been applied to data on the detritiation of trimethoxybenzene.[28] H/T exchange rates and partial rate factors have been measured for anthracene, coronene, and triphenylene in anhydrous trifluoroacetic acid. The 1- and 2-positions in anthracene are more reactive, but closer in reactivity, than the corresponding positions in naphthalene, as predicted by Hückel and CNDO/2 calculations. The reactivities of coronene and triphenylene are also correctly predicted.[29] Perdeuteriobenzo[a]pyrene has been prepared by heating the protio-compound with the product from reaction between $C_6D_6$, $Br_2$, and $AlBr_3$ at 100 °C for 2 h.[30] H/D exchange in triferrocenylphosphine and the phosphine oxide has been studied.[31]

The attack of $CH_5^+$ and $C_2H_5^+$ (from the $\gamma$-radiolysis of methane) on halogeno- and dihalogeno-benzenes leads to dehalogenation *via* two distinct pathways: protiodehalogenation is initiated by attack on the aromatic ring and methyldehalogenation by attack on the lone pairs of the halogen. The relative rates of these two processes depend primarily on the nature of the halogen.[32]

In 89.8% $H_2SO_4$, diacetylmesitylene undergoes two successive protodeacetylations and a final sulphonation to give mesitylenesulphonic acid. A kinetic analysis of the component reactions has been reported as an example of consecutive, irreversible, first-order reactions in which each of the three steps is much slower than the subsequent one.[33]

In the heterocyclic field, the basicities of 180 pyridines, pyrimidines, pyridazines, and pyrazines have been correlated with substituent effects, using the Hammett and Taft equations.[34] Kinetic data have been measured for H/D exchange in 3-hydroxypyridine at 161 °C in the pD range 2—11. H-2 exchanges first: H-4 more slowly, and only in the anionic form.[35] The monoprotonation of methyl- and amino-pyrimidines has been studied by [1]H and [13]C n.m.r., and the percentages of forms protonated at N-1 or N-3 have been determined from chemical shifts in TFA, DMSO, and water. For methyl-pyrimidines, a higher proportion (*ca.* 71%) of the form in which the protonated nitrogen is in the position *para* to the methyl group is observed. In 4-amino-6-methyl-pyrimidines, the influence of the amino-group is greater than that of the methyl, 94% protonation occurring at the nitrogen *para* to the amino-group.[36] The site of protonation in the conversion of thiamine into vitamin B₁ has been shown to be the pyrimidine N-1 by [15]N n.m.r. studies.[37] H/D exchange in pyrido[2,3-d]pyridazine (2) under acidic, basic, and neutral conditions has been examined. In

[28] A. J. Kresge, Y. Chiang, G. W. Koeppl, and R. A. More O'Ferrall, *J. Amer. Chem. Soc.*, 1977, **99**, 2245.

[29] H. V. Ansell, M. M. Hirschler, and R. Taylor, *J.C.S. Perkin II*, 1977, 353.

[30] J. C. Seibles, D. M. Bollinger, and M. Orchin, *Angew. Chem. Internat. Edn.*, 1977, **16**, 656.

[31] A. N. Nesmeyanov, D. N. Kursanov, V. N. Setkina, N. K. Baranetskaya, V. D. Vil'chevskaya, V. I. Losilkina, G. A. Panasyan, and A. I. Krylova, *Izvest. Akad. Nauk S.S.S.R., Ser. khim.*, 1977, 2263 (*Chem. Abs.*, 1978, **88**, 36 848).

[32] M. Speranza and F. Cacace, *J. Amer. Chem. Soc.*, 1977, **99**, 3051.

[33] J. Farooqi and P. H. Gore, *Tetrahedron Letters*, 1977, 2983.

[34] P. Tomaski and R. Zalewski, *Chem. Zvesti*, 1977, **31**, 246.

[35] V. P. Lezina, A. U. Stepanyants, L. D. Smirnov, and M. I. Vinnik, *Izvest. Akad. Nauk S.S.S.R., Ser. khim.*, 1978, 317.

[36] J. Riand, M. T. Chenon, and N. Lumbroso-Bader, *J. Amer. Chem. Soc.*, 1977, **99**, 6838.

[37] A. H. Cain, G. R. Sullivan, and J. D. Roberts, *J. Amer. Chem. Soc.*, 1977, **99**, 6423.

neutral or acidic media, H-8 and H-5 (only) exchange, the former more rapidly; in base, these protons exchange very rapidly, H-4 and H-3 more slowly.[38] The p$K$ values of twelve 1,5-naphthyridine derivatives (3) have been correlated with the sum of the $\sigma$ constants of the substituents.[39] In a study of H/D exchange in bromopurines, 6-bromopurine in $D_2O$ gave 63% of the 9-deuterio-compound; 8-bromoadenine gave mono-, di-, and tri-deuterio-derivatives.[40] N.m.r. studies demonstrate that 1-methylimidazo[4,5-$b$]pyrazines (4; R = H or Me) protonate first at N-3 and subsequently at N-4.[41]

(2)          (3)          (4)

Further studies have been published on the acid-catalysed H/D exchange of pyrrole and alkyl-pyrroles. Alkyl substituents activate adjacent positions, the influence of $N$-alkyl being less than $C$-alkyl substitution. The relative positional reactivities agree with INDO calculations.[42] Kinetic and equilibrium studies on the protonation of *meso*-tetraphenylporphyrin in DMSO–water show the presence of an equilibrium between the free base and the diprotonated species: the monoprotonated species could not be detected. Temperature-jump studies show that the equilibration is slow compared with other proton transfers from nitrogen acids, in agreement with earlier n.m.r. work.[43] The first-order rate constants for H/D exchange in tryptamine and the 5-hydroxy, 6-hydroxy, and 5-methyl derivatives, measured by 220 MHz n.m.r. spectroscopy, increase exponentially with increasing negative acidity function, $-H_o$, and correlate with the stabilities of the resonance structures of the conjugate acids.[44] The kinetics of the acid-catalysed decarboxylation of indole-3-carboxylic acid and of the 2-methyl- and 5-chloro-derivatives are consistent with an $A$-$S_E2$ mechanism involving a zwitterionic intermediate. Solvent deuterium isotope effects, estimated for reaction *via* both the neutral substrate and the carboxylate anion, have been used to assess the primary isotope effects for proton transfer from solvent to substrate.[45] The protonation of tetrazole and some $N$-aryl derivatives has been studied by $^1H$ and $^{13}C$ n.m.r.[46] The influence of phase-transfer catalysts ($Bu_4N^+$ $Br^-$ and cetyltrimethylammonium bromide) on H/D exchange rates of thiazole and some alkyl derivatives has been explored: attempts to obtain linear free-energy relationships with $\sigma$-parameters were unsuccessful.[47]

[38] D. Marchand, A. Turck, G. Queguiñer, and P. Pastour, *Bull. Soc. chim. France*, 1977, 919.
[39] I. V. Persianova, Yu. N. Sheinker, R. M. Titkova, and A. S. Elina, *Khim. geterotsikl. Soedinenii*, 1977, 965 (*Chem. Abs.*, 1977, **87**, 151 568).
[40] M. Kiessling, H. Beerbaum, and K. H. Grupe, *Z. Chem.*, 1977, **17**, 141.
[41] G. G. Dvoryantseva, T. Ya. Filipenko, I. S. Musatova, A. S. Elina, P. V. Petrovskii, E. V. Arshavskaya, and E. I. Fedin, *Izvest. Akad. Nauk S.S.S.R.*, *Ser. khim.*, 1977, 1060 (*Chem. Abs.*, 1977, **87**, 84 303).
[42] G. P. Bean and T. J. Wilkinson, *J.C.S. Perkin II*, 1978, 72.
[43] F. Hibbert and K. P. P. Hunte, *J.C.S. Perkin II*, 1977, 1624.
[44] S. Kang, T. H. Witherup, and S. Gross, *J. Org. Chem.*, 1977, **42**, 3769.
[45] B. C. Challis and H. S. Rzepa, *J.C.S. Perkin II*, 1977, 281.
[46] A. Könnecke, E. Lippmann, and E. Kleinpeter, *Tetrahedron*, 1977, **33**, 1399.
[47] T. Higgins, W. J. Spillane, H. J. M. Dou, and J. Metzger, *Compt. rend.*, 1977, **284**, *C*, 929.

**Heteroaromatic Tautomerism.**—A very timely and useful review has appeared on the problems associated with the study of heteroaromatic tautomerism, with particular reference to the 2-hydroxypyridine–2-pyridone equilibrium. The important point is made that the energy difference between, and the strength of association of, protomeric isomers can change by several kilocalories with change of medium.[48] This point is further stressed in a study of the 4-pyridone–4-hydroxypyridine system, in which the self-association energy (*ca.* 6 kcal mol$^{-1}$) shifts the position of the equilibrium in favour of the pyridone in CHCl$_3$ and cyclohexane, thereby calling into question previous interpretations of equilibria determined on associated material.[49] The conversion of 6-methoxy-2-pyridone into the hydroxy tautomer proceeds *via* concerted proton transfer involving self-association or association with water. In anhydrous media, the tautomerization is limited by diffusion-controlled dimerisation.[50] Photoelectron spectroscopic studies demonstrate, contrary to the situation in solution, that 3- and 4-hydroxy- and -mercapto-pyridines exist as such in the gas phase, with less than 5% of the oxo- and thioxo-tautomers: in the 2-substituted analogues, a similar situation obtains, although more substantial amounts of oxo- and thioxo-forms are present at equilibrium.[51] Ion cyclotron resonance experiments confirm these observations.[52] The effect of substitution at the 2-position on the tautomerism of 5-methoxy-4-pyridinethiones in solution has been studied. Electron-withdrawing substituents reduce the predominance of the thione form.[53] In an attempt to assess the structural requirements for direct proton transfer in tautomeric equilibria, the mechanism of tautomer interconversion in substituted pyridines, pyrimidines, and imidazoles has been studied by the temperature-jump method in aqueous solution. When the tautomeric groups are remote, isomer interconversion occurs *via* intermediate ionisation and dissociation followed by ion recombination; when the groups are close, a direct proton-transfer mechanism not involving intermediate ionic dissociation contributes to the interconversion rate.[54]

Several papers have appeared on tautomerism in pyrimidines, including the parent molecule,[55] pyrimidyl-2- and -4-cyanoacetic esters,[56] and *N*-substituted 4-aminopyrazolo[3,4-*d*]pyrimidines.[57] The last-mentioned system (5) is in equilibrium with (6) in aqueous solution. Interconversion is catalysed by acid and base, and proceeds *via* either an intermediate cation that is common to both neutral tautomers or through the anion. Other studies of condensed systems include

[48] P. Beak, *Accounts Chem. Res.*, 1977, **10**, 186.
[49] P. Beak, J. B. Covington, and J. M. Zeigler, *J. Org. Chem.*, 1978, **43**, 177.
[50] O. Bensaude and J. E. Dubois, *Compt. rend.*, 1977, **285**, *C*, 503.
[51] M. J. Cook, S. El-Abbady, A. R. Katritzky, C. Guimon, and G. Pfister-Guillouzo, *J.C.S. Perkin II*, 1977, 1652.
[52] C. B. Theissling, N. M. M. Nibbering, M. J. Cook, S. El-Abbady, and A. R. Katritzky, *Tetrahedron Letters*, 1977, 1777.
[53] H. Besso, K. Imafuku, and H. Matsumura, *Bull. Chem. Soc. Japan*, 1977, **50**, 3295.
[54] O. Bensaude, M. Dreyfus, G. Dodin, and J. E. Dubois, *J. Amer. Chem. Soc.*, 1977, **99**, 4438.
[55] R. Stolarski, M. Remin, and D. Shugar, *Z. Naturforsch.*, 1977, **32c**, 894.
[56] V. V. Lapachev, O. A. Zagulyaeva, and V. P. Mamaev, *Izvest. Akad. Nauk S.S.S.R., Ser. khim.*, 1977, 2633 (*Chem. Abs.*, 1978, **88**, 50 789); *Doklady Akad. Nauk S.S.S.R.*, 1977, **236**, 113 (*Chem. Abs.*, 1977, **87**, 183 640).
[57] G. Dodin, M. Dreyfus, O. Bensaude, and J. E. Dubois, *J. Amer. Chem. Soc.*, 1977, **99**, 7257.

4,8-dioxo-1,5-naphthyridines,[58] 2*H*-pyrazolo[3,4-*d*]pyridazines,[59] and dihydro-imidazo- and -pyrimido-[1,2-*a*]pyrimidines.[60] Phthalazin-1(2*H*)-one exists predominantly in the lactam form (8) with minor contributions of hydroxy (7) and zwitterionic forms (9). By contrast, the mobile monocationic species shows a tendency to prefer hydroxy-structure (11): there is no evidence for structure (10).[61]

Substituent effects in the tautomeric equilibrium between 4-oxo-2-pyrrolines and 3-hydroxy-pyrroles have been studied.[62] Photometric p*K* measurements on compounds containing the pyrrolinone partial structure as well as corresponding imino-ester derivatives permit the deduction of the position of equilibrium between lactam and lactim tautomers in bile pigments, pyrrolinones, pyrro-methenones, and pyrromethenes. In all cases, the lactam form is preferred by several orders of magnitude over the lactim form.[63] Recent [1]H and [13]C n.m.r. studies show that the hydroxy-form of 4-acyl-5-pyrazolones is favoured by factors which stabilise an intramolecular H-bond between the acyl and hydroxy-groups.[64] In a [1]H, [13]C, and [15]N n.m.r. study of tautomerically mobile 3-methyl-1-phenylpyrazolin-5-one, only [15]N n.m.r. (in [[2]H$_6$]DMSO) shows slow exchange between NH and OH sites: all three tautomers are present.[65] I.r. studies of a range of substituted pyrazolidine-3,5-diones suggest that the diketo tautomer pre-dominates both in the solid state and in CHCl$_3$ solution, whereas earlier u.v. work indicates a preference for the keto-iminol tautomer in methanol and cyclo-hexane.[66] The factors determining tautomeric preferences in histamine and

[58] S. B. Brown and M. J. S. Dewar, *J. Org. Chem.*, 1978, **43**, 1331.
[59] A. S. Shawali, *J. Heterocyclic Chem.*, 1977, **14**, 375.
[60] J. Clark and M. Curphey, *J.C.S. Perkin I*, 1977, 1855.
[61] M. J. Cook, A. R. Katritzky, A. D. Page, and M. Ramaiah, *J.C.S. Perkin II*, 1977, 1184.
[62] T. Momose, T. Tanaka, and T. Yokota, *Heterocycles*, 1977, **6**, 1827.
[63] H. Falk, S. Gergely, K. Grubmayr, and O. Hofer, *Annalen*, 1977, 565
[64] L. N. Kurkovskaya, N. N. Shapet'ko, A. S. Vitvitskaya, and I. Ya. Kvitko, *Zhur. org. Khim.*, 1977, **13**, 1750 (*Chem. Abs.*, 1977, **87**, 183 616).
[65] G. E. Hawkes, E. W. Randall, J. Elguero, and C. J. Marzin, *J.C.S. Perkin II*, 1977, 1024.
[66] M. Woodruff and J. B. Polya, *Austral. J. Chem.*, 1977, **30**, 421.

histidine have been discussed and pH-dependent [13]C n.m.r. chemical shifts used to demonstrate a 4:1 preference for the N[τ] tautomer of the imidazole ring in histamine.[67] Tautomerism in sixteen halogeno-1,2,4-triazoles has been studied by dipole moments,[68] and in the parent compound by n.m.r. In the latter case, activation parameters for the 1,2-prototropic isomerisation have been determined.[69]

Proton n.m.r. and i.r. studies on 2-mercapto-5-methyl- and 2,5-dimethyl-3-mercapto-furan, -thiophen, and -selenophen indicate the presence of mainly the thiol tautomers.[70] Carbon-13 n.m.r. has been used to distinguish thione and selenone tautomers[71] in appropriately substituted 1,3,4-thiadiazoles and -selenadiazoles, and tetrazole derivatives. N.m.r., i.r., and u.v. studies indicate that 4-aminoisothiazole exists as the amino- and not the imino-tautomer.[72]

## 3 Metallation

An exhaustive review on di- and poly-alkali-metal derivatives of heterofunctionally substituted organic molecules contains many useful references on aromatic ring metallation, particularly of nitrogen- and sulphur-containing heteroaromatics.[73] Other reviews include a description of the use of chromium tricarbonyl–arene complexes in organic synthesis[74] and the preparation and use in synthesis of highly reactive metal powders.[75]

**Lithium and Magnesium Substitutions.**—The competition between ether and thioether groups in directing benzene metallation has been investigated in the reactions between Bu[n]Li and *o*-, *m*-, and *p*-alkoxy-alkylthio-benzenes. In the *ortho*-isomers, ring lithiation occurs *ortho* to the alkoxy-group.[76] Treatment of an *o*-methoxyaryl-oxazoline with lithium alkylamides results in displacement of the methoxy-group by alkylamino, probably *via* nucleophilic addition and elimination of LiOMe, enhanced by strong chelation of the lithium cation.[77] Lithiation of 4,4-dimethyl-2-(2-thienyl)-2-oxazoline leads to 3- and 5-substitution of the thiophen nucleus, with the 3-pattern predominating when ether is the solvent. The *ortho*-directing effect thus decreases in the order: oxazoline > pyridyl > sulphonamides, carboxamides, and dimethylaminomethyl.[78] The *ortho*-lithiation of tertiary benzamides may be achieved without attack on the amide carbonyl by the use of s-butyl-lithium in THF at −78 °C.[79] *N*-Methyl- and *N*-phenyl-

[67]  W. F. Reynolds and C. W. Tzeng, *Canad. J. Biochem.*, 1977, **55**, 576.
[68]  A Bernardini, P. Viallefont, M. Gelize-Duvigneau, and H. Sauvaitre, *J. Mol. Structure*, 1976, **34**, 245.
[69]  I. D. Kalikhman, E. F. Shibanova, V. V. Makarskii, V. A. Pestunovich, V. A. Lopyrev, and M. G. Voronkov, *Doklady Akad. Nauk S.S.S.R.*, 1977, **234**, 1380 (*Chem. Abs.*, 1977, **87**, 133 630).
[70]  B. Cederlund, R. Lantz, A. B. Hörnfeldt, O. Thorstad, and K. Undheim, *Acta Chem. Scand.* (*B*), 1977, **31**, 198.
[71]  J. R. Bartels-Keith, M. T. Burgess, and J. M. Stevenson, *J. Org. Chem.*, 1977, **42**, 3725.
[72]  A. Avalos, R. M. Claramunt, and R. Granados, *Anales de Quim.*, 1976, **72**, 922.
[73]  E. M. Kaiser, J. D. Petty, and P. L. A. Knutson, *Synthesis*, 1977, 509.
[74]  G. Jaouen, *Ann. New York Acad. Sci.*, 1977, **295**, 59.
[75]  R. D. Rieke, *Accounts Chem. Res.*, 1977, **10**, 301.
[76]  S. Cabiddu, S. Melis, P. P. Piras, and M. Secci, *J. Organometallic Chem.*, 1977, **132**, 321.
[77]  A. I. Meyers and R. Gabel, *J. Org. Chem.*, 1977, **42**, 2653.
[78]  L. Della Vecchia and I. Vlattas, *J. Org. Chem.*, 1977, **42**, 2649.
[79]  P. Beak and R. A. Brown, *J. Org. Chem.*, 1977, **42**, 1823.

benzamide have been dilithiated (on N and at the *ortho*-C) with Bu$^n$Li and the products trimethylsilylated.[80] The reaction between phenyl isocyanide and Bu$^t$Li plus TMEDA at $-78\,^{\circ}$C yields $o$-LiC$_6$H$_4$N=C(CMe$_3$)Li, the product of $\alpha$-addition and *ortho*-lithiation. The isocyanide moiety thus provides a means of protecting an aromatic primary amine in order to achieve *ortho*-metallation.[81]

Conditions have been established for the high-yield syntheses of 2,5-dilithio-furan and -thiophen, and for 2,4- and 2,5-dilithiation of *N*-methylpyrrole. These conditions have been applied to the metallation of 2-methyl-furan and -thiophen, 2,5-dimethyl-furan and -thiophen, benzo[*b*]-furan and -thiophen, pyrrole, and indole.[82] Metallation of thiophen, blocked at the $\alpha$-positions with SO$_2$CMe$_3$ groups, by lithium di-isopropylamide yields ring-opened products, perhaps *via* a dilithio intermediate.[83] A full account of preliminary work on the inversion of Ar–Li reactivity *via* reaction with *trans*-chlorovinyliodosodichloride has now appeared.[84] The resulting iodonium chlorides (Ar = furan, thiophen, or seleno-phen) are strongly electrophilic and react readily with nucleophiles. The lithiation of 2- and/or 3-brominated benzofuran and its derivatives with Bu$^n$Li has been re-examined. $\beta$-Lithiobenzofuran is stable at $-115\,^{\circ}$C, but at $-75\,^{\circ}$C, after 10 minutes, it is 50% decomposed to 2-hydroxyphenylacetylene: at higher temperatures the proportion of ring-opened product increases. The 2,3-dilithio-compound may be prepared from the dibromo-compound and five equivalents of Bu$^n$Li at $-75\,^{\circ}$C.[85] Conditions for halogen–lithium exchange of the individual bromines in 2,7-dibromo-naphthalene and -anthracene have been published.[86] In the lithiation of 9-substituted fluorenes, alkyl groups on aryl substituents effectively block the reaction sites.[87] Other lithiations include attack on a pyridine derivative selectively at the 3-position,[88] 1,3-benzodioxole and -benzoxathiole derivatives,[89] 1-benzenesulphonylimidazole,[90] tetrathiafulvalene,[91] di($\eta^5$-ben-zene)chromium,[92] and phospha-, arsa-, and stiba-benzenes with MeLi to yield (12; X = P, As, and Sb).[93] The utility of $^{13}$C–$^1$H n.m.r. coupling constants for identifying reactive positions in $sp^2$-carbon lithiation has been recognised. The theory depends upon the stability of the carbanion depending qualitatively on the

(12)

[80] R. M. Sandifer, C. F. Beam, M. Perkins, and C. R. Hauser, *Chem. and Ind.*, 1977, 231.
[81] H. M. Walborsky and P. Ronman, *J. Org. Chem.*, 1978, **43**, 731.
[82] D. J. Chadwick and C. Willbe, *J.C.S. Perkin I*, 1977, 887.
[83] F. M. Stoyanovich, B. G. Chermanova, and Ya. L. Gol'dfarb, *Izvest. Akad. Nauk S.S.S.R., Ser. khim.*, 1977, 1367 (*Chem. Abs.*, 1977, **87**, 117 740).
[84] S. Gronowitz and B. Holm, *J. Heterocyclic Chem.*, 1977, **14**, 281.
[85] M. Cugnon de Sevricourt and M. Robba, *Bull. Soc. chim. France*, 1977, 142.
[86] G. Porzi and C. Concilio, *J. Organometallic Chem.*, 1977, **128**, 95.
[87] M. Nakamura, N. Nakamura, and M. Oki, *Bull. Chem. Soc. Japan*, 1977, **50**, 1097.
[88] G. R. Newkome, J. D. Sauer, and S. K. Staires, *J. Org. Chem.*, 1977, **42**, 3524.
[89] S. Cabiddu, A. Maccioni, P. P. Piras, and M. Secci, *J. Organometallic Chem.*, 1977, **136**, 139.
[90] R. J. Sundberg, *J. Heterocyclic Chem.*, 1977, **14**, 517.
[91] D. C. Green, *J.C.S. Chem. Comm.*, 1977, 161.
[92] C. Elschenbroich and J. Heck, *Angew. Chem. Internat. Edn.*, 1977, **16**, 479.
[93] A. J. Ashe, III, and T. W. Smith, *Tetrahedron Letters*, 1977, 407.

amount of $s$-character in the lone pair and upon a linear relation between $s$-character and the C–H coupling in the corresponding protonated carbon. The theory seems to hold for differences of $J$ values of as little as 2 Hz within molecules.[94]

Relevant Grignard reactions include those of nitrobenzothiazoles[95] with Grignards, indolemagnesium bromide with cyclohexanone,[96] and a range of *para*-substituted aryl-magnesium bromides with trifluoroacetonitrile.[97]

**Mercury Substitution.**—Perchloro- and periodo-naphthalene and -benzene derivatives have been prepared by halogenodemercuration of the appropriate permercurated arenes obtained by the reaction of an excess of molten mercury(II) trifluoroacetate with the arene.[98] Perbromobenzoic acids have been prepared analogously.[99] Arylmercuric acetates have been prepared in high yields by treatment of the appropriate $Ar_2SnEt_2$ compounds with mercury(II) acetate,[100] and by direct mercuration of arenes with the same reagent.[101] The rates of (and isomer distributions in) the mercuration and thallation of benzene and substituted benzenes by mercury(II) and thallium(III) trifluoroacetates in TFA have been reported.[102] Substituted ferrocenes have been permercurated with mercury(II) trifluoroacetate, thus providing a route to the perhalogenated compounds;[103] decaiodoferrocene has been prepared similarly.[104] A range of polybromophenyl-mercurials has been synthesized by decarboxylation of the mercury(II) salts of the corresponding polybromobenzoic acids. Substituent effects are consistent with an $S_E$i mechanism. The same paper also examines the cleavage reactions of the mercurials with $I_3^-$, $I_2$ in DMF, and mercury(II) halides (to poly-bromophenylmercuric halides).[105]

$Hg(OAc)_2$ catalyses the nitration of toluene with $HNO_3$ in AcOH at 80 °C, giving *o*-, *m*-, and *p*-nitrotoluenes in the ratios 3 : 1 : 4.3. Other mercury(II) salts give similar results. The reaction occurs in three steps: mercuration (rate-limiting), nitrosodemercuration, and oxidation of nitroso- to nitro-toluenes.[106]

**Other Metal Substitutions.**—Substituted arenes react with diborane (in THF) in the presence of Li (best), Na, K, or Ca to give aryl-boranes by transmetallation. Yields are superior to those of the two-step process *via* reaction between the pre-formed lithium aryl and diborane.[107] $Tl(O_3SCF_3)_3$ is a more reactive thalla-ting agent than the trifluoroacetate, and may be prepared effectively *in situ* by

[94] E. B. Pedersen, *J.C.S. Perkin II*, 1977, 473.

[95] G. Bartoli, R. Leardini, M. Lelli, and G. Rosini, *J.C.S. Perkin I*, 1977, 884.

[96] S.-H. Zee and K.-M. Kuo, *J. Chinese Chem. Soc. (Taipei)*, 1977, **24**, 57 (*Chem. Abs.*, 1977, **87**, 134 920).

[97] V. N. Fetyukhin, A. S. Koretskii, V. I. Gorbatenko, and L. I. Samarai, *Zhur. org. Khim.*, 1977, **13**, 271.

[98] G. B. Deacon and G. J. Farquharson, *Austral. J. Chem.*, 1977, **30**, 1701.

[99] G. B. Deacon and G. J. Farquharson, *Austral. J. Chem.*, 1977, **30**, 293.

[100] E. M. Panov, O. P. Syutkina, V. I. Lodochnikova, and K. A. Kocheshkov, *Zhur. obshchei Khim.*, 1977, **47**, 838.

[101] N. F. Chernov, T. A. Dekina, and M. G. Voronkov, *Zhur. obshchei Khim.*, 1977, **47**, 794.

[102] G. A. Olah, I. Hashimoto, and H. C. Lin, *Proc. Nat. Acad. Sci. U.S.A.*, 1977, **74**, 4121.

[103] V. I. Boev and A. V. Dombrovskii, *Izvest. V. U. Z., Khim. i khim. Tekhnol.*, 1977, **20**, 1789.

[104] V. I. Boev and A. V. Dombrovskii, *Zhur. obshchei Khim.*, 1977, **47**, 727.

[105] G. B. Deacon, G. J. Farquharson, and J. M. Miller, *Austral. J. Chem.*, 1977, **30**, 1013.

[106] L. M. Stock and T. L. Wright, *J. Org. Chem.*, 1977, **42**, 2875.

[107] F. G. Thorpe, G. M. Pickles, and J. C. Podesta, *J. Organometallic Chem.*, 1977, **128**, 305.

conducting the thallation with $Tl(O_2CCF_3)_3$ in TFA containing $CF_3SO_3H$: this approach avoids possible problems arising from the low solubility of $Tl(O_3SCF_3)_3$. By this means, *e.g.*, 2,3,5,6-tetrafluoroanisole, which does not react with $Tl(O_2CCF_3)_3$, has been thallated.[108] A full account has now appeared on the dithallation of monocyclic aromatics. Anisole, phenetole, and thiophen have been dithallated in high yields: partial dithallation has been achieved for benzene, toluene, and *m*-xylene.[109]

2,5-Di(trimethyl-silyl, -germanyl, and -stannyl)furan derivatives have been prepared by treatment of the appropriate lithiated precursors with the tri-methylmetal chlorides: yields range from 23 to 67%.[110] The optically active indenyl organotin compound (13) is obtained by treatment of (*S*)-(+)-1-methyl-indene with $Et_2NSnMe_3$. The compound is stereochemically unstable except in solvents of low polarity.[111] Treatment of the 9-silafluorene (14; M = $SiMe_2$) with $GeCl_4$ and $AlCl_3$ yields the germanium derivative (14; M = $GeCl_2$): $PCl_3$ similarly gives (14; M = $PCl$). Analogous reactions on the $SiCl_2$ compound yield *o*-$Cl_3GeC_6H_4$-$C_6H_4SiCl_3$-*o* and *o*-$Cl_2PC_6H_4$-$C_6H_4SiCl_3$-*o*.[112]

(13)                                    (14)

Biaryls can be synthesized in good yields by reduction (with Raney nickel) of diaryltellurium dichlorides. Thus, di(*p*-methoxyphenyl)tellurium dichloride, from anisole and $TeCl_4$, gives 4,4'-dimethoxy-1,1'-biphenyl in 78—90% yield.[113] Diaryltellurium dichlorides may also be synthesized, in high yields, by reaction between $TeCl_4$ and diaryl-mercury(II) compounds.[114]

The *ortho*-palladation of 3,4-dioxygenated benzylic tertiary amines by lithium tetrachloropalladate can be directed exclusively to either C-2 or C-6. Substitution at C-6 prevails when AcO, methylenedioxy, $PhCH_2O$, methoxymethyl ether, or HO substituents are attached to C-3, whereas palladation occurs exclusively at C-2 when C-3 bears methylthiomethyl ether or phenylthiomethyl ether substi-tuents. The resulting organopalladium compounds are crystalline solids, stable to air and moisture, and can readily be carbonylated, alkylated, arylated, *etc.*[115] Kinetic studies of the acetoxylation of arenes by potassium peroxydisulphate and acetic acid in the presence of (2,2'-bipyridyl)palladium(II) acetate catalyst have led to a revision of the mechanism. The reaction is now thought to proceed *via*

[108] G. B. Deacon and D. Tunaley, *J. Fluorine Chem.*, 1977, **10**, 177.
[109] G. B. Deacon, D. Tunaley, and R. N. M. Smith, *J. Organometallic Chem.*, 1978, **144**, 111.
[110] E. Lukevics and N. P. Erchak, U.S.S.R. P. 550 393 (*Chem. Abs.*, 1977, **87**, 23 501); E. Lukevics, N. P. Erchak, J. Popelis, and I. Dipans, *Zhur. obshchei Khim.*, 1977, **47**, 802.
[111] A. N. Kashin, V. A. Khutoryanskii, V. N. Bakunin, I. P. Beletskaya, and O. A. Reutov, *J. Organometallic Chem.*, 1977, **128**, 359.
[112] E. A. Chernyshev, T. L. Krasnova, E. F. Bugerenko, V. L. Rogachevskii, G. P. Metveicheva, and N. M. Babaeva, *Zhur. obshchei Khim.*, 1977, **47**, 2572.
[113] J. Bergman, R. Carlsson, and B. Sjöberg, *Org. Synth.*, 1977, **57**, 18.
[114] I. D. Sadekov, A. A. Maksimenko, and A. A. Ladatko, *Zhur. obshchei Khim.*, 1977, **47**, 2229.
[115] R. A. Holton and R. G. Davis, *J. Amer. Chem. Soc.*, 1977, **99**, 4175.

rapid, reversible formation of an aryl-palladium species; this is oxidized to the aryl acetate by the peroxydisulphate ion in the rate-limiting step.[116]

Ferrocene has been aurated with $[(Ph_3PAu)_3O]^+ BF_4^-$ and $HBF_4$ to give the mono-$AuPPh_3$-substituted derivative.[117]

### 4 Reactions with Electrophilic Carbon and Silicon

**Alkylation.**—Direct alkylation of pyridine and methyl-pyridines[118] and alkylation of benzofuran[119] have been reviewed, the latter in Russian.

*Gas-phase Alkylation.* In an attempt to establish the intrinsic reactivity of *bona fide* free methyl cations towards aromatic compounds, $CT_3^+$ (from the radio-chemical decay of $CT_4$) has been generated in the presence of benzene and toluene over a period of nine months. Products are mainly $PhCT_3$ (from benzene) and *o*-, *m*-, and *p*-$MeC_6H_4CT_3$ and $PhCH_2CT_3$, in ratios 32:21:26:4 (from toluene).[120] Previous work on gas-phase reactions of the t-butyl cation (reported in Volumes 5 and 6 of this series) has been extended to include the isopropyl cation (from the $\gamma$-radiolysis of propane). A comparison between gas-phase and solution reactions reveals no basic mechanistic differences, the reactivity of the cations in the gas phase representing the limit that is more or less closely approached by poorly solvated alkylating agents.[121] Full details of the reactions between the t-butyl cation and phenol and anisole have now appeared. The reactions appear to involve kinetic *O*-alkylation with subsequent intermolecular alkylation to the thermodynamically favoured *C*-alkylated products. There seems to be no evidence for *ortho*-alkylation *via* preliminary attack on the substituent and intramolecular rearrangement to the *ortho*-arenium ion.[122,123]

*Alkylation with Alkenes.* The alkylation of phenols by olefins and cyclo-olefins has been reviewed, in Russian.[124] Several papers on the alkylation of phenols have appeared. Thus, *p*-t-butylphenol has been prepared with good selectivity by treatment of phenol with isobutene and a Lewis acid catalyst followed by $HClO_4$ as an isomerization catalyst.[125] The *ortho*-alkylation of phenol by alkenes can be improved by using $Al(OPh)_3$ and an aluminosilicate catalyst.[126] Much greater *ortho*-selectivity in the alkylation of phenol with 4-bromostyrene is obtained if aluminium diphenylphosphorodithioate rather than $BF_3 \cdot OEt_2$ or $BF_3 \cdot H_3PO_4$ is used as catalyst.[127] Relative reactivities of the various ring positions towards

[116] L. Eberson and L. Jönsson, *Annalen*, 1977, 233.
[117] A. N. Nesmeyanov, E. G. Perevalova, V. P. Dyadchenko, and K. I. Grandberg, *Izvest. Akad. Nauk S.S.S.R., Ser. khim.*, 1976, 2844 (*Chem. Abs.*, 1977, **87**, 23 422).
[118] C. V. Digiovanna, P. J. Cislak, and G. N. Cislak, *A.C.S. Symp. Ser.*, 1977, **55**, 397.
[119] E. A. Karakhanov, A. V. Anisimov, S. K. Ermolaeva, and E. A. Viktorova, *Vestnik Moskov. Univ., Ser., 2: Khim.*, 1977, **18**, 610 (*Chem. Abs.*, 1978, **88**, 50 556).
[120] F. Cacace and P. Giacomello, *J. Amer. Chem. Soc.*, 1977, **99**, 5477.
[121] M. Attinà, F. Cacace, G. Ciranni, and P. Giacomello, *J. Amer. Chem. Soc.*, 1977, **99**, 2611.
[122] M. Attinà, F. Cacace, G. Ciranni, and P. Giacomello, *J. Amer. Chem. Soc.*, 1977, **99**, 5022.
[123] M. Attinà, F. Cacace, G. Ciranni, and P. Giacomello, *J. Amer. Chem. Soc.*, 1977, **99**, 4101.
[124] Ch.K. Rasulov, I. I. Sidorchuk, and Sh.G. Sadykhov, Deposited Document 1975, *VINITI* 3464 (*Chem. Abs.*, 1978, **88**, 62 062).
[125] Derivados Fenolicos S. A., Spanish P. 431 442 (*Chem. Abs.*, 1977, **87**, 134 544).
[126] N. L. Voloshin, G. I. Turyanchik, E. V. Lebedev, V. T. Sklyar, E. K. Bryanskaya, P. L. Klimenko, and N. I. Vykhrestyuk, U.S.S.R. P. 539 021 (*Chem. Abs.*, 1977, **87**, 39 097).
[127] D. A. Pisanenko, E. V. Alisova, and N. A. Goryachuk, *Izvest. V. U. Z., Khim. i khim. Tekhnol.*, 1977, **20**, 1717.

alkylation by cyclohexene and polyphosphoric acid in phenol and the cresols have been reported.[128] The effect of variation in size of alkyl substituents in 2,4- and 2,6-dialkyl-phenols on reactivity in alkylation by styrene has been studied, and the use of oxalic acid as the catalyst in the alkylation of phenol and alkyl-phenols by olefins has been reported by the same authors.[129] The kinetics of reaction between some pentadienyltricarbonyliron cations and $m$-dimethoxybenzene have been studied. Thus, reaction with (15) involves attack by (16) as the rate-limiting step, giving (17).[130] Details of the alkylation of $p$-chlorophenol by di-isobutene have been reported.[131]

(15)    (16)    (17)

The kinetics of alkylation of benzene[132] and $m$-xylene[133] by propene, and of benzene and toluene by propene, but-1-ene, and but-2-ene in the presence of $H_2SO_4$ have been studied.[134] Cyclohexylidene- and 1-cyclohexenyl-acetic acids react with benzene and $AlCl_3$ to yield mixtures of *trans*-2-phenyl-cyclohexaneacetic acid, *cis*-3-phenylcyclohexaneacetic acid, and *trans*-4-phenylcyclohexaneacetic acid in roughly the same proportions in both cases.[135] The effect of palladium(II) salts on the reactions between oct-1-ene, 1-phenyl-buta-1,3-diene, and 1,4-diphenylbuta-1,3-diene and benzene and toluene has been investigated.[136] The reaction between benzene and allyl alcohol in the presence of $AlCl_3$ yields n-propylbenzene and 1,1-diphenylprop-1-ene under mild conditions. Deuterium- and $^{13}C$-labelling studies show that the initial intermediate is 2-phenylpropan-1-ol, followed by a phenyl-assisted propyl cation. The final products are formed by intermolecular hydride shift and methyl migration.[137] A further study of the already much-studied cycloalkylation of benzene with isoprene and concentrated $H_2SO_4$ reveals 1,1-dimethylindane, the indacenes (18), (19; R = H), and (19; R = $CH_2CH_2CHMe_2$), and the tri-indane (20) in the complex mixture of products.[138]

[128] N. S. Kozlov, A. G. Klein, and Yu. A. Galishevskii, *Neftekhimiya*, 1977, **17**, 848 (*Chem. Abs.*, 1978, **88**, 89 265).
[129] Ya. A. Gurvich, S. T. Kumok, and E. L. Styskin, *Tezisy Vses. Simp. Org. Sint.: Benzoidnye Aromat. Soedin., 1st.*, 1974, 19 (*Chem. Abs.*, 1977, **87**, 5555); *ibid.*, p. 27 (*Chem. Abs.*, 1977, **87**, 5557).
[130] T. G. Bonner, K. A. Holder, P. Powell, and E. Styles, *J. Organometallic Chem.*, 1977, **131**, 105.
[131] E. A. Navarro, F. F. Muganlinskii, and M. A. Nizhnik, *Izvest. V. U. Z., Khim. i khim. Tekhnol.*, 1977, **20**, 1623 (*Chem. Abs.*, 1978, **88**, 74 149).
[132] R. Z. Akhmedova, I. A. Grishkan, V. V. Lobkina, and L. V. Arakelova, *Kinetics and Catalysis (U.S.S.R.)*, 1977, **18**, 786 (*Chem. Abs.*, 1977, **87**, 117 266).
[133] Yu. I. Kozorezov and A. N. Kuleshova, *Kinetics and Catalysis (U.S.S.R.)*, 1977, **18**, 813 (*Chem. Abs.*, 1977, **87**, 117 268).
[134] R. K. Tiwari and M. M. Sharma, *Chem. Eng. Sci.*, 1977, **32**, 1253.
[135] V. Dragutan, L. Stanescu, and A. M. Glatz, *Rev. Roumaine Chim.*, 1977, **22**, 1045.
[136] Y. Fujiwara, R. Asano, and S. Teranishi, *Israel J. Chem.*, 1977, **15**, 262.
[137] W. Ackermann and A. Heesing, *Chem. Ber.*, 1977, **110**, 3126.
[138] E. J. Eisenbraun, W. M. Harms, J. W. Burnham, O. C. Dermer, R. E. Laramy, M. C. Hamming, G. W. Keen, and P. W. Flanagan, *J. Org. Chem.*, 1977, **42**, 1967.

        (18)                 (19)  R = H or (CH$_2$)$_2$CHMe$_2$

                                                                    (20)

Mono- to tetra-substituted products have been isolated from the iso-propylation of naphthalene with propene and HF.[139] The utility of aluminium and titanium oxychlorides as catalysts in the alkylation of naphthalene by propene and but-2-ene has been explored.[140] The reactivities of benzene, bromobenzene, fluorobenzene, and toluene towards alkylation by propene, methylpropene, and but-1-ene in the presence of HF or H$_2$SO$_4$ catalyst have been measured.[141]

Two papers by the same authors have appeared on the alkylation of monosub-stituted benzene derivatives with vinyl triflates (*e.g.* R$_2$C=CPhO$_3$SCF$_3$, where R = Ph, Me, and 1-cyclohexenyl to cyclo-octenyl triflates). The reactions are carried out in the presence of 2,6-di-t-butyl-4-methylpyridine (a non-nucleo-philic base, to neutralise the triflic acid formed during reaction) and a large excess of the aromatic substrate. The mechanism is thought to involve a vinyl cation intermediate.[142,143]

In the deamination of aromatic amines with Bu$^t$ONO and CuCl$_2$ or CuBr$_2$ in the presence of CH$_2$=CHR (R = CN, CO$_2$Et, CONH$_2$, or Ph), yields of ArCH$_2$CHHalR are comparable or superior to those obtained by the Meerwein procedure.[144] Similar alkylations (in 15—35% yields) *via* aryldiazonium tetrafluoroborates, sulphates, and nitrates have been reported.[145]

The alkylation of 3-bromothiophen by allylic alcohols in the presence of Pd(OAc)$_2$ has been studied. Thus, 2-methyl-5-(3-thienyl)pentan-3-one is the major product from reaction between 3-bromothiophen, CH$_2$=CH-CH(OH)CHMe$_2$, Pd(OAc)$_2$, NaI, NaHCO$_3$, and Ph$_3$P in HMPT at 130 °C for 8 h. Thienyl aldehydes are similarly prepared from primary alcohols.[146]

*Alkylation with Alkyl Halides.* The effects of Zn, Fe, and Sn salts on the alkylation of some phenols and phenol ethers with a range of alkyl halides,[147] and of Fe and Zn salts on similar alkylations of chloro-phenols,[148] are reported. Selectivity

[139] N. I. Plotkina, N. V. Gein, M. I. Kachalkova, N. A. Shevchenko, and I. P. Kolenko, Deposited Document 1976, *VINITI* 92.
[140] K. Kagami, T. Masuda, and Y. Takami, *Yuki Gosei Kagaku Kyokaishi*, 1977, **35**, 909 (*Chem. Abs.*, 1978, **88**, 104 972).
[141] N. I. Plotkina, N. V. Gein, and M. I. Kachalkova, Deposited Document 1975, *VINITI* 1809 (*Chem. Abs.*, 1977, **87**, 84 338).
[142] P. J. Stang and A. G. Anderson, *Tetrahedron Letters*, 1977, 1485.
[143] P. J. Stang and A. G. Anderson, *J. Amer. Chem. Soc.*, 1978, **100**, 1520.
[144] M. P. Doyle, B. Siegfried, R. C. Elliott, and J. F. Dellaria Jr., *J. Org. Chem.*, 1977, **42**, 2431.
[145] N. I. Ganushchak, B. D. Grishchuk, V. A. Baranov, T. A. Shilo, and V. G. Nemesh, *Ukrain. khim. Zhur.*, 1977, **43**, 1301 (*Chem. Abs.*, 1978, **88**, 104 792).
[146] Y. Tamaru, Y. Yamada, and Z. I. Yoshida, *Tetrahedron Letters*, 1977, 3365.
[147] A. R. Abdurasuleva and K. N. Akhmedov, *Tezisy Vses. Simp. Org. Sint.: Benzoidnye Aromat. Soedin.*, *1st.*, 1974, 35 (*Chem. Abs.*, 1977, **87**, 5559).
[148] M. K. Turaeva, K. N. Akhmedov, and A. R. Abdurasuleva, Deposited Document 1975, *VINITI* 3172 (*Chem. Abs.*, 1978, **88**, 89 255).

towards *para*-substitution has been noted in aromatic alkylations (with, *e.g.*, $Me_3CCl$), acylations, and sulphonylations performed in the presence of $Mo(CO)_6$.[149]

Reactions of $RCH_2CHClCH_3$ (R = MeO or OH) with benzene and $AlCl_3$ give the unusual products PrPh and $Ph_2C=CHMe$ in addition to $RCH_2CHPhMe$ and 1,1- and 1,2-diphenylpropanes, whereas when R is $MeO_2C$ only methyl 3-phenylbutanoate is formed.[150] Alkylation of $PhCH_2CHMe_2$ with $F(CH_2)_3Br$ in the presence of $BF_3$ yields 80% of product that is 4-substituted by —$CHMeCH_2Br$.[151] Alkylation of benzene and derivatives with $ClCH_2NCO$ leads to substitution by —$CH_2NCO$ in *ca*. 50% yields.[152] The iodine in iodo-naphthalenes can be replaced by $CF_3$ *via* treatment with $CF_3I$ and Cu.[153]

Benzene has been amido-alkylated by reaction with $ArC_6H_4CHClNR^1COR^2$ and $AlCl_3$ ($R^1$ and $R^2$ are various alkyl and phenyl) to $ArCHPhNR^1COR^2$ in fair to good yields.[154] Anisole and phenetole have been alkylated with 1-chloro-2,3-epoxy-5-methylhex-5-ene and $AlCl_3$.[155] A dissertation on apparent anomalies in the Friedel–Crafts alkylation of 2,6-dimethylphenol and its alkyl ethers has been cited.[156]

Several papers have been published on allylation. The reaction between 2-alkoxy-phenols and allyl halides, catalysed by copper salts, is improved if carried out in a saturated aqueous solution of inorganic salts (*e.g.* NaCl) instead of in water.[157] $\beta$-Cyclodextrin catalyses the allylation of 2-methyl-1,4-naphthoquinone by allyl and crotyl bromides in aqueous media, providing one-step syntheses of vitamin $K_1$ and $K_2$ analogues respectively, in good yields.[158] Allylation of benzothiophen with allyl iodide in the presence of silver trichloroacetate yields the 3-allyl compound in minutes. Increased reaction time leads to the introduction of a second allyl group into the benzene ring; the use of allyl bromide leads to increased selectivity.[159] Polymethylated anthracenes, phenanthrenes, and naphthalenes are formed by reaction between suitably alkylated benzenes, $AlBr_3$, and 3,4-dichloro-1,2,3,4-tetramethylcyclobutene.[160]

Alkylation of thiophen at −70 °C with isopropyl chloride and $AlCl_3$ gives 2- and 3-isopropylthiophenium ions, which on treatment with water give 60:40 2- and 3-isopropylthiophens. Similar results are obtained with other alkyl halides, including t-butyl, where the 2- to 3-isomer ratio is 83:17.[161]

[149] F. M. Farona and J. F. White, U.S. P. 4 038 324 (*Chem. Abs.*, 1977, **87**, 184 685).
[150] H. Matsuda and H. Shinohara, *Chem. Letters*, 1978, 95.
[151] T. Sakakida, Japan. Kokai 76 122 032.
[152] V. A. Shokol, B. N. Kozhushko, and A. V. Gumenyuk, *Zhur. org. Khim.*, 1977, **13**, 664.
[153] K. Hosokawa and K. Inukai, *Nippon Kagaku Kaishi*, 1977, 1163.
[154] N. Mollov, A. Venkov, and M. Nikolova, *Doklady Bolg. Akad. Nauk.*, 1977, **30**, 253.
[155] S. I. Sadykh-Zade, N. M. Babaev, Sh. K. Kyazimov, and M. A. Akhmedov, *Azerb. khim. Zhur.*, 1977, 37 (*Chem. Abs.*, 1977, **87**, 151 896).
[156] M. P. McLaughlin, *Diss. Abs. Internat. B*, 1978, **38**, 3698.
[157] S. Umemura, N. Takamitsu, F. Iwata, and H. Hamada, Japan. Kokai 77 25 727 (*Chem. Abs.*, 1977, **87**, 67 974).
[158] I. Tabushi, K. Fujita, and H. Kawakubo, *J. Amer. Chem. Soc.*, 1977, **99**, 6456.
[159] A. V. Anisimov, N. Yu. Luzikov, V. M. Nikolaeva, Yu. N. Radyukin, E. A. Karakhanov, and E. A. Viktorova, *Khim. geterotsikl. Soedinenii*, 1977, 1625.
[160] A. P. Krysin, N. V. Bodoev, and V. A. Koptyug, *Zhur. org. Khim.*, 1977, **13**, 1290 (*Chem. Abs.*, 1977, **87**, 117 697).
[161] L. I. Belen'kii, A. P. Yakubov, and I. A. Bessonova, *Zhur. org. Khim.*, 1977, **13**, 364 (*Chem. Abs.*, 1977, **87**, 22 916).

*Other Alkylations.* The chloromethylation of aromatic and heteroaromatic compounds has been reviewed, in Russian.[162]

Experimental and calculated isomerisation rates for some alkyl-benzenes have been compared.[163] The relative rates of rearrangement of 1-phenyl-propane and -butane, on heating with AlCl₃ or AlBr₃, have been studied by $^{13}$C-labelling of C-1 of the side-chains. The butane fails to rearrange $^{13}$C to C-2 under conditions which lead to the complete equilibration of $^{13}$C between C-1 and C-2 in the propane.[164] A very detailed analysis of the effects of acidic and metallic functions on the isomerisation of ethylbenzene on platinum supported by a fluorinated alumina has been presented. The main product is *o*-xylene.[165] BF₃ and HF has been used as a catalyst in the transalkylation between isopropylbenzene and phenol to give nearly equal amounts of *o*- and *p*-isopropylphenol.[166] 2-t-Butyl-*p*-xylene, which is difficult to make in good yield, can be successfully prepared by the t-butylation of *p*-xylene with 2-t-butyl- or 2,6-di-t-butyl-*p*-cresol in the presence of AlCl₃ and MeNO₂.[167] The same authors have studied the transalkylation of some t-butyldiphenyl-methanes and -ethanes in benzene and toluene with a variety of Lewis acids; the use of the t-butyl function (removed by benzene and AlCl₃) as a positional protective group in this[168] and other aromatic systems[169] has been discussed.

Varying amounts of cycloalkyl-benzenes are formed in the reactions between benzene, chlorobenzene, or toluene and cyclohexane, methylcyclohexane, or decalin in the presence of AlCl₃ and a range of alkyl halides.[170] The alkylation of benzene or toluene by cyclohexane is promoted by the addition of Me₃CH, whose carbonium ion assists hydrogen abstraction from the cyclohexane.[171]

The alkylation of benzene by propan-1-ol and AlCl₃ in hexane or cyclohexane has been studied over the temperature range −40 to +100 °C by n.m.r. spectroscopy.[172] Structures of the several products of alkylation of mesitylene with 3,5-di-t-butyl-4-hydroxybenzyl alcohol and of resorcinol and its methyl ethers with α-butylbenzyl alcohol plus orthophosphoric acid have been elucidated.[173,174] Allylation of guaiacol by allyl alcohol and an acid catalyst is

[162] L. I. Belen'kii, Yu. B. Vol'kenshtein, and I. B. Karmanova, *Uspekhi Khim.*, 1977, **46**, 1698.
[163] V. A. Koptyug and V. I. Buraev, *Zhur. org. Khim.*, 1978, **14**, 18 (*Chem. Abs.*, 1978, **88**, 135 935).
[164] R. M. Roberts, T. L. Gibson, and M. B. Abdel-Baset, *J. Org. Chem.*, 1977, **42**, 3018.
[165] N. S. Gnep and M. Guisnet, *Bull. Soc. chim. France*, 1977, 429, 435.
[166] N. Yoneda, Y. Takahashi, C. Tajiri, and A. Suzuki, *Nippon Kagaku Kaishi*, 1977, 831 (*Chem. Abs.*, 1977, **87**, 134 237).
[167] M. Tashiro, T. Yamato, and G. Fukata, *J. Org. Chem.*, 1978, **43**, 743.
[168] M. Tashiro, T. Yamato, and G. Fukata, *J. Org. Chem.*, 1978, **43**, 1413.
[169] M. Tashiro, G. Fukata, T. Yamato, H. Watanabe, K. Oe, and O. Tsuge, *Org. Prep. Proced. Internat.*, 1976, **8**, 249.
[170] C. Ndandji, M. Desbois, R. Gallo, and J. Metzger, *Compt. rend.*, 1977, **285**, *C*, 591.
[171] R. Miethchen, A. Gaertner, and C. F. Kroeger, *Z. Chem.*, 1977, **17**, 443.
[172] V. G. Lipovich, L. E. Latysheva, N. G. Devyatko, T. G. Laperdina, G. A. Kalabin, V. I. Cherednichenko, and M. F. Polubentseva, *Zhur. obshchei Khim.*, 1978, **48**, 179 (*Chem. Abs.*, 1978, **88**, 169 264).
[173] I. G. Arzamanova, Ya. A. Gurvich, S. T. Kumok, R. M. Logvinenko, M. I. Naiman, and A. I. Rybak, *Tezisy Vses. Simp. Org. Sint.: Benzoidnye Aromat. Soedin., 1st.*, 1974, 24 (*Chem. Abs.*, 1977, **87**, 5556).
[174] A. Yusupov, A. R. Abdurasuleva, and G. Usmanov, Deposited Document 1976, *VINITI* 322 (*Chem. Abs.*, 1978, **88**, 89 260).

improved by the presence of copper(I) or copper(II) salts.[175] The unexpected *meta*-methylation of 2,6-xylenol by MeOH and highly active $Al_2O_3$ (to 2,3,6- rather than the expected 2,4,6-trimethylphenol) is rationalized by initial *ipso*-attack at a methyl-substituted position. In support of this, reaction with $CD_3OD$ leads to a product with nearly equal amounts of $CD_3$ in *ortho*- and *meta*-positions.[176] Alkylation of 1-naphthol by $PhCH_2OH$ and a cation-exchange resin gives a 45% yield of a *ca.* 2:1 mixture of 2- and 4-benzyl-1-naphthol. Similar results are obtained by alkylation with cyclohexene, giving the cyclohexyl analogues.[177] A wide range of 1-alkyl-2-naphthols has been prepared, in high yields, by the alkylation of naphthoxide ion with alcohols at *ca.* 270 °C. No isomerization of alkyl groups was detected.[178] The kinetics and product distributions in the alkylation of aniline[179] and benzylamine[180] by propan-2-ol–$H_2SO_4$–water have been studied.

Assorted methyl-benzenes have been alkylated to bicyclic products with a THF·$BF_3$ complex,[181] derivatives of PhOMgBr have been *ortho*-vinylated by substituted 1,3-dioxolans and 1,3-dioxan analogues,[182] normal and abnormal products have been reported in the $AlCl_3$-catalysed reaction between anisole and several epoxides,[183] and partial rate factors have been evaluated for the iso-propylation of toluene by isopropyl mesylate in the nitrobenzene–methanesul-phonic acid system.[184]

Whereas 2,4-dialkyl-phenols react readily with HCHO, the 2,6-analogues show little or no reactivity, depending on the size of the alkyl groups.[185] The kinetics and mechanism of the reaction between 4-t-butylphenol and formalde-hyde in the presence of alkaline hydroxide catalysts have been studied: the efficiencies of catalysts are in the order Ca < Ba < Na < Li.[186] The cyclo-condensation of veratrole with succindialdehyde yields 2,3-dimethoxynaph-thalene (21%).[187] Porphyrins that are unsubstituted in the *meso* positions have been obtained from the reaction between 3,4-disubstituted pyrroles and HCHO, with acid catalysis, in good yields.[188]

Other alkylations include those of pyrrole, indole, and *m*-(*NN*-dimethyl-amino)phenol with 3,3-dimethyl-3*H*-indole in AcOH (to indolinylated

[175] A. K. Moryashchev, V. I. Artem'ev, I. I. Sidorov, and V. A. Kovalenko, U.S.S.R. P. 536 157 (*Chem. Abs.*, 1977, **87**, 22 750).
[176] B. E. Leach, *J. Org. Chem.*, 1978, **43**, 1794.
[177] S. P. Starkov, N. L. Polyanskaya, N. A. Vozhzhova, and G. S. Leonova, *Izvest. V. U. Z., Khim. i khim. Tekhnol.*, 1977, **20**, 1099 (*Chem. Abs.*, 1977, **87**, 167 787).
[178] O. Kohki, *Yuki Gosei Kagaku Kyokai Shi*, 1976, **34**, 594; T. Kito and K. Ota, *J. Org. Chem.*, 1977, **42**, 2020.
[179] A. A. Zerkalenkov, V. G. Chekhuta and O. I. Kachurin, *Zhur. Vses. Khim. O-va.*, 1977, **22**, 464 (*Chem. Abs.*, 1977, **87**, 167 130).
[180] A. A. Zerkalenkov and O. I. Kachurin, *Kinetics and Catalysis* (*U.S.S.R.*), 1977, **18**, 1043 (*Chem. Abs.*, 1977, **87**, 183 679).
[181] A. P. Krysin and N. V. Bodoev, *Izvest. sibirsk. Otdel. Akad. Nauk, Ser. khim. Nauk*, 1977, 151 (*Chem. Abs.*, 1977, **87**, 117 711).
[182] G. Casiraghi, G. Puglia, G. Sartori, and G. Terenghi, *Chimica e Industria*, 1977, **59**, 458.
[183] M. Inove, T. Sugita, and K. Ichikawa, *Bull. Chem. Soc. Japan*, 1978, **51**, 174.
[184] O. Kachurin and N. Dereza, *Org. React.* (*Tartu*), 1976, **13**, 301 (*Chem. Abs.*, 1977, **87**, 38 437).
[185] Ya. A. Gurvich, O. F. Starikova, and E. L. Styskin, *Tezisy Vses. Simp. Org. Sint.: Benzoidnye Aromat. Soedin.*, *1st*, 1974, 28 (*Chem. Abs.*, 1977, **87**, 5558).
[186] B. S. R. Reddy, S. Rajadurai, and M. Santappa, *Indian J. Chem. Sect. A*, 1977, **15**, 424.
[187] A. Arcoleo, G. Fontana, G. Giammona, and S. Lo Curcio, *Chem. and Ind.*, 1977, 128.
[188] D. O. Cheng and E. LeGoff, *Tetrahedron Letters*, 1977, 1469.

products),[189] indole with N-acyl-imidazolium salts (to indolyl-imidazolines),[190] assorted benzene derivatives with $\beta$-oxymercuric compounds (*e.g.* AcOCH-MeCH$_2$HgBr) and Lewis acids,[191] and anisole and veratrole with the diacetate of but-2-yne-1,4-diol and AlCl$_3$.[192] 4-Amino-quinazolines have been prepared by the intramolecular cyclisation of suitably derivatised benzenes substituted with —NH—CR=N—C≡N.[193] In the first example of the intermolecular attack of intermediates in the Pummerer rearrangement on an aromatic ring, anisole and *p*-xylene are attacked by 4-BrC$_6$H$_4$S(O)CH$_2$CN *via* the formation of R$^1$S$^+$=CHR$^2$, yielding alkylated products.[194] Benzene has been alkylated (with AlCl$_3$) by the lactone (21) to (22), and the scope, limitations, and mechanism of amido-alkylation by the Nyberg procedure (F$_3$CCO$_2$H in CHCl$_3$) of aromatics (which may be regarded as a variant of the Mannich reaction, the amine function being replaced by amide or imide) have been discussed.[195,196] The polyphosphoric-acid-catalysed cyclization of crotonophenones and chalcones, without aryl migration, to indan-1-one derivatives has been shown to be a general reaction.[197]

CH$_2$CO$_2$H

Ph

(21)                    (22)

A procedure has now been published in *Organic Syntheses* for the *ortho*-alkylation of anilines by the Gassman method[198] (ylide 2,3-sigmatropic rearrangement). The previous approach has been extended to polycyclic aromatic amines,[199] modified,[200] and applied to the *ortho*-alkylation of anilides.[201]

The reaction between benzene and NN-dimethylaniline N-oxide in the presence of CF$_3$SO$_3$H gives mainly (64—76%) 4-(NN-dimethylamino)biphenyl, probably *via* protonation on oxygen and loss of water to the NN-dimethyl-immonium-benzenium ion.[202] Treatment of ferrocene with 2-carboxy-4,6-diphenylpyrylium perchlorate yields the 2-ferrocenylpyrylium salt with decarboxylation.[203]

*Heteroatom Alkylation.* Previous work (referred to in last year's Report) on the steric effect of benzo-fusion on the rates of quaternization of pyridine and thiazole

[189] V. Bocchi, R. Marchelli, and V. Zanni, *Synthesis*, 1977, 343.
[190] T. V. Stupnikova, A. K. Sheinkman, and A. I. Serdyuk, Deposited Document 1975, *VINITI* 746.
[191] J. Barluenga and A. M. Mastral, *Anales de Quim.*, 1977, **73**, 1032 (*Chem. Abs.*, 1978, **88**, 105 502).
[192] R. M. Lagidze, N. R. Loladze, Sh. D. Kuprava, G. G. Samsoniya, R. Sh. Kldiashvili, and D. G. Chavchanidze, *Soobshch. Akad. Nauk Gruz. S.S.R.*, 1977, **88**, 89 (*Chem. Abs.*, 1978, **88**, 89 420).
[193] K. Gewald, H. Schaefer, and K. Mauersberger, *Z. Chem.*, 1977, **17**, 223.
[194] D. K. Bates, *J. Org. Chem.*, 1977, **42**, 3452.
[195] D. N. Chatterjee and M. Sarkar, *Current Sci.*, 1977, **46**, 261.
[196] J. E. Barry, E. A. Mayeda, and S. D. Ross, *Tetrahedron*, 1977, **33**, 369.
[197] J. M. Allen, K. M. Johnston, J. F. Jones, and R. G. Shotter, *Tetrahedron*, 1977, **33**, 2083.
[198] P. G. Gassman and G. Gruetzmacher, *Org. Synth.*, 1977, **56**, 15.
[199] P. G. Gassman and W. N. Schenk, *J. Org. Chem.*, 1977, **42**, 3240.
[200] P. G. Gassman and R. L. Parton, *Tetrahedron Letters*, 1977, 2055.
[201] P. G. Gassman and R. J. Balchunis, *Tetrahedron Letters*, 1977, 2235.
[202] K. Shudo, T. Ohta, Y. Endo, and T. Okamoto, *Tetrahedron Letters*, 1977, 105.
[203] V. V. Krasnikov, Yu. P. Andreichikov, N. V. Kholodova, and G. N. Dorofeenko, *Zhur. org. Khim.*, 1977, **13**, 1566 (*Chem. Abs.*, 1977, **87**, 152 357).

has been extended through a study of the alkylation of salts of pyridinols, quinolinols, and isoquinolinols. $N$-Alkylation is generally favoured with EtI and $Et_2SO_4$ in aprotic solvents. The alkylation is subject to steric hindrance by adjacent methyl groups and less so by benzo-fusion.[204] Phase-transfer catalysis of alkylation of the anions from 2- and 4-pyridones yields $80:20$ $N$-: $O$-alkylation: the ratio is essentially unchanged by change of salts, solvents, temperature, or alkyl halide.[205] 2-Pyridone has also been alkylated on nitrogen (in *ca.* 85% yield) by treatment with $K_2CO_3$ in $MeOCH_2CH_2OMe$ and a range of alkyl halides.[206] $N$-Alkylation is preferred in the reaction between the sodium salt of 4-oxo-4,5-dihydrofuro[3,2-*c*]pyridine (23) and alkyl halides or sulphates, whereas the silver salt undergoes mainly $O$-alkylation. Nitration of the alkylation products occurs exclusively in the furan ring.[207]

(23)

An attempt has been made to find the optimum conditions for $C$- and $N$-alkylation of the pyrrolyl ambident anion. $N$-Alkylation is readily achieved by phase-transfer catalysis but, whereas it *is* possible to obtain almost total $C$-alkylation, the product is invariably an isomer mixture: from the synthetic standpoint, therefore, $C$-alkyl-pyrroles are best prepared *via* reduction of the acyl compounds.[208] A range of substituted pyrroles has been vinylated on nitrogen with $C_2H_2$–DMSO–KOH (30%) in yields up to 97%.[209] The methylation of *meso*-tetraphenylporphin with both MeI and methyl fluorosulphate gives $N$-methyl-, *trans*-$N_aN_b$-dimethyl-, and *trans*,*trans*-$N_aN_bN_c$-trimethyl-derivatives. In contrast to earlier findings in the octa-alkyl-porphin series, no *cis*- and *trans*-$N_aN_c$-dimethyl compounds are formed, this being ascribed to steric effects.[210] Indole, carbazole, and phenothiazine have been alkylated by polymer-supported 'solid' HMPT in high yields. The reactions are slower than with 'free' HMPT, though the yields are comparable.[211] Indole has also been alkylated with a range of aliphatic secondary alcohols in toluene containing $Al(OBu^i)_3$ and Raney nickel as catalyst, giving the $N$-alkylated products in good to excellent yields.[212]

Dialkyl phosphites $(RO)_2P(O)H$ ($R = Me$, Et, $Pr^i$, or Bu) are efficient alkylating agents for N-heterocycles, and have been applied particularly to imidazole analogues and pyridones.[213] Uracil, thymine, cytosine, adenine, and guanine have

[204] L. W. Deady, W. L. Finlayson, and C. H. Potts, *Austral. J. Chem.*, 1977, **30**, 1349.
[205] H. J.-M. Dou, P. Hassanaly, and J. Metzger, *J. Heterocyclic Chem.*, 1977, **14**, 321.
[206] C. S. Giam and A. E. Hauck, *Org. Prep. Proced. Internat.*, 1977, **9**, 5 (*Chem. Abs.*, 1977, **87**, 68 090).
[207] I. Ramos Raimundo, M. Blanco Jerez, J. Perez Chavez, and A. Macias Cabrera, *Sobre Deriv. Cana Azucar*, 1976, **10**, 30 (*Chem. Abs.*, 1977, **87**, 68 199).
[208] N.-C. Wang, K.-E. Teo, and H. J. Anderson, *Canad. J. Chem.*, 1977, **55**, 4112.
[209] B. A. Trofimov, A. I. Mikhaleva, S. E. Korostova, N. A. Vasil'ev, and L. N. Balabanova, *Khim. geterotsikl. Soedinenii*, 1977, 213 (*Chem. Abs.*, 1977, **87**, 22 934).
[210] H. M. G. Al-Hazimi, A. H. Jackson, A. W. Johnson, and M. Winter, *J.C.S. Perkin I*, 1977, 98.
[211] Y. Leroux and H. Normant, *Compt. rend.*, 1977, **285**, C, 241.
[212] F. De Angelis, M. Grasso, and R. Nicoletti. *Synthesis*, 1977, 335.
[213] M. Hayashi, K. Yamauchi, and M. Kinoshita, *Bull. Chem. Soc. Japan*, 1977, **50**, 1510.

been alkylated by triethyl phosphate in poor yields in aqueous media, and the reactivity order of the various alkylation sites has been determined. The work is of relevance to the mutagenic and carcinogenic properties of alkylating agents.[214] A range of substituted pyrimidines has been alkylated on nitrogen with triethyloxonium tetrafluoroborate,[215] and fluoroalkyl derivatives of uracil, uridine, and 2'-deoxyuridine have been prepared by treatment of the appropriate pyrimidinediones with perfluoroalkyl-copper compounds.[216]

Pyrazole and imidazole are alkylated in high yields on nitrogen with phase-transfer catalysis;[217] their N-alkyl derivatives, along with those of 1,2,4-triazole and benzotriazole, have been prepared by treatment of the N-tributyltin compounds with alkyl halides.[218] A range of N-alkyl-benzothiazolium salts has been synthesized from appropriate alkyl halides and benzothiazole,[219] N-alkyl-phenothiazines (24) are similarly prepared,[220] and some pyrazolo[1,5-a]pyrimidines (25) have been alkylated.[221] In the alkylation of 2,5-dimethyl-3-

(24)                              (25)

hydroxy-furan, -thiophen, and -selenophen, 'soft' MeI gives alkylation predominantly on carbon whereas 'hard' $Me_2SO_4$ leads mainly to O-alkylation. The thermodynamic equilibrium constants for the tautomeric equilibria of the three hydroxy-compounds have been compared.[222]

**Acylation.**—The acylation of benzene by $CO_2$–$AlCl_3$–Al is catalysed by transition-metal salts, *e.g.* $CuCl_2$, $WCl_6$, or $MoCl_5$.[223] A re-examination of the $AlCl_3$-catalysed reaction between benzene and 2-phenylbutanedioic anhydride has revealed the presence of 3-oxoindan-1-carboxylic acid among the products.[224] The reaction between benzene, $AlCl_3$, and thiophen-2,3-dicarbonyl chloride yields 2,3-dibenzoylthiophen, 3-benzoylthiophen-2-carbonyl chloride and -2-carboxylic acid, the naphthothiophen (26), and a thienofuran (27).[225] Benzo-, naphtho-, and pyrido-18-crown-6 have been benzoylated.[226]

[214] T. Tanabe, K. Yamauchi, and M. Kinoshita, *Bull. Chem. Soc. Japan*, 1977, **50**, 3021.
[215] E. A. Oostveen and H. C. Van der Plas, *Rec. Trav. chim.*, 1977, **96**, 64.
[216] D. Cech, R. Wohlfeil, and G. Etzold, East Ger. P. 122 381 (*Chem. Abs.*, 1977, **87**, 53 372).
[217] H. J.-M. Dou and J. Metzger, *Bull. Soc. chim. France*, 1976, 1861.
[218] R. Gassend, J. C. Maire, and J.-C. Pommier, *J. Organometallic Chem.*, 1977, **133**, 169.
[219] P. J. Nigrey and A. F. Garito, *J. Chem. and Eng. Data*, 1977, **22**, 451.
[220] J. Masse, *Synthesis*, 1977, 341.
[221] G. Auzzi, L. Cecchi, A. Costanzo, V. L. Pecori, and F. Bruni, *Farmaco, Ed. Sci.*, 1978, **33**, 14 (*Chem. Abs.*, 1978, **88**, 105 262).
[222] R. Lantz and A. B. Hornfeldt, *Chemica Scripta*, 1976, **10**, 126.
[223] M. Kh. Grigoryan and I. S. Kolomnikov, *Armyan. khim. Zhur.*, 1977, **30**, 357 (*Chem. Abs.*, 1977, **87**, 134 323).
[224] I. Hashimoto and R. Takatsuka, *Bull. Chem. Soc. Japan*, 1977, **50**, 2495.
[225] D. W. H. MacDowell and F. L. Ballas, *J. Org. Chem.*, 1977, **42**, 3717.
[226] J. C. Kauer, U.S. P. 4 024 158 (*Chem. Abs.*, 1977, **87**, 85 066).

(26)                    (27)

The effect of temperature, time, and amount of $AlCl_3$ on the acetylation of some alkyl-cyclohexyl-benzenes has been studied.[227] The product mixtures from the benzoylation of toluene with 2-, 3-, and 4-methylbenzoyl chlorides in the presence of $FeCl_3$ or $AlCl_3$ have been analysed.[228] Analogous studies on the xylenes and mesitylene have been published,[229] and 2-, 3-, and 4-chlorotoluenes have been chloroacetylated with $ClCH_2COCl$ and $AlCl_3$ or $FeCl_3$.[230]

The scope of the reaction between arenes and γ-butyrolactone has been investigated. Only activated aromatics can be annelated to tetralones: deactivated compounds react with rearrangement of the four-carbon fragment, yielding indanones. Thus, whereas benzene yields 1-tetralone, chlorobenzene gives a mixture of 1-indanones and -tetralones, and 1,3- and 1,4-dihalogeno-benzenes give only 1-indanones.[231] 1,3-Dichlorobenzene has been acylated with mono-and di-bromoacetyl chlorides[232] and a range of acetanilide derivatives similarly with dichloroacetyl chloride.[233] Spiro[4,5]decanes have been prepared *via* reaction between the anhydride of 1-carboxycyclopentane-1-acetic acid and chlorobenzene, t-butylbenzene, the xylenes, tetralin, and thiophen in the presence of $AlCl_3$ with subsequent reduction and cyclization.[234] Iron phosphate has been used to catalyse the benzoylation of a range of alkyl-benzenes and ethers,[235] and the catalysis of benzoylation of anisole, alkyl-benzenes, and halogeno-benzenes by a range of metal oxides has been studied.[236]

Naphthalene is diacetylated (with AcCl and $AlCl_3$) to give mainly the 1,5- and 1,6-products, small amounts of 1,3-, 1,7-, and 2,7-compounds also being formed.[237] Anthracene has been benzoylated with 4-methylbenzoyl[238] and 4-

[227] Sh. A. Ramazanova, R. A. Akhmedova, A. S. Suleimanov, E. M. Garyadyeva, and Z. I. Mamedova, *Azerb. khim. Zhur.*, 1976, 38 (*Chem. Abs.*, 1977, **87**, 134 295).
[228] K. D. Agzamova, Kh. Yu. Yuldashev, and N. G. Sidorova, *Zhur. org. Khim.*, 1977, **13**, 1452 (*Chem. Abs.*, 1977, **87**, 200 996).
[229] L. I. Leont'eva, Kh. Yu. Yuldashev, and N. G. Sidorova, *Zhur. org. Khim.*, 1977, **13**, 2178 (*Chem. Abs.*, 1978, **88**, 89 284); Kh. Yu. Yuldashev, *Zhur. org. Khim.*, 1977, **13**, 2369 (*Chem. Abs.*, 1978, **88**, 89 285).
[230] A. K. Abdushukurov, Kh. Yu. Yuldashev, and N. G. Sidorova, *Zhur. org. Khim.*, 1977, **13**, 2169 (*Chem. Abs.*, 1978, **88**, 89 283).
[231] C. A. Kerr and I. D. Rae, *Austral. J. Chem.*, 1978, **31**, 341.
[232] B. Sledzinski, L. Cieslak, and I. Missala, *Przemysl Chem.*, 1978, **57**, 23 (*Chem. Abs.*, 1978, **88**, 136 253).
[233] T. Chakrabortty, G. Podder, S. K. Desmukh, and N. N. Chakravarti, *Indian J. Chem., Sect. B*, 1977, **15**, 284.
[234] H. S. Bajaj and G. S. Saharia, *J. Indian Chem. Soc.*, 1976, **53**, 819.
[235] Kh. Yu. Yuldashev and N. G. Sidorova, Deposited Document 1975, *VINITI* 3325A (*Chem. Abs.*, 1978, **88**, 89 289).
[236] J. O. Morley, *J.C.S. Perkin II*, 1977, 601.
[237] P. H. Gore, M. Jehangir, and M. Yusuf, *Islamabad J. Sci.*, 1976, **3**, 16 (*Chem. Abs.*, 1978, **88**, 89 421).
[238] I. K. Buchina, A. I. Bokova, and N. G. Sidorova, Deposited Document 1976, *VINITI* 1351 (*Chem. Abs.*, 1978, **88**, 62 202).

methoxybenzoyl[239] chlorides with and without catalysts. The 9-substituted isomers are the major products along with small amounts of 1- and 2-substituted
compounds. Fluorene has been benzoylated with 4-bromo- and 2-methoxy-
benzoyl chlorides;[240] the 2-substituted compounds are the major products, along
with smaller amounts of the 4-isomers: FeCl$_3$ with MeNO$_2$ is superior to AlCl$_3$
with MeNO$_2$ as the catalyst for these reactions. Some 2-substituted fluorenyl
ethers have been acylated at the 7-position, following the usual pattern for
fluorene disubstitution.[241] Treatment of either 1-fluoro-9$H$-fluoren-9-one or the
3-fluoro-isomer with polyphosphoric acid at 140 °C for 4.5 h yields an equilibrium mixture of the two compounds (7:93), presumably *via* protonation,
deacylation, and intramolecular re-acylation. The authors regard this as the first
direct evidence of complete reversibility in an aromatic Friedel–Crafts acylation.[242] 2-Trifluoromethanesulphonyloxypyridine has been found useful in the
condensation of carboxylic acids with fluorene (to 2-acyl-fluorenes) and other
aromatics.[243] The utility of trifluoroacetylation of reactive aromatic and
heteroaromatic compounds as a route (*via* hydrolysis of the trifluoroacetyl
derivatives) to carboxylic acids has been stressed.[244] Deacylation has been
observed during attempted acetalization of some acetophenones bearing oxygen
functions at the 2- and/or 4-positions with ethylene glycol.[245]

The acetylation (with Ac$_2$O and HClO$_4$)[246] and Vilsmeier[247] formylation of
sterically hindered phenols have been investigated. Substituted *o*-hydroxy-
benzophenones have been prepared in 18–68% yields by treatment of the HMPT
complexes of bromomagnesium phenoxides with aromatic aldehydes.[248] Phenolic
*O*- *vs.* *C*-benzoylation has been studied, with particular reference to 3,4-
disubstituted phenols.[249] Previous work on the use of trifluoroacetic anhydride to
promote aromatic acylation has been extended to the preparation of symmetrical
and unsymmetrical benzophenones *via* reaction between the methyl and benzyl
ethers of orcinol and the same ethers of phloroglucinolcarboxylic acid.[250] Other
phenolic acylations include some chalcone syntheses[251] and the acetylation and
benzoylation of 2-hydroxy-4-methoxyacetophenone (peonol).[252]

Acylation of anisole with 3-oxocyclohexylacetyl chloride in CS$_2$ yields the
ketone (28) whereas in PhNO$_2$ the products are (4-MeOC$_6$H$_4$)$_2$C=CH$_2$ and the

239 A. I. Bokova, I. K. Buchina, N. G. Sidorova, and S. Teteneva, *Zhur. org. Khim.*, 1977, **13**, 1471 (*Chem. Abs.*, 1977, **87**, 134 801).
240 A. I. Bokova, N. G. Sidorova, and L. Grivtsova, Deposited Document 1975, *VINITI* 108 (*Chem. Abs.*, 1977, **87**, 22 841).
241 W. D. Jones, Jr., W. L. Albrecht, and F. P. Palopoli, *J. Org. Chem.*, 1977, **42**, 4144.
242 I. Agranat, Y. Bentor, and Y.-S. Shih, *J. Amer. Chem. Soc.*, 1977, **99**, 7068.
243 T. Keumi, H. Saga, R. Taniguchi, and H. Kitajima, *Chem. Letters*, 1977, 1099.
244 R. K. Mackie, S. Mhatre, and J. M. Tedder, *J. Fluorine Chem.*, 1977, **10**, 437.
245 K. Grözinger and F. Hess, *Synthesis*, 1977, 411.
246 M. V. Nekhoroshev, E. P. Ivakhnenko, and O. Yu. Okhlobystin, *Zhur. org. Khim.*, 1977, **13**, 662.
247 S. Morimura, H. Horiuchi, and K. Murayama, *Bull. Chem. Soc. Japan*, 1977, **50**, 2189.
248 G. Casnati, M. Colli, A. Pochini, and R. Ungaro, *Chimica e Industria*, 1977, **59**, 764.
249 I. N. Zemzina and N. G. Sidorova, *Tezisy Vses. Simp. Org. Sint.: Benzoidnye Aromat. Soedin.*, *1st*, 1974, 50 (*Chem. Abs.*, 1977, **87**, 5577).
250 E. G. Sundholm, *Tetrahedron*, 1977, **33**, 991.
251 S. P. Starkov and A. I. Panasenko, *Tezisy Vses. Simp. Org. Sint.: Benzoidnye Aromat. Soedin.*, *1st*, 1974, 37 (*Chem. Abs.*, 1977, **87**, 5569); A. S. Gupta, P. L. Trivedi, and J. R. Merchant, *Indian J. Chem.*, *Sect. B*, 1976, **14**, 903.
252 D. B. Jhaveri, V. M. Thakor, and H. B. Naik, *Vidya, B*, 1976, **19**, 149 (*Chem. Abs.*, 1977, **87**, 201 001).

bicyclo-octane (29).[253] Other acylations include those of alkyl phenyl ethers with benzoyl[254] and[255] 4-methylbenzoyl chlorides, phenols, phenol ethers, methyl-benzenes, and polycyclic aromatics with acid chlorides and anhydrides in the

(28)    (29)

presence of $FeCl_3$,[256] and of di- and tri-methoxybenzenes and alkyl-benzofurans with ring–chain tautomers, *e.g.* *o*-acetylbenzoyl chloride.[257] The cyclizations of some 3,5-disubstituted phenylalkanoic acids to tetralones and indanones with five different catalysts have been investigated in a study of the influence of MeO- *vs.* Cl-substitution and of the cyclizing agent on product distribution.[258]

Ferrocenecarboxylic acid has been prepared in high (74–83%) yield by benzoyl-ation of ferrocene with 2-chlorobenzoyl chloride plus $AlCl_3$, and subsequent cleavage with $KOBu^t$.[259] Phenylferrocene, and 22 derivatives substituted in the phenyl ring, have been acetylated ($Ac_2O–H_3PO_4$) and formylated (Vilsmeier) in a study of the effects of phenyl substituents on product distribution. The 1,1'- and 1,3-isomers are the main products.[260] In the acylation of methyl ferrocenecar-boxylate with $Ac_2O$ and $BF_3 \cdot Et_2O$ or AcCl, the main product is the 1'-acetyl derivative.[261] The previously reported diacetylation of cyclobutadieneiron tri-carbonyl has been used in the synthesis of 3,6-dimethylpyridazino[4,5-*a*]cyclo-butadieneiron tricarbonyl (30), this representing the first synthesis of a heteroareno-cyclobutadiene.[262]

Vilsmeier formylation of $\beta$-thienylacetonitrile gives the thieno[2,3-*c*]pyridine (31). With excess of reagent, N-$\beta$-thienyl and -seleno-phenylacetamides similarly

(30)    (31)

[253] V. Dabral, H. Ila, and N. Nanda, *Chem. and Ind.*, 1977, 952.
[254] Kh. Yu. Yuldashev, N. G. Sidorova, and S. U. Mukhamedalieva, Deposited Document 1975, *VINITI* 3171 (*Chem. Abs.*, 1978, **88**, 89 287).
[255] Kh. Yu. Yuldashev, N. G. Sidorova, and S. U. Mukhamedalieva, Deposited Document 1975, *VINITI* 3325 (*Chem. Abs.*, 1978, **88**, 89 291).
[256] N. G. Sidorova and Kh. Yu. Yuldashev, *Tezisy Vses. Simp. Org. Sint.: Benzoidnye Aromat. Soedin.*, *1st*, 1974, 33 (*Chem. Abs.*, 1977, **87**, 5568).
[257] W. Lonsky and H. Traitler, *Chem. Ber.*, 1977, **110**, 2601.
[258] M. S. Newman and J. O. Landers, *J. Org. Chem.*, 1977, **42**, 2556.
[259] P. C. Reeves, *Org. Synth.*, 1977, **56**, 28.
[260] Š. Toma, A. Maholanyiová, and E. Solčaniová, *Coll. Czech. Chem. Comm.*, 1977, **42**, 1013.
[261] M. N. Nefedova, V. N. Setkina, V. A. Kataev, T. V. Fisun, and D. N. Kursanov, *Doklady Akad. Nauk S.S.S.R.*, 1977, **237**, 352 (*Chem. Abs.*, 1978, **88**, 89 814).
[262] I. G. Dinulescu, E. G. Georgescu, and M. Avram, *J. Organometallic Chem.*, 1977, **127**, 193.

give thieno- and selenolo-[3,2-*b*]pyridines *via* initial attack at C-2.[263] A thorough re-examination of the much-neglected Vilsmeier aroylation of pyrroles has been published. Morpholides are, by and large, better than *NN*-dimethyl-amides. An excess of POCl$_3$ is advantageous since the viscosity of the reaction medium is thereby reduced and the rate of complex formation is enhanced. The reported conditions for the formylation of pyrrole are unnecessarily harsh and unsuitable for aroylation of pyrrole.[264] In a reinvestigation of the acetylation of pyrrole, 2,4-diacetylpyrrole has been isolated as the major product when trifluoroacetic anhydride was used as catalyst.[265] A useful study of acylation with agents of the C=O, C=N, and C≡N type of the pyrrolyl ambident anion has appeared, and the results have been rationalized on the HSAB principle. The metal cation, solvent, complexing agent, halide of the pyrrole Grignard, and temperature were varied; thus, magnesium and ethyl chloroformate give more *C*- than *N*-acylation, whereas lithium to potassium give only *N*-alkylation under comparable conditions: the decreasing hardness Mg > Li > Na > K implies a lowered association with the hard N-centre and therefore decreased hindrance to attack at that position. Addition of TMEDA increases *N*-acylation since increasing cation solvation weakens the attraction between the cation and hard N. Change of Grignard halide from Cl to I leads to decreased *N*-acylation: the softer the halide, the softer the magnesium becomes, thus weakening its co-ordinating ability.[266] Acetylation of 5-hydroxy-4-(2-hydroxyethyl)-3-methylpyrazole (for which there are eight possible tautomers) yields five acetyl derivatives.[267] Rate constants for Vilsmeier acetylation of a range of pyrroles, indoles, and carbazoles have been reported: for most substrates, *N*-acetylation occurs *via* rate-limiting direct attack on N, and not *via* initial attack at C and rearrangement.[268] Indoles can be *N*-acetylated conveniently with Ac$_2$O and KOH in Me$_2$SO.[269] Activated quinones have been used in the acylation of azulenes, benzofuran, and indoles.[270] Treatment of indole with 2-methylthio-2-phenyl-1,3-dithian and BF$_3$·Et$_2$O yields the indolyl-phenyl-dithian, which is a precursor for both 3-alkyl- and 3-aroyl-indoles.[271]

8-Azaindolizines (32) undergo Vilsmeier formylation and thioformylation preferentially at C-3, in agreement with M.O. calculations.[272] Acylation of 2-cyanomethylbenzimidazole under Schotten–Baumann conditions gives *N*-

(32)

[263] C. Paulmier and F. Outurquin, *J. Chem. Res. (S)*, 1977, 318.
[264] J. White and G. McGillivray, *J. Org. Chem.*, 1977, **42**, 4248.
[265] A. G. Anderson, Jr., and M. M. Exner, *J. Org. Chem.*, 1977, **42**, 3952.
[266] N.-C. Wang and H. J. Anderson, *Canad. J. Chem.*, 1977, **55**, 4103.
[267] H. Ochi, T. Miyasaka, and K. Arakawa, *Bull. Chem. Soc. Japan*, 1977, **50**, 2991.
[268] A. Cipiciani, S. Clementi, P. Linda, G. Marino, and G. Savelli, *J.C.S. Perkin II*, 1977, 1284.
[269] G. W. Gribble, L. W. Reilly, Jr., and J. L. Johnson, *Org. Prep. Proced. Internat.*, 1977, **9**, 271.
[270] J. N. Tsaklidis, A. Hofer, and C. H. Eugster, *Helv. Chim. Acta*, 1977, **60**, 1033.
[271] P. Stütz and P. A. Stadler, *Org. Synth.*, 1977, **56**, 8.
[272] R. Buchan, M. Fraser, and C. Shand, *J. Org. Chem.*, 1977, **42**, 2448.

acylation.[273] Uric acid undergoes cleavage and rearrangement with isobutyric anhydride to a new heterocyclic derivative (33), whose formation has been

(33)

rationalised in terms of cleavage of the rings by acyl exchange, giving a resonance-stabilized *NN'N'''*-tri-isobutyryl intermediate which undergoes ring closure and elimination of isobutyric acid.[274] Other heteroaromatic acylations include the carbamoylation of 5-fluoro-uracils,[275] the introduction of formyl, carboxyl, and other carbonyl functions into the $\beta$-pyridine positions *via* the use of $Fe(CO)_5$,[276] the acetylation[277] and aroylation of substituted benzofurans,[278] and the benzoylation of dibenzofuran.[279] 6-Hydroxy-1,2-benzisoxazole undergoes Reimer–Tiemann formylation at the 7-position.[280] Attempted aroylation of 5-amino-1,2,3,4-thiatriazole affords 2,5-diaryl derivatives of the hitherto unknown 1,6-dioxo-6a-thia-3,4-diazapentalene system (34). Acetylation, however, gives 3,5-diacetamido-1,2,4-thiadiazole.[281]

(34)

**Silicon Substitutions.**—The double Friedel–Crafts reaction of chloro-methyltrichlorosilane ($ClCH_2SiCl_3$) with benzene and $AlCl_3$ yields 75% of mainly *p*-(with some *o*- and *m*-)di(trichlorosilylmethyl)benzene, previously unknown.[282] Trimethylsilylbenzenes can be prepared directly by reaction between aryl chlorides, bromides, or iodides, hexamethyldisilane, and KOMe (or the Na or Li salts) in HMPT at 25 °C for 3 h. Yields are in the range 63–92%, depending on the nature of the aryl halide; the major side-product is ArH. The mechanism is not known, but may involve an Ar–K intermediate.[283] 1,8-Bis-(trimethylsilyl)- and -(trimethylstannyl)-naphthalenes have been prepared from the 1,8-dilithio-compound and $Me_3SiCl$ or $Me_3SnCl$. The n.m.r. spectra seem to imply that there is much greater hindrance to rotation about the C—Si than about the C—Sn bonds: thus the silicon compound displays three methyl resonances (incredibly)

[273] Y. Okamoto and T. Ueda, *Chem. and Pharm. Bull. (Japan)*, 1977, **25**, 3087.
[274] B. Coxon, A. J. Fatiadi, L. T. Sniegoski, H. S. Hertz, and R. Schaffer, *J. Org. Chem.*, 1977, **42**, 3132.
[275] S. Ozaki, Y. Ike, H. Mizuno, K. Ishikawa, and H. Mori, *Bull. Chem. Soc. Japan*, 1977, **50**, 2406.
[276] C.-S. Giam and K. Ueno, *J. Amer. Chem. Soc.*, 1977, **99**, 3166.
[277] J.-M. Clavel, J. Guillaumel, P. Demerseman, and R. Royer, *J. Heterocyclic Chem.*, 1977, **14**, 219.
[278] J. Astoin, P. Demerseman, N. Platzer, and R. Royer, *J. Heterocyclic Chem.*, 1977, **14**, 861.
[279] T. Keumi, S. Shimakawa, and Y. Oshima, *Nippon Kagaku Kaishi*, 1977, 1518 (*Chem. Abs.*, 1978, **88**, 36 840); A. G. Khaitbaeva, N. G. Sidorova, and Kh. Yu. Yuldashev, *Khim. geterotsikl. Soedinenii*, 1977, 895 (*Chem. Abs.*, 1977, **87**, 134 891).
[280] K. A. Thakar and B. M. Bhawal, *Indian J. Chem., Sect. B*, 1977, **15**, 1056.
[281] R. J. S. Beer and I. Hart, *J.C.S. Chem. Comm.*, 1977, 143.
[282] J. Dunoguès, N. Duffaut, and P. Lapouyade, *Bull. Soc. chim. France*, 1976, 1933.
[283] M. A. Shippey and P. B. Dervan, *J. Org. Chem.*, 1977, **42**, 2654.

even at 150 °C, whereas in the case of tin a singlet is observed at room temperature and down to −100 °C. The authors have rationalised these observations in terms of covalent radii and consequent out-of-planarity of *peri*-substituents.[284]

Uracil derivatives, alkylated on nitrogen, may be prepared by treatment of the free base with trimethylsilyl halides followed by alkylation of the products with, *e.g.*, 2,3-dihydrofuran or 2,3-dihydropyran.[285] Analogous alkylation *via* silylation at N-1 of 1,2,4-triazolo[4,3-*a*]pyridin-3(2*H*)-one (35) yields meso-ionic triazolopyridinium hydroxides (36).[286] Iodobenzenes may be prepared in excellent yield by the reaction of the corresponding trimethylsilyl-benzenes with $I_2$, ICl, or IBr.[287]

(35)                    (36)

## 5 Reactions with Electrophilic Nitrogen and Phosphorus

**Nitration.**—Whereas a plot of localisation energies *versus* partial rate factors for nitration of toluene, *m*- and *p*-xylene, and mesitylene is curved, other electrophilic substitutions (*e.g.* halogenation) yield linear correlations. The curvature has been ascribed to a shift of the nitration transition state from resemblance of the intermediate towards resemblance of the reactant with increasing ring substitution.[288]

The use of empirical linear relationships (for nitrations in moderately concentrated aqueous $H_2SO_4$ solutions) between the observed rates of aromatic nitration and assorted acidity functions as a criterion of mechanism has been criticised. In particular, the rate profiles for nitration of halogeno-benzenes used in the analysis display a breakdown of previous relationships either because of their incompatibility with the reaction mechanism ($H_0$ function) or because of the approximations intrinsic in the $H_R$ acidity function.[289] The utility of such relationships is also discussed in a reconsideration of the nitration (in $H_2SO_4$ over a range of concentrations) of acetanilide and some analogues. For weak bases, especially those containing oxygen functions, the occurrence of H-bonding can render the criterion ambiguous or anomalous. Acetanilide is nitrated as the free base and not as the cation, as previously reported: the marked increase in *ortho* : *para* ratio as the acidity of the reaction medium is lowered has been attributed to decreasing H-bonding. There is no evidence for the occurrence of a special mechanism for *ortho*-substitution of the form shown in Scheme 1.[290] In a further paper involving a reinvestigation of the nitration of benzene and halo-

[284]  D. Seyferth and S. C. Vick, *J. Organometallic Chem.*, 1977, **141**, 173.
[285]  K. Sakai and Y. Inamoto, Japan. Kokai 77  31 079.
[286]  A. Saito and B. Shimizu, *Bull. Chem. Soc. Japan*, 1977, **50**, 1596.
[287]  G. Félix, J. Dunoguès, F. Pisciotti, and R. Calas, *Angew. Chem. Internat. Edn.*, 1977, **16**, 488.
[288]  M. Sohrabi and T. Kaghazchi, *Pakistan J. Sci. Ind. Res.*, 1976, **19**, 133.
[289]  P. G. Traverso, N. C. Marziano, and R. C. Passerini, *J.C.S. Perkin II*, 1977, 845.
[290]  R. B. Moodie, P. N. Thomas, and K. Schofield, *J.C.S. Perkin II*, 1977, 1693.

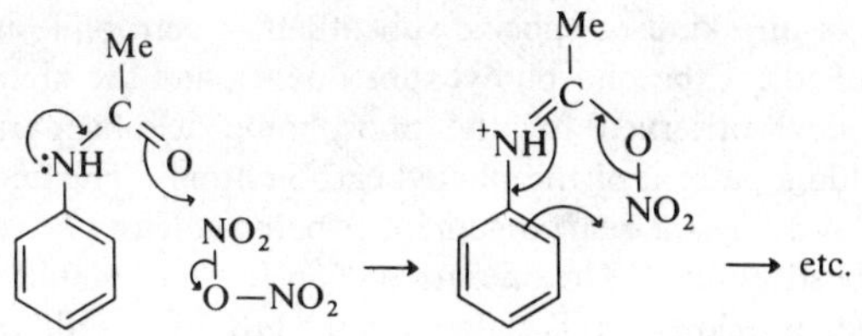

**Scheme 1**

genobenzenes in aqueous $H_2SO_4$ over a wide acidity range, Marziano *et al.* conclude that deviations from linearity of profiles of rate *versus* acidity function are not significant evidence for variation in mechanism, and that 'rate profiles are of little value for understanding mechanistic problems and appear in some cases not related to chemical behaviour of compounds'.[291]

In a study of the kinetics of nitration of benzene, toluene, *m*-xylene, mesitylene, and anisole by nitric acid in $Ac_2O$, the rate of nitration and the order with respect to acid are both sensitive to prior solvent treatment. Conductivity measurements show that the effect is due to some trace impurities in the solvent which on protonation give rise to $NO_3^-$ ions. Rates of nitration in purified solvent are sensitive to added $NO_3^-$ ions and water but almost unchanged by added AcOH. It has been deduced from arguments based on the true rate constant for electrophilic attack that reaction occurs through the $NO_2^+$ ion.[292]

In an attempt to solve the paradox that nitration of reactive aromatics displays intramolecular but not intermolecular selectivity, the necessity for a mechanism involving initial (diffusion-controlled) electron transfer to a radical pair [equation (1)] has been argued. Evidence in support of this contention includes polaro-

$$NO_2^+ + ArH \rightarrow \overline{NO_2^{\cdot} + ArH^{+\cdot}} \rightarrow HArNO_2^+ \qquad (1)$$

graphic data which demonstrate that electron transfer from reactive aromatics to $NO_2^+$ is exothermic and also the observation that electrochemical generation of the radical pair from naphthalene produces the same product mixture as that formed during conventional nitration.[293] The maintenance of regioselectivity of *ortho-*/*para-* over *meta*-substitution in the nitration of toluene, anisole, and *o*-xylene, regardless of the reactivity of the nitrating system, has been discussed by Olah and co-workers.[294]

The kinetics of nitration of 4-chloro-benzotrifluoride have been studied. The rate decreases in the order of solvents $H_2SO_4 > HNO_3 > MeNO_2 > AcOH \gg H_2O$ and passes through a maximum at a mixture corresponding to 34.6% free $SO_3$.[295] The substituent effects of the carboxyl and phosphonic groups have been compared through measurement of the partial rate factors for nitration (in $Ac_2O$ and in strongly acidic media) of $PhCH_2R$ and $PhCH=CHR$, where $R = CO_2H$ and $PO(OH)_2$. The two groups display a close similarity; in particular, there are

[291] N. C. Marziano, A. Zingales, and V. Ferlito, *J. Org. Chem.*, 1977, **42**, 2511.
[292] N. C. Marziano, R. Passerini, J. H. Rees, and J. H. Ridd, *J.C.S. Perkin II*, 1977, 1361.
[293] C. L. Perrin, *J. Amer. Chem. Soc.*, 1977, **99**, 5516.
[294] G. A. Olah, H. C. Lin, J. A. Olah, and S. C. Narang, *Proc. Nat. Acad. Sci. U.S.A.*, 1978, **75**, 545.
[295] J. Muchova and M. Paldan, *Chem. průmysl*, 1977, **27**, 187 (*Chem. Abs.*, 1977, **87**, 38 416).

no indications of any kind of specific substituent–electrophile interactions (as previously invoked for the phosphinyl substituent), and the high proportion of *ortho*-nitro-product observed for the phosphonic acid is paralleled by the $^{13}$C n.m.r. shielding pattern of the phenyl carbon atoms. The authors postulate *ipso*-attack followed by 1,2-rearrangement as being at least partially responsible for the *ortho*-substitution.[296] The operation of an $I_\pi$ effect (simultaneously with a $\sigma$-inductive and/or field effect) is claimed in the nitration (with fuming HNO$_3$ in excess of Ac$_2$O, at 25 °C) of toluene, t-butylbenzene, and various 4-substituted 1-phenylbicyclo[2.2.2]octanes. Log(partial rate factors) correlate with $\sigma_1$ values. Substituent effects on positions in the benzene ring decrease in the order $o > p > m$.[297]

The first detailed study of nitration in trifluoroacetic acid has been published. The kinetics of the quantitative mononitration of benzene, chlorobenzene, toluene, and di- and tri-methylbenzenes in CF$_3$CO$_2$H containing HNO$_3$ and water at 25 °C are reported. The rate of nitration under these conditions is limited either by the rate of formation of NO$_2^+$ or by a step which is insensitive to the nature of the reactive aromatic. It has been proposed that the nitrations occur by *ipso*-attack and subsequent migration of the NO$_2$ group to an unsubstituted position.[298] Direct evidence for bulk formation of a $\sigma$-complex from *ipso*-attack under normal nitration conditions has been provided by $^1$H and $^{13}$C n.m.r. spectra. Thus, the reaction of *NN*-dimethyl-*p*-toluidine with HNO$_3$ in 70–77% H$_2$SO$_4$ at 0 °C proceeds in two clearly separated stages: *ipso*-attack to the ion (37) followed by slow rearrangement (obeying approximately first-order kinetics) to (38).[299] The migratory aptitude of the NO$_2$ group in *ipso*-adducts has been

(37)          (38)

investigated in the 1,2-dimethyl-1-nitro-cyclohexadienyl cation (39) by studying the solvolysis (in 85% H$_2$SO$_4$, at 0 °C) of (40) to equal amounts of (41) and (42).

(39)          (40)          (41)          (42)

Given that an intramolecular 1,2-shift of NO$_2$ adequately summarizes the data, then the distribution of labelled products provides a measure of the rate of

[296] T. A. Modro, W. F. Reynolds, and E. Skorupowa, *J.C.S. Perkin II*, 1977, 1479.
[297] S. Sotheeswaran and K. J. Toyne, *J.C.S. Perkin II*, 1977, 2042.
[298] R. B. Moodie, K. Schofield, and G. D. Tobin, *J.C.S. Perkin II*, 1977, 1688.
[299] K. Fujiwara, J. C. Giffney, and J. H. Ridd, *J.C.S. Chem. Comm.*, 1977, 301.

migration of NO$_2$ to an equivalent *ipso* site relative to that for migration to an adjacent open site bearing hydrogen. The latter is *ca.* $\frac{1}{50}$th the former.[300] The shifts of nitro-groups in 4-methyl-4-nitrocyclohexa-2,5-dienones have been characterised through the preparation of a range of crystalline *ipso*-nitration products by nitrating ROAc (R = *p*-tolyl, 3,4-xylyl, or 3,4,5-Me$_3$C$_6$H$_2$) with HNO$_3$ and Ac$_2$O. The products decompose in a range of solvents by a clean first-order process to *o*-nitro-phenols *via* a radical dissociation–recombination mechanism (Scheme 2): the pathway for this formal 1,3-shift of NO$_2$ has a major intermolecular component.[301]

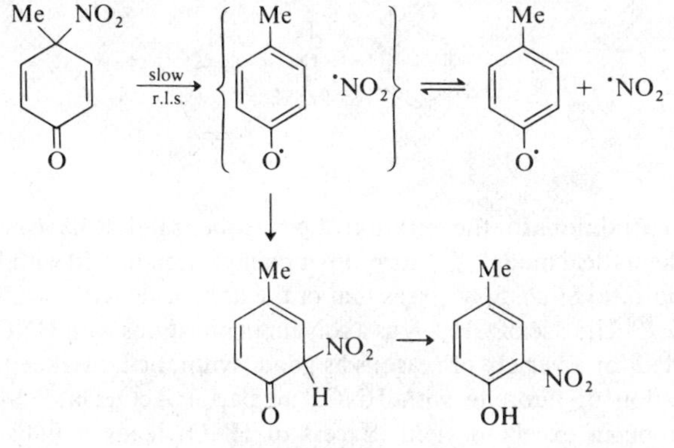

**Scheme 2**

The solvent effect on the aromatization of the *cis*- and *trans*-isomers of 4-nitro-3,4,5-trimethylcyclohexa-2,5-dienyl acetate (43) (prepared from acetyl nitrate and 1,2,3-trimethylbenzene) has been reported and two major pathways of solvolytic aromatization have been identified. In 50% H$_2$SO$_4$, loss of acetate yields the *ipso*-ion (44), which is then partitioned along two paths, one leading to oxidation (to 5-hydroxy-1,2,3-trimethylbenzene), the other to migration of the NO$_2$ group, eventually giving 4-nitro-1,2,3-trimethylbenzene. Aromatization in weakly acidic systems leads to loss of nitrite ion, formation of ion (45), and subsequent proton loss to yield 5-acetoxy-1,2,3-trimethylbenzene.[302]

*ipso*-Attack has been invoked in the nitration of aryl-cyclopropanes[303] and in the nitrative cyclisation of methyl $\beta$-arylisovalerate to the nitrohydrocoumarin system.[304]

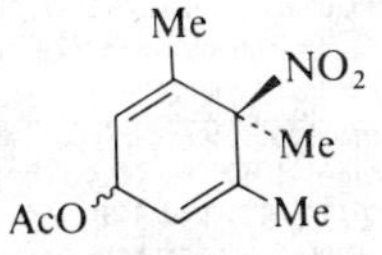

(43) 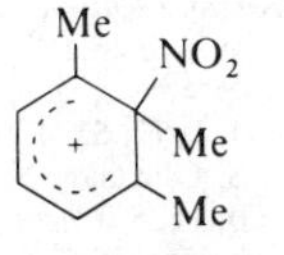           (44) 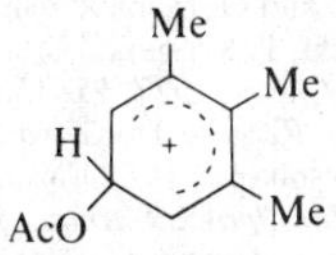           (45)

[300] C. E. Barnes and P. C. Myhre, *J. Amer. Chem. Soc.*, 1978, **100**, 975.
[301] C. E. Barnes and P. C. Muhre, *J. Amer. Chem. Soc.*, 1978, **100**, 973.
[302] T. Banwell, C. S. Morse, P. C. Myhre, and A. Vollmar, *J. Amer. Chem. Soc.*, 1977, **99**, 3042.
[303] S. S. Mochalov, N. B. Matveeva, I. P. Stepanova, and Yu. S. Shabarov, *Zhur. org. Khim.*, 1977, **13**, 1639 (*Chem. Abs.*, 1977, **87**, 184 123).
[304] M. Shinoda and H. Suzuki, *J.C.S. Chem. Comm.*, 1977, 479.

Because of the inefficiency of the nitrodeiodination reaction and the faster rate of nitration of *o*-xylene relative to 4-iodo-*o*-xylene, $I_2$ cannot be used as a catalyst to alter the substitution pattern in the nitration of *o*-xylene.[305] Mononitration of *m*-xylene with $HNO_3$ in a liquid mixture of *m*-benzenedisulphonic acid and phosphoric acid gives excellent yields and a 4-nitro-isomer content higher than that obtained by the usual method. In a typical experiment, the yield was 96% and the 4- to 2-isomer ratio *ca.* 7.5.[306] A white product obtained during the nitration of toluene has been identified as (46).[307] The catalytic *para*-nitration of substituted benzenes, especially toluene, has been reviewed, but in Japanese.[308]

(46)

Optimum conditions for the nitration of *p*-nitrobenzanilide have been deduced from a mathematical model.[309] Nitration of phenylsuccinic acid with $HNO_3$ gives a *para* : *ortho* ratio of *ca.* 5, whereas that of the anhydride with $Ac_2ONO_2$ gives a ratio of *ca.* 2.[310] The use of $CH_2Cl_2$ as a solvent in nitrations with $HNO_3$–$H_2SO_4$ is recommended for a variety of reasons as good 'synthetic housekeeping'.[311]

The nitration of fluorene with $HNO_3$ in glacial AcOH at 55–85 °C gives 2-nitrofluorene in excellent yield. Excess of $HNO_3$ leads mainly to the 2,7-dinitro-compound, with smaller amounts of the 2,5-isomer.[312] Conventional nitration of fluorenone shows poor selectivity, but nitration with $Cu(NO_3)_2$–AcOH–$Ac_2O$ for 90 minutes at 40 °C gives only 2-nitrofluorenone, in good yields.[313]

Increasing the temperature in the nitration of 1-nitronaphthalene increases the amount of 1,5-dinitro-isomer slightly, but the composition of the nitrating mixture appears to have no effect on the ratio of 1,5- to 1,8-isomers. The differences in activation energy and Arrhenius pre-exponential factor for formation of the 1,5- and 1,8-isomers are 0.5 kcal mol$^{-1}$ and 1.43 respectively.[314] Nitration of 8-chloro- and -bromo-naphthalene-1-sulphonic acids occurs *ortho* and *para* to the halogen: small amounts of 1-halogeno-8-nitro- and 1-halogeno-2,4,8-trinitro-naphthalenes are formed as side-products.[315] Naphthalene 1,8-

[305] A. Zweig, K. R. Huffman, and G. W. Nachtigall, *J. Org. Chem.*, 1977, **42**, 4049.
[306] T. Kameo and O. Manabe, *Nippon Kagaku Kaishi*, 1977, 691 (*Chem. Abs.*, 1977, **87**, 134 194).
[307] A. Kotarski, T. Krasiejko, D. Wasiak-Wisniewska, S. Galazka, T. Potocka-Smolka, and K. Gruber, *Roczniki Chem.*, 1977, **51**, 1357.
[308] T. Kameo, *Kagaku To Kogyo (Osaka)*, 1977, **51**, 136 (*Chem. Abs.*, 1977, **87**, 167 626).
[309] L. F. Myasnikova, I. L. Vaisman, S. S. Gluzman, N. I. Faingol'd, D. A. Novokhatka, and S. G. Panina, *Zhur. priklad. Khim.*, 1977, **50**, 1428 (*Chem. Abs.*, 1977, **87**, 134 380).
[310] F. Cuiban, A. Lupea, M. Silasi, and M. Sora, *Rev. Roumaine Chim.*, 1977, **22**, 869.
[311] G. Davis and N. Cook, *CHEMTECH*, 1977, **7**, 626.
[312] A. I. Levchenko and V. S. Efremova, Deposited Document 1974, *VINITI* 286 (*Chem. Abs.*, 1977, **87**, 22 843).
[313] J. Pielichowski and J. Obrzut, *Monatsh.*, 1977, **108**, 1163.
[314] I. K. Barvinskaya, *Izvest. V. U. Z., Khim. i khim. Tekhnol.*, 1977, **20**, 1457 (*Chem. Abs.*, 1978, **88**, 50 538).
[315] V. N. Lisitsyn and G. S. Stankevich, *Zhur. org. Khim.*, 1977, **13**, 1286 (*Chem. Abs.*, 1977, **87**, 117 710).

disulphide is nitrated initially at the 2-, 4-, 5-, and 7-positions, but all possible mono-, di-, tri-, and tetra-nitro-compounds can be isolated from the reaction mixture.[316]

Nitrations with urea nitrate in phosphoric acid (of bromobenzene, anisole, acetophenone, and 1,3-dimethoxyxanthene)[317] and in sulphuric acid (of anisole, acetophenone, diphenyl ether and benzyl methyl ether)[318] are reported: yields up to 66% are quoted. A statement in the literature that the value of nitrodesilylation is limited by the danger of explosions from the $HNO_3$–$Ac_2O$ mixture has been corrected: the mixture *can* be explosive, but no case is known of an explosion occurring under the conditions for nitrodesilylation. The reaction therefore remains very useful synthetically, as in, for example, the conversion of 1,4-di-trimethylsilyl-benzene into 4-nitrotrimethylsilylbenzene.[319] A substantial review has appeared on the side-reactions, particularly, reaction on substituent groups and addition reactions that can accompany aromatic nitration.[320]

A re-examination of the nitration of *N*-alkyl-phthalimides (with $HNO_3$–$H_2SO_4$) confirms earlier reports that the 4-nitro- is favoured over the 3-nitro-product. Yields are good except for long alkyl substituents (*e.g.* hexyl and octyl), when oxidation of the side-chain becomes significant.[321] In agreement with earlier work, nitration of 7-methyl- and -ethyl-quinoline yields only the 8-nitro-products; previous reports of minor products resulted from contamination of the starting material with 5-methylquinoline, which is nitrated at the 6- and 8-positions.[322] Nitration of isoquinoline *N*-oxide and of 1-cyanoisoquinoline *N*-oxide gives small amounts of 5-, 6-, and 8-nitro-derivatives.[323] The orientation of nitration of 2-phenylquinoline *N*-oxide with $KNO_3$–$H_2SO_4$ is principally governed by the concentration of $H_2SO_4$. Thus, whereas concentrated acid favours the 3'-nitro-product, the 4-nitro-derivative is preferred in 70—75% acid. Further nitration readily gives the 3',4-dinitro-product.[324]

Nitration of 1-methylpyrazole (with five-fold excess of 80% $H_2SO_4$, at 100 °C, for 18 h) gives 1-methyl-4-nitro- and 1-methyl-3,4-dinitro-pyrazoles in 4:1 ratio. The dinitro-compound (a new substance, which cannot be prepared by further nitration of 1-methyl-4-nitropyrazole) is formed by nitration of the 3-nitro-compound, which, although not isolated under the reaction conditions, can be detected by n.m.r. spectroscopy during reaction.[325] Significantly higher yields than in the conventional nitration of substituted thiazoles have been obtained using a mixture of $F_3CSO_3H$, $(F_3CSO_2)_2O$, and $KNO_3$ in 1,2-dichloro-ethane.[326] The kinetics of nitration of 1,4,5-trimethylimidazole 3-oxide and 1-methylpyrazole 2-oxide in $H_2SO_4$ have been studied: the compounds undergo

[316] B. I. Stepanov, V. Ya. Rodionov, and S. E. Voinova, *Zhur. org. Khim.*, 1977, **13**, 841 (*Chem. Abs.*, 1977, **87**, 39 339).
[317] V. B. Nabar and N. A. Kudav, *Indian J. Chem., Sect. B*, 1977, **15**, 89.
[318] M. P. Mujumdar and N. A. Kudav, *Indian J. Chem., Sect. B*, 1976, **14**, 1012.
[319] C. Eaborn, *J. Organometallic Chem.*, 1978, **144**, 271.
[320] H. Suzuki, *Synthesis*, 1977, 217.
[321] F. J. Williams and P. E. Donahue, *J. Org. Chem.*, 1978, **43**, 1608.
[322] P. A. Claret and A. G. Osborne, *Tetrahedron*, 1977, **33**, 1765.
[323] M. Hamana and H. Saito, *Heterocycles*, 1977, **8**, 403.
[324] M. Hamana, S. Takeo, and H. Noda, *Chem. and Pharm. Bull. (Japan)*, 1977, **25**, 1256.
[325] M. R. Grimmett and K. H. R. Lim, *Austral. J. Chem.*, 1978, **31**, 689.
[326] L. Grehn, *J. Heterocyclic Chem.*, 1977, **14**, 917.

mononitration as the free bases at C-2 and C-5 respectively. At high acidities, the pyrazole gives 1-methyl-3,5-dinitropyrazole 1-oxide.[327] The nitration of some 4-(4-alkoxyphenyl)imidazoles has also been reported.[328] Polynitro-indazoles have been prepared *via* nitration of mono- and di-nitroindazoles.[329]

Kinetic studies on the nitration of 3-methyl-1,2-benzisoxazole lead to the conclusion that the compound is nitrated as the free base in the 80—90% $H_2SO_4$ region, while at higher acidities the conjugate acid is nitrated. Like the parent benzisoxazole, the compound is nitrated exclusively at the 5-position, even though M.O. calculations indicate the highest charge density to be at position 7. The authors have concluded that the nitration is not charge-controlled, that the orbital term prevails over the coulombic term, and that the coefficients of the frontier orbitals are larger for position 5 than for 7.[330] Some ambiguities in structures of nitro-products from nitration of 1,2-benzisoxazoles have been resolved by synthesis from nitro-*o*-hydroxyaryl ketones and/or by n.m.r. spectroscopy.[331]

In the nitration of 2-alkenyl-furans and -thiophens with $N_2O_4$, the furans are substituted in the ring (5-position) whereas the thiophens undergo substitution on the alkenyl side-chain.[332] The nitration of furan, its 2-carbaldehyde and the diacetate, and methyl 2-furoate with fuming $HNO_3$–$Ac_2O$ has been studied over a range of temperatures. The nitrating mixture prepared between −20 and −40 °C appears to be a better nitrating agent than that prepared at 0 °C.[333] $CuNO_3$ in $Ac_2O$ at 8—10 °C nitrates 5-formyl-and -acetyl-2,2'-bithienyls at the 3'- and 5'-positions (mole ratios of products are 1 : 1.6 and 2 : 1 respectively). Increased acidity of the medium and lower temperatures lead to increasing amounts of 5'-isomers.[334] The same reagent [or $Al(NO_3)_3$] has been applied to the 2-nitration of thiophens.[335] Steric effects have been observed in the nitration of 1,2-dihydropyrrolizines (47; $R^1$ = H or Me; $R^2$ = H, Me, or Bu$^t$), from which 5-,

(47)

[327] I. J. Ferguson, K. Schofield, J. W. Barnett, and M. R. Grimmett, *J.C.S. Perkin I*, 1977, 672.
[328] M. A. Iradyan, A. G. Torosyan, R. G. Mirzoyan, and A. A. Aroyan, *Khim. geterotsikl. Soedinenii*, 1977, 1384 (*Chem. Abs.*, 1978, **88**, 62 333).
[329] M. S. Pevzner, N. V. Gladkova, G. A. Lopukhova, M. P. Bedin, and V. Yu. Dolmatov, *Zhur. org. Khim.*, 1977, **13**, 1300 (*Chem. Abs.*, 1977, **87**, 135 187).
[330] G. Bianchi, L. Casotti. D. Passadore, and N. Stabile, *J.C.S. Perkin II*, 1977, 47.
[331] K. A. Thakar and B. M. Bhawal, *Indian J. Chem.*, *Sect. B*, 1977, **15**, 1061.
[332] V. I. Klimenko, A. I. Sitkin, A. L. Fridman, and A. D. Nikolaeva, *Tezisy Vses. Soveshch. Khim. Nitrosoedinenii, 5th*, 1974, 42 (*Chem. Abs.*, 1977, **87**, 39 207).
[333] K. Venters and M. Trusule, *Latv. P.S.R. Zinat. Akad. Vestis, Kim. Ser.*, 1977, 230 (*Chem. Abs.*, 1977, **87**, 68 045).
[334] A. V. Zimichev, A. E. Lipkin, and T. M. Safargalina, *Khim. geterotsikl. Soedinenii*, 1977, 1047 (*Chem. Abs.*, 1978, **88**, 22 510).
[335] A. V. Zimichev and A. E. Lipkin, *Tezisy Vses. Soveshch. Khim. Nitrosoedinenii, 5th*, 1974, 33 (*Chem. Abs.*, 1977, **87**, 21 866).

6-, and 7-nitro-derivatives are obtained, the product ratios depending on the size of $R^2$.[336]

**Nitrosation.**—Copper(II) is claimed to retard the nitrosation[337] of phenol by $HNO_2$ and the subsequent oxidation of $o$-nitrosophenol to the nitro-compound by $HNO_3$ through formation of complexes with $HNO_2$ and the nitroso-phenol. The nitrosation of $N$-methylindole (a reaction studied almost a century ago by Fischer) has been re-examined and the products have finally been characterized. With equimolar amounts of substrate and $HNO_2$, a di-indole (48) (40%) and very little of the di-indole oxide (49) are formed, whereas with excess of $HNO_2$ the

(48)                                    (49)

latter is the sole indolic product (3.5%), the main product being $N$-methyl-anthranilic acid. The nitrosation of 3-phenylthio-$N$-methylindole (which gives five products) is also discussed. The products can be explained by assuming initial electrophilic attack on both free and 3-substituted indoles at the 3-position followed by further reactions at the open 2-position. When the latter is blocked, normal 3-nitrosation ensues.[338]

**Amination.**—Recent advances in the amination (by substitution and reduction) of aromatics have been reviewed, but in Japanese.[339] Benzene, toluene, chloro-benzene, nitrobenzene, anisole, and $NN$-dimethylaniline have been aminated with $NH_2OH$ and $FeSO_4$ and concentrated $H_2SO_4$ in AcOH, to produce the amino-derivatives in variable yields.[340] The mechanism of guanidination of benzene [with $BuNHC(=NH)NHOSO_3H$ and $AlCl_3$] has been studied by [15]N labelling, from which it was deduced that reaction occurs not *via* the nitrenium ion (50) but *via* the complex (51), followed by direct $S_N2$ attack on nitrogen.[341]

(50)                                    (51)

Phenols react with hexamethyldisilazane in the presence of diphenylseleninic anhydride, forming phenylseleno-imines, with a strong preference for *ortho* reaction. The resulting imines are easily reduced to amino-phenols.[342]

[336] L. N. Astakhova and I. M. Skvortsov, *Tezisy Vses. Soveshch. Khim. Nitrosoedinenii, 5th*, 1974, 36 (*Chem. Abs.*, 1977, **87**, 39 212).

[337] V. I. Trubnikova, I. Ya. Lubyanitskii, V. A. Preobrazhenskii, A. A. Miloradov, A. M. Gol'dman, and M. S. Furman, *Zhur. org. Khim.*, 1977, **13**, 1435 (*Chem. Abs.*, 1978, **88**, 22 270).

[338] A. H. Jackson, D. N. Johnston, and P. V. R. Shannon, *J.C.S. Perkin I*, 1977, 1024.

[339] K. Nara, *Kagaku To Kogyo (Osaka)*, 1977, **51**, 304 (*Chem. Abs.*, 1978, **88**, 89 408).

[340] F. Minisci, M. Stolfi, and S. Frigerio, Ger. Offen. 2 716 242 (*Chem. Abs.*, 1978, **88**, 37 403).

[341] A. Heesing and H. Šteinkamp, *Chem. Ber.*, 1977, **110**, 3862.

[342] D. H. R. Barton, A. G. Brewster, S. V. Ley, and M. N. Rosenfeld, *J.C.S. Chem. Comm.*, 1977, 147.

**Diazo-coupling.**—The kinetics of coupling of a range of substituted benzenediazonium ions with *NN*-dimethylaniline in a variety of non-aqueous solvents in the temperature range 13—39.5 °C have been studied.[343] In the coupling reactions between several 4-substituted benzenediazonium ions and eleven methyl-pyrroles, attack of the unprotonated pyrrole is favoured in an $S_E2$ mechanism with a steady-state intermediate (perhaps a protonated pyrrole). 3,4-Dimethylpyrrole and 2,3,4,5-tetramethylpyrrole display unusual reactions: within a few seconds of diazo-coupling with the former, the colour of the solution changes from intense yellow to pale blue, perhaps due to formation of a 2,2′-bipyrryl; the latter undergoes a very slow reaction (compared with pyrroles with free ring positions), giving a complex product mixture.[344] 4-Methyl- and 4,6-dimethyl-2-pyrimidone do not couple with diazotized 4-chloroaniline, as reported previously, but at the 4-methyl group, to give phenylhydrazones. 2,6-Dimethyl-4-pyrimidone and 1,4,6-trimethyl-2-pyrimidone show similar behaviour at the 6-methyl group.[345] 1-Aminophenazine fails to form a diazonium salt that will couple with $\beta$-naphthol (though it will couple with phloroglucinol) because the phenazine-1-diazonium cation reacts rapidly with water and is oxidized to a 1-diazo-phenazinone that has weak coupling ability. 2-Aminophenazine behaves similarly.[346]

5-Methyltropolone reacts with aryldiazonium salts to yield either 3-arylazo- or 3,7-bis(arylazo)- products, depending on the electrophilicity of the attacking ions.[347]

**Phosphorus Substitutions.**—The Friedel–Crafts reaction between benzene, sulphur, and $PCl_3$ yields $Ph_3PS$ in 71% yield (*via* $PSCl_3$): subsequent desulphurization gives triphenylphosphine, providing a good synthetic alternative to the established methods *via* organometallic precursors. In the absence of sulphur, only phenyldichlorophosphine and diphenylchlorophosphine are formed.[348]

Several dialkyl 2-(5-bromothienyl)phosphonites have been synthesised by the reaction between the mono-Grignard reagent from 2,5-dibromothiophen and $ClP(OR)_2$, where R = Et, Pr, or Bu.[349] The *N*-dimethylphosphonyl derivatives of pyrrole, indole, and 3-methylindole can be made by treatment of the heterocycles with $P(OMe)_3$ and $BrCCl_3$.[350]

## 6 Reactions with Electrophilic Oxygen and Sulphur

**Oxygen Substitutions.**—The electrophilic hydroxylation of benzene, alkyl-benzenes, and halogeno-benzenes with $H_2O_2$ in super-acids (*e.g.* $FSO_3H–SO_2ClF$ or $FSO_3H–SbF_5–SO_2Cl$) at low temperatures proceeds in good yields. The phenols so formed are protonated by the super-acidic medium and are therefore deac-

[343] I. L. Bagal, S. A. Skvortsov, and A. V. El'tsov, *Zhur. org. Khim.*, 1978 **14**, 361 (*Chem. Abs.*, 1978, **88**, 151 696).
[344] A. R. Butler, P. Pogorzelec, and P. T. Shepherd, *J.C.S. Perkin II*, 1977, 1452.
[345] D. T. Hurst and M. L. Wong, *J.C.S. Perkin I*, 1977, 1985.
[346] E. S. Olson, *J. Heterocyclic Chem.*, 1977, **14**, 1255.
[347] T. Ide, K. Imafuku, and H. Matsumura, *Chem. Letters*, 1977, 717.
[348] G. A. Olah and D. Hehemann, *J. Org. Chem.*, 1977, **42**, 2190.
[349] E. A. Krasil'nikova, A. I. Razumov, and O. L. Nevzorova, *Zhur. org. Khim.*, 1977, **47**, 1193 (*Chem. Abs.*, 1977, **87**, 85 097).
[350] R. P. Napier, U.S. P. 4 046 774 (*Chem. Abs.*, 1977, **87**, 201 310).

tivated against further electrophilic attack or secondary oxidation. The hydroxy-arenium intermediates in some cases undergo 1,2-shifts of the methyl group.[351] In the first report of the reactions of hypofluorous acid (prepared by the reaction of $F_2$ with ice at *ca.* −40 °C) with organic substrates, a range of monosubstituted benzenes, *p*-xylene, and naphthalene have been hydroxylated to phenols in poor to fair yields. The reagent spontaneously decomposes to $O_2$ and HF ($t_{1/2}$ *ca.* 30 min at 25 °C) and can detonate. Studies of isomer distribution and the absence of any fluorination suggest that the molecule is polarized in the reverse sense to HOCl.[352]

The structures and relative proportions of the *N*-oxides obtained from the oxidation of 2,4,6-trialkyl-pyrimidines with $H_2O_2$ and AcOH have been determined by hydrolysis and by n.m.r. studies in the presence of Eu(fod)$_3$.[353]

**Sulphur Substitutions.**—Several papers have appeared on the sulphonation of benzene and alkyl-benzenes. The kinetics of sulphonation of benzene, toluene, and *p*-xylene have been studied over a range of $H_2SO_4$ concentrations.[354] The effect of temperature on the distribution of monosulphonated products in the reactions between *o*- and *m*-nitrotoluenes and 30% oleum has been studied in the range 20—100 °C by $^1$H n.m.r. The *ortho*-isomer gives only the 4-sulphonic acid, whereas 4-, 5-, and 6-acids are formed from the *meta*-isomer. Addition of HgCl$_2$ leads to a slight increase in yields but has no significant effect on isomer composition.[355] In the sulphonation of *m*-xylene by 77% $H_2SO_4$, the ratio of rate constants for substitutions at the 5- and 4-positions is 0.011. The 5-sulphonic acid accumulates because of its greater stability towards hydrolysis.[356] The sulphonation of a range of monoalkylated benzenes under kinetic control has been investigated and the *para* : *ortho* ratios were found to be linearly correlated with steric parameters.[357] Isomer distributions and product stabilities in the sulphonation of polyethyl- and polyisopropyl-benzenes by aqueous $H_2SO_4$ have been studied.[358] The sulphonation of 2,5-dichloro-*p*-xylene in 30% oleum has been reported.[359] Sulphonation and sulphonylation of benzene, toluene, and chlorobenzene by ClSO$_3$H in MeNO$_2$ and in CH$_2$Cl$_2$ have been exhaustively investigated. Sulphonation in MeNO$_2$ in the initial stages proceeds *via* attack by MeNO$_2$·SO$_3$H$^+$ on the arene. In CH$_2$Cl$_2$, the observed kinetics may be explained in terms of ArS$_2$O$_6$H·ClSO$_3$H as the sulphonylating agent.[360] Aryl triflones have been synthesized in good yields from benzene and activated aromatics by sulphonylation with (CF$_3$SO$_2$)$_2$O and CF$_3$SO$_2$Cl in the presence of AlCl$_3$.[361] The

[351] G. A. Olah and R. Ohnishi, *J. Org. Chem.*, 1978, **43**, 865.
[352] E. H. Appelman, R. Bonnett, and B. Mateen, *Tetrahedron*, 1977, **33**, 2119.
[353] T. Sakamoto, S. Niitsuma, M. Mizugaki, and H. Yamanaka, *Heterocycles*, 1977, **8**, 257.
[354] M. Sohrabi, T. Kaghazchi, and C. Hanson, *J. Appl. Chem. Biotechnol.*, 1977, **27**, 453.
[355] H. Asakura and Y. Muramoto, *Nippon Kagaku Kaishi*, 1977, 1694 (*Chem. Abs.*, 1978, **88**, 89 276).
[356] B. G. Gnedin, T. A. Baranova, and A. S. Ocheretovyi, *Izvest. V. U. Z., Khim. i khim. Tekhnol.*, 1977, **20**, 1800 (*Chem. Abs.*, 1978, **88**, 88 753).
[357] E. N. Krylov and T. A. Khutova, *Zhur. obshchei Khim.*, 1977, **47**, 1601 (*Chem. Abs.*, 1977, **87**, 117 282).
[358] A. Koeberg-Telder and H. Cerfontain, *J.C.S. Perkin II*, 1977, 717; H. Cerfontain, A. Koeberg-Telder, and C. Ris, *J.C.S. Perkin II*, 1977, 720.
[359] N. S. Dokunikhin and S. A. Sokolov, *Tezisy Vses. Simp. Org. Sint.: Benzoidnye Aromat. Soedin.*, 1st, 1974, 43 (*Chem. Abs.*, 1977, **87**, 22 598).
[360] M. P. van Albada and H. Cerfontain, *J.C.S. Perkin II*, 1977, 1548; *ibid.*, p. 1557.
[361] J. B. Hendrickson and K. W. Bair, *J. Org. Chem.*, 1977, **42**, 3875.

reactions of sulphonyltriazines with benzene have been explored; *e.g.* 4-$MeC_6H_4O_2SNHN=NPh$ yields biphenyl and $H_2NSO_2C_6H_4Me$-4 with benzene and $AlCl_3$. The similar reaction of $PhNMeN=NSO_2C_6H_4Me$-4 yields $PhC_6H_4NHMe$, 4-$MeC_6H_4SO_2SC_6H_4Me$-4, and small amounts of $PhNMeSO_2C_6H_4Me$-4 and $MeNHC_6H_4SO_2C_6H_4Me$-4.[362] Mesitylene reacts with $SCl_2$ or $S_2Cl_2$ in $CHCl_3$ in the presence of iron powder to give the macrocycle (52).[363]

(52)

Phenylsulphamic acid (which has been postulated as an intermediate in the sulphonation of aniline) in a large excess of 86.8% $H_2SO_4$ at room temperature yields both $PhNH_3^+$ and $HO_3SC_6H_4NH_3^+$, by solvolysis and by sulphonation (followed by solvolysis) respectively. In aqueous $H_2SO_4$ (up to *ca.* 70%), at room temperature, only solvolysis occurs, whereas in >96% $H_2SO_4$ only sulphonation is observed.[364] Evidence has been presented for an intermolecular reaction pathway in the conversion of phenylsulphamic acid into *o*- and *p*-aniliniumsulphonic acids in a large excess of 96—100% $H_2SO_4$ at 25 °C.[365] The sulphonation of methanesulphonanilide ($PhNHSO_2Me$) is preceded by a protonation equilibrium. The substrate entity undergoing sulphonation is the unprotonated species, and at low $H_2SO_4$ concentrations the sulphonating agent is $H_3SO_4^+$: in >90% $H_2SO_4$, sulphonation occurs *via* $H_2S_2O_7$.[366] The sulphonation of anilinium sulphate in *ca.* 100% $H_2SO_4$ proceeds by two competing routes. At room temperature, in a large excess of fuming acid, sulphonation is exclusively *via* direct attack by $H_3S_2O_7^+$ on the ring; at elevated temperatures, and with less concentrated acid, reaction is predominantly *via* an indirect route involving sulphonation on nitrogen to produce phenylsulphamic acid.[367] Sulphonation of *m*-aminobenzenesulphonic acid in fuming $H_2SO_4$ at elevated temperatures yields a mixture of anilinium-2,5-di-, -2,4,5-tri-, and -2,3,4,6-tetra-sulphonic acids. The rate-limiting step is the conversion of the starting material into the intermediate sulphamic acid. The di:tri:tetra product ratios are thermodynamically controlled.[368] The isomeric composition of the products from the sulphonation of *NN*-diethylaniline with oleum varies with the amount of $SO_3$ present. The authors deduce that both protonated and non-protonated forms of the aniline

[362] R. Kreher and R. Halpaap, *Tetrahedron Letters*, 1977, 3147.
[363] F. Bottino, S. Foti, and S. Pappalardo, *Tetrahedron*, 1977, **33**, 337.
[364] P. K. Maarsen and H. Cerfontain, *J.C.S. Perkin II*, 1977, 929.
[365] P. K. Maarsen and H. Cerfontain, *J.C.S. Perkin II*, 1977, 921.
[366] P. K. Maarsen and H. Cerfontain, *J.C.S. Perkin II*, 1977, 1003.
[367] P. K. Maarsen and H. Cerfontain, *J.C.S. Perkin II*, 1977, 1008.
[368] P. K. Maarsen, R. Bregman, and H. Cerfontain, *J.C.S. Perkin II*, 1977, 1863.

participate in sulphonation.[369] In the chlorosulphonation of $N$-phenyl-$N'$-pyri-dyl-ureas and -thioureas, whereas, with excess of reagent, the urea gives a product that is disubstituted in the benzene ring, the thiourea gives only a mono-($para$-) substituted product. Neighbouring-group participation by the pyridine nitrogen in delivering —$SO_3H$ to a benzene $ortho$-position, this position being blocked by sulphur in the thiourea, has been invoked as the rationale.[370]

Other sulphonations include those of 2,3,5,6-tetramethylphenol and its methyl ether in $CF_3SO_3H$–$SO_2ClF$,[371] of $o$- and $p$-hydroxy-benzoic acids in $H_2SO_4$,[372] of fluoranthene by $H_2SO_4$ in nitrobenzene (major products being the 4- and 11-acids),[373] of ethylferrocene with $SO_3$ and dioxan (which is sulphonated on the ring that contains the ethyl group),[374] and of triptycene with sulphuryl chloride and $AlCl_3$ at $-40\,°C$, giving 2-triptycenesulphonyl chloride (54%) and 2,6-dichlorotriptycene (25%).[375] Toluene, aniline, dimethylaniline, anisole, and thiophen have been sulphonated in 23—79% yields by treatment at 100—200 °C with bis(trimethylsilyl) sulphate.[376]

The rates of (and isomer distribution in) the homogeneous monosulphonation of biphenyl with a large excess of aqueous $H_2SO_4$ (81.5–96.8%) at 25 °C have been studied. Sulphonation, which is first-order with respect to substrate, takes place mainly at the $para$-position, for steric reasons. At acid concentrations $\leqslant 86\%$, the sulphonating agent is $H_3SO_4^+$: $\rho^+$ is $-10.0 \pm 1.2$.[377]

In the heterocyclic field, sulphur substitutions include the sulphonation of quinoline to produce mainly the 8-acid, plus the 5-, 6-, and 7-acids,[378] the chlorosulphonation of 2-(3,4-dicholorophenyl)- and 2-(3-pyridyl)-imidazole (on the imidazole ring) and of 2-(4-chlorophenyl)imidazole on both aromatic rings,[379] the tosylation at N-3 of some 2-(monosubstituted amino)-1,3,4-thiadiazoles (which isomerise to products that are sulphonylated on the exocyclic nitrogen),[380] and the thiolation of 2-methylindolizine with diaryl disulphides to form (53).[381]

(53)

[369] R. N. Khelevin, *Zhur. priklad. Khim.*, 1977, **50**, 2035 (*Chem. Abs.*, 1977, **87**, 200 421).

[370] N. V. Badami, R. J. W. Cremlyn, and F. J. Swinbourne, *Austral. J. Chem.*, 1977, **30**, 1793.

[371] L. P. Kamshii, V. I. Mamatyuk, and V. A. Koptyug, *Zhur. org. Khim.*, 1977, **13**, 810 (*Chem. Abs.*, 1977, **87**, 38 421).

[372] L. P. Mel'nikova, *Uch. Zap., Yarosl. Gos. Pedagog. Inst.*, 1976, **154**, 91 (*Chem. Abs.*, 1977, **87**, 200 392).

[373] O. I. Kachurin and L. I. Velichko, *Ukrain. khim. Zhur.*, 1977, **43**, 387 (*Chem. Abs.*, 1977, **87**, 38 424).

[374] V. I. Boev, A. S. Osipenko, and A. V. Dombrovskii, *Zhur. obshchei Khim.*, 1977, **47**, 1573 (*Chem. Abs.*, 1977, **87**, 168 171).

[375] L. A. Knyazeva and V. R. Skvarchenko, *Vestnik Moskov. Univ., Ser. 2: Khim.*, 1977, **18**, 489 (*Chem. Abs.*, 1978, **88**, 89 426).

[376] M. G. Voronkov, S. V. Korchagin, and V. K. Roman, *Izvest. Akad. Nauk S.S.S.R., Ser. khim.*, 1977, 2340 (*Chem. Abs.*, 1978, **88**, 37 882).

[377] T. A. Kortekaas and H. Cerfontain, *J.C.S. Perkin II*, 1977, 1560.

[378] O. F. Sidorov, M. K. Murshtein, and V. V. Mochalov, *Khim.-Farm. Zhur.*, 1977, **11**, 107 (*Chem. Abs.*, 1978, **88**, 89 495).

[379] J. J. Baldwin, P. K. Lumma, G. S. Ponticello, F. C. Novello, and J. M. Sprague, *J. Heterocyclic Chem.*, 1977, **14**, 889.

[380] G. L'Abbe, G. Verhelst, L. Huybrechts, and S. Toppet, *J. Heterocyclic Chem.*, 1977, **14**, 515.

[381] M. Cardellini, F. Claudi, U. Gulini, and S. Martelli, *Synthesis*, 1977, 323.

## 7 Halogenation

**Fluorination.**—$XeF_2$ continues to find useful application as a selective fluorinating agent. Thus, $XeF_2$ with naphthalene gives 1-fluoro- (45%), 2-fluoro- (9%), and 1,4-difluoro- (15%) naphthalene,[382] (the last product was not detected in a previous study). With $XeF_2$ and HF, indane gives 5-fluoro- and 5,6-difluoro-indane, tetralin the 5- and 6-monofluoro-compounds and 6,7- and 5,8-difluoro-compounds, and *o*-xylene the 3- and 4-monofluoro-*o*-xylenes.[383] Intercalation of $XeF_6$ with graphite can alter the selectivity. Thus whereas anthracene and $XeF_6$ yield 1-fluoro- (30%), 2-fluoro- (4%), and 9-fluoro- (36%) anthracene, product yields with the intercalate $C_{19}XeF_6$ are 20, 10, and 18% and with $C_{8.7}XeOF_4$ (from $XeOF_4$ and graphite) are 6, 1, and 10% respectively.[384] Whereas dibenz[*a,h*]anthracene gives 7-fluoro- and 7,14-difluoro-products with $XeF_2$, the use of $C_{19}XeF_6$ allows isolation of the monofluoro-compound (16% yield). Similarly, benz[*a*]anthracene yields 7-fluorobenz[*a*]anthracene in 22% yield.[385]

The direct fluorination of uracil with $F_2$ in aqueous HF solution at *ca.* 0 °C yields 5-fluorouracil in 54–88% yield.[386] Fluorination of naphthalene, decalin, and tetralin with $CsCoF_4$ gives hexadecafluorobicyclo[4.4.0]dec-1(6)-ene (54) (50—63%) and dodecafluorotetralin (55) (27—31%): the reactions presumably involve addition/elimination, as established for other cases.[387]

(54)                                                    (55)

**Chlorination.**—Direct chlorination of benzofuran, known for almost a century to give the 2,3-dichloro-addition product, has been shown to produce an approximately 1:1 mixture of *cis*- and *trans*-isomers in $Et_2O$ at *ca.* 0 °C. A change of solvent to AcOH or $CCl_4$ has little effect on the proportion of isomers. Elimination of HCl (with NaOEt in EtOH, to produce 3-chlorobenzofuran) is much faster for the *cis*-than for the *trans*-compound. Acetolysis of the *trans*-isomer at 100 °C gives 2-chlorobenzofuran: the *cis*-isomer, at 80 °C, first isomerizes to the *trans*.[388] Direct chlorination and bromination of *N*-substituted (with assorted alkyl and aryl groups) 4-isothiazolin-3-ones has been studied.[389] Whereas chlorination even under mild conditions gives primarily the 4,5-dichloro-compound, the mono(4-)bromo-compound can be prepared in good yield by bromination. Chlorination and bromination of mono- and bis-(trifluoromethoxy)benzenes is also reported: no loss of fluorine was observed.[390]

[382] M. Rabinovitz, I. Agranat, H. Selig, and C.-H. Lin, *J. Fluorine Chem.*, 1977, **10**, 159.
[383] B. Šket and M. Zupan, *J. Org. Chem.*, 1978, **43**, 835.
[384] M. Rabinovitz, I. Agranat, H. Selig, C.-H. Lin, and L. Ebert, *J. Chem. Res. (S)*, 1977, 216.
[385] I. Agranat, M. Rabinovitz, H. Selig, and C.-H. Lin, *Synthesis*, 1977, 267.
[386] T. Takahara and S. Misaki, Ger. Offen. 2 719 245 (*Chem. Abs.*, 1978, **88**, 37 834).
[387] R. G. Plevey, I. J. Sallomi, D. F. Thomas, and J. C. Tatlow, *J.C.S. Perkin I*, 1976, 2270.
[388] E. Baciocchi, S. Clementi, and G. V. Sebastiani, *J. Heterocyclic Chem.*, 1977, **14**, 359.
[389] E. D. Weiler, R. B. Petigara, M. H. Wolfersberger, and G. A. Miller, *J. Heterocyclic Chem.*, 1977, **14**, 627.
[390] F. E. Herkes, *J. Fluorine Chem.*, 1977, **9**, 113.

The mechanism of catalysis of aromatic chlorination by pyridine has been examined by measurement of rates of chlorination of toluene in AcOH with and without added pyridine, pyridinium nitrate, or lithium chloride. All three additives show similar rate enhancements, from which it has been deduced that the catalysis caused by pyridine is simply a salt effect of pyridinium acetate.[391]

Chlorination of *o*- and *p*-toluidine by $Cl_2$ in concentrated $H_2SO_4$ is selective for positions *ortho* and *para* to the methyl groups: product mixtures are simpler than those from chlorination of amine–$AlCl_3$–HCl complexes.[392] Substituted pentamethylbenzenes (56; R = $CONH_2$, $CONMe_2$, or $NO_2$), on chlorination with $Cl_2$ in $FSO_3H$, undergo addition of $Cl^+$ *meta* to R. The resulting benzenonium ions (57) rearrange reversibly at $-20\,°C$ to the isomeric ions (58) and (59): at room temperature these are irreversibly converted into side-chain-halogenated products.[393]

(56)          (57)          (58)          (59)

The kinetics of chlorination (by $Cl_2$ and $FeCl_3$, at $50\,°C$) of chlorobenzene have been studied and the reactivities of isomers formed in the later stages of the reaction discussed in terms of the additivity principle and the partial rate factors ($f_o$ 0.111, $f_m$ 0.012, $f_p$ 0.354).[394] The intramolecular isomerisation of dichlorobenzenes on an $AlCl_3$–$MgSO_4$ catalyst at $160\,°C$ is reported. Under these conditions, *m*-dichlorobenzene, for example, yields 26.7% of *para*- and 6.6% of *ortho*-isomers.[395] Chlorination of indane under Friedel–Crafts-type conditions gives a mixture of the 4- and 5-chloroindanes. Further chlorination leads to di- and tri-substitution.[396] Addition of mercaptans to the reaction mixture for chlorination of 3,5-xylenol in the presence of $FeCl_3$, $AlCl_3$, *etc.* increases selectivity towards formation of the *para*-chloro-isomer; *e.g.*, with $C_{12}H_{25}SH$ a *para* : *ortho* ratio of 12.7 is obtained, whereas in the absence of mercaptan the quoted figure is 6.8.[397] A low *ortho* : *para* ratio is reported for the chlorination of alkyl-benzenes in the presence of a Lewis acid catalyst if halogeno- or acyl-thianthrenes (*e.g.* 2,3,7,8-tetrachlorothianthrene or dibenzoylthianthrene) or a sulphoxide (*e.g.* 2,7-dichlorothianthrene 5,10-dioxide) are used as the co-catalysts.[398]

[391] G. E. Dunn and J. A. Pincock, *Canad. J. Chem.*, 1977, **55**, 3726.
[392] C. W. Schimelpfenig, *J.C.S. Perkin I*, 1977, 1129.
[393] L. A. Ostashevskaya, M. M. Shakirov, I. S. Isaev, and V. A. Koptyug, *Zhur. org. Khim.*, 1977, **13**, 2362 (*Chem. Abs.*, 1978, **88**, 104 791).
[394] P. P. Alikhanov and K. I. Sakodynskii, *Zhur. org. Khim.*, 1977, **13**, 1703 (*Chem. Abs.*, 1977, **87**, 167 113).
[395] J. Suwinski and W. Zielinski, *Roczniki Chem.*, 1977, **51**, 455.
[396] J. Novročik and J. Poskočil, *Coll. Czech. Chem. Comm.*, 1977, **42**, 1374.
[397] T. Matsumoto, M. Matsuda, H. Mizokami, T. Kibamoto, and K. Hatta, Ger. Offen. 2 649 112 (*Chem. Abs.*, 1977, **87**, 134 542).
[398] J. C. Graham, Ger. Offen. 2 634 338 (*Chem. Abs.*, 1977, **87**, 39 074).

Chlorination of anthraquinone in the presence of $I_2$ or Pd salts at 100 °C in concentrated $H_2SO_4$ gives mixed chloro-products. With $I_2$ catalyst, the $\alpha : \beta$ ratio is 5 : 1; with Pd(OAc)$_2$ it is 10 : 1.[399] Chlorine fluorosulphate, with benzene, at −100 °C, in a Freon, yields $C_6Cl_6$ (77%).[400] Treatment of carbazole or bromo-carbazoles with $Cl_2SO_2$ at room temperature gives mixtures of mono- to tetra-chloro-derivatives.[401]

**Bromination.**—A range of substituted benzenes has been brominated by 'naked' $^{80}Br^+$ in the gas phase, the results of which indicate a mechanism similar to that for solvated ions. The author concludes that there is no need to establish a new reaction model for gas-phase aromatic substitution.[402]

Reactions of bromine in AcOH with aromatics display kinetics which vary with solvent and the initial $Br_2$ concentration. In a thorough paper, de la Mare and co-workers discuss the criteria for interpretation of such results in terms of the composition of the transition state, and in particular of the number of $Br_2$ molecules involved. In the same paper, more generally, the mechanisms available for bromination by molecular $Br_2$ are reviewed.[403] The role of pyridine and its HBr and HBr$_3$ salts in the catalysis of the bromination of mesitylene has been explored. At high concentrations, these additives inhibit halogenation. The authors state that this is *not* due to a simple salt effect.[404] Bromination of 3,4-dimethylphenol (or of the 2,6-dibromo-analogue) with $Br_2$ in AcOH gives a tribromo-dienone (60), and not the 2,5,6-tribromophenol. The dienone rear-ranges autocatalytically, over a period of days, to the bromomethyl compound (61).[405] Loss of the t-butyl group during bromination of substituted catechols has been noted: thus, 3,4-di-t-butylcatechol gives 3,4-dibromo-6-t-butylcatechol with $Br_2$ in $CCl_4$ whereas the monoacetate, dimethyl ether, or acetonide give the expected 4-bromo-derivatives.[406]

(60)                          (61)

The kinetics of bromination of substituted phenols and anisoles with pyri-dinium bromide perbromide are reported. The order in halogenating agent depends mainly on the position of a substituent: apparent fractional orders for some *ortho*-substituted substrates arise from concentration effects.[407] Cyclo-

[399] T. Ito, Y. Kindaichi, and Y. Takami, *Nippon Kagaku Kaishi*, 1977, 82 (*Chem. Abs.*, 1977, **87**, 5707).
[400] A. V. Fokin, Yu. N. Studnev, I. N. Krotovich, L. D. Kuznetsova, and A. M. Gukov, *Izvest. Akad. Nauk. S.S.S.R., Ser. khim.*, 1977, 2388 (*Chem. Abs.*, 1978, **88**, 89 228).
[401] J. Kyziol and J. Pielichowski, *Roczniki Chem.*, 1977, **51**, 815.
[402] E. J. Knust, *J. Amer. Chem. Soc.*, 1977, **99**, 3037.
[403] N. H. Briggs, P. B. D. de la Mare, and D. Hall, *J.C.S. Perkin II*, 1977, 106.
[404] S. Sharma, J. Rajaram, and J. C. Kuriacose, *Indian J. Chem., Sect. B*, 1977, **15**, 372.
[405] P. B. D. de la Mare, N. S. Isaacs, and P. D. McIntyre, *Tetrahedron Letters*, 1976, 4835.
[406] I. S. Belostotskaya, E. V. Dzhuaryan, and V. V. Ershov, *Izvest. Akad. Nauk. S.S.S.R., Ser. khim.*, 1977, 1100 (*Chem. Abs.*, 1977, **87**, 67 935).
[407] S. Sharma, J. Rajaram, and J. C. Kuriacose, *Indian J. Chem., Sect. B*, 1977, **15**, 274.

dextrin complexes of aromatics have been used as substrates in bromination and nitration. For example, the complex with chlorobenzene, on treatment with $Br_2$ in AcOH, yields *o*- and *p*-bromochlorobenzene in 1:4.5 ratio.[408] Bromination and nitration of 4-phenyl-isoxazoles give substitution exclusively at the *para*-position of the phenyl substituent.[409]

The bromination of 3-substituted indoles has been reviewed and the properties of 3-bromo-indolenines have been discussed.[410] Three polybromo-products (62), (63), and (64) have been identified from the bromination of indole with excess of

(62)          (63)          (64)

$Br_2$ in ice-cold AcOH. Similar products result from 2-methyl- and 3-methyl-indoles.[411] Bromination of gramine with $Br_2$ in AcOH yields 5- and 6-bromo-derivatives. 1-Methylgramine similarly gives the 2,6-dibromo-compound, and at higher $Br_2$ concentrations 2,5,6-tribromo-1-methylgramine. 2-Methylgramine (to the 5-bromo- and 5,6-dibromo-products) and 1,2-dimethylgramine (to the 6-bromo-derivative or 3,4,5,6,7-pentabromo-1,2-dimethylindole, depending on the $Br_2$ concentration) were similarly brominated.[412] Bromination of some pyrimidines[413] and (along with iodination and acetylation) of *m*-terphenyl,[414] and of some 9-substituted carbazoles (followed by KSCN)[415] is also reported.

Bromination of a range of *NN*-dimethyl-anilines with dibromoisocyanuric acid gives mixtures of products. Excess of the reagent readily leads to perbromination.[416] The kinetics of bromination of some aromatic amines (aniline, *N*-methyl- and *NN*-dimethyl-aniline, and some *para*-substituted anilines) by NBS in AcOH or AcOH–$H_2O$ with $HClO_4$ or $H_2SO_4$ and Hg(OAc)$_2$ (to fix liberated $Br^-$) have been studied. Reactions are first-order both in amine and NBS, and show an inverse acid dependence. The authors deduce that the mechanism involves direct transfer of $Br_2$ to the ring.[417] NBS has also been applied to the bromination of some spiro[benzopyran-indolines] and analogous -phenanthridines,[418] 1-acetyl- and 1- and 3-t-butyl-indoles,[419] and 2-bromo-, 2-ethylmercapto-, and 2-ethyl-sulphonyl-indoles.[420] 3-t-Butylindole yields 3-t-butyloxindole and 6-bromo-3-t-butylindole, but not the 2-bromo-compound.[419]

[408] K. Yamada, Japan. Kokai 77  42 828 (*Chem. Abs.*, 1977, **87**, 134 489).
[409] A. DeMunno, V. Bertini, and F. Lucchesini, *Chimica e Industria*, 1976, **58**, 880.
[410] T. Hino and M. Nakagawa, *Heterocycles*, 1977, **6**, 1680.
[411] A. Da Settimo, V. Santerini, G. Primofiore, G. Biagi, and C. Veneziano, *Gazzetta*, 1977, **107**, 367.
[412] A. Da Settimo, V. Santerini, G. Primofiore, and C. Veneziano, *Chimica e Industria*, 1977, **59**, 454.
[413] D. T. Hurst, K. Biggadike, and J. J. Tibble, *Heterocycles*, 1977, **6**, 2005.
[414] G. Rabilloud and B. Sillion, *Bull. Soc. chim. France*, 1977, 276.
[415] N. I. Baranova and V. I. Shishkina, *Izvest. V. U. Z., Khim. i khim. Tekhnol.*, 1977, **20**, 980 (*Chem. Abs.*, 1977, **87**, 151 948).
[416] J. Rosevear and J. F. K. Wilshire, *Austral. J. Chem.*, 1977, **30**, 1561.
[417] P. S. Radhakrishnamurti and S. N. Sahu, *Indian J. Chem., Sect. A*, 1977, **15**, 785.
[418] E. R. Zakhs, L. A. Zvenigorodskaya, N. G. Leshenyuk, and V. P. Martynova, *Khim. geterotsikl. Soedinenii*, 1977, 1320 (*Chem. Abs.*, 1978, **88**, 105 190).
[419] T. Hino, M. Tonozuka, Y. Ishii, and M. Nakagawa, *Chem. and Pharm. Bull.* (*Japan*), 1977, **25**, 354.

A 'new' brominating agent (PBr$_3$ in DMF) has proved useful in the conversion of methoxy-quinolines into 2- and 4-bromo-quinolines.[421]

**Iodination.**—The kinetics and mechanism of iodination of phenol and substituted phenols by I$_2$ at pH 6.85 and 11.7 in H$_2$O, aqueous Me$_2$SO (20%), and aqueous MeCN (20%) are reported. The reactions are first-order each in phenol and iodinating agent. The order of reactivity is *m*-cresol > *p*-cresol > phenol > *m*-bromophenol > *m*-chlorophenol > *p*-iodophenol > *p*-bromophenol > *p*-chlorophenol > *p*-nitrophenol and H$_2$O > aq. Me$_2$SO > aq. MeCN. The active halogenating species is deduced to be molecular I$_2$.[422]

Silver trifluoromethanesulphonate (plus halogen) is superior to the perchlorate or trifluoroacetate for aromatic halogenations since the respective dangers of formation of explosive complexes or CF$_3$I are avoided. Thus, Br$_2$ or I$_2$ with CF$_3$SO$_3$Ag in CHCl$_3$ and anisole gives the 4-halogeno-compound quantitatively. 2-Iodothiophen is similarly prepared. The method has also been applied to aromatics of lower nucleophilicity.[423] Iodonium nitrate (prepared by the reaction of ICl with AgNO$_3$ in CHCl$_3$–pyridine solution) poly-iodinates phenols, secondary and tertiary aromatic amines, and activated heteroaromatics in good to excellent yields. In the case of *NN*-dimethylaniline the reaction is regiospecific, only *para*- iodination occurring, in 94.5% yield.[424] Diethylstilboestrol diphosphate (65), which cannot be satisfactorily iodinated 'chemically', has been iodinated to (66) and (67) by the *N*-iodoacetonitrilium ion MeC≡NI$^+$, generated *in situ* by electro-oxidation of I$_2$ in MeCN and Et$_4$N$^+$ BF$_4^-$.[425]

(65) R$^1$ = R$^2$ = H
(66) R$^1$ = R$^2$ = I
(67) R$^1$ = I, R$^2$ = H

In a kinetic study of the halogenation of pyridine, quinoline, isoquinoline, 2-picoline, and 2,6-lutidine with ICl, linear Eyring plots are obtained. Dipole-moment calculations indicate that the charge-transfer state is the transition state.[426]

## 8  Miscellaneous Substitutions

This section contains details of more general studies, often on the more *recherché* aromatic systems, which usually span several of the substitution categories.

[420] T. Hino, M. Endo, M. Tonozuka, Y. Hashimoto, and M. Nakagawa, *Chem. and Pharm. Bull. (Japan)*, 1977, **25**, 2350.
[421] T. Yajima and K. Munakata, *Chem Letters*, 1977, 891.
[422] P. S. Radhakrishnamurti and C. Janardhana, *Indian J. Chem., Sect. A*, 1978, **16**, 142.
[423] Y. Kobayashi, I. Kumadaki, and T. Yoshida, *J. Chem. Res. (S)*, 1977, 215.
[424] J. W. Lown and A. V. Joshua, *Canad. J. Chem.*, 1977, **55**, 122.
[425] D. Maysinger, W. Wolf, J. Casanova, and M. Tarle, *Croat. Chem. Acta*, 1977, **49**, 123.
[426] A. Dasgupta and Rama-Basu, *J. Chim. Phys. Phys.-Chim. Biol.*, 1977, **74**, 1174.

The reactions between cyclopropenylidenemercury complexes and electrophiles have been studied; *e.g.*, with $I_2$ and $CH_2Cl_2$ the iodocyclopropenyl cation is formed in 55% yield.[427] The reactivity of indan-1-carboxylic acid towards chlorination, bromination, acylation, sulphonation, and nitration has been investigated: the 6-position is most reactive.[428] Bromination ($Br_2$ and $CCl_4$) and nitration ($HNO_3$ and AcOH) of 7$H$-benzo[$c$]fluorene (68) lead to 5-substituted products.[429] In the acetylation of 2a,3,4,5-tetrahydroacenaphthene and higher homologues (69; $m = 1$ or 2; $n = 1$–3) the *ortho-* : *meta-* : *para-* substitution ratios depend on the degree of strain and hyperconjugation from the non-aromatic rings.[430] Chlorination of 4,6,8-trimethylazulene gives the 2-chloro-derivative, but the 3-chloro-derivative is preferred if an electron-withdrawing substituent is present at the 1-position.[431] Vilsmeier formylation of the non-benzenoid aromatic 5-cyanoazuleno[1,2-$f$]azulene (70) yields the 1-aldehyde, in agreement with SCF-CI calculations.[432]

(68)  (69)  (70)

Acylation and metallation studies on monomers, dimers, trimers, and tetramers containing linked thiophen, pyridine, pyrimidine, furan, benzofuran, benzothiophen, and indole moieties have been published.[433] Deuteriation, halogenation, and diazo-coupling reactions of 2-oxo- and 2-thioxo-1,2-dihydropyrimidinium salts have been studied and compared with results for 2,2-dialkyl-1,2-dihydropyrimidinium and 2,3-dihydro-1,4-diazepinium salts in order to demonstrate the effect of an adjacent oxo- or thioxo-group on the properties of a 1,5-diazopentadienium system.[434] Vilsmeier formylation of, and tautomerism in, 2-hydroxypyrazolo[5,1-$b$]quinazolone and 1-phenylpyrazolo[5,1-$b$]-quinazoline-2,9-dione have been studied.[435] The pyrazolo[3,4-$c$]pyrazole (71) has been methylated and acetylated, the major products being (72).[436]

1,2,4-Triazole-5-thione derivatives have been alkylated, acylated, cyano-ethylated, hydroxymethylated, and aminomethylated.[437] The aromaticity of the 1-pyrindine system has been investigated by n.m.r. and chemical reactivity

[427] Z. Yoshida, H. Konishi, Y. Kamitori, and H. Ogoshi, *Chem. Letters*, 1977, 1341.
[428] T. Aono, S. Kishimoto, Y. Araki, and S. Noguchi, *Chem. and Pharm. Bull. (Japan)*, 1977, **25**, 3198.
[429] R. Bolton, *J. Chem. Res. (S)*, 1977, 149.
[430] R. Gruber, D. Cagniant, and P. Cagniant, *Bull. Soc. chim. France*, 1977, 773.
[431] Yu. N. Porshnev, V. I. Erikhov, N. A. Andronova, and M. I. Cherkashin, *Doklady Akad. Nauk. S.S.S.R.*, 1978, **238**, 112 (*Chem. Abs.*, 1978, **88**, 120 861).
[432] C. Jutz, H. G. Peuker, and W. Kosbahn, *Synthesis*, 1976, 673.
[433] T. Kauffman, J. König, D. Körber, H. Lexy, H. J. Streitberger, A. Vahrenhorst, and A. Woltermann, *Tetrahedron Letters*, 1977, 389.
[434] D. Lloyd, H. McNab, and K. S. Tucker, *J.C.S. Perkin I*, 1977, 1862.
[435] V. Purnaprajna and S. Seshadri, *Indian J. Chem., Sect. B*, 1977, **15**, 109.
[436] E. Gonzalez, R. Sarlin, and J. Elguero, *Bull. Soc. chim. belges*, 1976, **85**, 829.

(71)                    (72) R = Me or COMe

studies. Thus, (73; $R = CO_2Et$) undergoes substitution at the 5-position with trifluoroacetic anhydride. Complex formation with $Fe(CO)_8$ 'drains' the aromaticity.[438] Substitution reactions of the 2-methyl-2*H*-cyclopenta[*d*]pyridazine (74) system with nitrating agents, mercuric acetate, tetracyanoethene, and benzenediazonium tetrafluoroborate have been investigated. Because of the system's high reactivity, most nitrating agents are unsatisfactory: the 7- and 5- positions are the most easily substituted.[439] Pyrido[3,4-*d*]pyridazine undergoes bromination at the 8-position (75), addition of MeLi at the 1- and 4-positions, and deuteriation in $D_2O$ at the 1-, 4-, and 5-positions and, in $NaOD—D_2O$, additionally at the 8-position.[440] [1]Benzofuro[2,3-*d*]pyridazine (76) has been methylated on nitrogen by dimethyl sulphate and on nitrogen and oxygen by $CH_2N_2$.[441] Methylation of 2-methyl-1,3,6-triazacycl[3.3.3]azine (77) with MeI

(73)                    (74)

(75)                    (76)                    (77)

yields three salts quaternised on nitrogen, the major component of the mixture probably being the 6-methylated compound. Nitration with $Cu(NO_3)_2$ and $Ac_2O$ gives a mixture of the 4-, 7-, and 9-nitro-products.[442] The peripheral functionalisation of 'capped' porphyrins, in which conventional electrophilic nitration is unsuccessful, has been achieved by treatment of the compounds with $I_2$ and then $AgNO_2$, giving a *ca.* 1 : 1 mixture of the isomeric mono-nitro-derivatives in high yield.[443] Nitration and deuteriation of octaethylbilatriene-*abc* occur preferentially at the outer bridge positions (C-5 and C-15), in agreement with

[437] T. Bany and M. Dobosz, *Ann. Univ. M. Curie-Sklodowska, Sect. AA*, 1973, **28**, 111 (*Chem. Abs.*, 1978, **88**, 37 711); *ibid.*, p. 95 (*Chem. Abs.*, 1978, **88**, 37 713); *ibid.*, p. 103 (*Chem. Abs.*, 1978, **88**, 37 712); *ibid.*, 1975, **29–30**, 67 (*Chem., Abs.*, 1977, **87**, 135 170); *ibid.*, p. 75 (*Chem. Abs.*, 1977, **87**, 135 203).

[438] A. G. Anastassiou, E. Reichmanis, and S. J. Girgenti, *J. Amer. Chem. Soc.*, 1977, **99**, 7392.

[439] A. G. Anderson, Jr., D. M. Forkey, and L. D. Grina, *J. Org. Chem.*, 1978, **43**, 1602; A. G. Anderson, Jr., L. D. Grina, and D. M. Forkey, *ibid.*, p. 664.

[440] A. Decormeille, G. Queguiner, and P. Pastour, *Bull. Soc. chim. France*, 1977, 665.

[441] M. Robba, M. Cugnon de Sevricourt, and A. M. Godard, *Bull. Soc. chim. France*, 1977, 125.

[442] O. Ceder and K. Vernmark, *Acta Chem. Scand. (B)*, 1977, **31**, 239.

[443] J. E. Baldwin and J. F. DeBernardis, *J. Org. Chem.*, 1977, **42**, 3986.

calculations of the electron density.[444] Bromination, nitration, and H/D exchange take place exclusively at the 3-position in thieno[3,2-*c*]isoxazole (78; X = S), whereas in the selenium analogue (78; X = Se) the situation is more complex.[445] Electrophilic attack occurs preferentially at C-1 and then C-3 in 2-phenyl-benzo[*b*]cyclopenta[*e*]pyran (79).[446] Phenanthro[4,5-*bcd*]furan (80) displays substitution characteristics of both phenanthrene and dibenzofuran: thus, bromination occurs predominantly at C-9 and benzoylation at C-1.[447] 1-Thia-indan and a range of methylated derivatives have been brominated, nitrated, and chlorosulphonylated.[448] 2-Methylthiazolo[3,2-*a*]pyridinium perchlorate (81; R = H) has been nitrated with $HNO_3$ and $H_2SO_4$ to (81; R = $NO_2$) in 20% yield.[449]

(78)          (79)

(80)          (81)

Methylation of the pre-formed anion of quinoxalino[2,3-*b*]-[1,4]benzo-thiazine (82) with MeI gives 12- and 11-methyl derivatives in 2 : 1 ratio. Pyraz-ino[2,3-*b*]-[1,4]benzothiazine (83) similarly gives 10- and 1-methyl derivatives. Nitration of the two compounds leads to 2- and 3-substitution respectively.[450]

(82)          (83)

Attempted sulphochlorination of tetrazolo[5,1-*b*]benzothiazole (84) gave (85) *via* substitution and ring cleavage.[451] Acetylation and chloromethylation of benzo[*b*]thiophanthrene (86) give (probably) the 6-substituted products, and

[444] J. V. Bonfiglio, R. Bonnett, M. B. Hursthouse, K. M. A. Malik, and S. C. Naithani, *J.C.S. Chem. Comm.*, 1977, 829.
[445] S. Gronowitz, C. Westerlund, and A. B. Hornfeldt, *Chemica Scripta*, 1976, **10**, 165 (*Chem. Abs.*, 1977, **87**, 184 411).
[446] Yu. N. Porshnev, V. A. Churkina, N. A. Andronova, and V. V. Titov, *Khim. geterotsikl. Soedinenii*, 1977, 902 (*Chem. Abs.*, 1977, **87**, 151 962).
[447] T. Horaguchi, *Bull. Chem. Soc. Japan.*, 1977, **50**, 3329.
[448] I. M. Nasyrov and I. U. Numanov, *Izuch. Ispol'z. Geteroat. Komponentov Sernistykh Neftei, Plenarnye Dokl. Nauchn. Sess. Khim. Tekhnol. Org. Soedin. Sery Sernistykh Neftei, 13th*, 1974, 42 (*Chem. Abs.*, 1978, **88**, 50 574).
[449] V. P. Martynova, E. R. Zakhs, and A. V. El'tsov, *Zhur. org. Khim.*, 1978, **14**, 216 (*Chem. Abs.*, 1978, **88**, 170 022).
[450] S. D. Carter and G. W. H. Cheeseman, *Tetrahedron*, 1977, **33**, 827.
[451] V. N. Skopenko, L. F. Avramenko, V. Ya. Pochinok, and N. E. Kruglyak, *Ukrain. khim. Zhur.*, 1977, **43**, 518 (*Chem. Abs.*, 1977, **87**, 152 084).

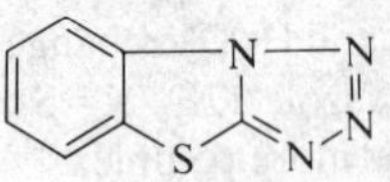

(84) 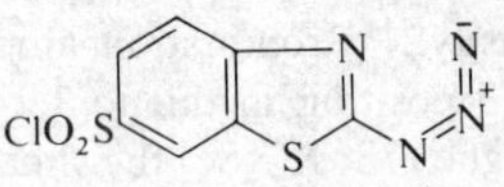                                    (85) 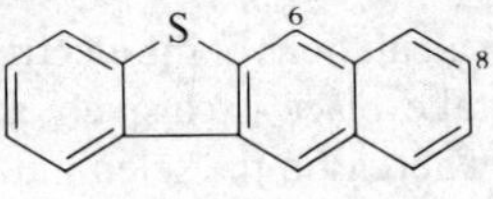                                    (86)

bromination gives the 6,8-dibromo-compound.[452] H/D exchange in some borazarothienopyridines and thienopyridines has been studied by n.m.r. spectroscopy: the comparable rates obtained for the two systems confirm the aromatic nature of the boron compounds.[453] Isomer distributions in the nitration and bromination of 2-(2'-thienyl)- and -(3'-thienyl)-pyrimidines have been measured in a study of the directing properties of the pyrimidyl group as a thiophen substituent. The pyrimidine ring acts as a *meta*-directing group in nitration in mixed acids but *ortho,para* for nitration with acetyl nitrate under neutral conditions.[454] A study of the nitration, bromination, and metallation of 2-(2-thienyl)thieno[2,3-*d*]pyrimidine has provided a comparison of the reciprocal influences of a thiophen ring substituted on to or fused to a pyrimidine ring (87).[455]

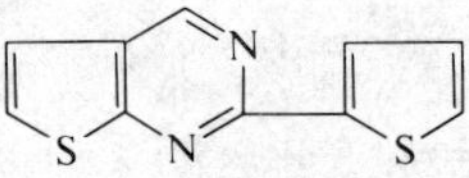

(87)

Other miscellaneous substitutions include the attempted bromination, acetoxymercuration, tritylation, idoination, lithiation, Vilsmeier, and diazo-coupling reactions on the 1,6-dioxa-6a$\lambda^4$-thiapentalene system (88): electrophilic reaction occurs at positions 3 (and 4).[456] The few attempts at electrophilic substitution in arsabenzene have been reviewed.[457]

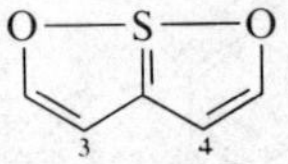

(88)

[452] J. N. Chatterjea and R. S. Gandhi, *J. Indian Chem. Soc.*, 1977, **54**, 719.
[453] S. Gronowitz, C. Roos, E. Sandberg, and S. Clementi, *J. Heterocyclic Chem.*, 1977, **14**, 893.
[454] S. Gronowitz and S. Liljefors, *Acta Chem. Scand.* (*B*), 1977, **31**, 771.
[455] J. Bourguignon, M. Moreau, G. Queguiner, and P. Pastour, *Bull. Soc. chim. France*, 1977, 676.
[456] D. H. Reid and R. G. Webster, *J.C.S. Perkin I*, 1977, 854.
[457] A. J. Ashe, *Accounts Chem. Res.*, 1978, **11**, 153.

# 8
# Nucleophilic Substitution Reactions

By G. M. BROOKE

## 1 Introduction

Since the publication of the last Report,[1] a further volume in the series 'Organic Reaction Mechanisms' has been published,[2] with a chapter on nucleophilic aromatic substitution.

## 2 Aromatic Systems

**Substituted Benzenes.**—The stability of aryldiazonium fluoroborates under a variety of conditions has been investigated. These compounds are rendered more thermally stable by sodium meta(octa)molybdate,[3] while complexes with 18-crown-6 polyether show enhanced stability over the uncomplexed salts in both thermal and photochemical de-diazoniations.[4] In the latter procedure, products from an ionic mechanism are obtained at 313 nm, but at >330 nm protodediazoniation of fluoroborates occurs by a radical mechanism.[5] The effect of solvents on the mechanism of nitrogen loss from diazonium salts has been reviewed.[6]

The conversion of $ArN_2^+$ salts into ArH takes place in hexamethylphosphoramide[7] (by a long radical chain process), and also by the use of the naphthalene anion radical $C_{10}H_8^{\cdot-}$ in THF.[8] Alternatively, deamination of $ArNH_2$ can be carried out with $Bu^tONO$ in DMF.[9] Unsymmetrical biaryls can be formed in a phase-transfer reaction between the diazonium fluoroborate (in an arene solvent) and KOAc complexed with 18-crown-6 polyether.[10] Meerwein arylation of fluorinated olefins with $ArN_2^+$ and copper(II) halides has been described,[11] but an alternative procedure for the overall addition of ArHal to an alkene involves the use of $ArNH_2$, $Bu^tONO$, and copper(II) halides in acetone or

[1] G. M. Brooke, in 'Aromatic and Heteroaromatic Chemistry', ed. H. Suschitzky and O. Meth-Cohn (Specialist Periodical Reports), The Chemical Society, London 1977, Vol. 6, p. 205.
[2] M. R. Crampton, in 'Organic Reaction Mechanisms 1976', ed. A. R. Butler and M. J. Perkins, Interscience, London 1977, p. 279.
[3] V. V. Kozlov, V. G. Smirnova, and V. P. Sagalovich, *Zhur. obshchei Khim.*, 1977, **47**, 2244 (*Chem. Abs.*, 1978, **88**, 61 814).
[4] R. A. Bartsch, N. F. Haddock, and D. W. McCann, *Tetrahedron Letters*, 1977, 3779.
[5] H. G. O. Becker, G. Hoffmann, and G. Israel, *J. prakt. Chem.*, 1977, **319**, 1021 (*Chem. Abs.*, 1978, **88**, 73 846).
[6] H. Zollinger, *Angew. Chem.*, 1978, **90**, 151.
[7] F. Troendlin and C. Ruechardt, *Chem. Ber.*, 1977, **110**, 2494.
[8] P. R. Singh, B. Jayaraman, and H. K. Singh, *Chem. and Ind.*, 1977, 311.
[9] M. P. Doyle, J. F. Dellaria, Jr., B. Siegfried, and S. W. Bishop, *J. Org. Chem.*, 1977, **42**, 3494.
[10] S. H. Korzeniowski, L. Blum, and G. W. Gokel, *Tetrahedron Letters*, 1977, 1871.
[11] C. S. Rondestvedt, Jr., *J. Org. Chem.*, 1977, **42**, 2618.

acetonitrile.[12] Phenolic compounds are formed, in yields superior to those obtained by the thermal process, by decomposing diazonium salts with $Cu_2O$ in a solution containing a large excess of $Cu(NO_3)_2$.[13] Diazotization of (1) and displacement of nitrogen gives (2), which undergoes an interesting thermolysis to produce (3), possibly *via* an aryloxenium ion $ArO^+$ (Scheme 1).[14] Displacement

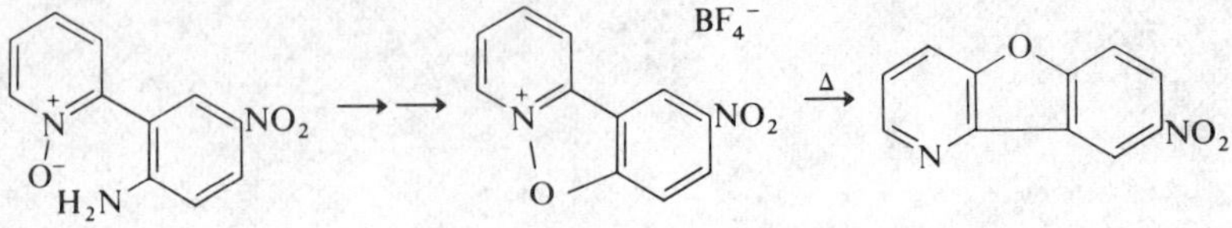

**Scheme 1**

of $N_2$ from diazonium compounds and the formation of C—S bonds has been accomplished with $CS_2$[15] and with 4-phenyl-1,3,4-thiadiazolidine-2,5-dithione;[16] the reaction of $ArN_2^+$ $BF_4^-$ with PhSeH in $CH_2Cl_2$ and acetone gives ArSePh.[17] Two alternative procedures to the Sandmeyer reaction for the preparation of aryl halides utilize arylhydrazines[18] or primary arylamines[19] with alkyl nitrites and copper(II) halides.

An anion radical has been shown to be a precursor in the nucleophilic substitution of one nitro-group of 1,4-dinitrobenzene by hydroxide ion in aqueous DMSO.[20] Spectroscopic evidence has shown that an electron transfer occurs which precedes or accompanies displacement of the nitro-group by 4-methylphenoxide ion in 3- and 4-nitrophthalate diesters.[21]

The mechanism of $S_N(AE)$ reactions involving nitrogen-containing nucleophiles continues to attract attention regarding the details of the base-catalysed product-forming step(s).[22] In protic solvents, examples have been found where deprotonation of (4) is rate-limiting,[23] as in the case when the leaving group is a

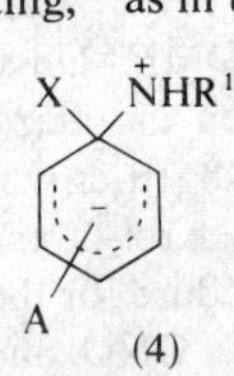

[12] M. P. Doyle, B. Siegfried, R. C. Elliott, and J. F. Dellaria, Jr., *J. Org. Chem.*, 1977, **42**, 2431.
[13] T. Cohen, A. G. Dietz, Jr., and J. R. Miser, *J. Org. Chem.*, 1977, **42**, 2053.
[14] R. A. Abramovitch and M. N. Inbasekaran, *J.C.S. Chem. Comm.*, 1978, 149.
[15] L. Benati and P. C. Montevecchi, *J. Org. Chem.*, 1977, **42**, 2025.
[16] E. P. Nesynov and M. M. Besprozvannaya, *Ukrain. khim. Zhur.* (*Russ. Edn.*), 1977, **43**, 1185 (*Chem. Abs.*, 1978, **88**, 62 343).
[17] F. G. James, M. J. Perkins, O. Porta, and B. V. Smith, *J.C.S. Chem. Comm.*, 1977, 131.
[18] M. P. Doyle, B. Siegfried, and W. F. Fobare, *Tetrahedron Letters*, 1977, 2655.
[19] M. P. Doyle, B. Siegfried, and J. F. Dellaria, Jr., *J. Org. Chem.*, 1977, **42**, 2426.
[20] T. Abe and Y. Ikegami, *Bull. Chem. Soc. Japan*, 1978, **51**, 196.
[21] H. M. Relles, D. S. Johnson, and J. S. Manello, *J. Amer. Chem. Soc.*, 1977, **99**, 6677.
[22] D. Ayediran, T. O. Bamkole, J. Hirst, and I. Onyido, *J.C.S. Perkin II*, 1977, 1580; D. Spinelli, G. Consiglio, and R. Noto, *ibid.*, p. 1316; G. D. Titskii, A. E. Shumeiko, and L. M. Litvinenko, *Doklady Akad. Nauk. S.S.S.R.*, 1977, **234**, 868 (*Chem. Abs.*, 1977, **87**, 84 088).
[23] C. F. Bernasconi, R. H. de Rossi, and P. Schmid, *J. Amer. Chem. Soc.*, 1977, **99**, 4090.

good one, *e.g.* X = PhO. This mechanism differs from that previously formulated for reactions in aprotic media, in which the rapid deprotonation of (4) is followed by a rate-limiting *general-acid-catalysed* expulsion of the leaving group X.[24]

Nucleophilic aromatic substitution reactions have been examined in terms of a model based on the combination of a cation with an anion.[25] The reactivities of 2,4-dinitro-halogenobenzenes with nucleophiles have been shown to be related to the basicity and polarisability of the nucleophile and the polarisability of the substrate: only atoms and bonds at or near the reaction centre are involved in producing the polarisability effects.[26] Hammett $\sigma^-$ values have been measured for unsaturated groups —CH=NX in the 4-position for nucleophilic displacement of chlorine in 2-nitrochlorobenzene derivatives and compared with those for established activating groups.[27]

A variety of synthetic reactions with substrates activated by the usual electron-attracting groups have been published since the last Report. Carbon–carbon bonds are formed in reactions involving base-catalysed reactions of active-methylene compounds; for example, of deoxybenzoins with halogeno-nitrobenzenes;[28] of ferrocenyl methyl ketone with 1,3,5-trinitrobenzene followed by hydride extraction with tropylium cation;[29] and of phenylacetonitrile derivatives with some chloronitrobenzophenones, in which the nitro-group is displaced from 3-chloro-4-nitro-compound and chloride ion from the 5-chloro-2-nitro-compound.[30] A nitrogen to carbon migration takes place following the capture of a heterocyclic carbene according to the reaction outlined in Scheme 2.[31] An Arbusov reaction occurs in the reaction of trialkyl phosphites with activated derivatives of chlorobenzene.[32] The $CF_3SO_2$ group is readily replaced by $MeO^-$

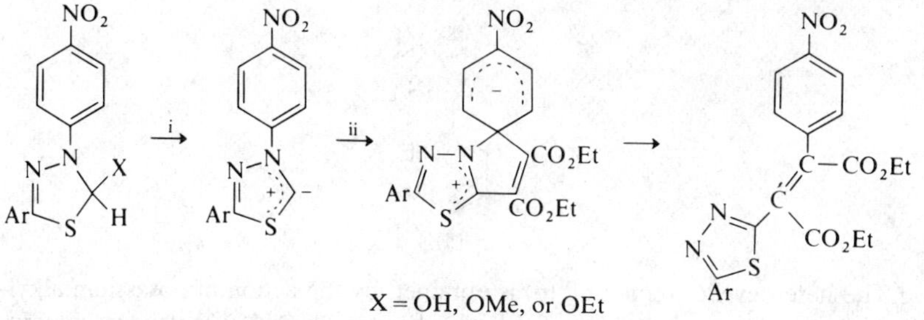

X = OH, OMe, or OEt

Reagents: i, benzene at 80 °C; ii, $EtO_2CC\equiv CCO_2Et$.

**Scheme 2**

[24] J. Orvik and J. F. Bunnett, *J. Amer. Chem. Soc.*, 1970, **92**, 2417.
[25] C. D. Ritchie and M. Sawada, *J. Amer. Chem. Soc.*, 1977, **99**, 3754.
[26] G. Bartoli, P. E. Todesco, and M. Fiorentino, *J. Amer. Chem. Soc.*, 1977, **99**, 6874.
[27] H. R. Freire and J. Miller, *J.C.S. Perkin II*, 1978, 108.
[28] M. Jawdosiuk, W. Wilczynski, and W. Cwikiewicz, *Roczniki Chem.*, 1977, **51**, 595 (*Chem. Abs.*, 1977, **87**, 134 294).
[29] M. I. Kalinkin, G. D. Kolomnikova, V. E. Puzanova, M. M. Gol'din, Z. N. Parnes, and D. N. Kursanov, *Izvest. Akad. Nauk. S.S.S.R., Ser. khim.*, 1977, 1852 (*Chem. Abs.*, 1977, **87**, 168 164).
[30] M. Makosza and M. Ludwikow, *Roczniki Chem.*, 1977, **51**, 829 (*Chem. Abs.*, 1977, **87**, 152 057).
[31] G. Scherowsky, K. Duennbier, and G. Hoefle, *Tetrahedron Letters*, 1977, 2095.
[32] V. A. Bondar, V. M. Sidorenko, and V. V. Malovik, *Zhur. obshchei Khim.*, 1977, **47**, 953 (*Chem. Abs.*, 1977, **87**, 39 604).

and PhS⁻ in 3,5-(CF$_3$SO$_2$)$_2$C$_6$H$_3$NO$_2$ and 1,3,5-(CF$_3$SO$_2$)$_3$C$_6$H$_3$;[33] phthalate ester groups activate 3- and 4-nitro-groups towards displacement by phenoxides,[34] as do *N*-substituted phthalimido-groups towards replacement of 3- and 4-nitro- or -halogeno-groups by phenoxides[35] and arylthiolates;[36] 3- and 4-nitro- and halogeno-phthalic anhydrides have been studied in similar reactions.[37] Acyl–oxygen *versus* aryl–oxygen bond fission has been investigated in the reaction between benzenethiolate and various nitrophenyl esters;[38] extreme examples are the nitrophenyl esters (100% acyl–oxygen fission) and 2,4,6-trinitrophenyl benzoate (0% acyl–oxygen fission). Methanethiolate effects polysubstitution of nitro- and halogen functions in benzoic acid and related derivatives:[39] for example, C$_6$(SMe)$_5$CONH$_2$ has been prepared. The overall replacement of NO$_2$ by —SCN is accomplished by treating 2-nitrobenzonitrile derivatives with HSCH$_2$CH$_2$CN and KOH in aqueous DMF followed by the reaction of the thiolate formed with ClCN.[40] A potential route for the preparation of aryl fluorides from sulphonates and the KF–18-crown-6 polyether complex was partially successful only in the case of the 2,4-dinitrophenyl compounds;[41] the major reaction was attack on sulphur with displacement of ArO⁻, which then gave diaryl ethers by further reaction with the starting material.

Traditional activating groups are not present in *o*-methoxyaryl-oxazolines, in which displacement of methoxide is readily accomplished by an alkyl, alkylidene, or aryl group, using the lithium or magnesium reagent.[42] The use of lithio-amides permits the formation of amines[43] (and thence *o*-aminobenzoic acid derivatives, by hydrolysis). Parallel reactions are followed by using *ortho*-fluorine-containing aryl-oxazolines,[44] and, with 2,6-difluoro-compounds (5), one or two substituents can be introduced into the aromatic ring.

(5)

The heterocyclic compound (6) is obtained by the action of potassium alkyl-dithiocarbonates with 4-chloro-3,5-dinitrobenzotrifluoride.[45] Nitrogen-containing anions from pyrroles, pyrazoles, and imidazoles have been used to displace

[33]  V. N. Boiko and G. M. Shchupak, *Zhur. org. Khim.*, 1977, **13**, 1042 (*Chem. Abs.*, 1977, **87**, 134 262).
[34]  F. J. Williams, H. M. Relles, J. S. Manello, and P. E. Donahue, *J. Org. Chem.*, 1977, **42**, 3419.
[35]  F. J. Williams and P. E. Donahue, *J. Org. Chem.*, 1977, **42**, 3414.
[36]  F. J. Williams and P. E. Donahue, *J. Org. Chem.*, 1978, **43**, 250.
[37]  F. J. Williams and P. E. Donahue, *J. Org. Chem.*, 1978, **43**, 255; F. J. Williams, H. M. Relles, P. E. Donahue, and J. S. Manello, *ibid.*, 1977, **42**, 3425.
[38]  G. Guanti, C. Dell'Erba, F. Pero, and G. Leandri, *J.C.S. Perkin II*, 1977, 966.
[39]  J. R. Beck and J. A. Yahner, *J. Org. Chem.*, 1978, **43**, 2052.
[40]  J. R. Beck and J. A. Yahner, *J. Org. Chem.*, 1978, **43**, 1604.
[41]  M. J. V. de Oliveira Baptista and D. A. Widdowson, *J.C.S. Perkin I*, 1978, 295.
[42]  A. I. Meyers, R. Gabel, and E. D. Mihelich, *J. Org. Chem.*, 1978, **43**, 1372.
[43]  A. I. Meyers and R. Gabel, *J. Org. Chem.*, 1977, **42**, 2653.
[44]  A. I. Meyers and B. E. Williams, *Tetrahedron Letters*, 1978, 223.
[45]  J. J. D'Amico, C. C. Tung, W. E. Dahl, and D. J. Dahm, *J. Org. Chem.*, 1977, **42**, 2896.

aromatic fluorine to yield precursors for 1,4-benzodiazepine syntheses.[46] A one-step route to quinoxaline $N$-oxides and related structures has been described by the annelations of amidines on halogeno-nitro-aromatic compounds.[47] Thus 2,4-dinitrofluorobenzene and $\alpha$-phenylacetamidine give (7).

(6)          (7)

A quantitative study of the photostimulated reaction of iodobenzene with diethylphosphite ion, a reaction which proceeds by a radical chain mechanism (designated $S_{RN}1$), did not give unequivocal identification of the initiation and termination steps.[48] Analogous reactions with isomeric iodo-anisoles and -toluenes and $(EtO)_2PO^-$ are suitable for preparative-scale work, while with isomeric halogeno-iodobenzenes, whether it is one or two halogen atoms that are replaced depends on the nucleophiles and the halogens involved, and on their orientation.[49] The conversion of $m$-$C_6H_4BrI$ into $m$-$C_6H_4[OP(OEt)_2]_2$ does not involve $m$-$C_6H_4Br[OP(OEt)_2]$ as an intermediate on the main route to the product,[50] an observation which can be rationalised on the basis of an $S_{RN}1$ propagation mechanism. This mechanism also satisfies quantitatively the partitioning between mono- and di-substitution products.[51]

A competing process has been found in the photostimulated reaction between iodobenzene and the enolate anion of di-isopropyl ketone in liquid ammonia:[52] in addition to the expected product $C_6H_5CMe_2COCHMe_2$, the 1,5-diketone $Me_2CHCOCMe_2CH_2CHMeCOCHMe_2$ was also identified, formed by hydrogen abstraction by $C_6H_5\cdot$ from the enolate anion to give the radical anion $\dot{C}H_2CMe{=}C(\bar{O})CHMe_2$, which in turn competed with $C_6H_5\cdot$ for reaction with more enolate anion. Other methods for promoting reactions by the $S_{RN}1$ process with aryl halides involve the use of hydrated electrons from Na–Hg and $H_2O$ (and reaction with phenoxide ion)[53] and the use of sunlight (with $\bar{C}H_2SOCH_3$).[54] Less clear are the mechanisms in which a nitro-group in 1,4-dinitrobenzene is replaced by $CMe_2NO_2$, using $Bu^n_4\overset{+}{N}[CMe_2NO_2]^-$ in benzene,[55] and in which the halogen in $p$-bromoacetophenone is replaced with benzenethiolate at the cathode of an electrolytic cell,[56,57] though some workers favour a non-radical chain pathway.[58]

[46] N. W. Gilman, B. C. Holland, G. R. Walsh, and R. I. Fryer, *J. Heterocyclic Chem.*, 1977, **14**, 1157.
[47] M. J. Strauss, D. C. Palmer, and R. R. Bard, *J. Org. Chem.*, 1978, **43**, 2041.
[48] S. Hoz and J. F. Bunnett, *J. Amer. Chem. Soc.*, 1977, **99**, 4690.
[49] J. F. Bunnett and R. P. Traber, *J. Org. Chem.*, 1978, **43**, 1867.
[50] J. F. Bunnett and S. J. Shafer, *J. Org. Chem.*, 1978, **43**, 1873.
[51] J. F. Bunnett and S. J. Shafer, *J. Org. Chem.*, 1978, **43**, 1877.
[52] J. F. Wolfe, M. P. Moon, M. C. Sleevi, J. F. Bunnett, and R. R. Bard, *J. Org. Chem.*, 1978, **43**, 1019.
[53] S. Rajan and P. Sridaran, *Tetrahedron Letters*, 1977, 2177.
[54] S. Rajan and K. Muralimohan, *Tetrahedron Letters*, 1978, 483.
[55] D. L. Burt, D. J. Freeman, P. G. Gray, R. K. Norris, and D. Randles, *Tetrahedron Letters*, 1977, 3063.
[56] W. J. M. van Tilborg, C. J. Smit, and J. J. Scheele, *Tetrahedron Letters*, 1977, 2113.
[57] W. J. M. van Tilborg, C. J. Smit, and J. J. Scheele, *Tetrahedron Letters*, 1978, 776.
[58] J. Pinson and J.-M. Saveant, *J. Amer. Chem. Soc.*, 1978, **100**, 1506.

Isoindoles and isoindolines have been prepared from tertiary amines derived from 2-chlorobenzylamine, *via* aryne intermediates.[59]

Nucleophilic aromatic photosubstitution has been reviewed.[60] Sodium arenesulphonates have been photodesulphonated in DMSO.[61]

A variety of reactions have been developed for introducing nucleophilic groups into an aromatic nucleus, some of which have only a superficial resemblance to better known methods of nucleophilic aromatic substitution. For example, the formation of aromatic C—H bonds from halogen-containing substrates can be accomplished by a variety of methods: by using NaH and amyl$^l$ONa, activated by transition-metal salts [*e.g.* $Ni(OAc)_2$];[62] by hydrogen transfer from an organic base, *e.g.* indoline, catalysed by $Pd^{II}$ salts;[63] by the use of $HCO_2^- Na^+$, catalysed by $(Ph_3P)_4Pd$[64] or $HCO_2^-$ $Et_3NH^+$ in the presence of Pd/C on a soluble $Ar_3P$–$Pd(OAc)_2$ catalyst;[65] and by using NaOMe in DMF, with $(PPh_3)_4Pd$ as catalyst.[66] Carbon–carbon bonds are formed from non-activated aryl halides and lithium ester enolates under the influence of a catalyst prepared by the addition of $Bu^nLi$ to $NiBr_2$–THF[67] [equation (1)]. A $\beta$-hydrogen in $\alpha\beta$-unsaturated acetals and

$$C_6H_5I + LiCH_2CH{=}CHCO_2Et \xrightarrow{\text{catalyst}} C_6H_5CH_2CH{=}CHCO_2Et \qquad (1)$$

ketals is replaced by an aryl group from an aryl halide, using $Et_3N$ and the catalyst $Pd(OAc)_2$–$(2\text{-}MeC_6H_4)_3P$.[68] Treatment of the *N*-chloro-derivatives of a 4-substituted acetanilide with $Me_2S$ at $-15\,°C$, and the reaction of the azasulphonium salts with $Et_3N$, gives the *ortho*-$CH_2SMe$-substituted acetanilide by a 2,3-sigmatropic shift.[69] Symmetrical biaryls are formed from arylcopper cluster compounds and copper(I) triflate in benzene[70] or by the self-coupling of aryl halides in the presence of zinc, $Ph_3P$, and a catalytic amount of $(Ph_3P)_2NiCl_2$, a reaction which tolerates a variety of carbonyl functions in the substrate.[71] A highly selective synthesis of unsymmetrical biaryls has been described[72] [equation (2)], and alkynyl-zinc halides may be used to prepare terminal and internal aryl-alkynes.[73]

$$ArBr(I) + Ar'ZnHal \xrightarrow[\substack{\text{or } (Ph_3P)_2PdCl_2 \\ \text{and } Bu^i_2AlH}]{Ni(Ph_3P)_4} ArAr' \qquad (2)$$

The overall formation of nucleophilically substituted *o*-phenylenediamines has been achieved according to the reactions shown in Scheme 3.[74]

[59] I. Ahmed, G. W. H. Cheeseman, B. Jaques, and R. G. Wallace, *Tetrahedron*, 1977, **33**, 2255.
[60] J. Cornelisse, G. P. De Gunst, and E. Havinga, *Adv. Phys. Org. Chem.*, 1971, **11**, 225; E. Havinga and J. Cornelisse, *Pure Appl. Chem.*, 1976, **47**, 1.
[61] N. Suzuki, K. Ito, A. Inoue, and Y. Izawa, *Chem. and Ind.*, 1977, 399.
[62] B. Loubinoux, R. Vanderesse, and P. Caubere, *Tetrahedron Letters*, 1977, 3951.
[63] H. Imai, T. Nishiguchi, M. Tanaka, and K. Fukuzumi, *J. Org. Chem.*, 1977, **42**, 2309.
[64] P. Helquist, *Tetrahedron Letters*, 1978, 1913.
[65] N. A. Cortese and R. F. Heck, *J. Org. Chem.*, 1977, **42**, 3491.
[66] A. Zask and P. Helquist, *J. Org. Chem.*, 1978, **43**, 1619.
[67] A. A. Millard and M. W. Rathke, *J. Amer. Chem. Soc.*, 1977, **99**, 4833.
[68] T. C. Zebovitz and R. F. Heck, *J. Org. Chem.*, 1977, **42**, 3907.
[69] P. G. Gassmann and R. J. Balchunis, *Tetrahedron Letters*, 1977, 2235.
[70] G. van Koten, J. T. B. H. Jastrzebski, and J. G. Noltes, *J. Org. Chem.*, 1977, **42**, 2047.
[71] M. Zembayashi, K. Tamao, J. Yoshida, and M. Kumada, *Tetrahedron Letters*, 1977, 4089.
[72] E. Negishi, A. O. King, and N. Okukado, *J. Org. Chem.*, 1977, **42**, 1821.
[73] A. O. King, E. Negishi, F. J. Villani, Jr., and A. Silveira, Jr., *J. Org. Chem.*, 1978, **43**, 358.
[74] A. M. Jefferson and H. Suschitzky, *J.C.S. Chem. Comm.*, 1977, 189.

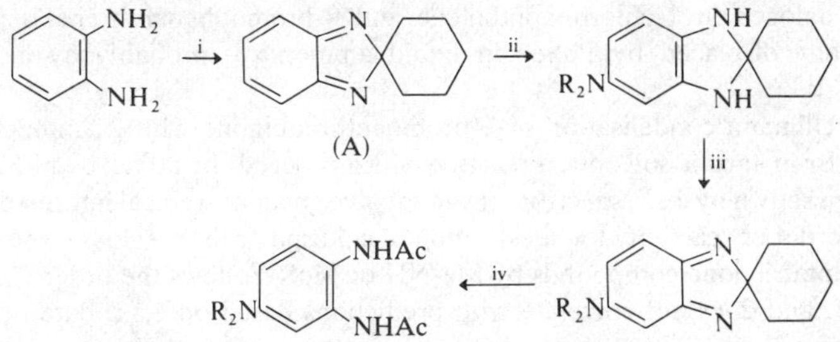

Reagents: i, cyclohexanone; ii, $R_2NH$; iii, (A); iv, $H_2$, Pd/C, $Ac_2O$, AcOH.

**Scheme 3**

Thiolates[75] and selenocyanates[76] displace halogen from ArHal with a palladium(0) compound and copper(I) iodide, respectively, as catalysts.

Aryl bromides can be converted into the iodo-compounds with KI and $NiBr_2$–Zn as catalyst under mild conditions.[77]

Cobalt(III) trifluoroacetate promotes overall nucleophilic displacement of hydrogen in ArH *via* intermediate $ArH^{\ddagger}$ cation radicals; in this way benzene has been converted into $C_6H_5X$ (X = Cl, Br, I, or CN), using LiX salts; the formation of nitrobenzene requires $LiNO_3$.[78]

**Fused Aromatic Systems.**—In an attempt to force nucleophilic substitution of a $\beta$-nitro-group in a 2,3-dinitro-naphthalene by blocking both $\alpha$-positions (which otherwise leads to a *cine*-substitution product at C-1), the reaction of the 1,4-dimethyl compound (8) with secondary amines gave an unusual *tele*-substitution product (9) (Scheme 4).[79]

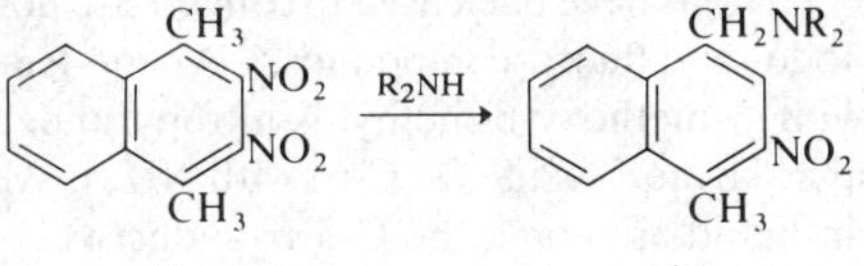

**Scheme 4**

1-Fluoro-2-methylnaphthalene and benzalaniline did not produce the expected 'stilbene' derivative with $Bu^tO^-$: following the initial C—C bond-forming reaction, an intramolecular displacement of fluoride ion by nitrogen occurred to give a 2,3-dihydro-1,2-diphenylindole derivative.[80] 1-Methoxy-2,4-dinitro-naphthalene undergoes an annelation reaction with an amidine, as does 2,4-$(NO_2)_2C_6H_3F$.[47]

[75] M. Kosugi, T. Shimizu, and T. Migita, *Chem. Letters*, 1978, 13 (*Chem. Abs.*, 1978, **88**, 135 848).
[76] H. Suzuki and M. Shinoda, *Synthesis*, 1977, 640.
[77] K. Tagaki, N. Hayama, and T. Okamoto, *Chem. Letters*, 1978, 191 (*Chem. Abs.*, 1978, **88**, 152 145).
[78] M. E. Kurz and G. W. Hage, *J. Org. Chem.*, 1977, **42**, 4080.
[79] G. Guanti, S. Thea, M. Novi, and C. Dell'Erba, *Tetrahedron Letters*, 1977, 1429.
[80] M. S. Newman, B. Dhawan, and S. Kumar, *J. Org. Chem.*, 1978, **43**, 524.

The halogen in 1-chloronaphthalene and 9-bromophenanthrene is photo-chemically displaced by PhSe⁻ in liquid ammonia – probably by an $S_{RN}1$ process.[81]

The Ullmann condensation of 1-bromoanthraquinone with 2-aminoethanol and CuBr in aprotic solvents, a reaction which proceeds by attack by nitrogen,[82] has been shown by e.s.r. spectroscopy to involve an anion-radical intermediate.[83]

The order of reactivity for substitution of chlorine or the tosyloxy-group from cycloheptatrienone compounds by $Me_2NH$ or MeS⁻ follows the order C-3 > C-2 ≈ C-4, and does not correlate with predictions based on e.s.r. data or M.O. treatments.[84]

## 3 Heteroaromatic Systems

$\pi$-Deficient Heteroaromatic Systems.—Substitution reactions by rearrangement of *N*-oxides have been reviewed.[85] Anion-radical mechanisms have been pro-posed to account for unexpected products in the reactions of some pyridine compounds with nucleophiles. For example, in a low-conversion experiment, treatment of 2-chloro-3-nitropyridine with isonicotinamide and $Bu^tO^- K^+$ in DMSO gave the 6-substituted product (by overall replacement of hydrogen),[86] and in the reaction of 2-amino-3,5-dihalogeno-pyridines with KOH, a by-product was the 2-amino-5-halogeno-pyridine, by a dehalogenation process.[87] The first light-induced substitution reactions of halogeno-pyridine compounds have been carried out, using the potassium enolates of a variety of ketones in liquid ammonia.[88] Anhydrous or hydrated $CuSO_4$ converts pyridine and related compounds into the 2-one compounds at 300 °C.[89]

3,6-Dichloropyridazine reacts with two equivalents of 6-bromo-3(2*H*)-pyri-dazinethione in slightly acidic methanol in stages, the second stage resulting in the unusual displacement of bromine rather than chlorine in the intermediate (10).[90] Halogen-exchange reactions have been used to convert 3-chloro-pyridazines into the corresponding iodo- and fluoro-compounds.[91] Aminodemethoxylation reac-tions take place with 3-methoxy-6-methyl-4-nitropyridine 1-oxide and 3,6-dimethoxy-4-nitropyridazine 1-oxide (at C-6) with $NH_3$ in MeOH, yet with the latter heterocycle in liquid ammonia the C-5 $\sigma$-adduct (11) is observed by ¹H n.m.r. spectroscopy.[92]

The conversion of 5-bromo-4-t-butylpyrimidine into the 6-amino-compound by $KNH_2$ in liquid $NH_3$, previously thought to proceed by an $S_N$(EA) process, has been re-investigated, using ¹⁵N-enriched materials. It has been shown that a

[81] A. B. Pierini and R. A. Rossi, *J. Organometallic Chem.*, 1978, **144**, C12.
[82] S. Arai, M. Hida, T. Yamagishi, and S. Ototako, *Bull. Chem. Soc. Japan*, 1977, **50**, 2982.
[83] S. Arai, M. Hida, and T. Yamagishi, *Bull. Chem. Soc. Japan*, 1978, **51**, 277.
[84] M. Cavazza, M. P. Colombini, M. Martinelli, L. Nucci, L. Pardi, F. Pietra, and S. Santucci, *J. Amer. Chem. Soc.*, 1977, **99**, 5997.
[85] R. A. Abramovitch and I. Shinkai, *Accounts Chem. Res.*, 1976, **9**, 192.
[86] K. Kurita and R. L. Williams, *J. Heterocyclic Chem.*, 1974, **14**, 673.
[87] G. Mattern, *Helv. Chim. Acta*, 1977, **60**, 2062.
[88] A. P. Komin and J. E. Wolfe, *J. Org. Chem.*, 1977, **42**, 2481.
[89] P. Tomasik and A. Woszczyk, *Tetrahedron Letters*, 1977, 2193.
[90] R. B. Phillips and C. C. Wamser, *J. Org. Chem.*, 1978, **43**, 1190.
[91] G. B. Barlin and C. Yoot Yap, *Austral. J. Chem.*, 1977, **30**, 2319.
[92] T. Sakamoto and H. C. van der Plas, *J. Heterocyclic Chem.*, 1977, **14**, 789.

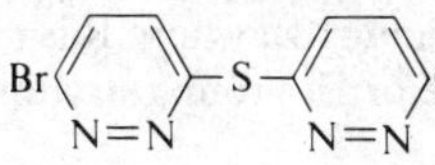

(10) 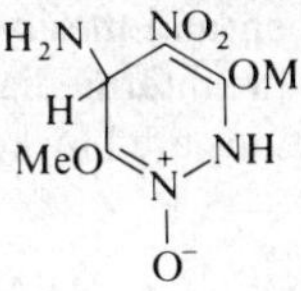          (11)

ring-opened intermediate is formed;[93] this new mechanism has been designated an $S_N(ANRORC)^{cine}$ process. A variety of 2-fluoro-pyrimidines have been prepared by halogen exchange with CsF on the chloro-compounds,[94] and 4-cyano-compounds are formed in a two-step process from the chloro-compounds by reaction first with Me₃N, followed by KCN in MeCONH₂.[95] Evidence for an intermediate 5,6-dihydropyrimidine intermediate has been obtained as a result of observing a secondary isotope effect in the cysteine-promoted dehalogenation of 5-bromo-2'-deoxyuridine.[96] 4-(6)-Ethoxy-1-ethyl-2-phenylpyridinium salts are de-*N*-ethylated in liquid ammonia by a ring-opening–ring-closing sequence.[97]

Multiheteromacrocycles possessing 2,6-pyrazino subunits connected by C—O and/or C—S linkages have been synthesised from 2,6-dichloropyrazine and dinucleophilic reagents.[98]

Polarographic reduction of chlorodiamino- and aminodichloro-*s*-triazines has been studied.[99] A unique deoxygenative 6-alkoxylation reaction of 1,2,4-triazine 2-oxide has been described[100] (Scheme 5).

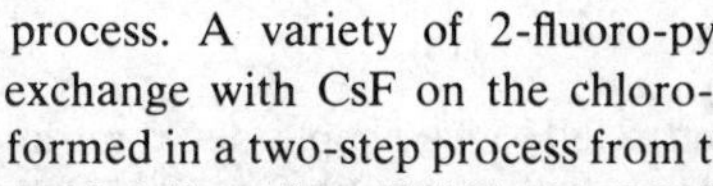

$$X = NH_2, SMe, etc.$$

**Scheme 5**

**π-Excessive Heteroaromatic Systems.**—The sequence of reactivity of the 2,5-dinitro-derivatives of the following heteroaromatic systems towards denitration by *p*-tolylthiolate has been established: furan > thiophen > *N*-methylpyrrole.[101] In contrast to piperidinodenitration, 1,4-dinitrobenzene is now less reactive than 2,5-dinitro-*N*-methylpyrrole as a result of entropy factors.

Genuine base catalysis has been found in the reaction of piperidine with 2-L-5-nitrothiophen (where L, the leaving group, is *p*-NO₂C₆H₄O or PhSO₂).[102]

[93] C. A. H. Rasmussen and H. C. van der Plas, *Rec. Trav. chim.*, 1977, **96**, 101.
[94] S. G. Baram, O. P. Shkurko, and V. P. Mamaev, *Izvest. Sibirsk. Otdel. Akad. Nauk S.S.S.R., Ser. khim. Nauk*, 1977, 106 (*Chem. Abs.*, 1977, **87**, 5900).
[95] D. T. Hurst, K. Biggadike, and J. J. Tibble, *Heterocycles*, 1977, **6**, 2005.
[96] Y. Wataye and D. V. Santi, *J. Amer. Chem. Soc.*, 1977, **99**, 4534.
[97] E. A. Oostveen and H. C. van der Plas, *Rec. Trav. chim.*, 1977, **96**, 183.
[98] G. R. Newkome and A. Nayak, *J. Org. Chem.*, 1978, **43**, 409.
[99] E. Yu. Khmel'nitskaya, *Zhur. obshchei Khim.*, 1977, **47**, 1158 (*Chem. Abs.*, 1977, **87**, 101 655).
[100] B. T. Keen, R. J. Radel, and W. W. Paudler, *J. Org. Chem.*, 1977, **42**, 3498.
[101] P. Mencarelli and F. Stegel, *J. Org. Chem.*, 1977, **42**, 3550.
[102] D. Spinelli, G. Consiglio, and R. Noto, *J. Heterocyclic Chem.*, 1977, **14**, 1325.

3,4-Dinitrothiophen and mesitylthiolate ultimately give 2-mesitylthio-4-nitro-thiophen through the intermediacy of (12) and (13), where R is mesityl.[103] When the 2- and 5-positions are blocked with Me or Et groups, arylthiolates give (14; R = H or Me).[104]

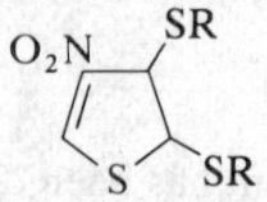

(12) 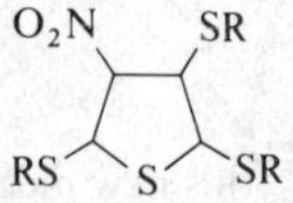                    (13) 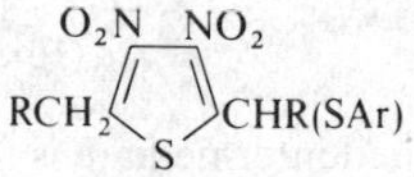                    (14)

1-Nitropyrazoles and $R_2NH$ give simple de-$N$-nitration products or de-$N$-nitration accompanied by substitution in the ring.[105] This latter reaction, favoured by having a nitro-group attached to C-4, is a *cine*-substitution, proceeding by attack at C-5 (Scheme 6).

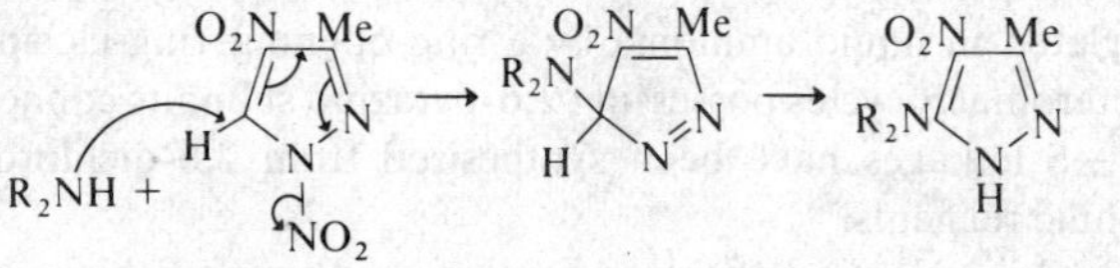

**Scheme 6**

When the 3- and 5-positions in 1,4-dinitropyrazole are blocked by methyl groups, reaction with piperidine or morpholine in simple alcohol solvents ROH gives (15).[106]

Nucleophilic photosubstitution of nitro-derivatives of 1-methylpyrazole and 1-methylimidazole strongly depends on the position of the nitro-group:[107] the 3-nitropyrazole compound and KCN give some hydrogen-substitution product at C-4 and an equal amount of 1-methyl-3-nitrosopyrazole, while 5-nitro-1-methylimidazole undergoes replacement of the nitro-group by cyanide, and by methoxide to give (16).

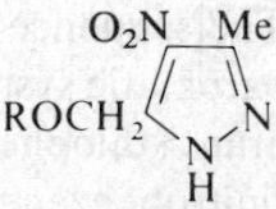

(15) 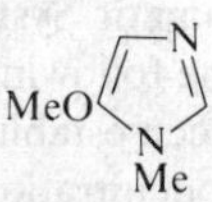                    (16)

**Fused Heteroaromatic Compounds.**—The halogen in 4-chloroquinoline is readily replaced by the enolate anions from isopropylidine alkylmalonates and acetic anhydride in the presence of the heterocycle;[108] acylation of the nitrogen provides the necessary activation of the system. Attempted overall methylation of 2-methylquinoline by $MeSO\bar{C}H_2$ and DMSO gives only a minute amount of the

[103] M. Novi, F. Sancassan, G. Guanti, C. Dell'Erba, and G. Lenadri, *Chimica e Industria*, 1977, **59**, 299.
[104] M. Novi, F. Sancassan, G. Guanti, C. Dell'Erba, and G. Lenadri, *Chimica e Industria*, 1977, **59**, 299.
[105] C. L. Habraken and E. K. Poels, *J. Org. Chem.*, 1977, **42**, 2893.
[106] C. L. Habraken and S. M. Bonser, *Heterocycles*, 1977, **7**, 259.
[107] C. Oldenhof and J. Cornelisse, *Rec. Trav. chim.*, 1978, **97**, 35.
[108] A. Scovill and F. X. Smith, *J. Heterocyclic Chem.*, 1977, **14**, 1081.

expected 2,4-dimethylquinoline;[109] the major product is the novel tricyclic system (17), which is the 1:1 adduct of the heterocycle and DMSO. 1-Nitro-7-bromoacridone is debrominated by MeOK, and with $NaNH_2$ it gives the 6-, 7-, and 8-amino-derivatives.[110] Attempted hydrolysis of 1-nitro-9-(2-methyl-aminoethylamino)acridine, at pH 7, gives the intramolecularly substituted compound (18).[111] A two-phase catalytic system promotes displacement of halogen from 9-chloroacridone by carbanions generated from PhCH(R)CN.[112] The chlorine in 1-chloro-4-nitroacridone is replaceable by $I^-$ and $NCS^-$.[113]

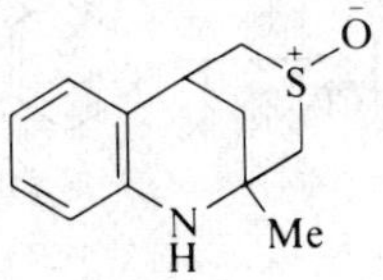

(17) 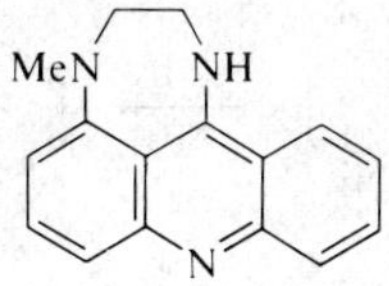                    (18)

2-Nitrophenazine 10-oxide (19) reacts with primary amines *via* σ-anionic intermediates (20) to give 1-alkylamino-2-nitrophenazine (21) and a trace of 2-alkylaminophenazine 10-oxide (22)[114] (Scheme 7), but both products are formed in respectable amounts with secondary amines.

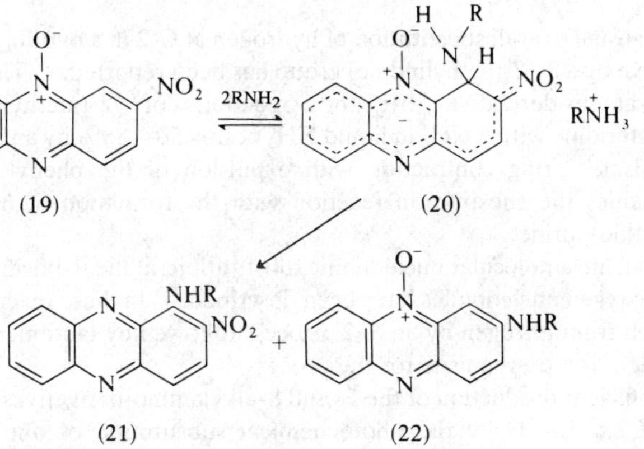

(19)                    (20)

(21)                    (22)

**Scheme 7**

Primary amines and (19), in photochemical reactions, give 2-alkylamino-phenazine 10-oxide, and reduction to 2-nitrophenazine occurs.[115]

[109] H. Kato, I. Takeuchi, Y. Hamada, M. Ono, and M. Hirota, *Tetrahedron Letters*, 1978, 135.
[110] A. Ledochowski, S. Skonieczny, A. Glowacki, and J. Mogielnicki, *Roczniki Chem.*, 1977, **51**, 357 (*Chem. Abs.*, 1977, **87**, 134 983).
[111] S. Skonieczny and A. Ledochowski, *Roczniki Chem.*, 1977, **51**, 2279 (*Chem. Abs.*, 1978, **88**, 120 242).
[112] W. Wilczynski, M. Jawdosiuk, and M. Makosza, *Roczniki Chem.*, 1977, **51**, 1643 (*Chem. Abs.*, 1978, **88**, 62 277).
[113] J. Romanowski and Z. Eckstein, *Roczniki Chem.*, 1977, **51**, 2455 (*Chem. Abs.*, 1978, **88**, 169 927).
[114] G. Minoli, A. Albini, G. F. Bettinetti, and S. Pietra, *J.C.S. Perkin II*, 1977, 1661.
[115] S. Pietra, G. F. Bettinetti, A. Albini, E. Fasani, and R. Oberti, *J.C.S. Perkin II*, 1978, 185.

2-Halogeno-1,7-naphthyridines undergo interesting *tele*-amination reactions with $KNH_2$ in liquid $NH_3$.[116] With the chloro-compound, in addition to the 2-amino-compound and the parent hydrocarbon, the 4-amino- and 8-amino-compounds are formed (by 1,3- and 1,4-*tele*-substitution reactions respectively) and a ring-transformation product. $^1H$ n.m.r. data on the reacton system at $-50\,^\circ C$ showed C-4, C-6, and C-8 adducts; the 1,3-*tele*-amination was formulated as proceeding *via* the intermediate (23) (Scheme 8).

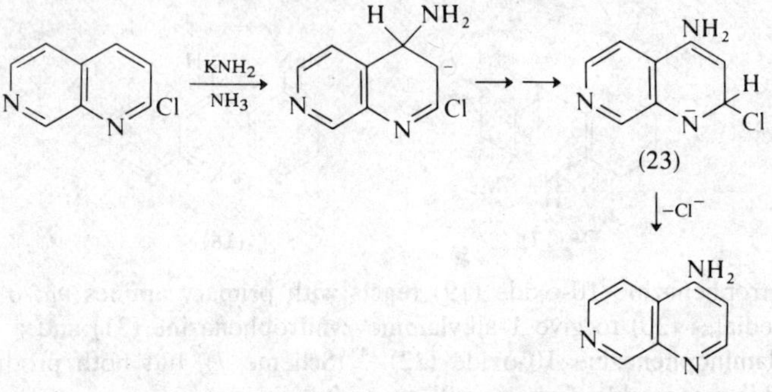

**Scheme 8**

An unusual overall substitution of hydrogen at C-2 in a pyrido[2,3-*b*]pyrazine derivative by a 3-(2-methylindole) group has been reported.[117] The formation of the 2-amino-derivative by the reaction of 2-(methylthio)-4,6,7-tri-phenylpteridine with $KNH_2$ in liquid $NH_3$ occurs 50—85% by an $S_N$(ANRORC) mechanism;[118] ring contraction with expulsion of the phenyl group at C-7 accompanies the substitution reaction with the formation of 6,8-diphenyl-2-(methylthio)purine.

Overall intramolecular nucleophilic substitutions at the $\beta$-position of an indole ring by oxygen nucleophiles have been described.[119] In these reactions $PhSO_2^-$ is displaced from nitrogen by an $S_N2'$ process followed by tautomerism; an $\alpha$-acyl substituent is a prerequisite for the process.

The efficient production of the 5- and 8-alkylamino-derivatives (formed in the ratio of *ca.* 1.8:1) by the photochemical substitution of one methoxide in compound (24) by primary amines can be explained in terms of the predicted $\pi$-electron densities at the 5- and 8-positions in the first excited state.[120]

The action of Grignard reagents on nitrobenzothiazoles followed by work-up with $BF_3$ or HCl effects alkylation of the six-membered ring and reduction of the nitro-group to the nitroso-group [with the exception of the 4-nitro-compound, which gives the 7-alkyl-4-nitro-derivative (25)].[121] The 5-nitro-isomer gives a

[116] H. C. van der Plas, M. Wozniak, and A. van Veldhuizen, *Rec. Trav. chim.*, 1977, **96**, 151.
[117] O. N. Chupakhin, V. N. Charushin, I. Ya. Postovskii, N. A. Klyuev, and E. N. Istratov, *Zhur. org. Khim.*, 1978, **14**, 431 (*Chem. Abs.*, 1978, **88**, 152 561).
[118] J. Nagel and H. C. van der Plas, *Heterocycles*, 1977, **7**, 205.
[119] M. M. Cooper, G. J. Hignett, R. F. Newton, J. A. Joule, M. Harris, and J. D. Hinchley, *J.C.S. Chem. Comm.*, 1977, 432.
[120] G. Green-Buckley and J. Griffiths, *J.C.S. Chem. Comm.*, 1977, 396.
[121] G. Bartoli, R. Leardini, M. Lelli, and G. Rossini, *J.C.S. Perkin I*, 1977, 884.

(24)          (25)

4-alkylated 5-nitroso-compound; with the 6- and 7-nitro-compounds, the former gives the 7- and the latter both the 4- and 6-alkylated compounds. A kinetic study of the displacement of the nitro-group by amines in 2-nitrobenzothiazole in the presence of DABCO (1,4-diazabicyclo[2.2.2]octane) has indicated the occurrence of both base-catalysed and uncatalysed reactions.[122] The effect of solvents in the thioalkoxydehalogenation of halogeno-benzofurazans has been studied.[123]

The bromine in (26) is replaced by alkoxides, thiolates, and carbanions derived from active-methylene compounds $R^1R^2CH_2$;[124] in the latter case, the product is the tautomeric 2-alkylidene compound.

(26)          (27)

The chlorine at C-8 in (27; R = H) and in (27; R = Cl) has been replaced with a variety of N-, O-, and S-containing nucleophiles;[125] the use of *p*-thiocresol gives substitution of both chlorine substituents. A closely related compound, 2-chloro-3-ethoxycarbonyl-cyclohepta[*b*]pyrrole, with enolate anions gives a variety of substitution and addition products, dependent upon the reaction conditions.[126]

## 4 Meisenheimer and Related Compounds

**Meisenheimer Complexes from Aromatic Systems.**—Stopped-flow rapid-scan spectrophotometry has been used to study the colour changes produced during the reaction of 1,3,5-trinitrobenzene with acetophenone under Janovsky conditions in different media.[127] In MeOH, the changes result from conversion of (28; $R^1 = OH$) into (28; $R^1 = CH_2COPh$). In MeOH–$H_2O$ mixtures, as the concentration of $H_2O$ increases, a succession of complexes is formed: (28; $R^1 = OH$); (29; $R^1 = R^2 = OH$); (29; $R^1 = CH_2COPh$, $R^2 = OH$), and (29; $R^1 = R^2 = CH_2COPh$). A Janovsky σ-complex has been allowed to react with $ArN_2^+$

[122] S. Di Pietro, L. Forlani, and P. E. Todesco, *Gazzetta*, 1977, **107**, 135 (*Chem. Abs.*, 1978, **88**, 61 728).
[123] L. Di Nunno and S. Florio, *Tetrahedron*, 1977, **33**, 1523.
[124] R. Neidlein and K. F. Cepera, *Chem. Ber.*, 1977, **110**, 2388.
[125] K. Yamane, K. Fujimore, J.-K. Sin, and T. Nozoe, *Bull. Chem. Soc. Japan*, 1977, **50**, 1184.
[126] N. Abe and T. Nishiwaki, *Bull. Chem. Soc. Japan*, 1978, **51**, 667.
[127] K. Kohashi, T. Kabeya, and Y. Ohkura, *Chem. and Pharm. Bull* (*Japan*), 1977, **25**, 50 (*Chem. Abs.*, 1977, **87**, 67 513).

salts.[128] The reaction of picryl chloride[129] and bromide[130] with sodium diethyl-malonate has been interpreted on the basis of the formation of a C-1 rather than a C-3 complex. Stopped-flow kinetics have been carried out on a mixture of methoxide and dimethylmalonate in MeOH or MeOH–DMSO (4:1) with 1,3,5-trinitroanisole, which ultimately forms the MeO-displacement product [by $CH(CO_2Me)_2$] or the conjugate base of the latter.[131]

The exclusive formation of the C-1 adduct between 2,4,6-trinitroanisole and n-butylamine has actually been detected, as a transient intermediate *en route* to *N*-(n-butyl)picramide, by a low-temperature flow $^1H$ n.m.r. experiment.[132] Further kinetic data for the reversible reaction between the 1,3,5-tri-nitrobenzene–methoxide $\sigma$-complex and aniline in DMSO–MeOH solutions, yielding the 1,3,5-trinitrobenzene-anilide $\sigma$-complex, mentioned in last year's Report, have been reported[133] and are best fitted by a dissociative mechanism involving rate-determining interconversion of free TNB and the protonated anilide complex. The failure of *aromatic* primary amines by themselves to form stable $\sigma$-complexes with TNB is due to an unfavourable overall equilibrium represented by equation (3), while the additional presence of the strong base

$$TNB + 2RNH_2 \rightleftharpoons TNBNH_2R^\pm + RNH_2 \rightleftharpoons TNBNHR^- + RNH_3^+ \qquad (3)$$

DABCO causes the initial pre-equilibrium formation of the zwitterionic complex (30) to be followed by its rate-determining deprotonation.[134] Furthermore, a primary kinetic isotope effect has been observed in this latter reaction with $PhND_2$, the relevance of which, in relation to nucleophilic aromatic substitution, has been discussed.[135] 3-Dimethylamino-2,2-dimethyl-2*H*-azirine and 2,4-dinitrofluorobenzene react initially through the ring nitrogen to give (31) as the first of a series of reaction intermediates which with $H_2X$ finally give $ArNHCMe_2C(X)NMe_2$ [X = O, S, or $NCH_2Ph$].[136] 3,5-Di-adducts, dianionic materials, are formed from *N*-phenyl-2,4,6-trinitroaniline and $NH_2OH$–NaOMe.[137] It has been concluded from an ion cyclotron resonance study that

[128] A. Ya. Kaminskii, S. S. Gitis, I. L. Bagal, Yu. D. Grudtsyn, T. P. Ikher, L. V. Illarionova, and M. D. Stepanov, *Zhur. org. Khim.*, 1977, **13**, 803 (*Chem. Abs.*, 1977, **87**, 39 040).
[129] K. T. Leffek and A. E. Matinopoulos-Scordou, *Canad. J. Chem.*, 1977, **55**, 2656.
[130] K. T. Leffek and A. E. Matinopoulos-Scordou, *Canad. J. Chem.*, 1977, **55**, 2664.
[131] J. Kavalek, V. Machacek, M. Pastrnek, and V. Sterba, *Coll. Czech. Chem. Comm.*, 1977, **42**, 2928 (*Chem. Abs.*, 1978, **88**, 61 739).
[132] C. A. Fyfe, A. Koll, S. W. H. Damji, C. D. Malkiewich, and P. A. Forte, *J.C.S. Chem. Comm.*, 1977, 335.
[133] E. Buncel, J. G. K. Webb, and J. F. Wiltshire, *J. Amer. Chem. Soc.*, 1977, **99**, 4429.
[134] E. Buncel and W. Eggimann, *J. Amer. Chem. Soc.*, 1977, **99**, 5958.
[135] E. Buncel, W. Eggimann, and H. W. Leung, *J.C.S. Chem. Comm.*, 1977, 55.
[136] P. Parthasarathi, H. Heimgartner, and H. Schmid, *Helv. Chim. Acta*, 1977, **60**, 2270.
[137] Yu. D. Grudtsyn and S. S. Gitis, *Sint., Anal. Strukt. Org. Soedin.*, 1976, **7**, 3 (*Chem. Abs.*, 1978, **88**, 135 979).

(30)                    (31)

Meisenheimer adducts are formed in the gas phase by the reaction of nitrite ions (and chloride ions) with *o*-dinitrobenzene.[138] Trialkyl phosphites form the C-2 $\sigma$-complexes with TNB,[139] and give products of halogen substitution [by —P(O)(OR)$_2$] *via* the C-1 adduct with picryl chloride.[140] The relative stabilities of complexes formed between TNB and a variety of phosphorus(III) derivatives have been determined.[141]

Equilibrium and kinetic data have been obtained for the addition of sodium ethoxide, in ethanol, to a series of 2,4-dinitro-6-X-phenetoles (X = NO$_2$, CO$_2$Et, Cl, or H) to give 1,1-$\sigma$-complexes, and for 2,4,6-trinitrophenetole the 1,3-isomer also.[142] The 1,1-complexes associate strongly with the sodium ions through the two oxygen atoms of the alkoxy-groups and the oxygen atoms of the *ortho*-substituents. 2-Fluoro-4,6-dinitroanisole and MeO⁻ in MeOH–DMSO give the C-1 adduct when the solvent contains <70% DMSO, but at higher concentrations of DMSO the C-3 adduct is formed, which reverts to the C-1 adduct.[143] The 1 : 2 complex between TNB and HO⁻ did not revert to the 1 : 1 complex when neutralized or acidified.[144] An *X*-ray analysis on the Meisenheimer adduct formed from methyl 2,4,6-trinitrobenzoate and MeOK shows that the ring is non-planar and that the latter group has the *s-cis* configuration.[145] Raman[146] and ¹³C n.m.r.[147] spectroscopic studies on carbon–oxygen $\sigma$-adducts have been carried out.

Independent studies have confirmed that the stabilities of spiro-Meisenheimer complexes decrease with increasing size of the spiro-ring.[148] The rates of Smiles' rearrangement of 1-(2-*N*-acetylaminoethoxy)-2-X-4-nitrobenzene with base depend only on the decomposition process of the anionic spiro $\sigma$-complexes.[149]

[138] J. H. Bowie and J. B. Stapleton, *Austral. J. Chem.*, 1977, **30**, 795.
[139] P. P. Onys'ko and Yu. G. Gololobov, *Zhur. obshchei Khim.*, 1977, **47**, 2480 (*Chem. Abs.*, 1978, **88**, 61 752).
[140] Yu. G. Gololobov, P. P. Onys'ko, and V. P. Prokopenko, *Doklady Akad. Nauk S.S.S.R.*, 1977, **237**, 105 (*Chem. Abs.*, 1978, **88**, 50 974); *Zhur. obshchei Khim.*, 1977, **47**, 2632 (*Chem. Abs.*, 1978, **88**, 61 753).
[141] P. P. Onys'ko, L. F. Kasukhin, and Yu. G. Gololobov, *Zhur. obshchei Khim.*, 1978, **48**, 342 (*Chem. Abs.*, 1978, **88**, 189 536).
[142] M. R. Crampton, *J.C.S. Perkin II*, 1977, 1442.
[143] F. Millot and F. Terrier, *Compt. rend.*, 1977, **284**, C. 979 (*Chem. Abs.*, 1977, **87**, 151 305).
[144] K. Kohashi, Y. Tsuruta, M. Yamaguchi, and Y. Ohkura, *Chem. and Pharm. Bull.* (*Japan*), 1977, **25**, 1103 (*Chem. Abs.*, 1977, **87**, 67 559).
[145] E. G. Kaminskaya, S. S. Gitis, A. I. Ivanova, N. V. Margolis, A. Ya. Kaminskii, and N. V. Grigoreva, *Zhur. strukt. Khim.*, 1977, **18**, 386 (*Chem. Abs.*, 1977, **87**, 67 884).
[146] E. G. Kaminskaya, S. S. Gitis, and A. Ya. Kaminskii, *Zhur. priklad. Spektroskopii*, 1977, **26**, 1053. (*Chem. Abs.*, 1977, **87**, 133 322).
[147] M. P. Simonnin, M.-J. Pouet, and F. Terrier, *J. Org. Chem.*, 1978, **43**, 855.
[148] C. F. Bernasconi and J. R. Gandler, *J. Org. Chem.*, 1977, **42**, 3387.
[149] K. Okada and S. Sekiguchi, *J. Org. Chem.*, 1978, **43**, 441.

Stable adducts containing the oxazolidine ring are formed with 2,4,6-tri-nitroaniline derivatives that have Ph and $CH_2CH_2OH$ on nitrogen;[150] replacing Ph by $PhCH_2$ gives a cyclized product, resulting from displacement of the nitro-group by alkoxide when the reaction is conducted under the conditions necessary for a Smiles' rearrangement. Treatment of 1-(2-*N*-substituted amino-ethylthio)-2,4,6-trinitrobenzene compounds with base gives thiazolidine deriva-tives for the less nucleophilic nitrogen substrates, but more generally cyclized products result by displacement of a nitro-group by the nitrogen and the sulphur (*via* a Smiles' rearrangement).[151]

**Meisenheimer Complexes from Heteroaromatic and Fused Ring Systems.**—2-Methoxy-3,5-dinitropyridine and $MeO^-$ gives, as the stable adduct (32), the '1,3-complex' relative to the MeO group, and not the '1,1-complex' (33) that is expected by analogy with 2,4,6-trinitroanisole.[152]

A *meta*-bridging reaction across the 2,6-positions in 3,5-dinitropyridine takes place with dibenzyl ketone and $Et_3N$ to give (34), but, with the amidine $PhCH_2C(=NH)NMe_2$, the compound (35) is formed by 2,4-bridging.[153] The spirocyclic compound (36) is formed when diethyl pyridazine-4,5-dicarboxylate is treated with diphenylguanidine and NaH.[154]

The first examples of formation of a $\sigma$-complex between a polyfluoro-aromatic compound and $F^-$ have been reported:[155] trifluoro-*s*-triazine and CsF in tetra-methylene sulphone formed a $\sigma$-complex which could be isolated and charac-terized by $^{19}F$ n.m.r. spectroscopy. Analogous reactions with perfluoro-mono-and -di-isopropyl-*s*-triazines gave analogous complexes containing —$CF_2$—, but they were not isolated.

[150] V. N. Drozd, V. N. Knyazev, and V. M. Minov, *Zhur. org. Khim.*, 1977, **13**, 396 (*Chem. Abs.*, 1977, **87**, 22 021).
[151] V. N. Knyazev, V. N. Drozd, V. M. Minov, and N. P. Akimova, *Zhur. org. Khim.*, 1977, **13**, 1255 (*Chem. Abs.*, 1977, **87**, 152 098).
[152] A. P. Chatrousse, F. Terrier, and R. Schaal, *J. Chem. Res. (S)*, 1977, 228.
[153] R. Bard, M. J. Strauss, and S. A. Topolsky, *J. Org. Chem.*, 1977, **42**, 2589.
[154] G. Adembri, S. Chimichi, R. Nesi, and M. Scotton, *J.C.S. Perkin I*, 1977, 1020.
[155] R. D. Chambers, P. D. Philpot, and P. L. Russell, *J.C.S. Perkin I*, 1977, 1605.

The overall nucleophilic displacement of hydrogen at C-9 in 10-methyl-acridinium salts by 4-$R^1R^2NC_6H_4$, using the substituted aniline $C_6H_5NR^1R^2$, has been shown by $^1H$ n.m.r. spectroscopy to involve an intermediate having a saturated carbon at C-9.[156] 2-Quinoxalone and *NN*-dimethylaniline give overall addition of 4-$Me_2NC_6H_4$ and H across the C-3–N-4 bond, respectively.[157] Some $^1H$ and $^{13}C$ n.m.r. studies on the adducts of 1,$n$-naphthyridines ($n$ = 5, 6, or 8) with $KNH_2$ in liquid $NH_3$ showed the formation of the C-2 adduct in each case (a 2-amino-1,2–dihydro-1,$n$-naphthyridinide ion); 1,7-naphthyridine gave C-6 and C-8 adducts as well as the C-2 adduct.[158]

Transformation of the kinetically controlled C-5 adduct formed from 4-nitrobenzofurazan 1-oxide and $MeO^-$ into the C-7 adduct is subsequently followed by conversion into 7-hydroxy-4-nitrobenzofurazan (as the anion) and 7-methoxy-4-nitrobenzofurazan (as the methoxide C-7 adduct).[159] Meisenheimer complexes formed from both 4-nitro-2,1,3-benzothiadiazole and 4-nitro-7-halogeno-2,1,3-benzothiadiazole with methoxide ion are much less stable than those formed from related benzofurazans.[160]

Replacement of the ethoxy-group in 1-ethoxy-2,4-dinitronaphthalene by $Bu^nNH$ has been shown by flow $^1H$ n.m.r. spectroscopy to proceed through the C-1 Meisenheimer adduct.[132,161] However, 1-piperidino-2,4-dinitronaphthalene and MeOK in DMSO form a transient 1,3-anionic adduct and, by a much slower process, a 1,1-adduct, which was detected by a stopped-flow photometric method.[162]

A spirocyclic intermediate was detected in the base-catalysed Smiles' rearrangement of 2-(2-*N*-acetylaminoethoxy)-5-nitropyridine ether.[149]

### 5 Substitution in Polyhalogeno-compounds

**Aromatic Polyhalogeno-compounds.**—A recent molecular orbital study of orientational selectivity in polyhalogenated benzenes has been criticized and an extension of an older 'I$\pi$-repulsion' method has been presented to rationalise substitution reactions in perfluoro-polycycloaromatic compounds.[163]

Displacement of fluorine by isopropoxide in the 2-keto-compound (37) occurs at a significantly increased rate compared with (38) [a factor of 100 is involved at 25 °C], but, since F-6 and F-7 are replaced at similar rates, homoconjugative interaction in the transition state of the type shown in (39) is excluded; the field effect of the carbonyl group on the energy of the ground state in (37) must be responsible for the observed reactivity.[164]

[156] V. N. Charushin, O. N. Chupakhin, E. O. Sidorov, J. Beilis, and I. A. Terent'eva, *Zhur. org. Khim.*, 1978, **41**, 140 (*Chem. Abs.*, 1978, **88**, 169 271).
[157] O. N. Chupakhin, E. O. Sidorov, A. L. Kozerchuk, and Yu. I. Beilis, *Khim. geterotsikl. Soedinenii*, 1977, 684 (*Chem. Abs.*, 1977, **87**, 84 946); E. O. Sidorov and O. N. Chupakhin, *Zhur. org. Khim.*, 1978, **14**, 134 (*Chem. Abs.*, 1978, **88**, 189 520).
[158] H. C. van der Plas, A. Van Veldhuizen, M. Wozniak, and P. Smit, *J. Org. Chem.*, 1978, **43**, 1673.
[159] E. Buncel, N. Chuaqui-Offermans, B. K. Hunter, and A. R. Norris, *Canad. J. Chem.*, 1977, **55**, 2852.
[160] L. Di Nunno and S. Florio, *Tetrahedron*, 1977, **33**, 855.
[161] C. A. Fyfe, A. Koll, S. W. H. Damji, C. D. Malkiewich, and D. A. Holden, *Canad. J. Chem.*, 1977, **55**, 1468.
[162] S. Sekiguchi, T. Takei, T. Aizawa, and K. Okada, *Tetrahedron Letters*, 1977, 1209.
[163] J. Burdon and I. W. Parsons, *J. Amer. Chem. Soc.*, 1977, **99**, 7445.
[164] G. M. Brooke, R. S. Matthews, and A. C. Young, *J.C.S. Perkin I*, 1977, 1411.

(37)                    (38)                    (39)

Hexafluorobenzene reacts with 1-dialkylamino-cyclohexenes to give the 2-perfluorophenyl enamine, which on prolonged heating gives $N$-alkyl tetrafluoro-tetrahydrocarbazoles by intramolecular displacement of fluorine and subsequent dealkylation of the quaternary ammonium fluoride.[165] Pentafluorobenzoic acid undergoes replacement of the *ortho*-fluorine by ArMgBr.[166] The alcohols $C_6F_5CH(OH)R$ (R = H, Me, Ph, or $C_6F_5$) undergo replacement of the *ortho*-fluorine by H with $LiAlH_4$.[167]

In spite of being highly hindered, the fluorine at C-2 in 2,4,6-trifluoro-1,3-dinitrobenzene is displaced by ammonia, indicating that there is a high degree of hydrogen-bonding between the incoming nucleophile and the adjacent nitro-groups.[168] Treatment of $N$-lithio-$N$-substituted anilides with $C_6F_6$ gives the corresponding perfluorophenyl compounds; surprisingly, of these compounds, $C_6F_5N(Ph)SiMe_3$ was particularly unreactive towards further displacement of the 4-fluorine by $C_6H_5NHLi$.[169] The phosphinimines $C_6F_5N=PR_3$ are prepared from $C_6F_6$ and $Me_3SiN=PR_3$.[170] Triethyl phosphite undergoes Arbusov reactions with $C_6F_5CN$ and $C_6F_5C_6F_5$ to give mono- and di-substituted compounds containing the $P(O)(OEt)_2$ group in *para*-positions.[171] The orientation reactivity pattern for displacement of fluorine in $C_6F_5P(O)Ph_2$ with $MeNH_2$ changes from 93:7 *ortho* : *para*-substitution in benzene at 50 °C for 24 h to <2 : >98 *ortho*- : *para*-substitution for the solid–gas reaction at 55 °C over 2 days.[172]

A low yield of a perfluorobenzodioxin is formed from $C_6F_5O^-$ $Cs^+$ and hexafluoropropene oxide.[173] The reactions of fluorinated anilines, phenyl-hydrazines,[174] and carboxylic acids[175] with thiolates have been reported. All the fluorine in $C_6F_6$ is replaced by chlorine under mild conditions, using the reagent BMC ($S_2Cl_2$–$AlCl_3$–$SO_2Cl_2$).[176]

[165] C. Wakselman and J.-C. Blazejewski, *J.C.S. Chem. Comm.*, 1977, 341.
[166] T. V. Fomenko, T. N. Gerasimova, and E. P. Fokin, *Izvest. Akad. Nauk. S.S.S.R., Ser. khim.*, 1977, 1880 (*Chem. Abs.*, 1977, **87**, 167 666).
[167] T. N. Gerasimova, N. V. Semikolenova, and E. P. Fokin, *Zhur. org. Khim.*, 1978, **14**, 100 (*Chem. Abs.*, 1978, **88**, 152 156).
[168] M. E. Sitzmann, *J. Org. Chem.*, 1978, **43**, 1241.
[169] R. Koppang, *J. Fluorine Chem.*, 1977, **9**, 449.
[170] D. Dahmann and H. Rose, *Chem.-Ztg.*, 1977, **101**, 401 (*Chem. Abs.*, 1978, **88**, 22 257).
[171] L. N. Markovskii, G. G. Furin, Yu. G. Shermolovich, and G. G. Yakobson, *Izvest. Akad. Nauk. S.S.S.R., Ser. khim.*, 1977, 2839 (*Chem. Abs.*, 1978, **88**, 105 470).
[172] T.-W. Lin and D. G. Naae, *Tetrahedron Letters*, 1978, 1653.
[173] S. A. Lopyreva, T. I. Ryabtseva, V. P. Sass, A. V. Tumanov, and S. V. Sokolov, *Zhur. org. Khim.*, 1977, **13**, 1122 (*Chem. Abs.*, 1977, **87**, 68 257).
[174] W. J. Frazee, M. E. Peach, and J. R. Sweet, *J. Fluorine Chem.*, 1977, **9**, 377.
[175] I. Cervena, K. Sindelar, Z. Kopicova, J. Holubek, E. Svatek, J. Metysova, M. H. Hrubantova, and M. Protiva, *Coll. Czech. Chem. Comm.*, 1977, **42**, 2001 (*Chem. Abs.*, 1977, **87**, 201 469).
[176] C. Glidewell and J. C. Walton, *J.C.S. Chem. Comm.*, 1977, 915.

**Heteroaromatic Polyhalogeno-compounds.**—Rate constants for the reactions of various polyfluoro-[177] and chloropolyfluoro-pyridines[178] with ammonia in dioxan–water have enabled the separate effects of fluorine and chlorine atoms that are *ortho*, *meta*, and *para* to the reaction centre to be determined. In each case the order is *ortho* > *meta* > *para*; for fluorine, 31:23:0.26; and for chlorine, 86:24:6.9, both sequences being relative to a hydrogen atom at the same position. Both pentafluoro-[171] and pentachloro-pyridine[179] undergo Arbusov reactions at C-4 with $(EtO)_3P$, the chloro-compound also giving the 2,4-bis(diethoxyphosphinyl) derivative. 3-Chloro-2,4,5,6-tetrafluoro- and 3,5-dichloro-2,4,6-trifluoro-pyridines are readily converted into the corresponding 4-iodo-compounds with NaI and DMF.[180]

The chlorine in 6-chloro-1,2,3,4-tetrafluorophenanthridine is replaced by piperidine and hydrazine.[181] All the halogens are replaced by nucleophiles in 2,2',3,3'-tetrachloro-6,6'-biquinoxaline.[182]

## 6 Other Reactions

The formation of carbon–hydrogen bonds from halogen-containing compounds has been carried out by a number of procedures: reductive dehalogenation of halogeno-phenol with Raney alloys in alkaline $D_2O$ provides a route to deuteriated phenols;[183] treatment of dichlorobenzene with Na or K in liquid ammonia ('solvated electrons') gives benzene, and, in addition, aniline *via* benzyne.[184]

Nitrobenzene is monomethylated in *ortho*- and *para*-positions (32% and 20% respectively) and dimethylated (7% 2,4-$Me_2$) by MeLi followed by $Br_2$ and $Et_3N$ or by DDQ;[185] 1-nitronaphthalene is n-butylated by an analogous process in the 2- and 4-positions. But-1-en-3-ol undergoes thienylation at C-1 and C-2 with 5-substituted 2-bromo-thiophens in the presence of $Pd(OAc)_2$, $Ph_3P$, and $Na_2CO_3$, the final products being ketones.[186]

One bromine atom undergoes metal–halogen exchange in the reactions of 2,7-dibromo-naphthalene and -anthracene with $Bu^nLi$ in ether at 20 °C; both halogens are exchanged in THF at −35 °C.[187]

[177] R. D. Chambers, J. S. Waterhouse, and D. L. H. Williams, *J.C.S. Perkin II*, 1977, 585.
[178] R. D. Chambers, D. Close, W. K. R. Musgrave, J. S. Waterhouse, and D. L. H. Williams, *J.C.S. Perkin II*, 1977, 1774.
[179] S. D. Moshchitskii, L. S. Sologub, A. F. Pavlenko, and V. P. Kukhar, *Zhur. obshchei Khim.*, 1977, **47**, 1263 (*Chem. Abs.*, 1977, **87**, 85 100).
[180] R. E. Banks, R. N. Haszeldine, and E. Phillips, *J. Fluorine Chem.*, 1977, **9**, 243.
[181] T. V. Fomenko, T. N. Gerasimova, and E. P. Fokin. *Izvest. Sibirsk. Otdel. Akad. Nauk S.S.S.R.*, *Ser. khim. Nauk*, 1977, 99 (*Chem. Abs.*, 1977, **87**, 5778).
[182] V. V. Titov and L. F. Kozhokina, *Khim. geterotsikl. Soedinenii*, 1977, 414 (*Chem. Abs.*, 1977, **87**, 39 419).
[183] M. Tashiro, A. Iwasaki, and G. Fukata, *J. Org. Chem.*, 1978, **43**, 196.
[184] R. A. Rossi, A. B. Pierini, and R. H. de Rossi, *J. Org. Chem.*, 1978, **43**, 1276.
[185] F. Kienzle, *Helv. Chim. Acta*, 1978, **61**, 449.
[186] Y. Tamaru, Y. Yamada, and Z. Yoshida, *Tetrahedron Letters*, 1978, 919.
[187] G. Porzi and C. Concilio, *J. Organometallic Chem.*, 1977, **128**, 95.

Oxidative additions of triethylphosphine–nickel(O) complexes,[188] or of a specially active slurry of nickel in the presence of $Et_3P$,[189] to aryl halides give $ArNi(PEt_3)_2Hal$; $C_6F_5Cl$ and $C_5F_5Br$ react very rapidly. Highly reactive palladium and platinum slurries have also been prepared which undergo analogous reactions.

[188] D. R. Fahey and J. E. Mahan, *J. Amer. Chem. Soc.*, 1977, **99**, 2501.
[189] R. D. Rieke, W. J. Wolf, N. Kujundzic, and A. V. Kavaliunas, *J. Amer. Chem. Soc.*, 1977, **99**, 4159.

# 9

# Aromatic Substitution by Free Radicals, Carbenes, and Nitrenes

BY R. S. ATKINSON

## 1 By Free Radicals

Rate constants for homolytic alkylation of several protonated heterocyclic bases with alkyl radicals have been measured and the values obtained ($10^5$— $10^8 \, \text{l mol}^{-1} \, \text{s}^{-1}$) contrasted with those from alkylation of benzene derivatives ($10^2 \, \text{l mol}^{-1} \, \text{s}^{-1}$).[1] This great reactivity of protonated heteroaromatic bases towards alkyl radicals has been put to preparative use in the reactions of 1-acetyl-2-pyrrolidinyl and 1-formyl-2-pyrrolidinyl radicals with pyridazine, giving nicotine analogues (Scheme 1).[2]

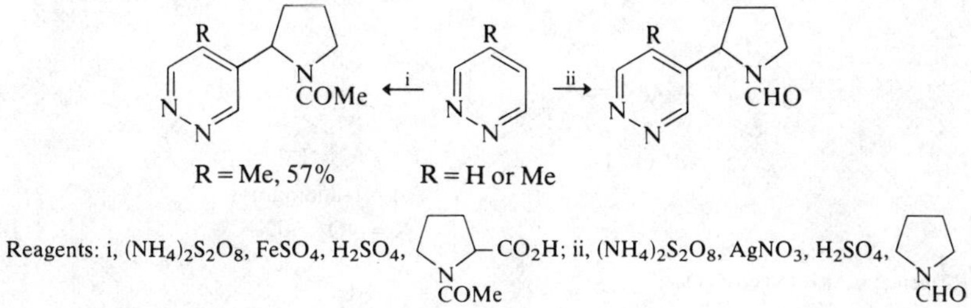

R = Me, 57%     R = H or Me

Reagents: i, $(NH_4)_2S_2O_8$, $FeSO_4$, $H_2SO_4$, [structure]—$CO_2H$; ii, $(NH_4)_2S_2O_8$, $AgNO_3$, $H_2SO_4$, [structure]

**Scheme 1**

Oxidation of pent-4-en-1-ol by sodium persulphate and silver ion in the presence of protonated lepidine leads to (1) and (2) in a 9 : 1 ratio (Scheme 2). 2-Allyloxyethanol, however, gives (3) by preliminary $\beta$-fission to an alkoxyalkyl radical before attack on the protonated lepidine.[3]

Details of the reaction of 1-adamantyl and other radicals with 2-acyl-benzo-thiazoles, effecting displacement of the acyl group, have appeared. The reaction is facilitated by electron-withdrawing groups in the 5- and 6-positions of the benzothiazole nucleus.[4] A related homolytic aromatic *ipso*-substitution occurs in the reactions of nitrobenzenes bearing electron-withdrawing *para*-substituents

[1] A. Citterio, F. Minisci, O. Porta, and G. Sesana, *J. Amer. Chem. Soc.*, 1977, **99**, 7690.
[2] G. Heinisch, A. Jentzsch, and I. Kirchner, *Tetrahedron Letters*, 1978, 619.
[3] A Clerici, F. Minisci, K. Ogawa, and J.-M. Surzur, *Tetrahedron Letters*, 1978, 1149.
[4] M. Fiorentino, L. Testaferri, M. Tiecco, and L. Troisi, *J.C.S. Perkin II*, 1977, 1679.

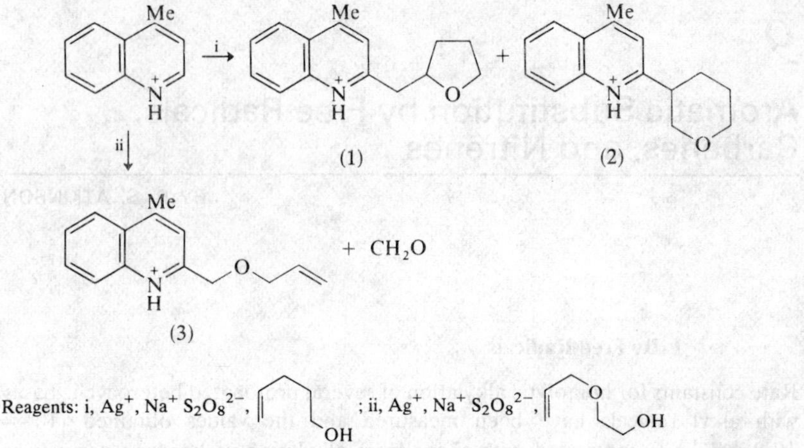

**Scheme 2**

with adamantyl radicals giving *para*-substituted 1-adamantylbenzenes in surprisingly good yields (Scheme 3). $MeCHEt$ and $Me_2CHCH_2$ radicals can also be used, but the yields are lower.[5]

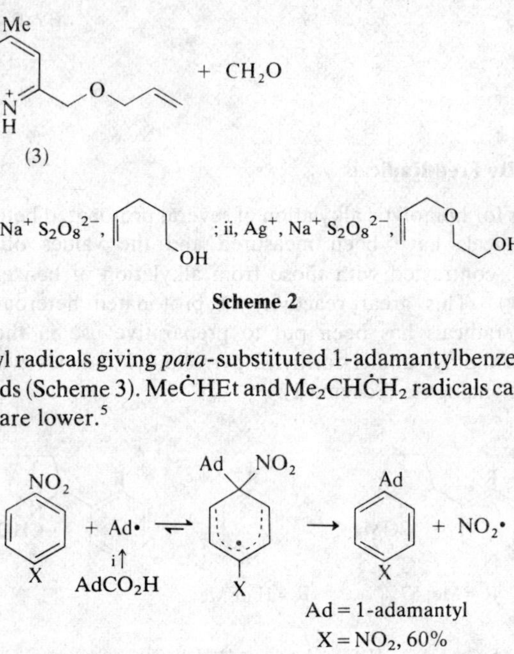

Reagents: i, $Ag^+$, $(NH_4)_2S_2O_8$

**Scheme 3**

Yields in the arylation of aromatic substrates with diaroyl peroxides are known to be increased by the presence of small quantities of aromatic nitro-compounds. It has now been found that nitro-compounds bearing additional electron-withdrawing groups are even more effective.[6] An oxidation of the intermediate arylcyclohexadienyl radicals by an electron-transfer mechanism has been proposed (Scheme 4).

$$Ar'NO_2 + [Ar—PhH]^{\cdot} \rightarrow Ar'NO_2^{\bar{\cdot}} + [Ar—PhH]^+$$

$$[Ar—PhH]^+ \rightarrow Ar—Ph + H^+$$

$$Ar'NO_2^{\bar{\cdot}} + (PhCO_2)_2 \rightarrow Ar'NO_2 + PhCO_2{\cdot} + PhCO_2^-$$

**Scheme 4**

[5] L. Testaferri, M. Tiecco, M. Tingoli, M. Fiorentino, and L. Troisi, *J.C.S. Chem. Comm.*, 1978, 93.
[6] R. Henriquez and D. C. Nonhebel, *J. Chem. Res. (S)*, 1977, 253.

The term 'radiationless deactivation' refers to all processes through which electronically excited states undergo relaxation without the emission of photons, and the triplet state of benzophenone is known to undergo deactivation more rapidly in aromatic solvents than in, for example, carbon tetrachloride at room temperature. Evidence for addition of triplet benzophenone to an aromatic solvent has now been adduced in the case of diphenyl ether (Scheme 5). The authors conclude that triplet excited benzophenone adds to an aromatic nucleus to give a diradical which either undergoes chemical reaction (path *b*) or dissociates to ground state benzophenone (path *a*) depending upon the nature of the substituents.[7]

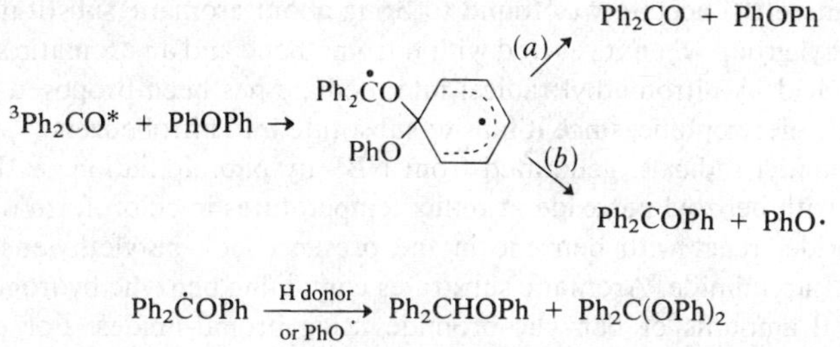

$$Ph_2CO + PhOPh$$

$$^3Ph_2CO^* + PhOPh \rightarrow$$

$$(a)$$

$$(b)$$

$$Ph_2\dot{C}OPh + PhO\cdot$$

$$Ph_2\dot{C}OPh \xrightarrow[\text{or PhO}\cdot]{\text{H donor}} Ph_2CHOPh + Ph_2C(OPh)_2$$

**Scheme 5**

Tributylstannane-induced reactions of 1-arylsulphonyl-2-iodomethyl-piperidines (4) yield products which include substituted piperidines (5) *via* radical *ipso*-substitution on the aromatic ring and thiazine *S*-dioxides (6) by

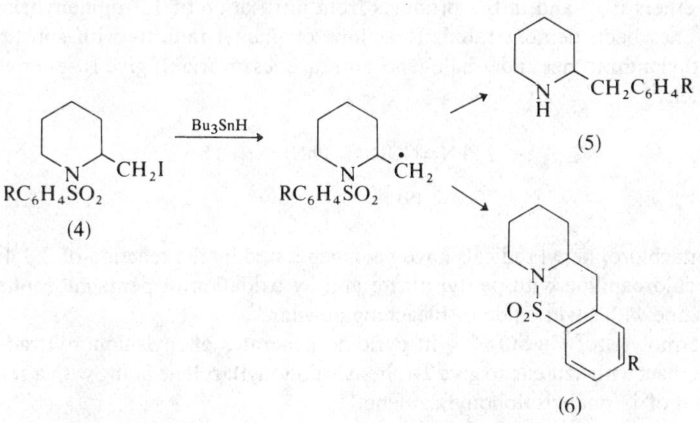

homolytic substitution in the *meta*-position. Whereas a Hammett plot for formation of the latter is linear, with $\rho = -0.31$, two intersecting straight lines are obtained for *ipso*-substitution, with $\rho = -0.98$ for R = OMe, Me, F, and H and $\rho = +0.92$ for R = H, Cl, Br, $CO_2Et$, COMe, and CN. To explain this result requires a change in the polar character of the methyl group – a switch in the importance of the resonance contributors (7) and (8) to the transition state.

[7] K. Nowada, M. Hisaoka, H. Sakuragi, K. Tokumaru, and M. Yoshida, *Tetrahedron Letters*, 1978, 137.

$$[RCH_2 \cdot + YPhX] \leftrightarrow [R\overset{+}{C}H_2 - Y\overset{-}{P}hX] \leftrightarrow [R\overset{-}{C}H_2 - Y\overset{+}{P}hX]$$

$$(7) \qquad\qquad\qquad (8)$$

Alternatively, it may indicate a change from singly occupied–highest occupied to singly occupied–lowest unoccupied orbital overlap, in frontier-orbital parlance.[8]

A wide variety of mixed biaryls has been prepared, in good to excellent yield, by a modified Gomberg–Bachmann–Hey reaction. Potassium acetate is solubilised in an arene solvent by 18-crown-6 and reacts with aryldiazonium tetrafluoroborate, with the subsequent decomposition leading to coupling products between the aryl radical and arene solvent.[9]

Manganese(III) acetate was found to bring about aromatic substitution by a nitromethyl group when it reacted with nitromethane and an aromatic substrate in acetic acid. A nitromethyl radical intermediate has been proposed which is apparently electrophilic, since it fails to substitute into nitrobenzene.[10]

Succinimidyl radicals, generated from NBS by photoinitiation or thermally initiated with benzoyl peroxide at reflux temperatures in chloroform or carbon tetrachloride, react with benzene in the presence of t-butylethylene to give $N$-phenylsuccinimide. Aromatic substrates containing benzylic hydrogens yield substantial amounts of benzylic bromide using bromo-imides. For example, toluene and $N$-bromo-3,3-dimethylglutarimide (NBDMG) at 65 °C yielded a product mixture consisting of 65% benzyl bromide and 35% imidation product (48% *ortho*, 31% *meta*, and 21% *para*). However, at − 20 °C the product mixture contained more than 80% imidation product.[11]

The intermediacy of phenyl radicals in the thermal decomposition of aryldiazo alkyl ethers $(9)$[12] and in the products from nitrosation of 1,3-diphenyltriazene $(10)$[13] has been demonstrated. Reactions of phenyl radicals with substituted 9-methylanthracenes and 9-halogeno-anthracenes invariably give 10-phenylated products.[14]

$$PhN{=}NOR \qquad PhN{=}NNHPh$$

$$(9) \qquad\qquad (10)$$

Pentachlorophenyl radicals have been generated by the reaction of 2,3,4,5,6-pentachloroaniline with pentyl nitrite and by oxidation of pentachlorophenyl-hydrazine with silver oxide or bleaching powder.[15]

Thermolysis of $Cu(SO_2Ph)_2$ in pyridine generates phenylsulphonyl radicals, which react with azulene to give 2-(phenylsulphonyl)azulene along with a smaller amount of 1-(phenylsulphonyl)azulene.[16]

[8]  W. N. Speckamp and J. J. Köhler, *J.C.S. Chem. Comm.*, 1978, 166.
[9]  S. H. Korzeniowski, L. Blum, and G. W. Gokel, *Tetrahedron Letters*, 1977, 1871.
[10]  M. E. Kurz and T. R. Chen, *J. Org. Chem.*, 1978, **43**, 239.
[11]  J. C. Day, M. G. Katsaros, W. D. Kocher, A. E. Scott, and P. S. Skell, *J. Amer. Chem. Soc.*, 1978, **100**, 1950.
[12]  R. M. Paton and R. U. Weber, *J.C.S. Chem. Comm.*, 1977, 769.
[13]  J. I. G. Cadogan, R. G. M. Landells, and J. T. Sharp, *J.C.S. Perkin I*, 1977, 1841.
[14]  F. M. Cromarty, R. Henriquez, and D. C. Nonhebel, *J. Chem. Res. (S)*, 1977, 309.
[15]  R. Bolton, E. P. Mitchell, and G. H. W. Williams, *J. Chem. Res. (S)*, 1977, 223.
[16]  V. A. Nefedov, L. V. Kryuchkova, and L. K. Tarygina, *Zhur. org. Khim.*, 1977, **13**, 1735.

Hydroxide radicals are conveniently produced in known concentration by radiolysis of water, and the effect of metal ions on the hydroxylation of benzonitrile, anisole, and fluorobenzene has been studied. Some metal salts, *e.g.* $[Fe(CN)_6]^{3-}$, $Fe^{3+}$, and $Cr_2O_7^{2-}$, oxidize the intermediate hydroxycyclohexadienyl radicals and give high conversion of OH radicals into phenols.[17]

Homolytic aromatic chlorination has been studied in simple monosubstituted benzenes, using recoil atoms produced by the $^{37}Cl(n, \gamma)^{38}Cl$ reaction.[18]

### 2 By Carbenes

Rearrangements and interconversions of aromatic carbenes and nitrenes have been reviewed.[19]

Carbene–carbene rearrangements are believed to proceed *via* bicyclo-[4.1.0]heptatrienes (Scheme 6).

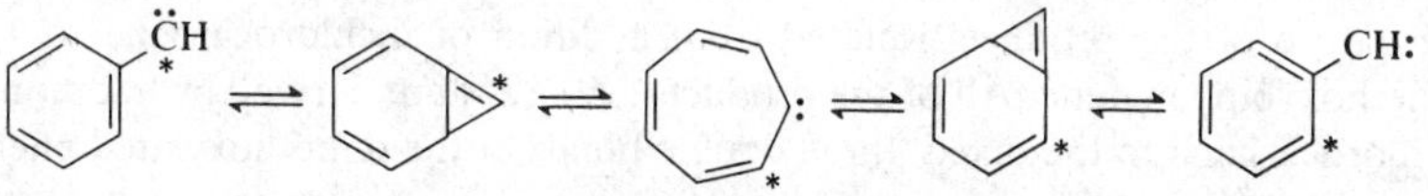

**Scheme 6**

Isolation of the t-butyl ethers (11) and (12) from treatment of the dichlorocarbene adduct (13) with potassium t-butoxide is considered to be the result of such a carbene–carbene rearrangement.[20]

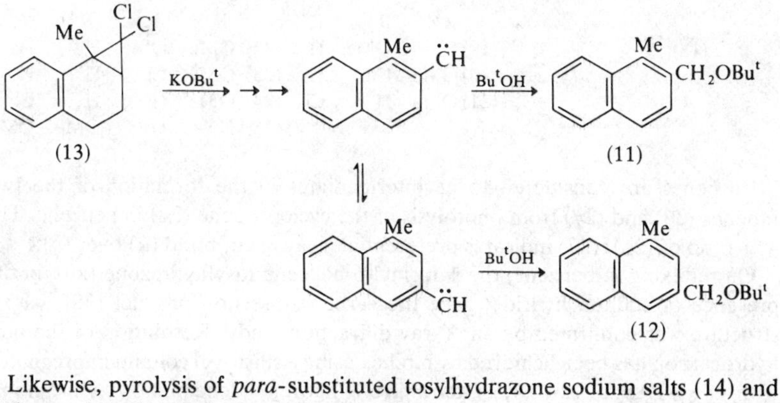

Likewise, pyrolysis of *para*-substituted tosylhydrazone sodium salts (14) and (15) gave 1,1-dimethylindan (16) and the sila-indan (17), in 23% and 15% yield respectively. Two carbene–carbene rearrangements are required before insertion of the carbene into a saturated C—H bond can occur.[21]

[17] M. Eberhardt, *J. Phys. Chem.*, 1977, **81**, 1051.
[18] H. H. Coenen, H.-J. Machalla, and G. Stöcklin, *J. Amer. Chem. Soc.*, 1977, **99**, 2892.
[19] C. Wentrup, *Topics Current Chem.*, 1976, **62**, 173.
[20] W. E. Billups and L. E. Reed, *Tetrahedron Letters*, 1977, 2239.
[21] A. Sekiguchi and W. Ando, *Bull. Chem. Soc. Japan*, 1977, **50**, 3067. See also E. B. Norsoph, B. Coleman, and M. Jones, *J. Amer. Chem. Soc.*, 1978, **100**, 994 and refs. therein.

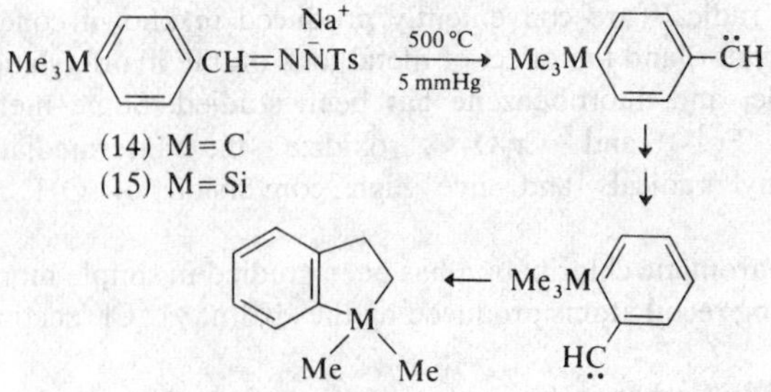

(14) M = C
(15) M = Si

(16) M = C
(17) M = Si

Biphenylene has long been considered to have predominant bond fixation as in the canonical structure (18), in which anti-aromaticity in the central four-membered ring is mitigated. This bond-fixation is thought to be responsible for the nature of the products isolated from addition of dichlorocarbene to 1,2-dimethoxybiphenylene. All of the products (19)–(25) are derived by addition of dichlorocarbene to the 'fixed' three double bonds of the dimethoxylated phenyl ring in (18).[22]

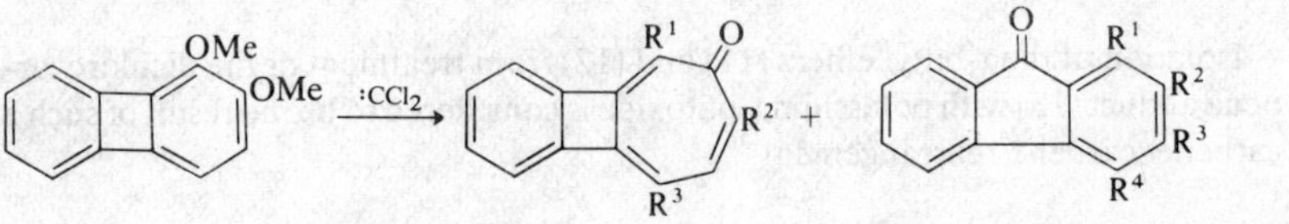

|        | $R^1$ | $R^2$ | $R^3$ |        | $R^1$ | $R^2$ | $R^3$ | $R^4$ |
|--------|-------|-------|-------|--------|-------|-------|-------|-------|
| (18)   |       |       |       |        |       |       |       |       |
|        | (19) Cl | OMe | H |        | (22) OMe | OMe | H | H |
|        | (20) OMe | Cl | H |        | (23) OMe | Cl | H | H |
|        | (21) OMe | H | Cl |        | (24) OMe | H | H | Cl |
|        |       |       |       |        | (25) H | H | OMe | OMe |

Carbenes are considered to be intermediates in the formation of the two indenes (26) and (27) from photolysis of the cyclopropene (28) in benzene. The 4:1 ratio of (26):(27) indicates preferential cleavage of bond (a) over (b).[23]

On refluxing in benzene, the 1-formyl[6]helicene tosylhydrazone (29), in the presence of sodium hydride, gave the carbene-insertion product (30), whose structure was confirmed by an X-ray diffraction study. Resolution of the new hydrocarbon has been achieved by h.p.l.c., using a silica-gel column impregnated with (−)-TAPA [2-(2,4,5,7-tetranitro-9-fluorenylideneamino-oxy)propionic acid].[24]

The latent carbenoid character of isocyanides is manifest on heating (31) in refluxing n-tetradecane; it gives the indole derivative (32) and the phenanthridine (33), in addition to some naphthalene (34).[25]

[22] M. Sato, A. Uchida, J. Tsunetsugu, and S. Ebine, *Tetrahedron Letters*, 1977, 2151.
[23] A. Padwa, R. Loza, and D. Getman, *Tetrahedron Letters*, 1977, 2847.
[24] J. Jespers, N. Defay, and R. H. Martin, *Tetrahedron*, 1977, **33**, 2141.
[25] J. H. Boyer and J. R. Patel, *J.C.S. Chem. Comm.*, 1977, 855.

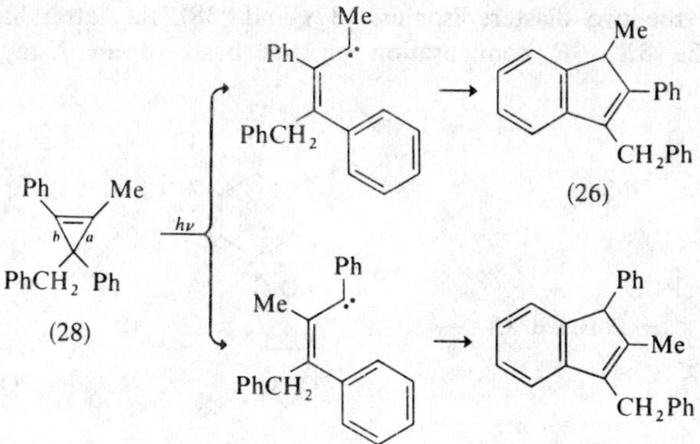

(26)

(27)

(28)

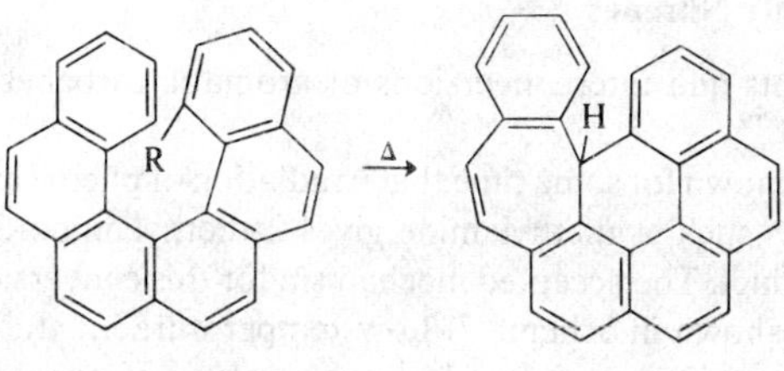

(29) R = CH=NN̄Ts Na⁺ (30)

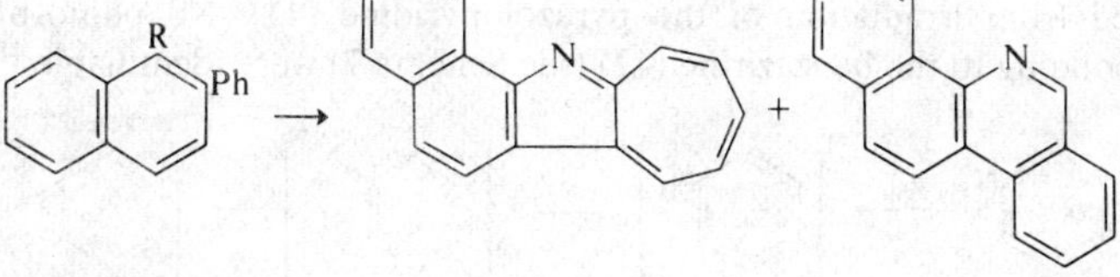

(31) R = NC (32) (33)
(34) R = CN

Indole-3-acetic acids, functionalized at position 2, are obtained by decomposition of $N_2CHCO_2Et$ in the presence of the appropriate indole derivative at elevated temperatures. Using *N*-acyl-indoles, the cyclopropane adducts, *e.g.* (35), are isolated.[26]

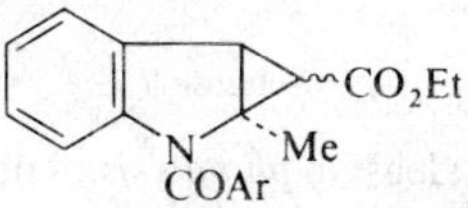

(35)

Reactions of dihalogeno-carbenes and :$CHCO_2Et$ with various uracil and uridine derivatives give adducts from addition to the 5,6-bond. Thus (36) gave a

[26] H. Keller, E. Langer, and H. Lehner, *Monatsh.*, 1977, **108**, 123.

mixture of the two diastereoisomers (37) and (38), the latter having been assigned the 5R, 6R configuration on the basis of an X-ray structure determination.[27]

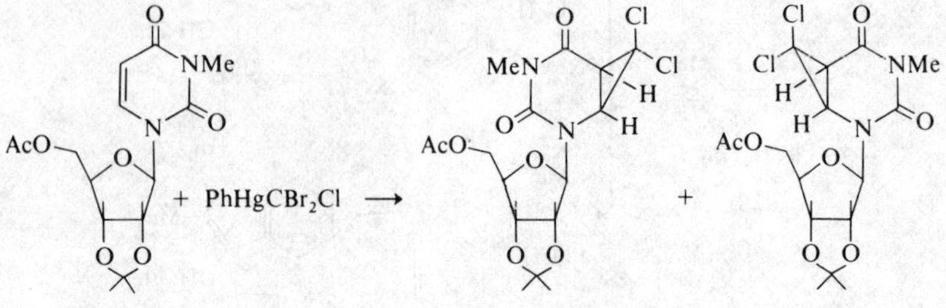

(36)                                    (37)                        (38)

### 3 By Nitrenes

Rearrangements and interconversions of aromatic carbenes and nitrenes have been reviewed.[19]

It has been known for some time that irradiation of phenyl azide in the presence of nucleophiles such as diethylamine gives 2-diethylamino-3*H*-azepine (39) as the major product. The accepted mechanism for this conversion has involved the intermediates shown in Scheme 7. Low-temperature i.r. studies, however, have now revealed a primary product in the photolysis of phenyl azide in an argon matrix at 8 K which has an intense absorption at 1895 cm$^{-1}$. The same photo-product, which is believed to be 1-aza-1,2,4,6-cycloheptatetraene (40), is obtained from irradiation of the pyrazolopyridine (41). No photo-products corresponding to the benzazirine (42) (see Scheme 7) were identifiable.[28]

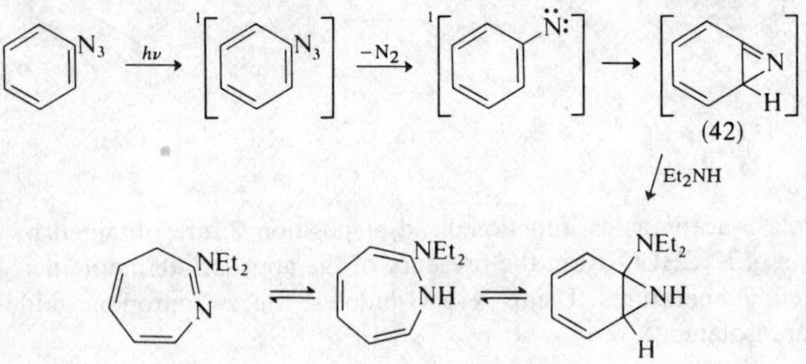

(39)

**Scheme 7**

It appears therefore that, at least in *photolysis* of phenyl azide, the intermediate that reacts with the diethylamine present is probably the ketenimine (40). It remains for experiment to show how far this result of Chapman and his co-workers can be extended both to the photolysis at room temperature and the

[27] H. P. M. Thiellier, G. J. Koomen, and U. K. Pandit, *Tetrahedron*, 1977, **33**, 1493; see also U. K. Pandit, *Heterocycles*, 1977, **6**, 1520.
[28] O. L. Chapman and J.-P. Le Roux, *J. Amer. Chem. Soc.*, 1978, **100**, 282.

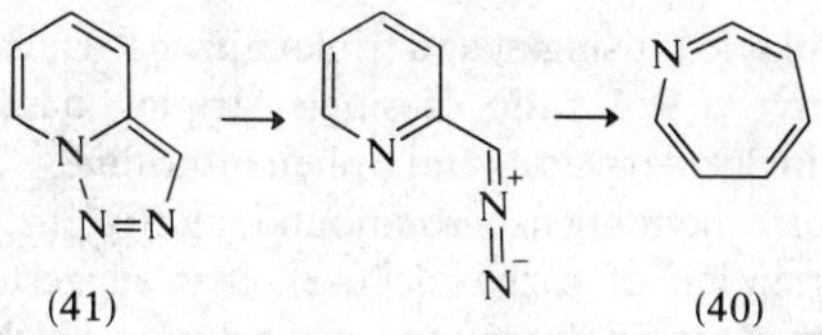

(41)  (40)

thermolysis of other aromatic azides, including five-membered aromatic ring azides.

Photolysis of phenyl azide in ethanethiol gave *o*-thioethoxyaniline in 39% yield. Decomposition of *p*-tolyl azide under similar conditions gave 2-thioethoxy-5-methylaniline (15%), the mechanism in Scheme 8 being suggested by the authors.[29] This mechanism could easily be modified to accommodate a keten-imine intermediate analogous to (40) rather than the benzazirine (43). However,

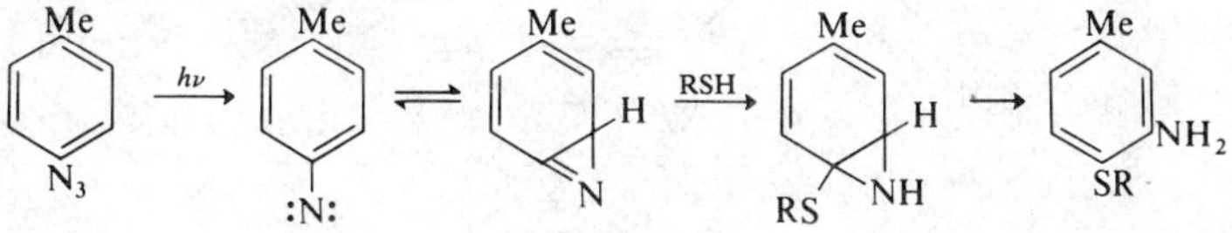

(43)

**Scheme 8**

the formation of 2-thioethoxy-1-naphthalene (44) (40%) from photolysis of β-azidonaphthalene requires the intermediacy of (45) rather than (the more stable?) (46). Presumably the benzazirine could still be an intermediate *en route* to a ketenimine intermediate.

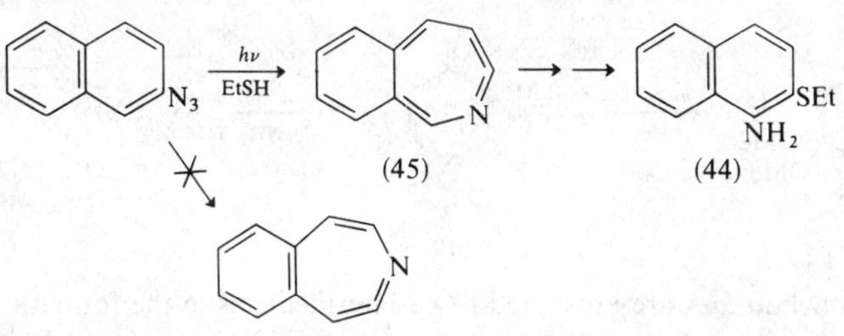

(45)  (44)

(46)

In the photolysis of phenyl azide, the yield of 2-ethylamino-3*H*-azepine can be raised from 30% to 60% by using triethylamine instead of diethylamine as the solvent.[30]

It is now thought that the conversion of 2-azidobiphenyl into carbazole involves the reaction of a singlet nitrene, doubt having been cast on the interpretation of earlier flash photolysis results suggesting a triplet nitrene. To probe the multiplicity of this species, some model systems have been examined in which separate

[29] S. E. Carroll, B. Nay, E. F. V. Scriven, H. Suschitzky, and D. R. Thomas, *Tetrahedron Letters*, 1977, 3175.
[30] B. Nay, E. F. V. Scriven, H. Suschitzky, D. R. Thomas, and S. E. Carroll, *Tetrahedron Letters*, 1977, 1811.

decay routes are available for singlet and triplet nitrene. Thus direct irradiation of (47) appeared to give a 9:1 ratio of singlet:triplet, based on the yields of carbazole (48) (singlet-derived) and phenanthridine, 2-amino-2'-methyl-biphenyl, and the corresponding azo-compound (all triplet-derived), as shown in Scheme 9. The formation of carbazole (48), it is suggested, is the result of concerted cyclization of the singlet nitrene, a conclusion which is supported by the effects of sensitizers and solvents that contain heavy atoms.[31]

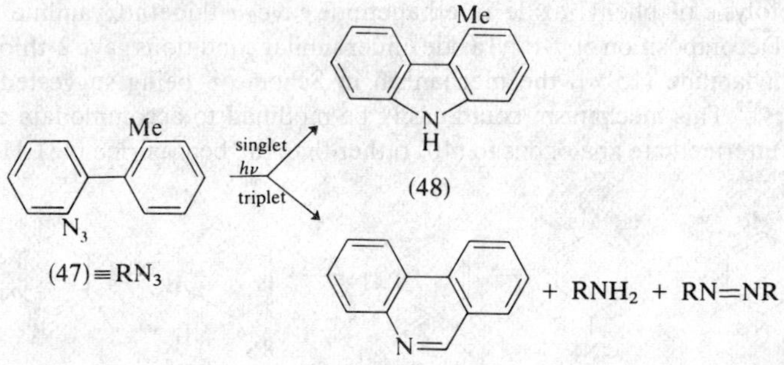

**Scheme 9**

Other reactions involving azide-derived nitrenes include the photolysis of 6-azido-1,3-dimethylthymine (49), which produced the 1,3,5-triazepinedione (50) in methylamine but the dihydrouracil (51) in methanol. The formation of both products was rationalized by the usual nitrene → azirine conversion followed by nucleophilic attack.[32]

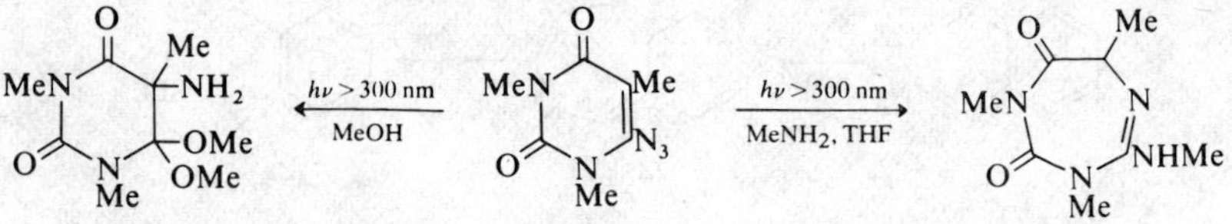

Sulphonylnitrenes are presumed to be intermediates in the formation of the oxathiazepine 5,5-dioxide (52) and the sultam (53) from thermolysis of the corresponding azides.[33]

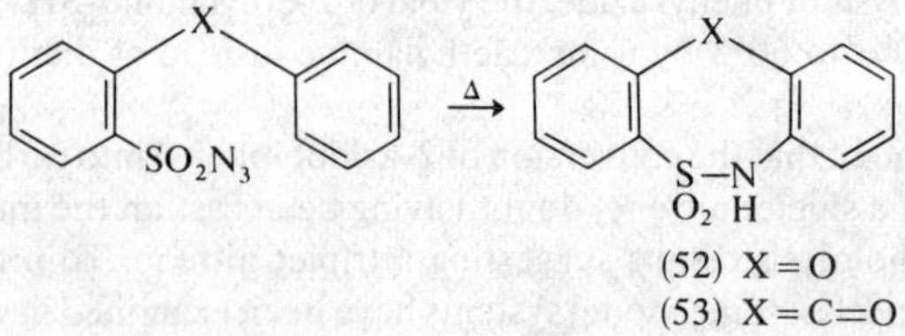

(52) X = O
(53) X = C=O

[31] J. M. Lindley, I. M. McRobbie, O. Meth-Cohn, and H. Suschitzky, *J.C.S. Perkin I*, 1977, 2194.
[32] S. Senda, K. Hirota, T. Asao, K. Maruhashi, and H. Kitamura, *Tetrahedron Letters*, 1978, 1531.
[33] R. A. Abramovitch, C. I. Azogu, I. T. McMaster, and D. P. Vanderpool, *J. Org. Chem.*, 1978, **43**, 1218.

Thermolysis of the *o*-azido-chalcones (54) in boiling toluene yielded the 2,1-benzisoxazoles (55) (70—80%) by intramolecular 1,3-dipolar addition of the azide to the carbonyl group followed by nitrogen loss. A minor product (5—7%) was identified as 3-aryl-4-quinolone (56), which is believed to be formed by direct attack on the side-chain followed by rearrangement (Scheme 10). High-temperature thermolysis (245 °C) of the (*p*-methoxystyryl)anthranil (55; Ar = *p*-MeO C₆H₄) gave the 3-*p*-methoxyphenyl-4-quinolone (65%); ring-opening back to the nitrene is probably involved at this temperature. Intermediate in these transformations is (57), which resembles the spiro-species postulated in other arylnitrene rearrangements and which presumably is also involved in the formation of the 2-arylideneindoxyl (58; Ar = *p*-MeOC₆H₄) on heating (54; Ar = *p*-MeOC₆H₄) at an intermediate temperature.[34]

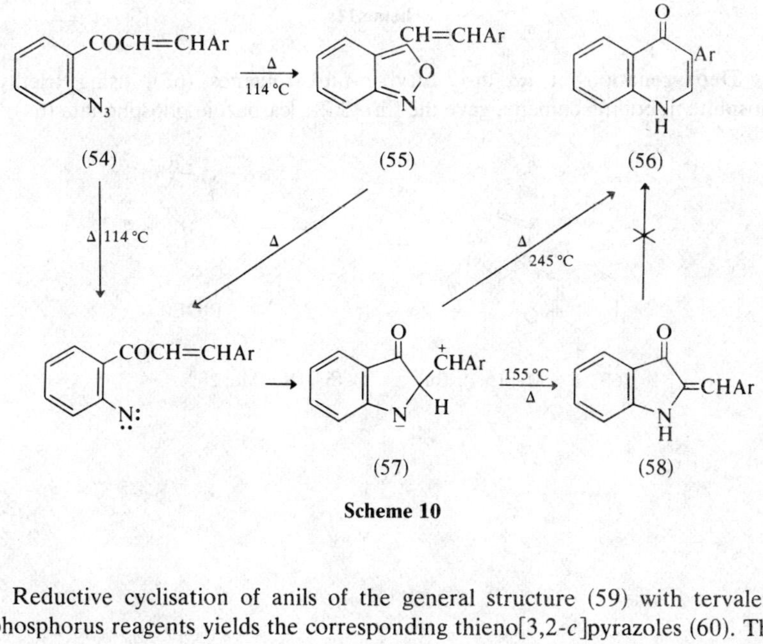

**Scheme 10**

Reductive cyclisation of anils of the general structure (59) with tervalent phosphorus reagents yields the corresponding thieno[3,2-*c*]pyrazoles (60). The isomeric anil (61), however, gave 1-phenyl-3-carbonitrile (62) in 55% yield by the route suggested in Scheme 11.[35]

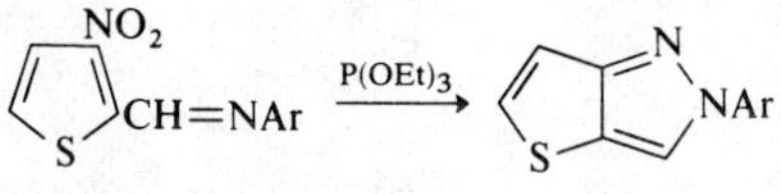

[34] R. K. Smalley, R. H. Smith, and H. Suschitzky, *Tetrahedron Letters*, 1978, 2309.
[35] V. M. Colburn, B. Iddon, H. Suschitzky, and P. T. Gallagher, *J.C.S. Chem. Comm.*, 1978, 453.

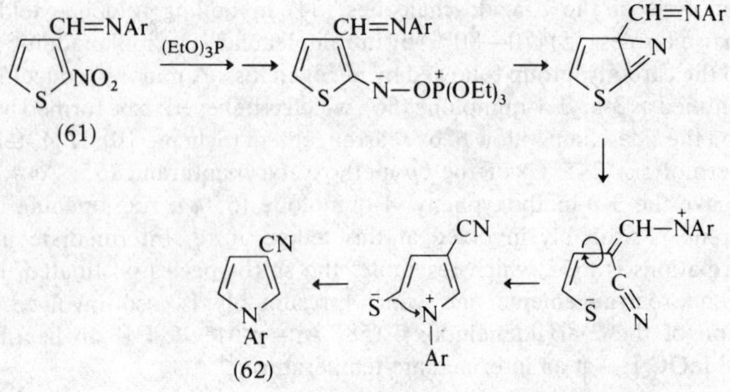

Ar = Ph or C₆H₄NMe₂-*p*

**Scheme 11**

Deoxygenation of αα-di-(2-furyl)-*o*-nitro-toluenes (63), using triethyl phosphite in boiling cumene, gave the furo[3,2-*c*]carbazole phosphonate (64).[36]

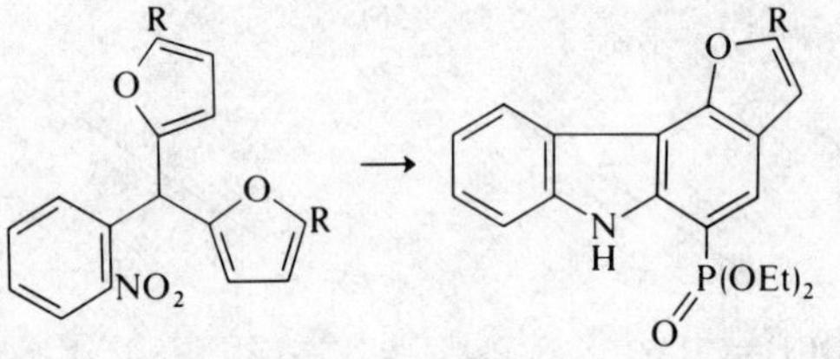

(63)  R = Me, Et, or Buᵗ          (64)  R = Me, 28%

[36] G. Jones and W. H. McKinley, *Tetrahedron Letters*, 1977, 2457.

# 10
## Porphyrins and Related Compounds

BY A. H. JACKSON

## 1 Introduction

A similar format and order of topics will be adopted as in previous years, to facilitate cross-referencing. The pace of research in this field has increased substantially this year, the number of papers published in some sections increasing over two-fold, so that the Reporter has of necessity had to be more selective than previously.

## 2 Reviews

The publication of a seven-volume treatise[1] on the chemistry and biochemistry of porphyrins, edited by D. Dolphin, was a major landmark this year, and this will provide an authoritative reference work for many years to come. A new book concerned with the chemistry of pyrroles has also appeared,[2] and pyrrole pigments have been reviewed in the new edition of 'Rodd'.[3]

The biosynthesis and degradation of haem has been reviewed,[4a] and also the pathobiochemistry of the porphyrias,[4b] whilst high-performance liquid chromatography (h.p.l.c.) has been used to differentiate[5] between the various types of porphyria (abnormalities of porphyrin biosynthesis). Stereochemical aspects of metalloporphyrin chemistry have been discussed;[6] reviews on the organisation of chlorophyll *in vivo*[7] and its role in photosynthesis[8] have also appeared. Accounts of the structure and function of the vitamin $B_{12}$ coenzyme and its mode of action,[9] as well as its synthesis and biosynthesis, have been presented,[10] and revised

[1] 'The Porphyrins', ed. D. Dolphin, Academic Press, New York, 1978, Vol. I, 'Structure and Synthesis, Part A'; Vol. II, 'Structure and Synthesis, Part B'; Vol. III, 'Physical Chemistry, Part A'; Vol. IV, 'Physical Chemistry, Part B'; Vol. V, 'Physical Chemistry, Part C'; Vol. VI, 'Biochemistry, Part A'; Vol. VII, 'Biochemistry, Part. B'.

[2] R. A. Jones and G. P. Bean, 'The Chemistry of Pyrroles', Academic Press, London, 1977.

[3] K. M. Smith, 'Pyrrole Pigments', in "Rodd's Chemistry of Carbon Compounds", 2nd edn., ed. S. Coffey, Elsevier, Amsterdam, 1977, Vol. IVB, p. 237.

[4] (a) G. H. Tait, *Handbook Exp. Pharmacol.*, 1978, **44**, 1; (b) M. Doss, *Med. Klin.* (*Munich*), 1977, **72**, 1501.

[5] C. H. Gray, C. K. Lim, and D. C. Nicholson, *Clinica Chim. Acta*, 1977, **77**, 167.

[6] R. W. Scheidt, *Accounts Chem. Res.*, 1977, **10**, 339.

[7] J. P. Thornber and R. S. Alberte, *Encycl. Plant Physiol.* (*New Series*), 1977, **5**, 574.

[8] D. Mauzerall, *Encycl. Plant Physiol.* (*New Series*), 1977, **5**, 117.

[9] T. Toraya, *Vitamins*, 1977, **51**, 87; J. S. Krouwer and B. M. Babior, *Mol. Cell Biochem.*, 1977, **15**, 89.

[10] (a) A. Eschenmoser and C. W. Wintner, *Science*, 1977, **196**, 1410; (b) A. W. Johnson, *Chem. and Ind.*, 1978, 27; (c) M. H. Georgopapakou and A. I. Scott, *J. Theor. Biol.*, 1977, **69**, 381; (d) A. I. Scott, *Accounts Chem. Res.*, 1978, **11**, 29.

IUPAC rules[11] for corrinoid nomenclature have been published. A useful and timely review of the properties of phytochrome, the photoreactive tetrapyrrolic plant pigment, has also appeared.[12] Other more specialised reviews are mentioned in the various sections below.

## 3 Porphyrins

**Synthesis.**—The classical Fischer synthesis of porphyrins from dipyrromethenes has been markedly improved[13] by using bromine in refluxing formic acid instead of the much harsher conditions previously employed (*i.e.* heating in molten succinic acid); substantially higher yields of centrosymmetric porphyrins (2) were obtained in this way by self-condensation of 5-bromo-5'-methyl-pyrromethenes (1), and of other less symmetrical porphyrins (5) by condensation of 5,5'-dimethyl-pyrromethenes (3) with 5,5'-dibromo-pyrromethenes (4). The synthesis

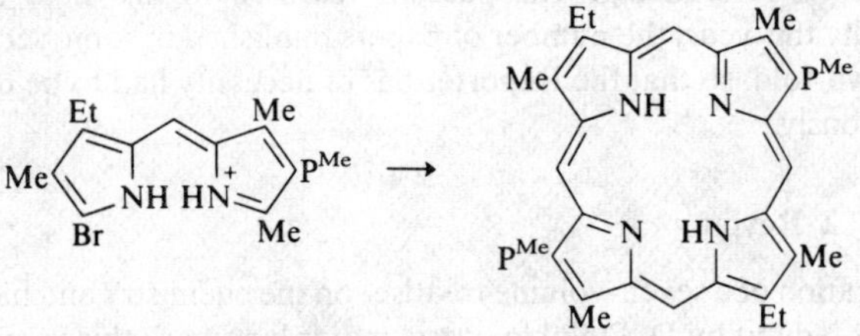

(1)   $P^{Me} = CH_2CH_2CO_2Me$                                  (2)

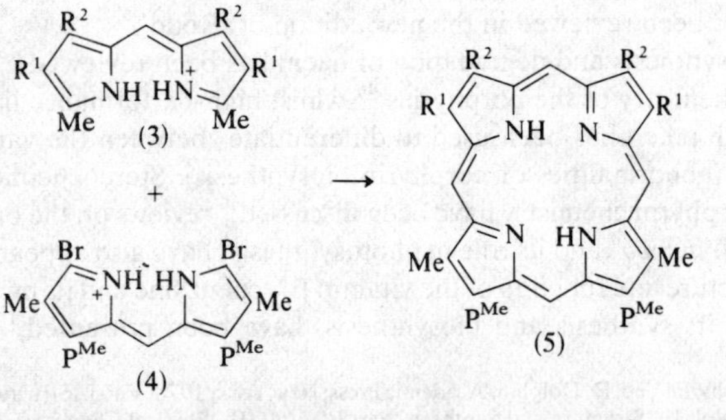

(3)

+

(4)

(5)

$R^1 = Me, R^2 = Et; R^1 = R^2 = Et;$ or $R^1 = n\text{-}C_{12}H_{25}, R^2 = Me$

of symmetrical octa-alkyl-porphyrins (7) from monopyrrolic precursors has also been facilitated,[14] both by the development of a new synthesis of the appropriate 2,5-unsubstituted monopyrroles, *e.g.* (6), from toluene-*p*-sulphonylmethyl iso-cyanide and unsaturated ketones, and by modifying the cyclisation conditions.

A series of *meso*-mono- and di-aryl-octaethylporphyrins (9) has been synthesized by condensation of 3,3',4,4'-tetraethylpyrromethane (8) with

[11] *Pure Appl. Chem.*, 1976, **48**, 495.
[12] L. H. Pratt, *Photochem. and Photobiol.*, 1978, **27**, 81.
[13] J. B. Paine, C. K. Chang, and D. Dolphin, *Heterocycles*, 1977, **7**, 831.
[14] D. O. Cheng and E. Legoff, *Tetrahedron Letters*, 1977, 1469.

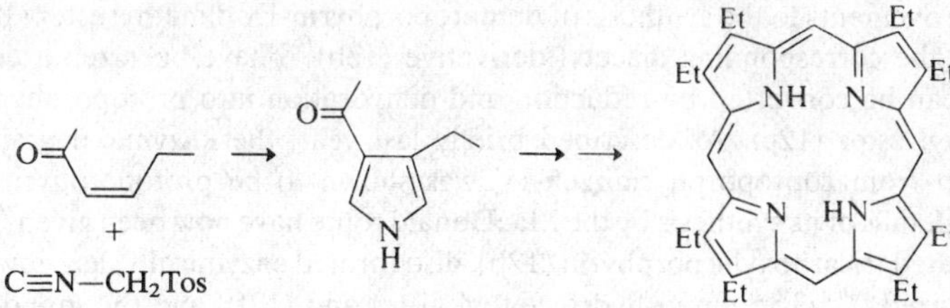

(6)

(7)

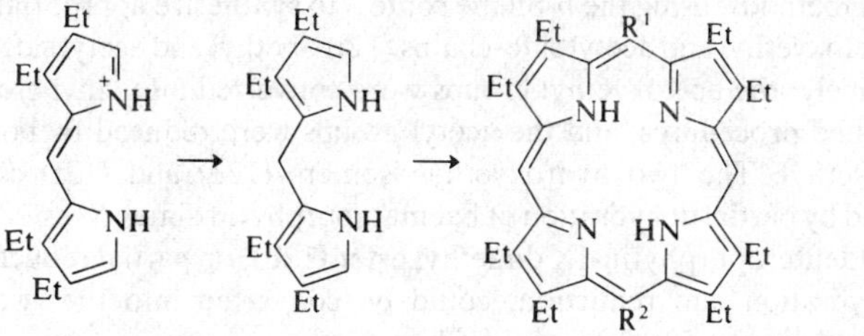

(8)

(9) a; $R^1 = H, R^2 = $ aryl
b; $R^1 = R^2 = $ aryl

aryl = Ph, naphthyl, 2,3,4-$Me_3C_6H_2$, or 2,3,4-$(OMe)_3C_6H_2$

aromatic aldehydes in a variety of acidic media.[15] The *meso*-formylation products of the complex of copper with aetioporphyrin-I undergo intramolecular cycliza-tion[16] to form porphyrins of type (10). *meso*-Methyl porphyrins (11) can be synthesized[17] from the appropriate *meso*-formyl derivatives by a three-stage process; reduction with borohydride, acetylation of the resulting hydroxymethyl compound, and reduction with borohydride or hydrogen over palladium/char-coal.

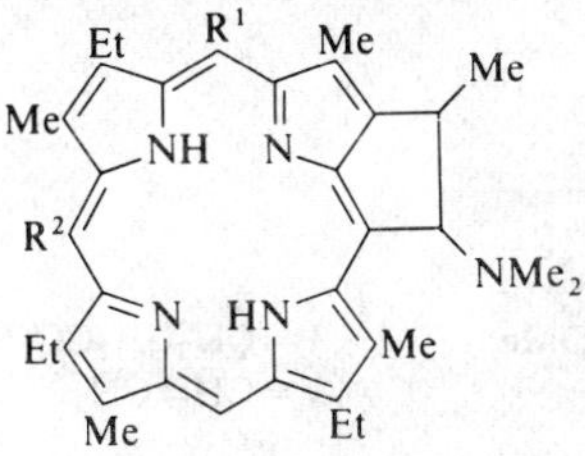

(10) a; $R^1 = H, R^2 = CH_2OH$ or CHO
b; $R^1 = CH_2OH$ or CHO, $R^2 = H$

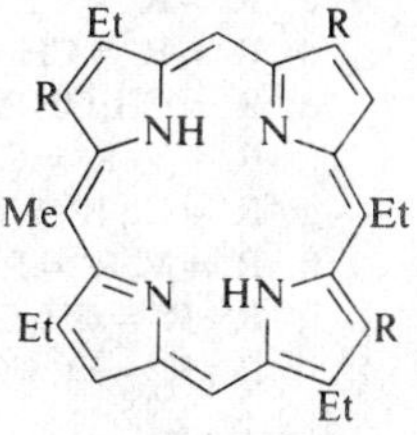

(11) R = Me or Et

A useful ring synthesis of *meso*-substituted porphyrins has been developed.[18] It involves the condensation of *ac*-biladienedicarboxylic acids with aldehydes or acetals in the presence of nickel(II) acetate followed by removal of the nickel from the resulting complexes.

[15] H. Ogoshi, H. Sugimoto, T. Nishiguchi, T. Watanabe, Y. Matsuda, and Z. Yoshida, *Chem. Letters*, 1978, 29.
[16] G. V. Ponomarev, *Khim. geterotsikl. Soedinenii*, 1977, 1691.
[17] M. J. Bushell, S. Evans, G. W. Kenner, and K. M. Smith, *Heterocycles*, 1977, 7, 67.
[18] D. Harris and A. W. Johnson, *J. C. S. Chem. Comm.*, 1977, 771.

Improvements to the synthesis of deuteroporphyrin-IX dimethyl ester (12a)[19a] and of the corresponding diacetyl derivative (12b)[19b] have been reported; the latter can be converted by reduction and dehydration into protoporphyrin-IX dimethyl ester (12c). As described briefly last year, the enzymic degradation product from coproporphyrinogen-IV was shown to be protoporphyrin-XIII (13a); details of its synthesis by the MacDonald route have now been given[20a] and the related tricarboxylic porphyrin (12b), also formed enzymically, has now been synthesized.[20b,c] The mono-hydroxyethyl (12e) and (12f), and the mono-ethyl (12g) and (12h) analogues of protoporphyrin-IX (12c) have been prepared by rational methods, using the *b*-bilene route[21] to synthesize appropriate precursors with acetoxyethyl and acetyl side-chains (12i) or ethyl and acetyl side-chains (12j) respectively; the acetoxyethyl groups were converted into vinyl groups by well-established procedures, and the acetyl groups were reduced by borohydride to hydroxyethyl. The two hydroxyethyl isomers (12e) and (12f) could also be obtained by partial dehydration of haematoporphyrin dimethyl ester (12d), whilst diacetyldeuteroporphyrin-IX dimethyl ester (12b), on partial reduction followed by dehydration and reduction, could be converted into the two monoethyl monovinyl porphyrins (12g) and (12h).

(12) a; $R^1 = R^2 = H$
b; $R^1 = R^2 = COMe$
c; $R^1 = R^2 = V$
d; $R^1 = R^2 = CH(OH)Me$
e; $R^1 = CH(OH)Me, R^2 = V$
f; $R^1 = V, R^2 = CH(OH)Me$
g; $R^1 = Et, R^2 = V$
h; $R^1 = V, R^2 = Et$
i; $R^1, R^2 = CH_2CH_2OAc, COMe$
j; $R^1, R^2 = Et, COMe$

(13) a; $R = V$
b; $R = P$

$P = CH_2CH_2CO_2H$
$V = CH=CH_2$

Perhaps the most significant achievement in porphyrin synthesis during the past year has been the total synthesis of porphyrin-*a* dimethyl ester (14a), the demetallated pigment obtained from cytochrome oxidase.[22] The main features of the structure of porphyrin-*a* had been established beyond doubt by earlier work,

[19] (a) A. D. Adler, D. L. Ostfeld, and E. H. Abbott, *Bioinorg. Chem.*, 1977, **7**, 187; (b) V. N. Luzgina, E. I. Filippovich, N. K. Doan, M. N. Moskvin, and R. P. Evstigneeva, *Zhur. obshchei Khim.*, 1977, **47**, 1416.
[20] (a) G. Buldain, L. Kiaz, and B. Frydman, *J. Org. Chem.*, 1977, **42**, 2957; (b) G. Buldain, J. Hurst, R. B. Frydman, and B. Frydman, *ibid.*, p. 2953; (c) P. S. Clezy, C. J. R. Fookes, and S. Sternhell, *Austral. J. Chem.*, 1978, **31**, 639.
[21] P. S. Clezy, C. J. R. Fookes, and T. T. Hai, *Austral. J. Chem.*, 1978, **31**, 365.
[22] M. Thompson, J. Barrett, E. McDonald, A. R. Battersby, C. J. R. Fookes, I. A. Chaudhry, and P. S. Clezy, *J. C. S. Chem. Comm.*, 1977, 278.

but some doubts remained about the precise nature of the $C_{17}$ side-chain, although n.m.r. evidence obtained by Caughey[23] favoured the structure shown as (14a) rather than related, reduced, or cyclic forms. By use of h.p.l.c. and field desorption (f.d.) mass spectrometry[22,24] it was shown that the porphyrin ester obtained after demetallation of beef heart cytochrome $c$ oxidase was a mixture of two components, the less polar of which had the previously accepted structure (14a) for porphyrin-$a$ dimethyl ester; the structure of the more polar component has not yet been described. The key intermediate in the new synthesis was the formyl-porphyrin ester (15a), which had been prepared previously by Clezy's group in Australia;[25] this was converted into the corresponding acid chloride (15b), which was used to acylate the magnesium enolate (16) prepared from *trans-trans*-farnesyl bromide and malonic ester. The resulting $\beta$-keto-ester (15c) was demethylated and decarboxylated with lithium iodide to give the corresponding ketone (15d), and the formyl group of the latter was selectively protected by formation of an acetal. Reduction of the ketone with borohydride,

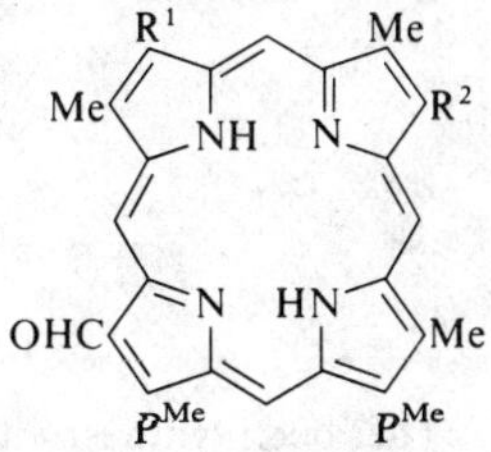

(14) a; $R^1 =$ [HO—...], $R^2 = V$

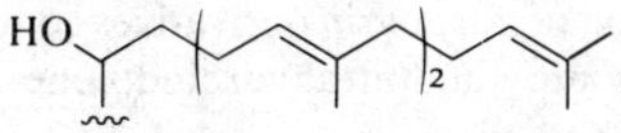

b; $R^1 =$ [HO—...], $R^2 = V$

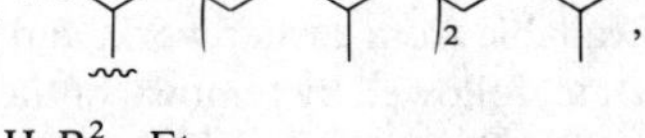

c; $R^1 = H, R^2 = Et$

(15) a; $R^1 = R^2 = CO_2Me$

b; $R^1 = R^2 = COCl$

c; $R^1 = R^2 = -COCH$—[...], $CO_2Me$

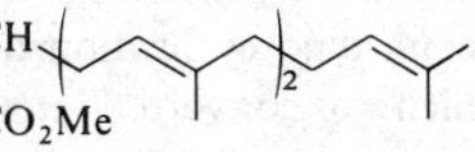

d; $R^1 = R^2 = -COCH_2$—[...]

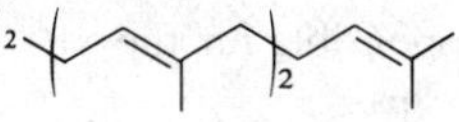

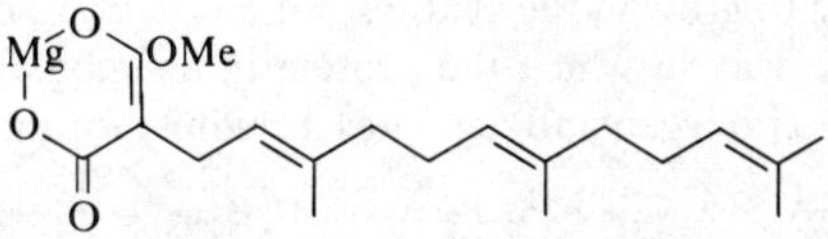

(16)

[23] W. S. Caughey, G. A. Smythe, D. H. O'Keefe, J. E. Maskasky, and M. L. Smith, *J. Biol. Chem.*, 1975, **250**, 7602.
[24] L. J. Defilippi and D. E. Hultquist, *Biochim. Biophys. Acta*, 1977, **498**, 395.
[25] P. S. Clezy and C. J. R. Fookes, *Austral. J. Chem.*, 1977, **30**, 1799.

followed by hydrolysis of the acetal and re-esterification, then gave $(R, S)$-porphyrin-*a* dimethyl ester (14a), identical with the natural product {h.p.l.c., n.m.r., u.v., and 'mixed n.m.r.',[26] using $[Eu([^2H_9]fod)_3]$}. A closely isomeric porphyrin (17) was also prepared, and this was readily distinguished spectroscopically and chromatographically from porphyrin-*a* dimethyl ester (14a). This route to porphyrin-*a* was developed from earlier model experiments[25,27] leading to the synthesis of the hexahydro-derivative (14b) and of an analogue (18a), elaborated from the more readily accessible porphyrin acid (18b). A new synthesis of 4-ethyl-8-formylcytodeuteroporphyrin (14c) has also been reported.[28]

(17) R =

(18) a; R =

b; R = CO₂H

A monobenzoporphyrin (23) has been synthesized by two routes involving the construction of porphyrin derivatives containing a fused cyclohexanone ring.[29] In the first route the tetrahydroindolone (19) was prepared by a Knorr-type synthesis, and was incorporated into a dipyrrylmethane (20a) as shown in Scheme 1. Condensation of the formyl analogue (20b) of the latter with the imine salt (21), available from earlier work, and oxidation of the derived *b*-bilene with copper acetate, followed by removal of the metal, gave the oxotetrahydrobenzo-porphyrin (22). Reduction, dehydration, and dehydrogenation then afforded the desired benzoporphyrin (23). The second route involved synthesis of the porphyrin (24) by the MacDonald method from appropriate dipyrromethanes. Dieckmann-type cyclization of the latter then gave the unstable $\beta$-keto-ester (25), which was converted into the same benzoporphyrin (23). The spectroscopic properties of this compound supported an earlier proposal[30] that such a species was responsible for the 'rhodo'-type spectra exhibited by some types of petro-porphyrins.

Interest continues to grow in the synthesis of model compounds which can be utilised to study the way in which the protein in haemoglobin and myoglobin controls the binding of oxygen and regulates its affinity for the central iron atom.

[26] *cf.* A. H. Jackson, H. A. Sancovich, A. M. Ferramola, N. Evans, D. E. Games, S. A. Matlin, G. H. Elder, and S. G. Smith, *Phil. Trans. Roy. Soc.* 1976, **B273**, 191.

[27] *cf.* also R. V. H. Jones, G. W. Kenner, T. Lewis, and K. M. Smith, *Chem. and Ind.*, 1971, 129; R. V. H. Jones, G. W. Kenner, and K. M. Smith, *J. C. S. Perkin I*, 1974, 531.

[28] M. A. Kulish, A. F. Mironov, and R. P. Evstigneeva, *Trady Moskov. Inst. Tonkoi Khim. Tekhnol.*, 1976, **6**, 71 (*Chem. Abs.* 1977, **87**, 152 159).

[29] P. S. Clezy, C. J. R. Fookes, and A. H. Mirza, *Austral. J. Chem.*, 1977, **30**, 1337.

[30] *cf.* J. R. Maxwell, C. T. Pillinger, and G. Eglinton, *Quart. Rev.*, 1971, **25**, 571.

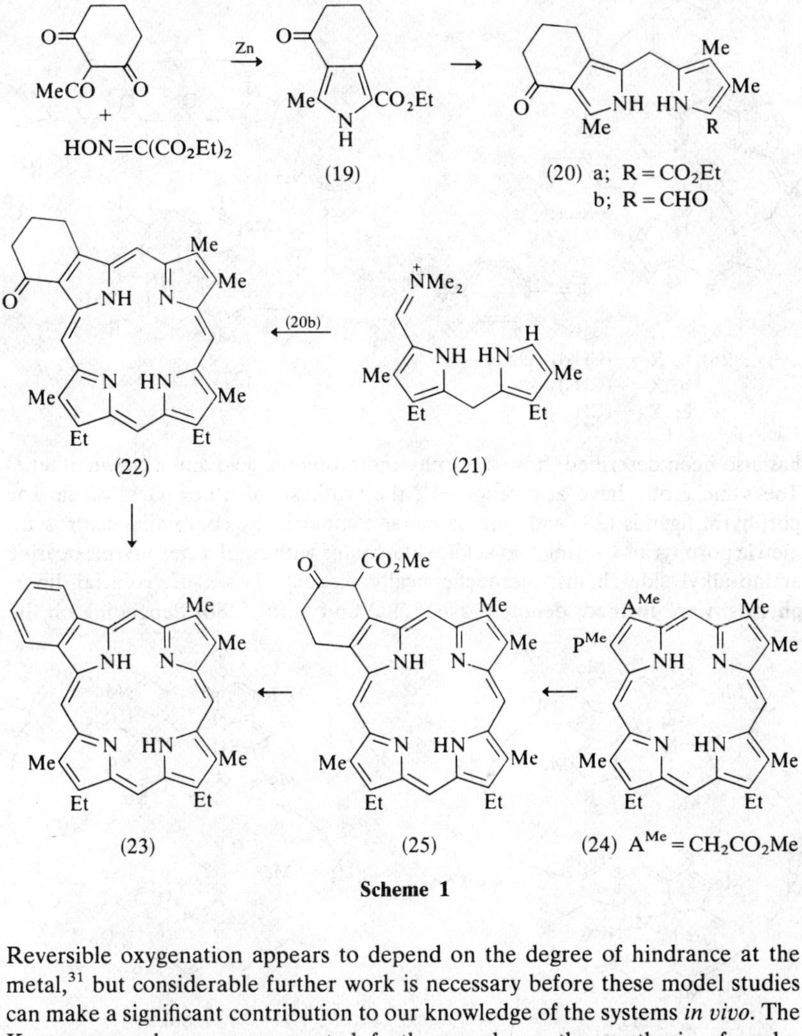


(19)

(20) a; R = $CO_2Et$
b; R = CHO

(21) (22) (23) (25)

(24) $A^{Me} = CH_2CO_2Me$

**Scheme 1**

Reversible oxygenation appears to depend on the degree of hindrance at the metal,[31] but considerable further work is necessary before these model studies can make a significant contribution to our knowledge of the systems *in vivo*. The Kyoto group have now reported further work on the synthesis of cyclophaneporphyrins[32] related to those described in last year's Report, and the peripheral functionalization of copper porphyrins has also been achieved.[33] A new type of 'looping-over' porphyrin has been synthesized[34] from a cobalt *meso*-tritolyl-*o*-hydroxyphenyl-porphyrin by attachment of a pyridine ring containing side-chains to the phenolic hydroxy-group; the resulting cobalt porphyrin (26), however, exhibited little enhancement of oxygen affinity compared with other porphyrins. The preparation of a 'crowned' porphyrin (27)

[31] *cf.* J. P. Collman, *Accounts Chem. Res.*, 1977, **10**, 265.
[32] H. Ogoshi, H. Sugimoto, and Z. Yoshida, *Tetrahedron Letters* 1977, 1515.
[33] J. Baldwin and J. E. DeBernardis, *J. Org. Chem.*, 1977, **42**, 3986.
[34] F. S. Molinaro, R. G. Little, and J. A. Ibers, *J. Amer. Chem. Soc.*, 1977, **99**, 5628.

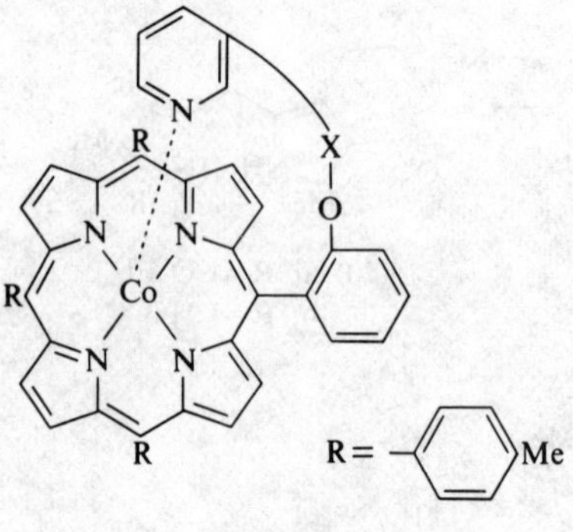

(26) a; X = -(CH$_2$)$_4$CONH-
     b; X = -(CH$_2$)$_3$-
     c; X = -(CH$_2$)$_2$-

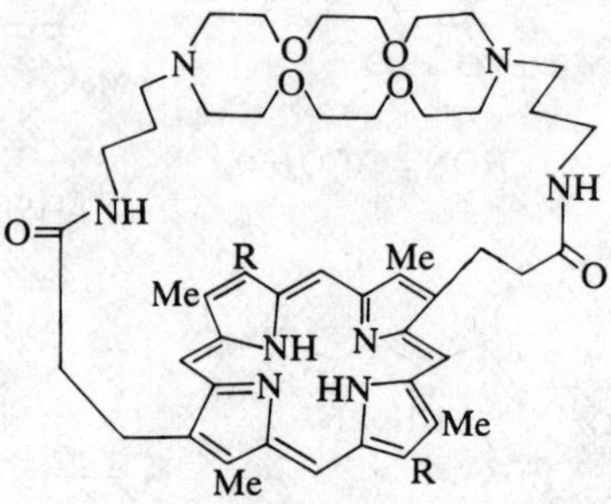

R = —⟨ ⟩—Me

(27) R = n-hexyl

has also been described, from a porphyrindipropionic acid and a crown ether.[35] The same group have also reported[36] the synthesis of a new class of dimeric porphyrin ligands (28) and various metal complexes by combining centrosymmetric porphyrins bearing two acidic side-chains with similar porphyrins bearing amino-alkyl side-chains; stereochemically distinct classes of 'cofacial-diporphyrins' were formed, denoted '*syn*' (28a) and '*anti*' (28b) depending on the

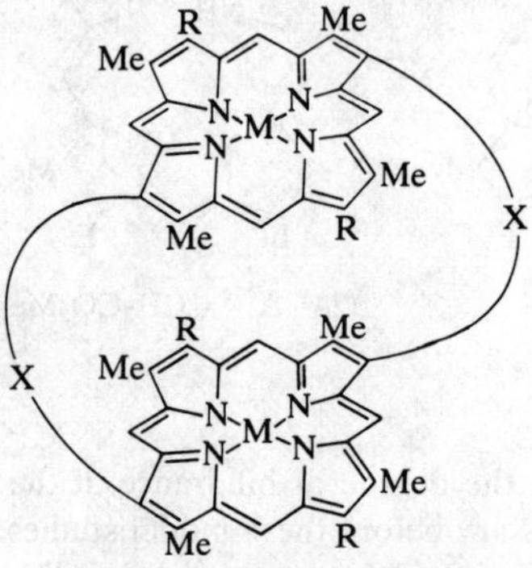

(28a) 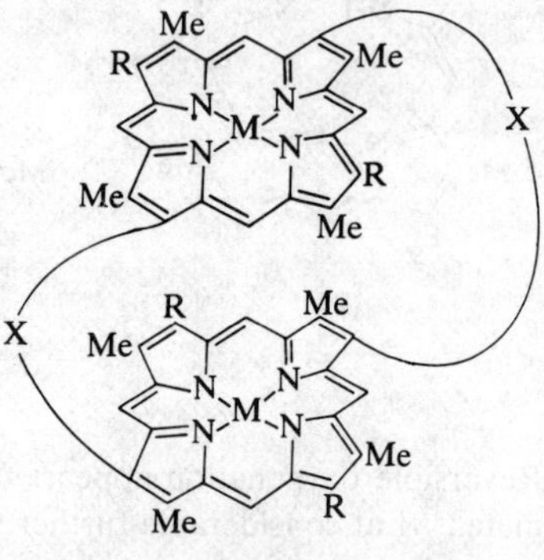                                    (28b)

R = n-hexyl, X = (CH$_2$)$_n$CON(Bu$^n$)(CH$_2$)$_m$   (n = 1 or 2; m = 2 or 3)
M = Mg, Fe, Cu, Co, or Zn

arrangements of the side-chains. Three homologous series were studied, differing in the number of methylene groups in the bridges interconnecting the macrocyclic rings; the interplanar differences ranged from 4.2 to 6.4 Å, and their absorption and emission spectra showed strong exciton coupling. A dicobalt complex with a separation of 4.2 Å reacts to give a sandwiched μ-superoxo-dicobalt complex.[37]

[35] C. K. Chang, *J. Amer. Chem. Soc.*, 1977, **99**, 2819.
[36] C. K. Chang, M. S. Kuo, and C. B. Wang, *J. Heterocyclic Chem.*, 1977, **14**, 943; C. K. Chang, *ibid.*, p. 1285.
[37] C. K. Chang, *J. C. S. Chem. Comm.*, 1977, 800.

Another type of bisporphyrin derivative (29) has been prepared[38] by coupling a *meso*-tetra-*p*-(2-hydroxyethoxy)phenyl-porphyrin with *p*-formylbenzyl chloride (4 moles) and then condensing the product with pyrrole under acidic conditions. Yet further types of face-to-face porphyrins (30) have been prepared by the

(29)  X = —⟨  ⟩—O(CH₂)₂OCO⟨  ⟩— 

(30)  Y = –(CH₂)₂CONH⟨  ⟩

Stanford group,[39] whilst haemoglobin models with conformationally linked haems (31) have been synthesized,[40] and the preparation and visible spectra of bis- and tris-porphyrins, *e.g.* (32), and their metal complexes have also been studied.[41]

**Reactions and Properties.**—Treatment of octaethylporphyrin with hypofluorous acid gives the corresponding *N*-oxide;[42] this novel finding has been discussed in connection with a possible role for metal porphyrin *N*-oxides in the biological degradation of porphyrins to bile pigments. The ozonisation of octaethylporphyrin has also been described.[43] A new water-soluble porphyrin, *meso*-tetrakis-(*p*-trimethylammoniumphenyl)-porphyrin, has been prepared which does not appear to aggregate in solution at concentrations below $10^{-2}$ mol l$^{-1}$ at pH 6.9, and which forms a dication in two distinct steps.[44a] Temperature-jump studies of the protonation of the parent tetraphenyl-porphyrin have also been described.[44b]

Further studies of the $^{15}$N n.m.r. spectra of porphyrins have appeared;[45] reproducible $^{1}$H chemical shifts of porphyrin substituents can be obtained by measuring the spectra of the zinc complexes in the presence of an excess of

[38] N. E. Kagan, D. Mauzerall, and R. B. Merrifield, *J. Amer. Chem. Soc.*, 1977, **99**, 5484.
[39] J. P. Collman, M. C. Elliott, T. R. Halbert, and B. S. Torrog, *Proc. Nat. Acad. Sci. U.S.A.*, 1977, **74**, 18.
[40] T. G. Traylor, Y. Tatsuno, D. W. Powell, and J. B. Cannon, *J. C. S. Chem. Comm.*, 1977, 732.
[41] K. Ichimura, *Chem. Letters*, 1977, 641.
[42] R. Bonnett, R. J. Ridge, and E. H. Appelman, *J. C. S. Chem. Comm.*, 1978, 310.
[43] A. M. Shul'ya, I. M. Bytera, I. F. Gurinovich, L. A. Grubina, and G. P. Gurinovich, *Biofizika*, 1977, **22**, 771.
[44] (a) M. Krishnamurthy, *Indian J. Chem., Sect. B.*, 1977, **15**, 964; (b) F. Hibbert and K. P. P. Hunte, *J. C. S. Perkin II*, 1977, 1624.
[45] D. Gust and J. D. Roberts, *J. Amer. Chem. Soc.*, 1977, **99**, 3637.

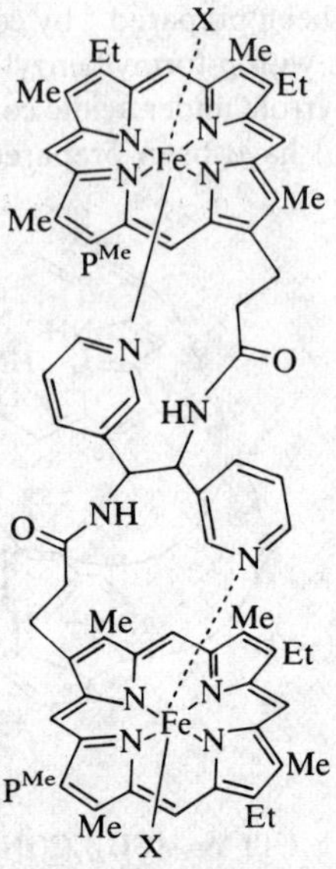

(31)

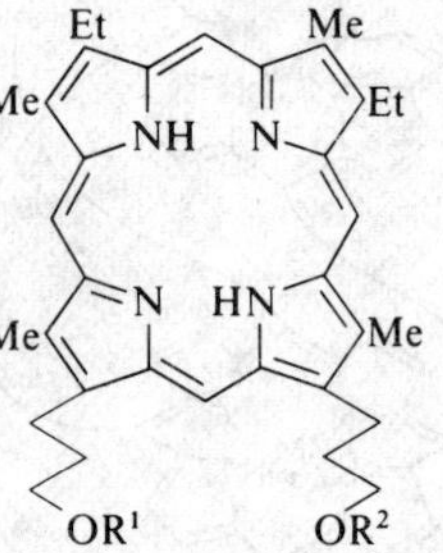

(32) $R^1$, $R^2$ = H or the
acyl group derived from
the monomethyl ester
of mesoporphyrin-IX

pyrrolidine, which inhibits aggregation of porphyrins.[46] The resonance Raman
spectra of [15]N-enriched metallo-octaethylporphyrins have been studied to help in
the characterization of the oxidation-state marker bands in haemoproteins.[47]
CNDO M.O. calculations on the ground and excited states of porphyrins have
been reported,[48] and the effect of structure on the photosensitizing efficiency of
porphyrins has been discussed.[49]

An interesting regioselective exchange of the protons of the methyl groups of
rings A and B in protoporphyrin-IX occurs in basic media.[50] Further mass spectral
studies of porphyrinogens indicate the possibility of localising the positions of
*meso*-substituents.[51]

**Biosynthesis.**—Further details of the enzymic synthesis of porphobilinogen
(PBG) (33) using immobilized ALA dehydratase have been described,[52] as well as
the possible use of immobilized subunits for enzyme isolation;[53] the role of SH
groups in the mechanism of action of enzyme has also been discussed.[54] The

[46] R. J. Abraham, F. Eivazi, R. Nayyirmazhir, H. Pearson, and K. M. Smith, *Org. Magn. Resonance*,
1978, **11**, 52.
[47] T. Kitagawa, M. Abe, Y. Kyoguku, H. Ogoshi, H. Sugimoto, and Z. Yoshida, *Chem. Phys. Letters*,
1977, **48**, 55.
[48] S. J. Chantrell, C. A. McAuliffe, R. W. Munn, A. C. Pratt, and R. F. Weaver, *Bioinorg. Chem.*,
1977, **7**, 283.
[49] G. Canizzo, G. Gennasi, G. Julio, and J. D. Spikes, *Photochem. and Photobiol.*, 1977, **25**, 389.
[50] B. Evans, K. M. Smith, G. N. La Mar, and D. B. Viscio, *J. Amer. Chem. Soc.*, 1977, **99**, 7070.
[51] H. Budzikiewicz and W. Neuenhaus, *Heterocycles*, 1977, **7**, 251.
[52] D. Gurne and D. Shemin, *Methods Enzymol.*, 1976, **44**, 844; A. M. Stella, E. Wider de Xifra, and
A. M. del C. Batlle, *Mol. Cell Biochem.* 1977, **16**, 97.
[53] G. F. Barnard, R. Itoh, L. H. Hohberger, and D. Shemin, *J. Biol. Chem.*, 1977, **252**, 8964.
[54] D. Gurne, J. Chen, and D. Shemin *Proc. Nat. Acad. Sci. U.S.A.*, 1977, **74**, 1383.

purification and modification of PBG oxygenase[55] (which has been implicated in
the regulation of porphyrin biosynthesis) and its induction by AIA* have been
reported.[56] New syntheses of lactams derived by oxidation of hemopyrrole and
kryptopyrrole have been described;[57] it was thought that the latter might be
responsible for neurological symptoms in certain types of porphyria, but their
presence in both normal and schizophrenic urine has not been substantiated.[58]

Several groups[59—63] have reported results concerned with the manner and
timing of the reversal of ring D of uroporphyrinogen-III (36) during its biosynthesis from four PBG units, and the results are shown in Scheme 2. Work in this
area has been complicated by the propensity of PBG and polypyrrylmethanes to
undergo non-enzymic conversions into porphyrinogens; furthermore, the enzymic polymerization of PBG probably involves a rapid 'zipping-up' process on the
enzyme surface, with relatively little leakage of intermediates into the cytoplasm,
and only low incorporations of potential intermediates have been obtained
hitherto. The Cambridge group,[64] in a definitive series of three papers,[62] have,
however, now shown that the [14]C-labelled 'AP–AP' pyrromethane (34) and the
[13]C-labelled bilane (35) corresponding to head-to-tail linkage of four PBG units
are both incorporated intact and in relatively high yields into uroporphyrinogen-
III (36) by the deaminase–cosynthetase system. The ratios of the products formed
from the bilane after aromatization were 70% uroporphyrinogen-II and 30%
uroporphyrinogen-I; non-enzymic conversions led only to uroporphyrinogen-I
(37). The bilane (35) was synthesized in a stepwise manner[65] from a pyrromethane (corresponding to the BC rings) *via* a tripyrrene and an *ac*-biladiene;
the synthesis of a related biladiene with the D ring reversed has also been
reported.[66] Incorporation of the unrearranged bilane (35) into uroporphyrinogen-III by the enzyme system from *P. shermanii* has also been reported by
the Stuttgart group,[63] 84% of the product being the type-III isomer and 16%
type-I; the difference between the relative amounts of the uroporphyrinogen
isomers obtained in Stuttgart and Cambridge is probably due to differences in the
enzyme systems used and their concentrations. These results lead to the
conclusion that the unrearranged bilane (35) is the normal intermediate in both
uroporphyrinogen-III and -I biosynthesis, and that the cosynthetase present in

* AIA is allylisopropylacetamide

[55] M. L. Tomaro, R. B. Frydman, and B. Frydman, *Arch. Biochem. Biophys.*, 1977, **180**, 239.
[56] M. L. Tomaro and R. B. Frydman, *F.E.B.S. Letters*, 1977, **84**, 29.
[57] T. A. Wooldridge and D. A. Lightner, *J. Heterocyclic Chem.*, 1977, **14**, 1283.
[58] P. L. Gendler, H. A. Duhan, and H. Rapoport *Clin. Chem.*, 1978, **24**, 230.
[59] A. M. del C. Batlle and M. V. Rossetti, *Internat. J. Biochem.*, 1977, **8**, 25; M. V. Rossetti and A. M.
   del C. Batlle, *ibid.*, p. 277; M. V. Rossetti, A. A. Jaknat de Geralnik, and A. M. del C. Batlle, *ibid.*,
   p. 78.
[60] R. B. Frydman, E. S. Levy, A. Valasinas, and B. Frydman, *Biochemistry*, 1978, **17**, 110, 115.
[61] B. Franck, G. Fels, G. Ufer, and A. Rowold, *Angew. Chem.*, 1977, **89**, 676; B. Franck, G. Fels, and
   G. Ufer, *ibid.*, p. 677.
[62] A. R. Battersby, E. McDonald, D. C. Williams, and H. K. W. Wurziger, *J. C. S. Chem. Comm.*, 1977,
   113; A. R. Battersby, D. G. Buckley, E. McDonald, and D. C. Williams, *ibid.*, p. 115; A. R.
   Battersby, D. W. Johnson, E. McDonald, and D. C. Williams, *ibid.*, p. 117.
[63] H.-O. Dauner, G. Gunzer, I. Heger, and G. Müller, *Z. physiol. Chem.*, 1976, **357**, 147.
[64] A. R. Battersby, *Experientia*, 1978, **34**, 1.
[65] *cf.* J. A. P. Baptista de Almeida, G. W. Kenner, J. Rimmer, and K. M. Smith, *Tetrahedron*, 1976, **32**,
   1793.
[66] A. Gossauer and J. Engel, *Annalen*, 1977, 225.

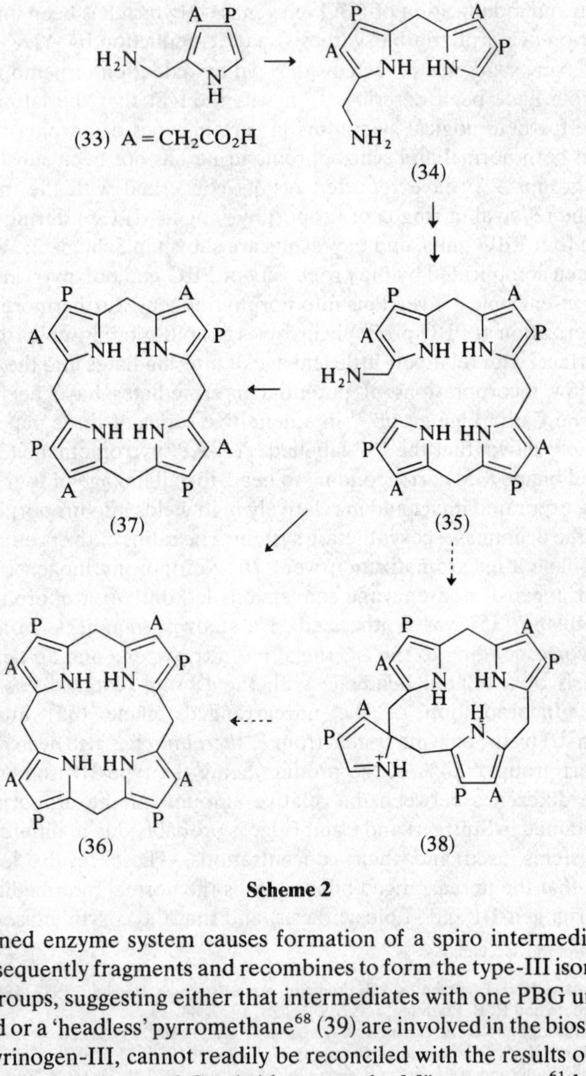

(33) A = CH₂CO₂H

(34)

(37)

(35)

(36)

(38)

**Scheme 2**

the combined enzyme system causes formation of a spiro intermediate[67] (38), which subsequently fragments and recombines to form the type-III isomer. Work by other groups, suggesting either that intermediates with one PBG unit already rearranged or a 'headless' pyrromethane[68] (39) are involved in the biosynthesis of uroporphyrinogen-III, cannot readily be reconciled with the results obtained by the workers in Stuttgart and Cambridge; e.g., the Münster group[61] have shown that the labelled tripyrranes (40) and (41) are incorporated into uroporphyrinogen-III. However, the incorporations obtained in these and other experiments were rather low, due perhaps in part to differences in the enzyme preparations; it may well be that the combined enzyme system will also accept and transform rearranged substrates (albeit in lower yield) as well as head-to-tail combinations of two, three, or four PBG units, and further clarification is awaited with interest.

[67] cf. J. H. Mathewson and A. H. Corwin, *J. Amer. Chem. Soc.*, 1961, **83**, 135.
[68] A. I. Scott, K. S. Ho, M. Kajiwara, and T. Takahashi, *J. Amer. Chem. Soc.*, 1976, **98**, 1589.

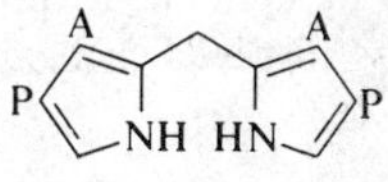

(39)

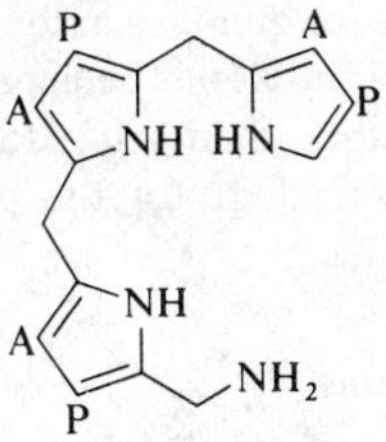

(40) 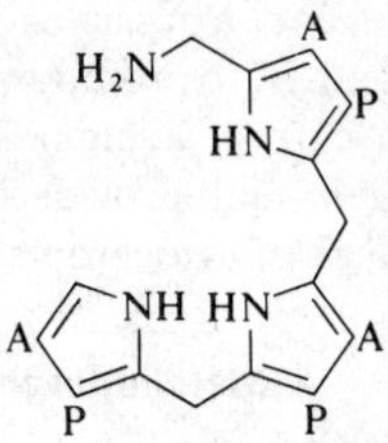          (41)

The preferred 'clockwise' pathway for the enzymic decarboxylation of uro-porphyrinogen-III to coproporphyrinogen-III was reported last year, and it is interesting to note that uroporphyrinogen-I (37) is non-specifically decarboxyl-ated enzymically by both possible routes (Scheme 3) into coproporphyrinogen-I (42).[69] This was shown by synthesis of heptacarboxylic, hexacarboxylic, and pentacarboxylic type-I porphyrins by the *b*-oxobilane route and of a second hexacarboxylic porphyrin by the Fischer method, and by incorporation of the corresponding porphyrinogens (see Scheme 3) with a chicken haemolysate

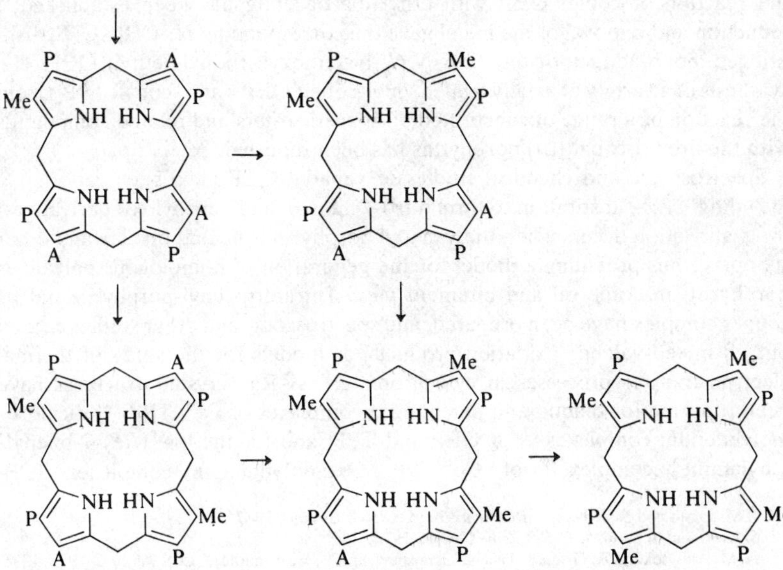

**Scheme 3**

preparation; the heptacarboxylic porphyrinogen gave rise to both hexacarboxylic porphyrinogens in equal amounts, and the corresponding porphyrins were also identified in the urine of human and bovine porphyrics.[69]

[69] A. H. Jackson, K. R. N. Rao, D. M. Supphayen, and S. G. Smith, *J. C. S. Chem. Comm.*, 1977, 696.

Evidence for a possible role for qunones in the enzymic oxidation of protopor-phyrinogen-IX to protoporphyrin-IX has been presented;[70] the direct synthesis[71] of haem-*c* from haemin, rather than *via* addition of cysteine to protopor-phyrinogen, may provide an analogy for the process involved *in vivo* in the biosynthesis of cytochrome-*c*.

## 4 Metalloporphyrins

**Synthesis.**—Further studies of oxidative addition reactions of the bis-dicarbonyl-rhodium(I) complexes of porphyrins have been carried out; the products were of the type R–Rh$^{III}$–porphyrin, with an axial ligand R (R = aryl, acyl, aroyl, alkoxy-carbonyl, *etc.*). Stable acyl-rhodium(III) complexes were formed from the rhodium(I) complexes and aldehydes, and an improved method for pre-paring rhodium(III) porphyrins was described.[72] The preparation of a hydridorhodium(III) porphyrin and a rhodium(II) porphyrin dimer have also been reported.[73] Cobalt complexes of *meso*-tetraphenylporphyrin (TPP) react with ethyl diazoacetate (in a similar manner to those of octaethylporphyrin, OEP) to form an *N*-ethoxycarbonylmethyl cobalt(III) complex, a $N_a$-CH(CO$_2$Et)-$N_b$-bridged cobalt(II) complex, and an $N_a$-CH(CO$_2$Et)-$N_b$-bridged metal-free porphyrin.[74] The *X*-ray crystal structure of the nitrato-bis-adduct derived from the reaction of cobalt OEP with ethyl diazoacetate has been established;[75] reduction and removal of the metal gave the corresponding $N_a$-CH(CO$_2$Et)-$N_b$-bridged mono-adduct, or the *trans*-$N_aN_b$-bis(ethoxycarbonylmethyl)-OEP. The synthesis of a variety of axially ligated vinyl-cobalt OEP's and -cobalt TPP's from the reaction of a range of substituted diazoacetic esters and related compounds with the bromo-cobalt(III) porphyrins has been reported.[76]

Spectroscopic and chemical studies of vanadyl OEP have been reported,[77] including *meso*-substitution to form nitro-, chloro-, and benzoyloxy-derivatives; *trans*-alkylation occurs when the vanadyl porphyrin is heated on clay and other supports, thus providing a model for the generation of homologous polyalkyl-porphyrins in crude oil and bitumen. *meso*-Tetraferrocenyl-porphyrin and its copper complex have been prepared, and spectroscopic and other studies carried out on mixed-valence oxidation products, as models for the study of thermal electron-transfer processes in cytochrome-*c*.[78] *X*-Ray crystal structures have been reported for diaquo and perchlorate complexes of Fe$^{III}$ TPP,[79a] dichloro-molybdenum complexes[79b] of OEP and TPP, and for the bis-[(*RS*)-1-phenyl-ethylamine] complex[79c] of Co$^{III}$TPP. Oxomolybdenum complexes TPP-

[70] J. M. Jacob and N. J. Jacobs, *Biochem. Biophys. Res. Comm.*, 1977, **78**, 429.

[71] S. Kojo and S. Sano, *J. C. S. Chem. Comm.*, 1977, 249.

[72] A. M. Abeysekera, R. Grigg, J. Trocha-Grimshaw, and V. Viswanatha, *J. C. S. Perkin I*, 1977, 1395.

[73] H. Ogoshi, J. Setsune, and Z. Yoshida, *J. Amer. Chem. Soc.*, 1977, **99**, 3869.

[74] A. W. Johnson and D. Ward, *J. C. S. Perkin I*, 1977, 720.

[75] P. Batten, A. L. Hamilton, A. W. Johnson, M. Mahendran, D. Ward, and T. J. King, *J. C. S. Perkin I*, 1977, 1623.

[76] H. J. Callot and E. Schaeffer, *Tetrahedron Letters*, 1977, 239.

[77] R. Bonnett, P. Brewer, K. Noro, and T. Noro, *Tetrahedron*, 1978, **34**, 379.

[78] R. G. Wollman and D. N. Hendrickson, *Inorg. Chem.*, 1977, **16**, 3079.

[79] (*a*) M. E. Kastner, W. R. Scheidt, T. Mashiko, and C. A. Reed, *J. Amer. Chem. Soc.*, 1978, **100**, 666; (*b*) T. Diebold, B. Chevrier, and R. Weiss, *Angew. Chem.*, 1977, **89**, 819; (*c*) C. Riche, A. Chiarmi, and M. Gonedan, *J. Chem. Res. (S)*, 1978, 32.

Mo(O)(OEt), OEP-Mo(O)(OEt), and OEP-Mo(O)Cl were prepared from the corresponding porphyrins by their reaction with molybdenum pentachloride;[80] e.s.r. and electronic spectra indicate that these complexes are monomeric in dichloromethane but associate in aromatic solvents.

**Properties and Reactions.**—*meso*-Substituted zinc or cadmium porphyrins (43) undergo oxidation by thallium trifluoroacetate at the neighbouring position and after acidic work-up they form the corresponding β-substituted oxophlorin (44);

(43) R¹ = Me or Et        (44)
R² = OCOCF₃, OCOCH₃, OMe, CHO, CN, or Cl

in contrast, enolisable *meso*-hydroxy or -amino zinc porphyrins (which come from the corresponding oxo-phlorins, or from imino-phlorins) are oxidized to β-chloro-α-oxo-phlorins or the corresponding α-amino-porphyrins after work-up with hydrochloric acid.[81a] The reactions involve intermediate π-cation radicals or π-dications, and zinc OEP itself also undergoes novel *meso*-substitution reactions on treatment with oxidizing agents [*e.g.* 1-chlorobenzotriazole, or (*p*-BrC₆H₄)₃N⁺ SbCl₆⁻] and nucleophiles, to form, for example, *meso*-chloro-, -thiocyanato-, and -cyano-porphyrins.[81b] *meso*-Dimethylaminomethyl- and *meso*-hydroxymethyl-copper OEP can be readily transformed (under acidic conditions) into the corresponding *meso*-methoxymethyl-porphyrin, and this fact, together with spectral and kinetic data, has been interpreted as evidence for the intermediacy of a stabilized carbonium ion.[82]

The ethylaluminium complex of TPP undergoes a novel reaction with carbon dioxide in visible light (but not in the dark) in the presence of *N*-methylimidazole to form the corresponding carboxyethyl-aluminium complex.[83] Some new organo-germanium and tri-TPP derivatives TPP–M–R₂ (M = Ge or Sn) were prepared and found to undergo photo-oxidation to form mono- and di-peroxides TPP–M–(OOR)R and TPP–M(OOR)₂.[84] Iron–oxygen complexes formed in the autoxidation of iron(II) porphyrins have been characterized,[85] and the well-known rapid oxidation of iron(II) porphyrins to iron(III) porphyrins has also been investigated.[86] Superoxide anion radical reacts with iron(III) porphyrins to form

---

[80] Y. Matsuda, F. Kubota, and Y. Murakami, *Chem. Letters*, 1977, 1281.

[81] (a) B. Evans and K. M. Smith, *Tetrahedron*, 1977, **33**, 629; (b) B. Evans and K. M. Smith, *Tetrahedron Letters*, 1977, 3079.

[82] G. V. Ponomarev, *Khim. geterotsikl. Soedinenii*, 1977, 90.

[83] S. Inoue and N. Takeda, *Bull. Chem. Soc. Japan*, 1977, **50**, 984.

[84] C. Cloutour, D. Lafargue, J. A. Richards, and J. C. Pommier, *J. Organometallic Chem.*, 1977, **137**, 157.

[85] D.-H. Chin. J. Del Gaudio, G. N. La Mar, and A. L. Balch, *J. Amer. Chem. Soc.*, 1977, **99**, 5486.

[86] J. H. Ong and C. E. Castro, *J. Amer. Chem. Soc.*, 1977, **99**, 6740.

low-spin $(O_2^-)_2$–Fe$^{II}$ complexes, which are stable in organic solvents but rapidly decompose in water to form $\mu$-oxo-dimers.[87] Kinetic and mechanistic studies of the well-known demetallation of iron(III) porphyrins by iron(II) salts in acidic media have confirmed that reduction to the iron(II) porphyrin occurs, followed by acid-catalysed demetallation, and either step can be rate-limiting, depending on the conditions.[88]

A wide variety of spectroscopic investigations have been reported recently. ENDOR studies of a series of axially ligated proto- and deutero-haemins have been correlated with n.m.r. shifts of *meso*-protons,[89] and n.m.r. spectra of ligated iron(II) octa-alkyl- and tetraphenyl-porphyrins have been analysed.[90] Aggregation in high-spin iron(III) TPP complexes has also been studied by intermolecular electron nuclear dipolar relaxation,[91] and self-aggregation of *meso*-trifluoroacetoxy zinc OEP has been investigated by n.m.r. spectroscopy.[92] The $^{13}$C and $^1$H n.m.r. spectra of zinc complexes of the coproporphyrin isomers have been assigned in relation to aggregation effects,[93] and the spectra of the 'monomeric' forms obtained on complexing with pyrrolidine[46] readily allow the 'type' isomers to be distinguished by symmetry arguments.[93] The effect of the axial ligand on $^{15}$N chemical shifts and $^{15}$N–$^{57}$Fe coupling constants in bis-amine complexes of iron(II) low-spin TPP has been studied.[94] Clear evidence for 'sitting atop' porphyrin complexes in non-aqueous media has now been obtained,[95] whilst resonance Raman studies have provided evidence for interactions between the porphyrin ring and conjugated substituents in metalloporphyrins and porphyrin dications.[96]

## 5 Chlorophylls and Chlorins

**Structure and Synthesis.**—Methyl 10-*epi*-phaeophorbide-*a* (45a) shows an unusual epimeric stability[97] compared with chlorophyll-*a* or -*a'*. Magnesium complexes of the $\beta$-keto-ester system (46) of methyl phaeophorbides-*a* and -*b*, methyl bacteriophaeophorbide-*a*, and methyl bacteriophaeophytin-*b* have been prepared,[98] whilst the isolation and identification of protochlorophylls (47) obtained by oxidation of chlorophylls has been reported.[99] Several improvements in the methods used for the extraction and separation of chlorophylls have been described,[100] including t.l.c.[101] and h.p.l.c.[102] New methods for the introduction of

[87] I. B. Afanas'ev, S. V. Prigoda, A. M. Khenkin, and A. A. Shteinman, *Doklady Akad. Nauk S.S.S.R.*, 1977, **236**, 3.
[88] J. H. Espenson and R. J. Christensen, *Inorg. Chem.*, 1977, **16**, 2561.
[89] H. L. Vancamp. C. P. Scholes, C. F. Mulks, and W. S. Caughey, *J. Amer. Chem. Soc.*, 1977, **99**, 8283.
[90] H. Goff, G. N. La Mar, and C. A. Reed, *J. Amer. Chem. Soc.*, 1977, **99**, 3641.
[91] R. V. Snyder and G. N. La Mar, *J. Amer. Chem. Soc.*, 1977, **99**, 7178.
[92] R. J. Abraham, G. H. Bonnett, G. E. Hawkes, and K. M. Smith, *Tetrahedron*, 1976, **32**, 2949.
[93] R. J. Abraham, F. Eivazi, H. Pearson, and K. M. Smith, *Tetrahedron*, 1977, **33**, 2277.
[94] I. Morishima, T. Inubushi, and M. Sato, *J. C. S. Chem. Comm.*, 1978, 106.
[95] E. B. Fleischer and F. Dixon, *Bioinorg. Chem.*, 1977, **7**, 129.
[96] W. M. Fuchsman, Q. R. Smith, and M. M. Stein, *J. Amer. Chem. Soc.*, 1977, **99**, 4190; T. Kitagawa, Y. Kyogoku, and Y. Orii, *Arch. Biochem. Biophys.*, 1977, **181**, 228.
[97] P. A. Ellsworth and C. B. Storm, *J. Org. Chem.*, 1978, **43**, 281.
[98] H. Scheer and J. J. Katz, *J. Amer. Chem. Soc.*, 1978, **100**, 561.
[99] M. V. Sarzhevskaya and A. P. Losev, *Biokhimiya (Moscow)*, 1977, **42**, 2105.
[100] P. H. Hynninen, *Acta Chem. Scand. (B)*, 1977, **31**, 829; N. Sato and N. Murata, *Biochim. Biophys. Acta*, 1978, **501**, 103.
[101] M. Shiraki, M. Yoshiura, and K. Iriyama, *Chem. Letters*, 1978, 103.
[102] T. R. Jacobsen, *Mar. Sci. Comm.*, 1978, **4**, 33 (*Chem. Abs.*, 1978, **88**, 133 416); M. Yoshiura, K. Iriyama, and M. Shiraki, *Chem. Letters*, 1978, 281.

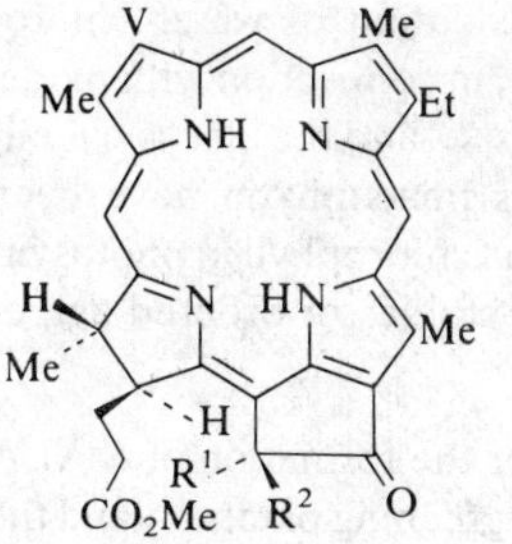

(45) a; $R^1 = H$, $R^2 = CO_2Me$
b; $R^1 = CO_2Me$, $R^2 = H$
c; $R^1 = R^2 = H$

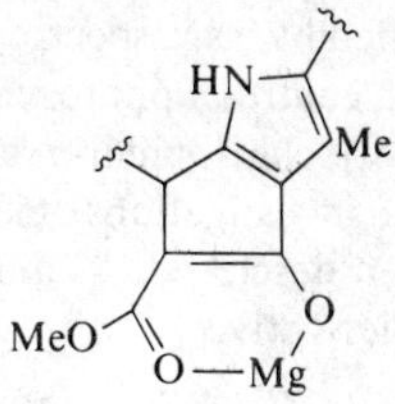

(46)

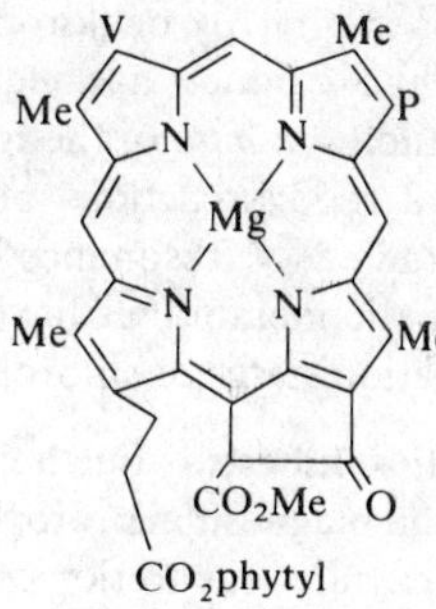

(47)

metals (Cr, Cd or Mn) into chlorins have been reported; of particular interest was the formation of the magnesium complex of methyl phaeophorbide-*a* by treatment with magnesium acetate in acetone and dimethyl sulphoxide.[103]

Models for 'reaction centre' and 'antenna' chlorophyll have been reviewed,[104] and a new model (48) for special pair chlorophyll and bacteriochlorophyll-*a* has been synthesised.[105] A chlorophyll derivative (49) which does not aggregate has been synthesised by replacing the phytyl ester group with an *N*-imidazolylpropyl residue,[106] the basic nitrogen of which can complex internally with the magnesium.

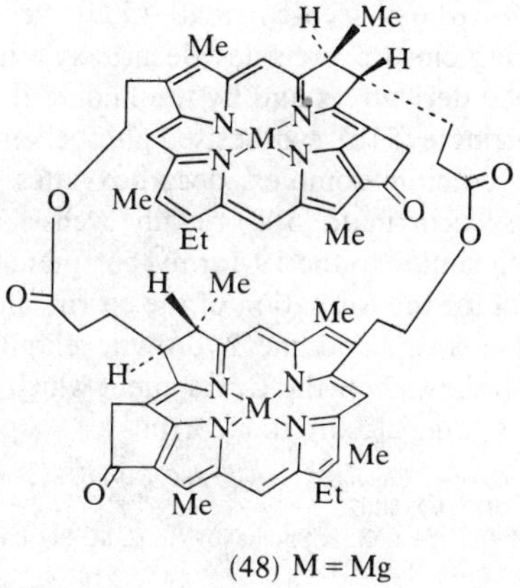

(48) M = Mg

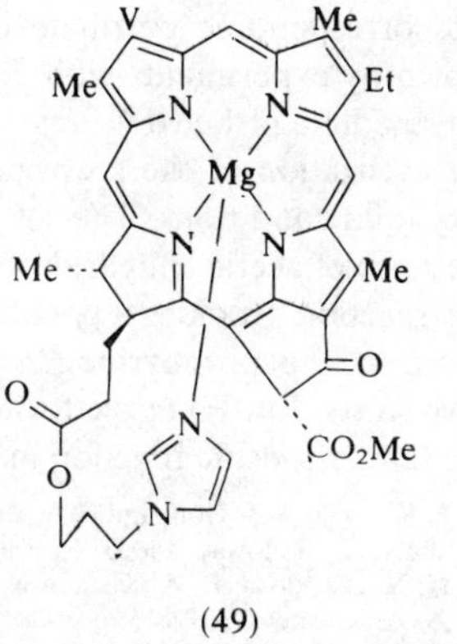

(49)

**Properties and Reactions.**—The chlorophyll-*a* cation radical has been studied by ENDOR spectroscopy,[107] and coupling constants have been assigned by comparison with the magnesium complexes of specifically deuteriated derivatives

[103] M. Strell and T. Crumow, *Annalen*, 1977, 970.
[104] J. J. Katz, J. R. Norris, and L. L. Shipman, *Brookhaven N. L. Reports*, 1977, 10—55 (*Chem. Abs.* 1978, **88**, 71 379).
[105] M. R. Wasiliewski, U. N. Smith, B. T. Cope, and J. J. Katz, *J. Amer. Chem. Soc.*, 1977, **99**, 4172; M. R. Wasiliewski, W. A. Svec, and B. T. Cope, *ibid.*, 1978, **100**, 1961.
[106] I. S. Denniss and J. K. M. Sanders, *Tetrahedron Letters*, 1978, 295.
[107] H. Scheer, J. J. Katz, and J. R. Norris, *J. Amer. Chem. Soc.*, 1977, **99**, 1372.

of methyl pyrochlorophyllide-*a* (45c). The photo-oxidation of water–chlorophyll-*a* dimers has been studied by e.s.r. spectroscopy, in connection with model studies *in vitro* of the reaction centre of photosynthesis,[108] and the electrochemical redox reactions of bacteriochlorophyll-*b* and its phaeophytin have been studied.[109] Resonance Raman investigations of antenna chlorophyll in photosynthetic membranes have beeen described,[110] and also studies of ordered aggregation states of chlorophyll derivatives.[111]

**Biosynthesis.**—Further evidence has been adduced for the formation of δ-ALA and magnesium protoporphyrin-IX from glutamate in greening plants,[112] and the enzymic preparation of the monomethyl ester of magnesium protoporphyrin has been described.[113] Intermediates in the last stages of chlorophyll biosynthesis in which the geranylgeranyl side-chain is converted into phytyl have been studied;[114] 1-$^{13}$C-labelled acetate is incorporated into the expected alternate positions in the phytyl ester side-chain (as shown by $^{13}$C n.m.r. spectral studies), but only low incorporations of mevalonate were obtained, presumably because of poor transport into the cell.[115] The biosynthesis of chlorophyll-*b* has also been studied.[116]

## 6 Vitamin B$_{12}$, Corrins, and Corroles

**Structure and Synthesis.**—Recent progress in the synthesis of corrins has been reviewed,[117] and the isolation and identification of cobalamins has been studied. using h.p.l.c.[118] and reverse affinity chromatography.[119] Further work on the ring closure of A/D-seco-corrins to corrins (described in last year's Report) has now been reported;[120] a 1-methylidene-19-carboxy-seco-corrinate (50a) cyclises to the corresponding corrinate (51a); ring closure precedes decarboxylation, as shown by experiments with deuteriated derivatives and by the finding that the intermediate nickel(II) 19-carboxy-corrinate (51b), synthesised photochemically *via* cyclization of the corresponding cadmium complex, decarboxylates under very mild conditions. The 19-formyl secocorrinate (50b) readily cyclises in the presence of acetic anhydride and triethylamine to the 19-formyl corrinate (51c); this reaction provides a possible model for the formation of the corrin chromophore at the porphyrinogen oxidation level, and the hydrolytic elimination (2M-KOH, EtOH) of the formyl group shows how the C$_1$ fragment which is lost can fulfil a specific function in the (A–D) ring closure to a corrin.[120b]

[108] F. K. Fong, A. J. Hoff, and F. A. Brinkman, *J. Amer. Chem. Soc.*, 1978, **100**, 619; F. K. Fong, J. S. Polles, L. Galloway, and D. R. Fruge, *ibid.*, 1977, **99**, 5802.
[109] N. N. Drozdova, B. A. Kuznetsov, N. M. Mestechkina, G. P. Shumakovich, E. M. Pushkina, and A. A. Krasnovskii, *Doklady Akad. Nauk. S.S.S.R.*, 1977, **235**, 6.
[110] M. Lutz, *Biochim. Biophys. Acta*, 1977, **460**, 408.
[111] C. Kratky and J. D. Dunitz, *J. Mol. Biol.*, 1977, **113**, 431.
[112] J. D. Weinstein and P. A. Castelfranco, *Arch. Biochim. Biophys.*, 1978, **186**, 376; P. A. Castelfranco and S. Schwartz, *ibid.*, p. 365.
[113] R. K. Ellsworth and S. J. Murphy, *Photosynthetica*, 1978, **12**, 81.
[114] S. Schoch, V. Lampert, and W. Rüdiger, *Z. Pflanzenphysiol.*, 1977, **83**, 427.
[115] E. H. Ahrens, D. C. Williams, and A. R. Battersby, *J. C. S. Perkin I*, 1977, 2540.
[116] S. Aronoff and E. Kwok, *Canad. J. Biochem.*, 1977, **55**, 1091.
[117] A. Eschenmoser, *Chem. Soc. Rev.*, 1976, **5**, 377.
[118] F. Pellerin, J. F. Letavenier, and N. Chaudon, *Ann. Pharm. France*, 1977, **35**, 413.
[119] J. F. Kolhouse and R. H. Allen, *Analyt. Biochem.*, 1978, **84**, 486.
[120] (*a*) V. Rasetti, B. Krautler, A. Pfaltz, and A. Eschenmoser, *Angew. Chem.*, 1977, **89**, 475; (*b*) A. Pfaltz, N. Buhler, R. Neier, K. Hirai, and A. Eschenmoser, *Helv. Chim. Acta*, 1977, **60**, 2653.

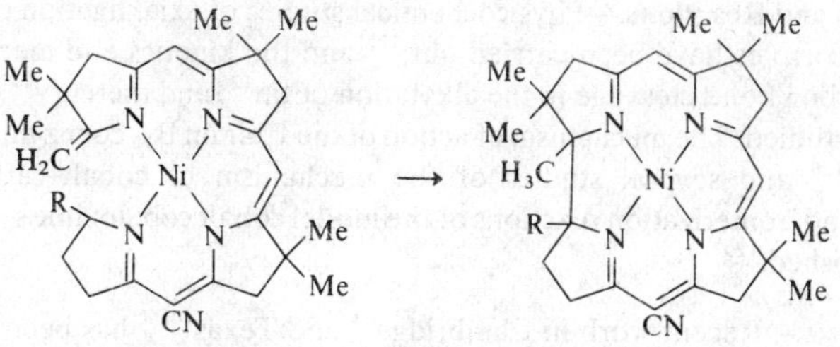

(50) a; R = CO₂H
b; R = CHO

(51) a; R = H
b; R = CO₂H
c; R = CHO

Dicyanocobyrinic acid heptamethyl ester (52) undergoes oxidation[121] at the α-bridge by molecular oxygen in the presence of ascorbic acid to form the *meso*-hydroxy-lactone (53); the dicyanocobyrinic acid lactone (54a) can be transformed into a variety of β-*meso*-substituted derivatives, *e.g.* (54b), and reduction of the bromo- and iodo-lactones (54b; R = Br or I) led to dicyano-cobyrinic acid heptamethyl ester (52).[122] Nickel complexes of 5-methyl derivatives of 1,19-dimethyl-octadehydrocorrins have been prepared,[123] and the synthesis of 1,19-diethoxycarbonyl-octamethylplatinum(II) tetradehydrocorrin bromide by oxidative ring closure of the corresponding 1,19-diethoxycarbonyl-bilatriene in the presence of platinum chloride has been reported.[124]

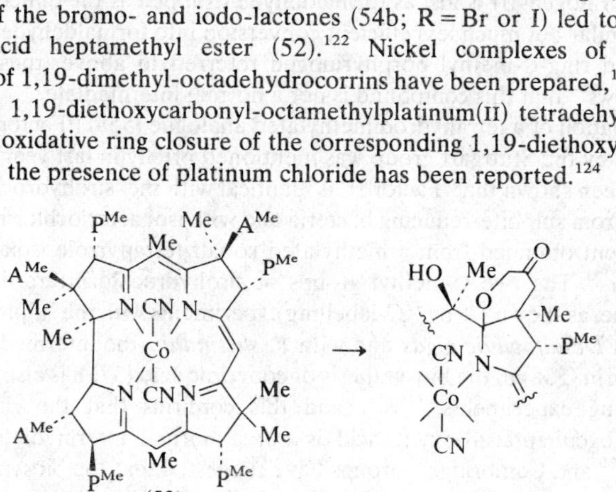

(52)

(53)

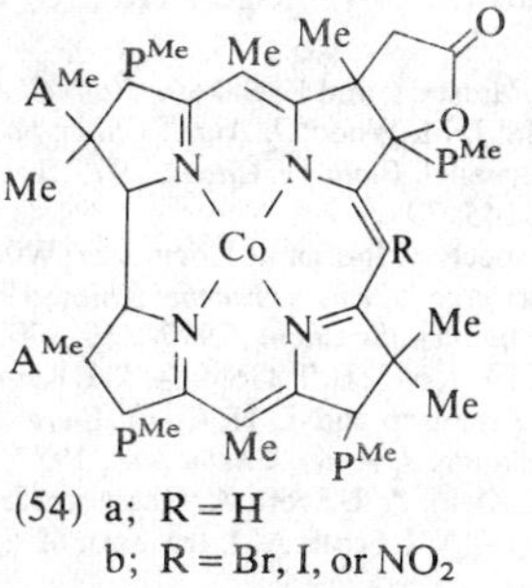

(54) a; R = H
b; R = Br, I, or NO₂

[121]. A. Gossauer, B. Gruning, L. Ernst, W. Becker, and W. S. Sheldrick, *Angew. Chem.*, 1977, **89**, 486.
[122] A. Gossauer, K. P. Heise, H. Gotze, and H. H. Inhoffen, *Annalen*, 1977, 1480.
[123] T. A. Melenteva, N. S. Genokhova, and V. M. Berezovskii, *Zhur. obshchei Khim.*, 1977, **47**, 2797.
[124] J. Engel and H. H. Inhoffen, *Annalen*, 1977, 767.

**Properties and Reactions.**—Physicochemical studies of axial ligation of pyridine to cobalt corroles have been carried out,[125] and the kinetics and mechanism of cobalt–carbon bond cleavage in the alkylation of tin[126] and mercury[127] ions have also been studied. The mechanism of action of the vitamin $B_{12}$ coenzyme has been reviewed,[128] and several studies of the mechanism of cobalt–carbon bond cleavage and isomerization reactions of the model cobalt cobaloxime system have been published.[129]

**Biosynthesis.**—Recent work in Cambridge[64] and Texas[10c,d] has been reviewed, and it has become clear that uroporphyrinogen-III (36) is the branch point where the corrinoid and porphyrin pathways diverge.[130,131] As mentioned in last year's Report, the related ring-C-methyl porphyrinogenheptacarboxylic acid is converted into cobyrinic acid, but in much lower yield, whereas the normal porphyrin pathway (*cf.* ref. 26) starts with the decarboxylation of the D-ring acetic acid residue of uroporphyrinogen-III. The Texas group have now shown,[130a] in experiments with a cell-free system from *P. shermanii*, that the δ-carbon atom of uroporphyrinogen-III is lost as formaldehyde (trapped as the dimedone derivative); a similar but much less efficient conversion into formaldehyde occurs with the related ring-C-methyl porphyrinogen referred to above, thus confirming earlier views[64] that this compound is not a normal intermediate.

The isolation of a tetrahydrodimethylated analogue (55a) ('Factor II') from *P. shermanii* by the Stuttgart group was mentioned briefly in last year's Report. It has now been shown that 'Factor II' is identical with the 'sirohydrochlorin'[130,131] obtained from sulphite-reducing bacteria and with isobacteriochlorin, the metal-free pigment obtained from a methylated cobalt tetrapyrrole isolated from *P. shermanii*.[131] The two C-methyl groups of sirohydrochlorin are derived from methionine, as shown[131c] by $^{13}C$-labelling experiments with the sulphite-reducing bacterium *Desulfovibrio gigas* and with *P. shermanii*; the intermediacy of sirohydrochlorin (55a) in the biosynthesis of cobyrinic acid (57) has also been shown by labelling experiments,[130b,131c] and this confirms that the ring-C-methyl porphyrinogenheptacarboxylic acid is not a normal intermediate. Both the Texas[10d,130] and Cambridge[64] groups have suggested that the biosynthetic intermediate is a dihydro-derivative (56) of sirohydrochlorin, which could be derived by direct methylation of uroporphyrinogen-III (36) on rings A and B. Interes-

[125] Y. Murakami, S. Yamada, Y. Matsuda, and K. Sakata, *Bull. Chem. Soc. Japan*, 1978, **51**, 123.

[126] L. J. Dizikes, W. P. Ridley, and J. M. Wood, *J. Amer. Chem. Soc.*, 1978, **100**, 1010.

[127] V. C. W. Chu and D. W. Gruenwedel, *Bioinorg. Chem.*, 1977, **7**, 169; P. J. Craig and S. F. Morton, *J. Organometallic Chem.*, 1978, **145**, 79.

[128] R. H. Abeles in 'Biological Aspects of Inorganic Chemistry' Wiley, New York, 1977, p. 245.

[129] E. A. Parfenov and T. G. Chervyakova, *Zhur. obshchei Khim.*, 1977, **47**, 923; R. Dreos Garlatti, G. Tauzher, and G. Costa, *J. Organometallic Chem.*, 1977, **139**, 179; B. T. Golding, C. S. Sell, and P. J. Sellers, *J. C. S. Chem. Comm.*, 1977, 693; B. T. Golding, T. J. Kemp, P. J. Sellers, and E. Nocchi, *J. C. S. Dalton*, 1977, 1266; J. H. Espenson and T. H. Chao, *Inorg. Chem.*, 1977, **16**, 2553; D. Dodd, M. D. Johnson, and B. L. Lockman, *J. Amer. Chem. Soc.*, 1977, **99**, 3664.

[130] (*a*) M. Kajiwara, K. S. Ho, H. Klein, A. I. Scott, A. Gossauer, J. Engel, E. Newman, and H. Zilch, *Bio-org. Chem.*, 1977, **6**, 397; (*b*) A. I. Scott, A. J. Irwin, L. M. Siegel, and J. N. Shoolery, *J. Amer. Chem. Soc.*, 1978, **100**, 316.

[131] (*a*) A. R. Battersby, K. Jones, E. McDonald, J. A. Robinson, and H. R. Morris, *Tetrahedron Letters*, 1977, 2213; (*b*) A. R. Battersby, E. McDonald, H. R. Morris, M. Thompson, D. C. Williams, V. Y. Bykhovsky, N. I. Zaitseva, and V. N. Bukin, *ibid.*, p. 2217; (*c*) A. R. Battersby, E. McDonald, M. Thompson, and V. Y. Bykhovsky, *J. C. S. Chem. Comm.*, 1978, 150.

tingly, the Stuttgart group have recently reported[132] the isolation of an analogue (55b) of sirohydrochlorin containing an additional methyl group (as shown by f.d. mass spectrometry and visible spectroscopy). As this compound was incorporated into cobyrinic acid (57) by *P. shermanii*, the extra methyl group (derived, like the ring-A- and-B-methyl groups, from methionine) was assigned to the α-*meso*-position, as shown in (55b).[132]

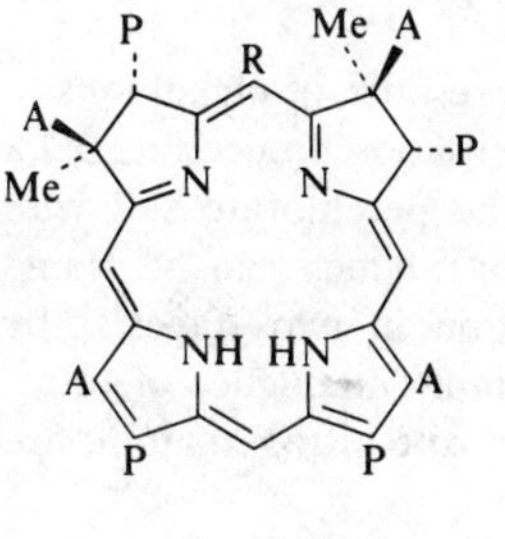

(55) a; R = H
    b; R = Me

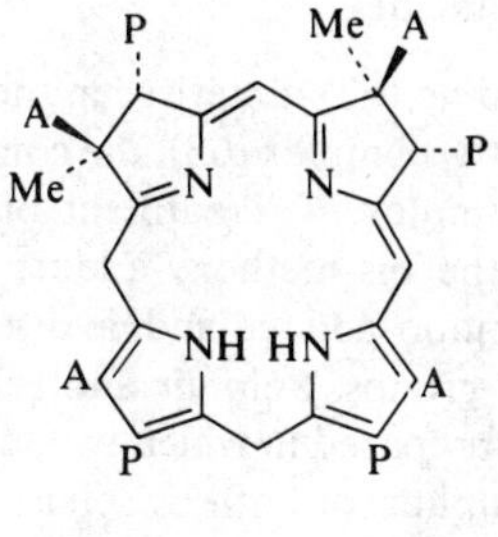

(56)

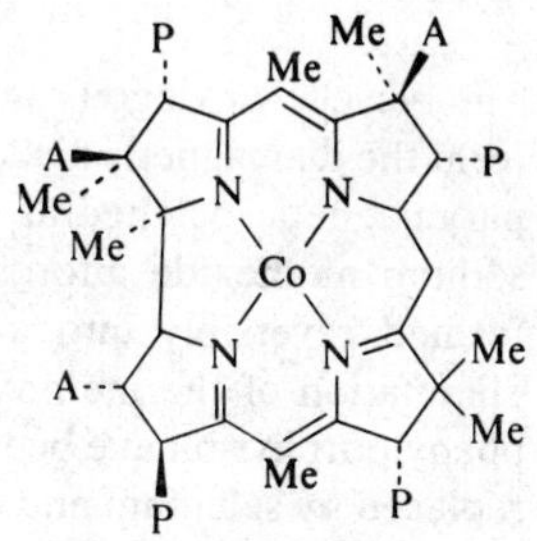

(57)

## 7 Prodigiosins

Details of the synthesis of metacycloprodigiosin (61) have been given;[133] cyclododecanone was converted (by a series of reactions) into the formyl derivative (58) and hence into the pyrrole (59), which was condensed with the known methoxy-bipyrrole aldehyde (60). 3-Hydroxy-pyrromethenes and 3-hydroxy-pyrroles can be alkylated on oxygen with trialkyloxonium tetrafluoroborate, whereas other reagents effect alkylation on nitrogen,[134a] and these reactions were

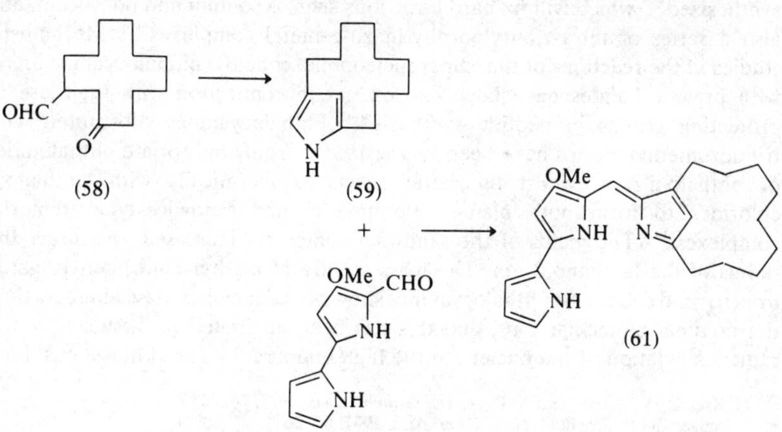

(58)    (59)

+

(60)

(61)

[132] K. H. Bergmann, R. Deeg, K. D. Gneuss, H. P. Kriemler, and G. Müller, *Z. physiol. Chem.*, 1977, **358**, 1315.
[133] H. H. Wasserman, D. D. Keith, and J. Nadelson, *Tetrahedron*, 1976, **32**, 1867.
[134] (*a*) H. Berner, G. Schulz, and H. Reinshagen, *Monatsh.*, 1977, **108**, 915; *ibid.*, 1978, **109**, 137.

utilized in the synthesis of a pyrromethane analogue of metacycloprodigiosin. A phenyl analogue of metacycloprodigiosin was prepared *via* condensation of a 3,5-cyclononyl-pyrrole with 3-acetoxy-2-formyl-5-phenypyrrole, the acetoxy function having been introduced by oxidation with lead tetra-acetate.[134b] Other analogues of prodigiosin have also been synthesized.[135]

## 8 Porphyrin Analogues

The bicyclic vinylogous amidine (62) dimerises in the presence of nickel ions to form the diamagnetic nickel(II) complex (63), the condensation proceeding *via* a paramagnetic octahedral complex.[136] Treatment of the perchlorate salt with sodium methoxide affords the bis-methoxy-adduct (64), which can be transformed reversibly into a mono-adduct and a di-cation in two stages[136] by elimination of the methoxy-groups. Sulphur and selenium analogues of tetraphenylporphyrin have been prepared in which two, opposite nitrogen atoms are replaced by selenium and sulphur, or both by selenium.[137]

(62)                         (63) X = Cl or $ClO_4$                      (64)

A new macrocyclic system, 1,6,11,16-tetra-azaporphyrinogen (65), has been synthesized[138] which will fix hard basic ions such as sodium and potassium, and also a series of tetra-t-butylporphyrinazine–metal complexes.[139] Mechanistic studies of the reactions of the 'supernucleophile' cobalt(I) phthalocyanine anion with organic halides have been carried out in connection with their use as protecting groups in peptide synthesis.[140] Phthalocyanines substituted with trifluoromethyl groups have been synthesized[141] from appropriate phthalimides or phthalonitriles, whilst the latter reacts exothermically with lanthanoid chlorides to form both planar (monomeric) and sandwich-type (dimeric) complexes.[142] The yields of the sandwich complexes increased, the larger the radius of the lanthanoid ion. The mass spectra of methyl- and phenyl-ligated iron(III) and cobalt(III) phthalocyanines show peaks at higher mass numbers than the nominal molecular ions, and this has been attributed to dissociation and radical alkylation of the nucleus at the high temperatures used in the m.s. inlet

[135] G. Kresze, M. Morper, and A. Bijer, *Tetrahedron Letters*, 1977, **26**, 2259.
[136] J. Loliger and R. Scheffold, *Helv. Chim. Acta*, 1977, **60**, 2644.
[137] A. Ulman, J. Manassen, F. Frolow, and D. Rabinovich, *Tetrahedron Letters*, 1978, 167.
[138] J. Fifani, A. Ramdani, and G. Tarrago, *Nouveau J. Chim.*, 1977, **1**, 52.
[139] V. N. Kopranenkov, L. S. Goncharova and E. A. Lukyanets, *Zhur. obshchei Khim.*, 1977, **47**, 2143.
[140] H. Eckert, I. Lagerlund, and I. Ugi, *Tetrahedron*, 1977, **33**, 2243.
[141] I. G. Oksengendler, N. V. Kondratenko, E. A. Lukyanet, and L. M. Yagupolskii, *Zhur. org. Khim.*, 1977, **13**, 1554.
[142] M. Yamana, *J. Chem. Soc. Japan, Chem. Ind. Chem.*, 1977, 144.

system.[143] Further studies of homoporphyrins derived from tetraphenylporphyrin, and the contraction of anion radicals of their metal complexes to
porphyrins, have been reported.[144,145] *meso*-Aza-, -oxonia-, and -thionia-*meso*-
porphyrin– and -protoporphyrin–metal complexes (66) have been prepared by
cyclization of the zinc complexes of the corresponding 1-substituted biliverdins.

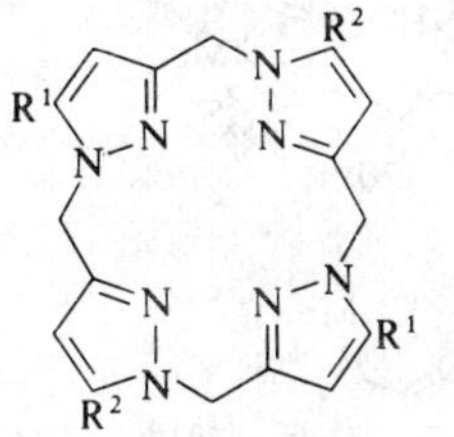

(65) a; $R^1 = R^2 = Me$
b; $R^1 = R^2 = Ph$
c; $R^1 = Me, R^2 = Ph$

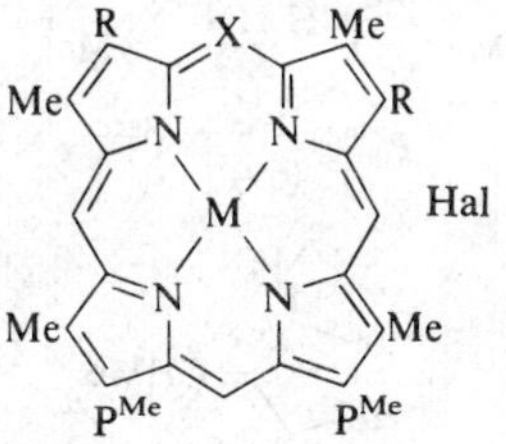

(66) R = Et or V    X = $O^+$, N, or $S^+$
M = Zn, VO, Cu, Pd, $Mn^{III}$, or $H_2$
Hal = $Cl^-$ (when X = $O^+$ or $S^+$)

The reactivity of the heteroporphyrin analogues, however, was similar to that of
the parent biliverdin; the rings were readily opened by strong acid or alkali and
the compounds underwent addition reactions on the γ-bridge.[146]

## 9 Bile Pigments

**Structure and Synthesis.**—Spectral studies and correlations with other bile pigments have shown that the blue butterfly bile pigment, sarpedobilin,[147] has the
novel structure (67). Total syntheses of diastereoisomeric mesobilirhodins and
isomesobilirhodins (68),[148a] $(4R)(16R)$- and $(4R)(16S)$-18-vinyl analogues (69)
of mesourobilin,[148b] and racemic phycocyanobilin dimethyl ester[148c] (70) have
been reported by the Braunschweig group. Perhaps the most novel feature of this
very substantial effort was the use of a phosphorus ylide (71) in the preparation of
one of the dipyrromethenones (72) that is required for the synthesis of the

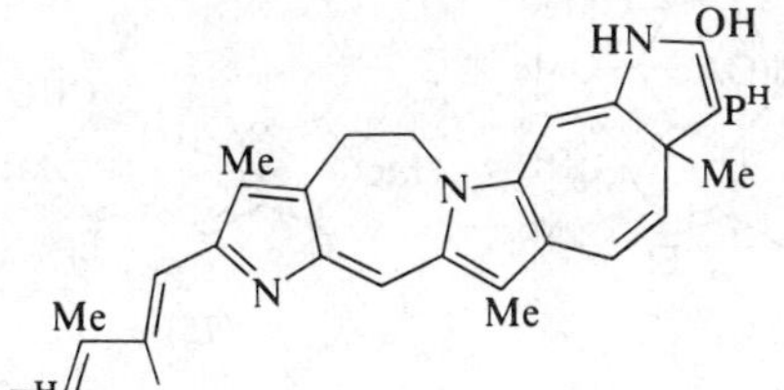

(67)

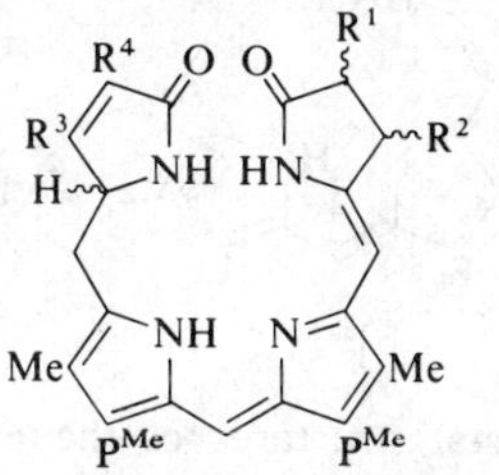

(68) a; $R^1 = R^3 = Et, R^2 = R^4 = Me$
b; $R^1 = R^3 = Me, R^2 = R^4 = Et$

[143] E. C. Muller, R. Kraft, G. Etzold, H. Drevs, and R. Taube, *J. prakt. Chem.*, 1978, **320**, 49.
[144] A. Louati, E. Schaeffer, H. J. Callot, and M. Gross, *Nouveau J. Chim.*, 1978, **2**, 163.
[145] H. J. Callot and E. Schaeffer, *J. Chem. Res.(S)*, 1978, 690.
[146] J. H. Fuhrhop, P. Kruger, and W. S. Sheldrick, *Annalen*, 1977, 339.
[147] M. Choussey and M. Barbier, *Experientia*, 1977, **33**, 1407.
[148] (a) A. Gossauer and G. Kohne, *Annalen*, 1977, 664; (b) A. Gossauer and J. P. Weller, *Chem. Ber.*,
1978, **111**, 486; (c) A. Gossauer and R. P. Hinze, *J. Org. Chem.*, 1978, **43**, 283.

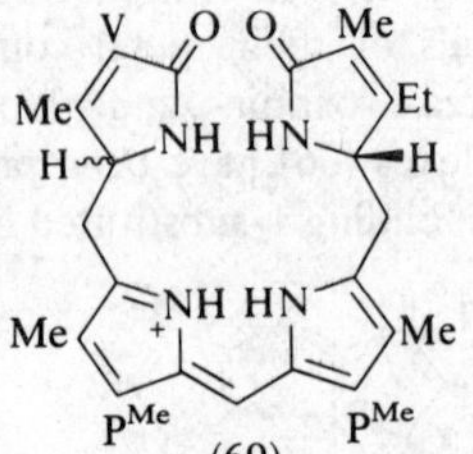

(69) 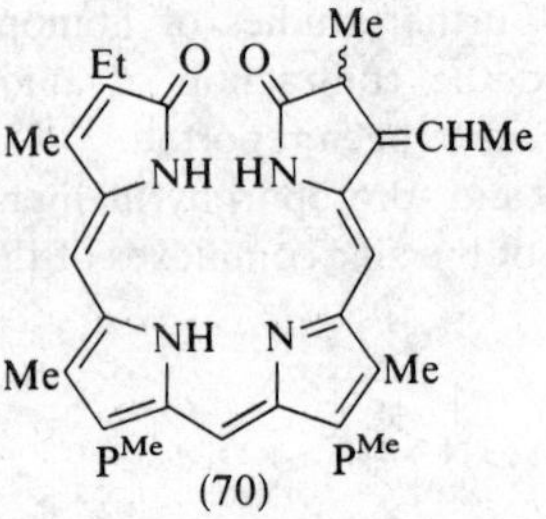      (70)

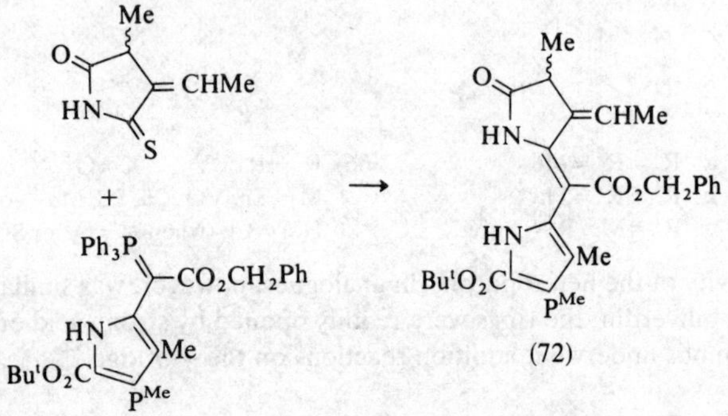

(71)      (72)

phycocyanobilin.[148c] Thiol- and acylthio-adducts of bilirubin have been synthesized to assist in labelling the bilirubin-binding sites of proteins.[149]

**Properties and Reactions.**—The unstable thallium complex of the bilindione (73) is oxidized by air in methanol to a dimethoxy-derivative (74); the latter undergoes photo-oxidation to a tripyrrinone (75) and ethylmethyl maleimide.[150] The

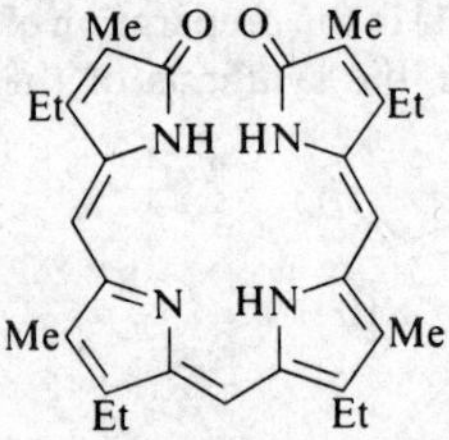

(73) 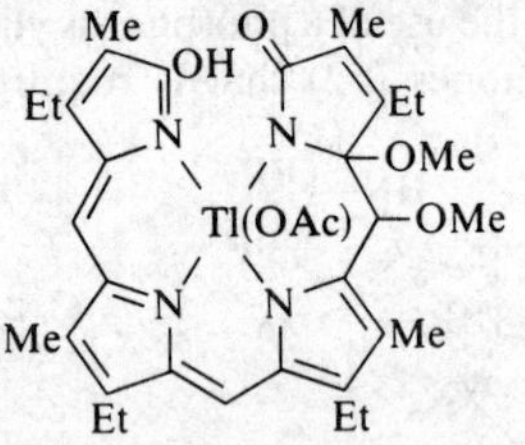      (74) 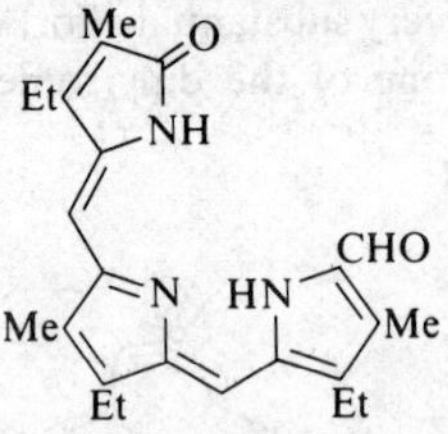      (75)

crystal structures of the nickel complex of octaethylbiliverdin (76a) and a 5-nitro-derivative (76b) have been determined; the former has a helical configuration, but the C-4—C-5 bond in the latter has the (*Z*)-configuration, with ring A at approximately 90° to ring B.[151] The ¹³C n.m.r. spectra of bilirubin and its

149 P. Manitto and D. Monti, *Gazzetta*, 1977, **107**, 573.
150 F. Eivazi, W. M. Lewis, and K. M. Smith, *Tetrahedron Letters* 1977, 3083; F. Eivazi, M. F. Hudson, and K. M. Smith, *Tetrahedron*, 1977, **33**, 2959.
151 J. V. Bonfiglio, R. Bonnett, M. B. Hursthouse, K. M. Abdulmalik, and S. C. Naithani, *J. C. S. Chem. Comm.*, 1977, 829; J. V. Bonfiglio, R. Bonnett, M. B. Hursthouse, and K. M. Abdulmalik, *ibid.*, p. 83.

(76) a; R = H
b; R = NO$_2$

(77) R = Et or V; Y = O; M = Zn or H$_2$

X = OH, OMe, OEt, NH$_2$, N⌂O, NH(CH$_2$)$_2$Me, or SH

and isomers with substituents X and Y
on rings A and D transposed

dimethyl ester have been determined and spin–lattice relaxation times determined for some carbon atoms,[152] and the photoreactivity of bilirubin has been reviewed.[153] The oxonia- and thionia-porphyrin–zinc complexes (66) referred to above undergo ring opening with amines, sulphides, or methoxides to form the corresponding biliverdinates (77); the amino-biliverdinates yield very stable $\pi$-radicals.[154]

Intensive studies of the structure and stereochemistry of bile pigments have been reported during the past year[155—162] involving both the synthesis of model bilatrienes, tripyrrolinones, and pyrromethenones, and wide-ranging spectroscopic investigations.[159] Mono-oxygen and -sulphur analogues[155] of pyrromethenones undergo photochemical (Z)–(E) isomerization, as do the parent compounds,[159c,161,162] and the crystal structure of a pyrromethenone (78) shows that it is nearly planar and has the (Z)-configuration, as shown.[156] Bilatrienes normally have the (Z, Z, Z)-configuration, but an (E, Z, Z)-geometrical isomer

(78)

[152] D. Kaplan, R. Panigel, and G. Navon, *Spect. Meth.*, 1977, **10**, 881.
[153] D. A. Lightner, *Photochem. and Photobiol.*, 1977, **26**, 427.
[154] J. H. Fuhrhop and P. Kruger, *Annalen*, 1977, 360.
[155] J. A. Vankoeveringe and J. Lugtenburg, *Rec. Trav. chim.*, 1977, **96**, 55.
[156] D. L. Cullen, P. S. Black, E. F. Meyer, D. A. Lightner, G. D. Quistad, and C. S. Dak, *Tetrahedron*, 1977, **33**, 477.
[157] H. Falk and K. Grubmayr, *Angew. Chem.*, 1977, **89**, 487.
[158] H. Falk and K. Grubmayr, *Synthesis*, 1977, 614.
[159] (a) H. Falk, S. Gergely, K. Grubmayr, and O. Hofer, *Annalen*, 1977, 565; H. Falk and K. Grubmayr, *Monatsh.*, 1977, **108**, 625; (c) H. Falk, K. Grubmayr, G. Hollbacher, O. Hofer, A. Leodolter, F. Neufingerl, and J. M. Ribo, *ibid.*, p. 1113.
[160] H. Falk, A. Leodolter, and G. Schade, *Monatsh.*, 1978, **109**, 183.
[161] D. A. Lightner and Y. T. Park, *J. Heterocyclic Chem.*, 1977, **14**, 415.
[162] H. Heinz, K. Grubmayr, and F. Neufingerl, *Monatsh.*, 1977, **108**, 1185.

(79)

(79) has been prepared by photochemical isomerization on alumina.[157] Photometric p$K$ measurements of the lactam–lactim tautomerism of bile pigments, supported by $X$-ray photoelectron studies and theoretical calculations, show that the former is favoured by at least four to more than ten orders of magnitude.[159a] Electrochemical oxidations of a series of pyrromethenes and pyrromethenones show that reversible one-electron oxidations occur with arylmethylene-pyrrolinones and pyrromethenones that are unsubstituted in the 5-position of the pyrrole ring, whereas the 5-methoxy-analogues of the latter undergo a two-step oxidation, only the first stage of which is reversible.[160]

The major diazo-positive bile pigments in the bile of homozygous Gunn rats have been characterized,[163a] and studies of the structure of azodipyrroles from mesobilirubin have been reported.[163b] The enzymic conversion of bilirubin monoglucuronide into the diglucuronide has been studied,[164a] and it has been shown that the bilirubin aglycone in the monoglucuronide can migrate to the 2-, 3-, and 4-hydroxyl functions of the glucuronic acid.[164b] Further reports of the detection of non-$\alpha$-isomers of bilirubin have appeared,[165] and of the reaction of biliverdin with singlet oxygen.[166] The role of iron in the degradation of haem to bile pigments has been discussed, and this is also of interest in relation to the mechanism of the formation of isomeric bile pigments from abnormal haemoglobins.[167]

The oxidative photodimerization of a model pigment for phytochrome $P_r$ has been described;[168a] this probably occurs at the 4,5-double bond, and the product has spectral similarities to phytochrome $P_{fr}$, suggesting that in the latter a photo-reaction has occurred at the 4,5-double bond of the $P_r$ form. Theoretical models for the phytochrome chromophore have been discussed,[169] and further investigations of the structure and mode of linkage of the protein have been carried out.[170] The role of phycobilin pigments in photosynthesis[168b] and the biological properties of phytochrome[12] have both been reviewed.

[163] (a) N. Blanckaert, J. Fevery, K. P. M. Heirwegh, and F. Compernolle, *Biochem. J.*, 1977, **164**, 237; (b) M. Salmon, *J. Heterocyclic Chem.*, 1977 **14**, 1101.
[164] (a) P. L. M. Jansen and J. R. Chowdhury, *J. Biol. Chem.*, 1977, **252**, 2710; (b) F. Compernolle, N. Blanckaert, and K. P. M. Heirwegh, *Biochem. Soc. Trans.*, 1977, **5**, 317.
[165] K. P. M. Hierwegh, N. Blanckaert, F. Compernolle, J. Fevery, and Z. Zaman, *Biochem. Soc. Trans.*, 1977, **5**, 316.
[166] I. B. C. Matheson and M. M. Toldo, *Photochem. and Photobiol.*, 1977, **25**, 243.
[167] S. B. Brown, J. C. Docherty, and T. B. Bradley, *Biochem. Soc. Trans.*, 1977, **5**, 1020; S. B. Brown and M. S. Grundy, *ibid.*, p. 1017.
[168] (a) H. Scheer and C. Krauss, *Photochem. and Photobiol.*, 1977, **25**, 311; (b) O. D. Bekasova and W. B. Evstigneeva, *Biofizika*, 1977, **22**, 429.
[169] T. Sugimoto, M. Oishi, and H. Suzuki, *J. Phys. Soc. Japan*, 1977, **43**, 619.
[170] G. Klein, S. Grombein, and W. Rüdiger, *Z. physiol. Chem.*, 1977, **358**, 1077.

# Author Index

354

*Author Index*